PERSPECTIVES IN GENE EXPRESSION

Krishnarao Appasani, PhD, MBA
GeneExpression Systems™, Inc.
P.O. Box 540170
Waltham, MA 02454, USA

Library of Congress Cataloging-in-Publication Data

Perspectives in gene expression / edited by Krishnarao Appasani ;
foreword by Phillip A. Sharp.
p. ; cm.
ISBN 1-881299-16-3
1. Gene expression--Research--Methodology.
[DNLM: 1. Gene Expression. QH 450 P467 2003] I. Appasani,
Krishnarao, 1959-
QH450.P47 2003
572.8'65--dc21

2003007089

ISBN 1-881299-16-3

Printed in the United States of America

9 8 7 6 5 4 3 2 1

Eaton Publishing
One Research Drive
P.O. Box 1070
Westborough, MA 01581-6070
www.BioTechniques.com

James D. Emerton: *Publisher*
Christine Emerton: *Managing Editor*
Patrice Mailloux: *Copy Editor*
Sandy Lamont: *Production Manager*
Anne G. Hann: *Asst. Production Manager*
Robin Davis: *Marketing Projects Coordinator*
Leigh Linsley: *Sales and Marketing Assistant*

Perspectives in Gene Expression

Edited by

Krishnarao Appasani
GeneExpression Systems, Inc.
Waltham, MA, USA

Foreword by

Phillip A. Sharp
Massachusetts Institute of Technology

BioTechniques® Press
Eaton Publishing
Part of the Informa Life Sciences Group

Contributors

Krishnarao Appasani
GeneExpression Systems, Inc.
Waltham, MA, USA
(DrAppasani@aol.com)

Kevin G. Becker
DNA Array Unit
National Institute on Aging
National Institutes of Health
Baltimore, MD, USA

Debajit K. Biswas
Dana-Farber Cancer Institute
Boston, MA, USA

Hidemasa Bono
Laboratory for Genome Exploration Research Group
RIKEN Genomic Sciences Center
RIKEN Yokohama Institute
Yokohama, Japan

Andrea D. Branch
Mount Sinai School of Medicine and
Recanati/Miller Transplantation Institute
New York, NY, USA

Kurt Brunden
Athersys, Inc.
Cleveland, OH, USA

Jeffrey H. Charlap
Department of Pathology, Anatomy and Cell Biology
Developmental Biology and Teratology Program
Jefferson Medical College
Thomas Jefferson University
Philadelphia, PA, USA

Rakesh Chaterjee
Department of Biomedical Engineering
Boston University
Boston, MA, USA

Chris Cheadle
DNA Array Unit
National Institute on Aging
National Institutes of Health
Baltimore, MD, USA

Jianjun Chen
Department of Medicine
University of Chicago Medical Center
Chicago, IL, USA

Georgiana Cheng
Department of Ophthalmology and
The Visual Sciences Research Center
Case Western Reserve University and
University Hospitals of Cleveland
Cleveland, OH, USA

Rita Cheng
Texas Children's Cancer Center
Baylor College of Medicine
Houston, TX, USA

Yee Sook Cho
Cellular and Biochemistry Section
Basic Research Laboratory, CCR
National Cancer Institute
National Institutes of Health
Bethesda, MD, USA

Yoon S. Cho-Chung
Cellular Biochemistry Section,
Basic Research Laboratory, CCR
National Cancer Institute
National Institutes of Health
Bethesda, MD, USA
(chochung@helix.nih.gov)

Terry Clark
Department of Computer Science
University of Chicago Medical Center
Chicago, IL, USA

Rina Das
Department of Molecular Pathology
Walter Reed Army Institute of Research
Silver Spring, MD, USA

Charles DeLisi
Bioinformatics Graduate Program
Department of Biomedical Engineering
Boston University
Boston, MA, USA
(delisi@bu.edu)

David R. Dorris
Ambion, Inc.
The Woodlands, TX, USA

Evan A. Facher
Athersys, Inc.
Cleveland, OH, USA
(efacher@athersys.com)

Stanley Falkow
Department of Microbiology and Immunology
Stanford University School of Medicine
Stanford, CA, USA

Kurt Fellenberg
Department of Theoretical Bioinformatics and Functional Genome Analysis
German Cancer Research Center
Heidelberg, Germany
(K.Fellenberg@DKFZ-Heidelberg.de)

E.J. Fernandez
Mount Sinai School of Medicine
New York, NY, USA

Heide L. Ford
Departments of Obstetrics and Gynecology and
Biochemistry and Molecular Genetics
University of Colorado Health Sciences Center
Denver, CO, USA

Xiaohu Gao
Departments of Biomedical Engineering, Chemistry, Hematology, Oncology, and
The Winship Cancer Institute
Georgia Institute of Technology and
Emory University School of Medicine
Atlanta, GA, USA

Rhonda Harrison-Tang
Bioinformatics Graduate Program
Boston University
Boston, MA, USA

Nicole C. Hauser
Department of Functional Genome Analysis
German Cancer Research Center
Berlin, Germany

Jörg D. Hoheisel
Department of Functional Genome Analysis
German Cancer Research Center
Berlin, Germany

Marti Jett
Department of Molecular Pathology
Walter Reed Army Institute of Research
Silver Spring, MD, USA

Hiroaki Kawasaki
Department of Chemistry and Biotechnology
School of Engineering
The University of Tokyo
Tokyo, and
Gene Function Research Center
National Institute of Advanced Industrial Science and Technology
Tskuba Science City, Japan

Sangeeta Khanna
Department of Ophthalmology and The Visual Sciences Research Center
Case Western Reserve University and University Hospitals of Cleveland
Cleveland, OH, USA

Jae-Hoon Kim
Department of Obstetrics and Gynecology
Saint Vincent Hospital and Catholic University of Korea
Kyong-Ki-Do, Korea

Meyoung-Kon Kim
Cellular and Biochemistry Section
Basic Research Laboratory, CCR
National Cancer Institute
National Institutes of Health
Bethesda, MD, USA

Thomas Knudsen
Department of Pathology, Anatomy and Cell Biology
Developmental Biology and Teratology Program
Jefferson Medical College
Thomas Jefferson University
Philadelphia, PA, USA
(Thomas.Knudsen@mail.tju.edu)

Tomoko Kohno
Division of Cytokine Signaling
Department of Molecular Microbiology and Immunology
Nagasaki University Graduate School of Medical Sciences
Nagasaki, Japan

James J. Kovach
Athersys, Inc.
Cleveland, OH, USA

Patrick Leahy
The Gene Expression Array Core Facility and The Comprehensive Cancer Center
Case Western Reserve University and University Hospitals of Cleveland
Cleveland, OH, USA

Sanggyu Lee
Department of Medicine
University of Chicago Medical Center
Chicago, IL, USA

Qi-Xiang Li
Immusol, Inc.
San Diego, CA, USA

Suxing Liu
Department of Tumor Biology
Schering-Plough Research Institute
Kenilworth, NJ, USA
(suxing.liu@spcorp.com)

Randall Lockner
Amersham Biosciences
Piscataway, NJ, USA

Katherine J. Martin
Dana-Farber Cancer Institute
Boston, MA, USA

Shelby A. Martensen
Quantitative Expression Group
Department of Cellular Genomics
GlaxoSmithKline Pharmaceuticals
Research Triangle Park, NC, USA

Ann-Marie Martoglio
Department of Pathology
University of Cambridge
Cambridge, UK
(amm53@cam.ac.uk)

Toshifumi Matsuyama
Division of Cytokine Signaling
Department of Molecular Microbiology and Immunology
Nagasaki University Graduate School of Medical Sciences
Nagasaki, Japan
(tosim@net.nagasaki-u.ac.jp)

Chanaka Mendis
Department of Molecular Pathology
Walter Reed Army Institute of Research
Silver Spring, MD, USA
(chanaka.mendis@na.amedd.army.mil)

D. Scott Merrell
Department of Microbiology and Immunology
Stanford University School of Medicine
Stanford, CA, USA
(dmerrell@leland.stanford.edu;
dmerre01@yahoo.com)

Makoto Miyagishi
Department of Chemistry and Biotechnology
School of Engineering
The University of Tokyo
Hongo, Tokyo, and
Gene Function Research Laboratory
National Institute of Advanced Industrial Science and Technology
Tsukuba Science City, Japan

Samuel C. Mok
Laboratory of Gynecological Oncology
Brigham and Women's Hospital
Harvard Medical School
Boston, MA, USA
(scmok@rics.bwh.harvard.edu)

Catherine Neary
Cellular and Biochemistry Section
Basic Research Laboratory, CCR
National Cancer Institute
National Institutes of Health
Bethesda, MD, USA

Maria Nesterova
Cellular and Biochemistry Section
Basic Research Laboratory, CCR
National Cancer Institute
National Institutes of Health
Bethesda, MD, USA

Shuming Nie
Departments of Biomedical Engineering, Chemistry, Hematology, and Oncology, and
the Winship Cancer Institute
Georgia Institute of Technology and
Emory University School of Medicine
Atlanta, GA, USA
(snie@emory.edu)

Yasushi Okazaki
Laboratory for Genome Exploration Research Group
RIKEN Genomic Sciences Center
RIKEN Yokohama Institute
Yokohama, Japan
(okazaki@gsc.riken.go.jp)

Arthur B. Pardee
Department of Cancer Biology
Dana-Farber Cancer Institute and
Departments of Biological Chemistry and Molecular Pharmacology
Harvard Medical School
Boston, MA, USA
(Pardee@mbcrr.harvard.edu;
Arthur_pardee@dfci.harvard.edu)

Leigh N. Pierce
Immusol, Inc.
San Diego, CA, USA

John D. Porter
Departments of Ophthalmology, Neurology, and Neuroscience, and
The Visual Sciences Research Center
Case Western Reserve University and
University Hospitals of Cleveland
Cleveland, OH, USA
(jdp7@po.cwru.edu)

Teresa J. Raich
Amersham Biosciences
Piscataway, NJ, USA

Janet D. Rowley
Department of Medicine
University of Chicago Medical Center
Chicago, IL, USA

Kimberly Rozett Nemeth
Department of Pathology, Anatomy and Cell Biology
Developmental Biology and Teratology Program
Jefferson Medical College
Thomas Jefferson University
Philadelphia, PA, USA

Myron E. Schwartz
Recanati/Miller Transplantation Institute
New York, NY, USA

Phillip A. Sharp
The McGovern Institute
Massachusetts Institute of Technology
Cambridge, MA, USA
(sharppa@mit.edu)

Avrum Spira
Bioinformatics Graduate Program and
Division of Pulmonary Medicine
Boston University
Boston, MA, USA

Rakesh Srivastava
Cellular and Biochemistry Section
Basic Research Laboratory, CCR
National Cancer Institute
National Institutes of Health
Bethesda, MD, USA

Jay Strum
Quantitative Expression Group
Department of Cellular Genomics
GlaxoSmithKline Pharmaceuticals
Research Triangle, NC, USA
(jay.c.strum@gsk.com)

Shinichi Sugita
Division of Cytokine Signaling
Department of Molecular Microbiology and Immunology
Nagasaki University Graduate School of Medical Sciences, and
Department of Obstetrics and Gynecology
Nagasaki University School of Medicine
Nagasaki, Japan

Kazunari Taira
Department of Chemistry and Biotechnology
School of Engineering
The University of Tokyo
Tokyo and
Gene Function Research Center
National Institute of Advanced Industrial Science and Technology
Tsukuba Science City, Japan
(taira@chembio.t.u-tokyo.ac.jp;
kawasaki@chembio.t.u-tokyo.ac.jp)

Philip K. Tan
Immusol, Inc.
San Diego, CA, USA

Swan N. Thung
Mount Sinai School of Medicine
New York, NY, USA

Martina L. Veigl
The Gene Expression Array Core Facility and The Comprehensive Cancer Center
Case Western Reserve University and University Hospitals of Cleveland
Cleveland, OH, USA

Martin Vingron
Department of Theoretical Bioinformatics
German Cancer Research Center and
Department of Computational Molecular Biology
Max-Planck Institute for Molecular Genetics
Berlin, Germany

José L. Walewski
Division of Liver Diseases
The Mount Sinai Hospital and Mount Sinai School of Medicine
New York, NY, USA
(walewj01@doc.mssm.edu)

Karolyn A. Wanat
Amersham Biosciences
Piscataway, NJ, USA

Luquan Wang
Department of Discovery Technology
Schering-Plough Research Institute
Kenilworth, NJ, USA

San Ming Wang
Department of Medicine
University of Chicago Medical Center
Chicago, IL, USA
(swang1@midway.uchicago.edu)

Kwong-Kwok Wong
Texas Children's Cancer Center
Baylor College of Medicine
Houston, TX, USA
(KKWong@txccc.org)

Flossie Wong-Staal
Immusol, Inc.
San Diego, CA, USA
(fwongstaal@immusol.com)

Jie Wu
Department of Biomedical Engineering
Boston University
Boston, MA, USA

Itai Yanai
Bioinformatics Graduate Program
Boston University
Boston, MA, USA

Kazuo Yamamoto
Division of Cytokine Signaling
Department of Molecular Microbiology and Immunology
Nagasaki University Graduate School of Medical Sciences
Nagasaki, Japan

Guolin Zhou
Department of Medicine
University of Chicago Medical Center
Chicago, IL, USA

Contents

FOREWORD

The revolutionary advance brought by the sequencing of the human genome has empowered the development of numerous new techniques for analysis of changes in gene expression corresponding to biological states and transitions. This technology has evolved from the identification of a few genes whose expression differs between two cellular states to the ability to assay the level of expression of thousands of genes. It is this new and emerging power to confront problems with experiments reporting on thousands of genes that will transform biology and medicine. In some cases, it will simply be a descriptive tool to distinguish subgroups of cellular states such as subgroups of cancers, which share specific features and prognoses. In other cases, it will be the basis for analysis of the network of molecular interactions controlling the cellular state such as a developmental state or cell type. The ability to describe the expression status of thousands of genes will demand the advancement of a science that explains these parameters. This is a very exciting prospect given that most of the interesting biology of multicellular organisms reflects changes in the expression of genes.

This volume provides a wonderful summary of the status of the science of measuring and interpreting changes in gene expression. It begins with the early technology of differential display pioneered by scientists such as Arthur Pardee. In addition, there are several chapters demonstrating the power of this method in lymphoid cells, leukemia cells, and disease states such as colitis. The new technologies for quantitating the expression of thousands of genes—SAGE, cDNA arrays, and oligochips—are described with numerous examples of analysis of normal and disease conditions. These are rapidly evolving technologies where investigators must closely compare the results of current experiments to design future experiments. These chapters are invaluable for this comparison.

Bioinformatics is central to all experiments that generate large amounts of data in parallel. This is perhaps the biggest challenge in the study of the expression of large numbers of genes. Bioinformatics is the handmaiden of the gene expression technology in this volume but is nicely represented.

Finally, several emerging technologies such as the silencing of genes by siRNA and the use of mutational methods to identify the genotype/phenotype relationship are presented. The marriage of this technology to modulate the expression of specific genes in mammalian cells to that of the ability to measure the expression of thousands of genes in the first sections of this book will produce a rich and exciting future of cell biology and medicine.

Phillip A. Sharp, PhD
Institute Professor and Director of the McGovern Institute
Massachusetts Institute of Technology
Cambridge, MA, USA

INTRODUCTION

Never put off doing something useful for the fear of evil that may never arrive.

–James D. Watson

Although this planet has an immense variety of life forms, all living organisms share a master plan of structural and functional organization. This book is about the expression of the building blocks—genes—that constitute the unity of the living world, and will provide the reader with insights into the expression of these building blocks in normal and diseased states. It follows my philosophy that molecular biology must emphasize both molecules and biology, not only at the molecular level but also at the chemical and physical. Cell and molecular biology have become the basic pillars on which all biological and medical sciences are based. The vocabulary of molecular biology is vast because it is drawn from so many disciplines, such as genetics, physiology, biochemistry, cell biology, developmental biology, physics, organic chemistry, and biophysical chemistry. Gene expression (genomics) is a new discipline within molecular biology that narrates the functional organization of genes.

Perspectives in Gene Expression is primarily intended for readers in the genomics field but may be useful in more advanced graduate level courses. I assume that most readers have at least introductory biology knowledge, but I have attempted organize this book so that even a stranger to biology could follow it easily. I hope this text will be useful for working scientists that are in search of a guide to help them choose a particular gene expression strategy. Much of the text should be of interest to those in the applied sciences such as medicine, toxicology, and biotechnology.

This book is organized around the spectacular information provided by the contributors. The chapters are written by top scientists from academia and biotechnology and pharmaceutical industries around the world and provide a timeless collection of theory and applications in various fields. The objective of this book is to provide a meaningful interpretation and critical review of those aspects of the molecular biology of disease that reflect on the role of a cell's genomic information. The study of differences in gene expression patterns is one of the most promising approaches for understanding mechanisms of differentiation and development. In addition, the identification of disease-related target molecules opens new avenues for rational pharmaceutical intervention. Recent technical advances and improvements are accelerating the analysis of gene expression profiles at the transcript level. The knowledge and comprehension of currently applied methods is one of the central criteria for an efficient and successful screening approach. Therefore, several such expression strategies are compiled with several applications in basic and medical sciences. The various themes of the book are summarized below.

Differential Display

Differential display was the first reliable, one-tube method for comparing differential gene expression of two or more cell populations or tissues. This section describes the basic methodology and its applications in the identification of novel genes in various disease processes including breast cancer, leukemia, enterotoxin-stimulated lymphoid cells, and ulcerative colitis. Using this method, some of the investigators have postulated an approach for the development of "fingerprint" or "barcode" methods to classify cell types. Most importantly, some of the markers identified by differential display have been tested for effectiveness as cancer therapies as well as to assess gene function. The chapters contain the latest improvements in differential display PCR and have been extensively tested for general applicability.

SAGE and Validation

Serial analysis of gene expression (SAGE) is a sequence-based approach for the identification of differentially expressed genes through comparative analyses. This method does not depend on prior transcript information and allows the simultaneous analysis of sequences that derive from different cell populations or tissues. This section includes an interesting chapter in which SAGE has been successfully utilized in the genome-wide analysis of gene expression in hematopoietic differentiation. The data obtained from this method will provide a basis for further studies on stem cell biology and hematopoiesis. Another chapter in this section describes a systematic way of procuring human tissue specimens for gene discovery projects. In addition, this chapter also covers validation of genes identified by various expression strategies by using real-time quantitative PCR method.

cDNA Microarrays

Microarray technology provides insight into the transcriptional state of the cell, measuring RNA levels for thousands of genes at once. The design and manufacture of tailored cDNA arrays are beneficial for researchers trying to understand developmental processes or the molecular basis of disease, and to identify potential diagnostic and therapeutic targets for various diseases. Two chapters in this section detail the use of cDNA arrays in the understanding of pathogenetic pathways in ovarian cancer. A separate chapter describes the use of commercial cDNA array formats in the study of embryonic-transcriptome based gene expression patterns in developmental toxicology and teratology. High-density mouse cDNA arrays were developed by the RIKEN group and utilized in the study of metabolic pathways (metabolome) of 49 mouse tissues; this study provides in-depth information about the coordinated regulation of gene expression in mammalian tissues. Another chapter in this section describes the applications of cDNA arrays in the field of antisense research for the elucidation of signal transduction signatures.

Oligochips

Genome-wide expression profiling using commercially available oligonucleotide-based chips has revolutionized the understanding of basic biology and drug discovery in the pharmaceutical industry. The first two chapters in this section highlight the applications of such arrays in the elucidation of molecular pathways involved in skeletal muscle phenotypic diversity and in the discovery of p53-target genes. Another chapter in this section provides the application of a new commercial microarray platform in the identification of genes involved in hepatocellular (liver cancer) carcinoma. The age of genomics has ushered in a wealth of new possibilities for discovery of the repertoire of genes employed by a pathogen as it interacts with and establishes disease in a particular host. In a separate chapter, the initial utilization of these oligochips to gain a more thorough understanding of the intimate host/pathogen interactions employed by *Helicobacter pylori* during colonization of the host is described. Tools such as DNA microarrays and other relevant technology are likely to go a long way toward improving the divining rods currently used to determine cancer type and treatment. Drug researchers, too, take advantage of the "guilt-by-association method" to discover proteins not previously known to operate in biological pathways involved in diseases. Once those proteins are found, they can be enlisted as targets for the development of new and better medicines.

The gene expression technologies described in this book provide clinically relevant and practically cogent information. One of the most important uses of microarrays is to study changes in gene expression that accompany changes in cell physiology, such as during development, cell-cycle progression, drug treatment, or disease progression. In addition, this technology has led to a broad spectrum of applications such as drug target validation, pathway dissection, and annotation of the human genome.

Microarray technology has become one of the workhorses of functional genomics; however, the technology results in large amounts of data, the interpretation of which is still a major bottleneck. Addressing this arena of computational genomics, two chapters in this section describe the details of correspondence analysis and domain fusion analysis. Correspondence analysis is a sensitive and insightful method for microarray data analysis, especially for clustering genes or measurements of discrete groups with similar expression profiles. Most importantly, this method is capable of simultaneous visualization of data in pictorial form for ease of interpretation. The domain fusion analysis described in a separate chapter is a cluster of orthologous groups (COGs) database approach that has been adopted systematically to analyze 30 microbial genomes.

Emerging Technologies

Molecular biotechnology is still progressing toward new technologies that satisfy the demands of gene expression analyses. This section features two chapters on the RNA interference and gene silencing mechanisms, a new frontier in the RNA world. Double-stranded RNA-mediated interference (RNAi) has recently emerged as a powerful reverse-genetic tool for silencing gene expression in organisms including plants, fungi, viruses, *Xenopus laevis*, trypanosomes, Drosophila, *Caenorhabditis elegans,* and in other higher organisms. One natural function of RNAi seems to be protection of the genome against endogeneous transposable elements. The use of RNAi in human cultured cells should dramatically facilitate the dissection of signaling pathways and the study of cell growth and division to understand the biology of cancer. Independent of biomedical applications, sequence-specific tools that interfere with gene expression will be of great demand as tools for functional genomics and as therapeutics, and RNAi will undoubtedly grow to be one of the leading methodologies in the gene expression field. Two chapters described in this book detail the identification of RNAi and gene repression phenomenon in mammalian cells with the help of a novel engineered siRNA expression vector system.

Another chapter in this section describes the applications of the Inverse Genomics platform that is based on ribozymes—a method to identify genes that are specifically linked to a phenotype. Using this platform technology, more rapid gene discovery is possible because the ribozyme target recognition sequence can be utilized to directly identify a gene, forgoing the much slower positional cloning approaches used in traditional forward genetics. This approach will also be useful in the development of therapeutic genes. Another group uses random activation of gene expression (RAGE) to create normalized, genome-wide protein expression libraries; thus, novel genes can be efficiently identified by virtue of their overexpression imparting phenotypes upon a host cell. Likewise, because of its method for up-regulating gene/protein expression on a random basis, RAGE enables researchers to conduct small molecule screening programs against previously validated drug target(s) of interest without the need for cDNA or other cloned and isolated gene sequences. A very interesting chapter in this section describes the development of a new series of color dyes, called semiconductor quantum dots, for applications in gene expression studies. This technology platform will be ideal for real-time detection of single-base mutations and expression profiling studies.

The beauty of these emerging technologies is that they help us understand the function of the genome and provide a unifying mechanism for targeted restructuring of the epigenome. The bridging of transcriptome, proteome, and RNome data will be the next challenge to the post-sequence functional and systematic genomics era; before we face that challenge, though, single-cell gene expression profiling, or stochastic gene expression, in single cells will be crucial to the understanding of coordinated transcription events and organization of gene expression. This can be achieved on the cellular genomics level. The ability to collect a binary "snapshot" of genes that are "on" or "off" in a single cell not only offers new insights into transcript regulation but also allows the understanding of causes and effects of expression. We are entering into an age of "whole-istic biology," where we need to study the biological pathways and systems involved in physiology and disease. Systems biology approaches will integrate and analyze biological information across the drug discovery disciplines of pharmacology, toxicology, molecular biology, chemistry, and clinical development.

A new age of predictive medicine lies before us. The expression strategies and emerging technologies described in this book will serve as examples of the possibilities for integrated healthcare and personalized medicine. One day gene expression profiling performed on DNA arrays and other technology platforms could replace the classical histological examination of patients' tissues in pathology laboratories.

Many people have contributed to making my involvement in this project possible. Born a Hindu, I always remember the slogan from Sanskrit, the ancient language, "Maatru Devo Bhava, Pitru Devo Bhava and Guru Devo Bhava" (One should treat mother, father, and teacher as the gods for giving the existence in the world). It is my privilege to thank my parents for my existence in the educational enterprise. I thank my teachers who have taught me the wonder of science and helped me become a scientist. Thanks to my scientific mentors Debi P. Burma and Maharani Chakravorty-Burma (associates of Nobel laureate Severo Ochoa and Bernard L. Horecker) who trained me in the wonderful saga of science during my PhD studies at Banarus Hindu University, India. I give my special thanks to my postdoctoral mentors Edward B. Goldberg (associate of Nobel laureate A.D. Hershey), Robert E. Hausman, and Nobel laureate Har Gobind Khorana for their excellent guidance and brilliant ideas, all which I absorbed.

Many people have had a hand in the launch of this book. Each chapter has been passed back and forth between the authors for criticism and revision, so that each chapter represents a joint composition. I am indebted to the staff at BioTechniques Press, Eaton Publishing, and in particular to Christine Emerton for her kindness, generosity, and efficiency throughout, which have made the enterprise of editing a pleasure as well as an education for me. And last but not least, I thank Nobel laureate Phillip A. Sharp for his kindness and support in writing the Foreword to this book.

Krishnarao Appasani
March 2003

To my wife Shyamala and my sons Raakish and Raghu

Section I

DD-PCR

1 Discovery of Expressed Genes by Differential Display and Their Applications

Heide L. Ford[1], Debajit K. Biswas[2], Katherine J. Martin[2], and Arthur B. Pardee[2]
[1]Departments of Obstetrics and Gynecology and Biochemistry and Molecular Genetics, University of Colorado Health Sciences Center, Denver, CO, and [2]Dana-Farber Cancer Institute, Boston, MA, USA

"*Expression genetics is a conceptually different approach to the identification of cancer-related genes than the search for mutations at the genomic level.... Cancer geneticists have ignored the productive potential of examining downstream events based on screening for differential gene expression.*"

—Ruth Sager (58)

EXPRESSION GENETICS

The 30 000 to 100 000 genes in each human cell provide a library of blueprints for all the possible cellular structures and functions. In a given cell, only some of these genes, possibly 20 000, are expressed as their mRNAs. These depend on the type of tissue, cell, and environmental conditions that determine the properties of that cell type. Furthermore, expressions of a few percent of the genes are modified in abnormal cells, as in cancers. Thus, gene expression profiles provide fundamental information regarding the molecular basis of both normal processes such as differentiation and of diseases. The first problem is to identify the expressed genes.

Expression genetics has numerous applications. In cancer biology, these include: (*i*) identification of markers useful for diagnosis (1); (*ii*) discovery of relevant gene expressions and determinations of their functions; (*iii*) identification of markers that determine the characteristics of a cancer; (*iv*) determination of potential therapeutic responses; (*v*), characterization of the metastatic capability of a tumor; (*vi*) early tumor detection; and (*vii*) assessment of response and recurrence following treatments, including surgery. We will illustrate these applications primarily with examples from our research as well as several others in the literature.

Perspectives in Gene Expression
Edited by K. Appasani

Techniques invented in the past decade to identify gene expression patterns have replaced prior useful but laborious methods (40). Gene expression patterns determine the amounts of gene-specific mRNAs or proteins in cells. The most frequently applied method is cDNA differential display (DD) (33) and the similar technique, arbitrary primed PCR fingerprinting (69). DD can simultaneously display the products obtained from a considerable number of samples. It is widely applied to the discovery of genes that are differently expressed under various conditions. For example, comparisons are often made between gene expressions in cancer versus normal cells. In the PubMed database, there are approximately 2500 references to "differential display + gene," of which 200 had appeared after mid-2002.

DD has been described in detail elsewhere (33,36). The process consists of several steps. First, total RNAs are isolated from the cells of interest, and then 3′-ends of several hundred mRNAs are reverse-transcribed using a 5′-T_{12}-XY-3′ primer; X and Y are other bases at its 3′-end. These products are then are amplified as cDNAs by PCR, with the original primer plus a selected arbitrary primer of known sequence. The products are separated by size on a sequencing gel, and the cDNAs in bands of interest, generally those differing in intensity between samples, are confirmed by an independent method. The selected cDNAs are then sequenced and investigated further by comparison with databases and other methods, such as the expression of the corresponding full-length cDNA into the appropriate cells.

There are advantages and disadvantages to DD. An advantage is that it allows for the detection of sets of differentially expressed genes, which are found without bias, that are related to the phenomenon under investigation. Many genes will be previously unknown, and others will be recognized when identified by sequencing and by comparison with databases. This process for discovery is in contrast to the increasingly applied DNA microarrays, which generally permit the search for genes that are expressed within a set of known genes (21). The DD technique is fundamentally simpler and quicker than other methods, and it requires equipment that is commonly available in molecular biology laboratories, such as for PCR and gel electrophoresis. Very little starting material is needed.

An initial problem with DD was the high frequency of false positives obtained. This frequency varied greatly between investigations, probably depending largely on the meticulousness of the applied technique. Recent improvements, such as in primer design (48), have reduced false positives to a very acceptable level of about 5% (37,38), yet it is very important to verify a candidate gene's differences with an alternative technique such as Northern blot analysis. Another problem is that the number of primer pair reactions that need to be applied for the saturation of an mRNA pool is estimated at several hundred. However, only a few pairs are sufficient to reveal several differentially expressed genes. The determination of function of any one of these genes can thus be a formidable task.

DISCOVERY OF EXPRESSED GENES AND THEIR FUNCTIONS

In all gene discovery techniques, the most difficult hurdles include choosing which differentially expressed gene(s) to pursue and determining the relevance of the newly discovered gene(s) to the process of interest. Because thousands of genes have been identified by DD, there are many more than we can address in this review. Here we will provide examples from our laboratory and others, in which genes initially discovered through DD were further characterized to determine their functions. However, we will first address how DD may itself allow one to decipher gene function.

DD as a Tool to Assess Gene Function

An outstanding example of how DD can be utilized to identify gene function was recently published by Ryo et al. (56). This group was interested in understanding the function of Pin1, a peptidyl-prolyl *cis-trans* isomerase that isomerizes phosphorylated Ser/Thr-Pro bonds (52,72) of a protein, thereby affecting its activity (59,64,71). Since overexpression of Pin1 is observed in breast cancer, and previous work demonstrated that this overexpression was accompanied by changes in gene expression, a system using a modified DD approach was employed to examine gene expression changes before and after the induction of Pin1.

Of the 12 genes identified as up-regulated by Pin1 overexpression, four were downstream of β-catenin/TCF, leading the authors to conclude that Pin1 might regulate β-catenin activity and therefore affect its transcriptional targets. Indeed, they demonstrated that Pin1 directly binds to a phosphorylated Ser-Pro motif on β-catenin, inhibiting its interaction with the adenomatous polyposis coli (APC) protein, a protein that normally promotes the nuclear exclusion and degradation of β-catenin (26,29,43,54,55). Therefore, the inhibition of the APC/β-catenin interaction by Pin1 leads to the stabilization of β-catenin and increases its presence in the nucleus. This, in turn, allows it to activate transcription. Thus, in this case, DD identified a novel, previously undescribed function for Pin1.

Discovery of a Novel Function for a Previously Cloned Gene Identified by DD

The human Six1 gene (HSIX1) was identified in our search for genes that are differentially expressed toward the end of the S-phase in a comparative study between normal and human breast cancer cells (19). The expression of such genes might therefore be preparative for events that occur late in the cell cycle. We reasoned that some genes identified in this manner may be important for the transition into mitosis; an event that, if improperly regulated, could affect genomic stability and thus tumorigenesis.

We carried out DD after isolating RNA from 21PT mammary carcinoma cells that had been synchronized at the G1/S boundary and then released into the cell cycle. With a particular primer pair, several genes were observed whose expression increased in these tumor cells relative to normal controls as they exited the S-phase. It should be noted that because DD is a gene discovery tool, relevant genes can be discovered without examination of all changes, and thus the laboriousness of successive DD experiments is often unnecessary. Therefore, although the number of primer pairs utilized in this experiment was far less than needed to saturate the mRNA pool, interesting gene changes were observed, and the decision was made to pursue these changes rather than to attempt to discover every gene change.

The Choice to Pursue a Particular Gene Identified by DD (HSIX1)

Upon sequencing the several cDNAs of interest, only one, HSIX1, had a known match in the database. The others were novel cDNAs. HSIX1 had previously been cloned from human adult skeletal muscle (9) by virtue of its homology to the human myotonic dystrophy-associated homeodomain protein (DMAHP) (8). The mouse homolog of HSIX1, *six1*, is implicated in limb development and is 62% similar to a *Drosophila* gene, *sine oculis* (*so*) (46). The choice to pursue the HSIX1 function arose from several clues in the literature. First, the *so* gene plays a role in the development of the fly visual system, which, interestingly, involves coordinate regulation of cell cycle progression. *So* expression precedes a wave of mitosis, suggesting that it could be involved in this transition (11). Because HSIX1 expression increases as cells exit the S-phase, it is possible that, as suggested of *so*, it may be involved in the transition into mitosis. The choice to pursue this particular gene was made because we were interested in cell cycle transitions.

The second rationale for pursuing the role of HSIX1 in cell cycle control and tumorigenesis revolved around its belonging to the homeobox gene superfamily, which encodes proteins that act as transcription factors to regulate the coordinated expression of genes involved in development and differentiation. Both tumorigenesis and normal development involve alterations in cell proliferation and differentiation, cell death, neovascularization, cell motility, and invasion of surrounding tissue (31). Therefore, homeobox genes involved in normal developmental processes may contribute to tumorigenesis if mis-expressed. Indeed, a large number of homeobox genes have been implicated in tumorigenesis, as extensively reviewed (12,13,18).

Taken together, we believed that HSIX1 might be an important gene in regulating cell cycle control, thereby affecting tumorigenesis. Proving or disproving this theory would lie ahead. We then decided to set aside the unknown genes to focus specifically on HSIX1. The pursuit of a completely novel gene with no known homologies in the database could pose a much more difficult question and depend on the robustness of the expression changes and if the gene contains relevant domains (e.g., putative kinase domains and DNA binding domains), which may be deemed important to the process of interest.

How to Determine Relevance and Function of the Differentially Expressed Gene

There are many ways to determine the function of a differentially expressed gene. After the comparison with databases to assess the novelty of the gene and to determine if it contains any conserved domains, the next obvious step is to overexpress and/or knock out the gene of interest in a tissue culture system. We chose to stably overexpress HSIX1 in the mammary carcinoma cell line MCF7, in which endogenous levels of the mRNA are low, and then to examine cell cycle profiles (19). Overexpression of HSIX1 in this system attenuated the DNA damage-induced G2 cell cycle checkpoint. When these cells were injected into the flanks of nude mice, HSIX1 overexpression led to a statistically significant increase in tumor burden (our unpublished data). The observation of a phenotype in multiple clones with two separate assay systems suggests that overexpression of HSIX1 occurs in these cells. Together, these data imply that the role of HSIX1 in tumorigenesis may be due to its effect on the cell cycle.

But what is the relevance of this gene to human breast cancer? To answer this question, we examined the level of expression of HSIX1 by Northern blot analysis in breast tumor biopsies as compared to adjacent normal breast (Figure 1). Of the 35 breast biopsies examined, HSIX1 was overexpressed in 44% of primary breast cancers and 90% of metastatic lesions compared to controls (19). This underscored our belief that this gene plays an important role in the etiology of breast cancer.

Our Job Is Not Done

While the above-mentioned data suggest a role by HSIX1 in cell cycle control and tumorigenesis, there are still many questions to be answered. These include how HSIX1 is regulated in the cell cycle, both transcriptionally and post-translationally (20). As HSIX1 is a transcription factor, it is important to identify its transcriptional targets. Identifying targets of transcription factors can be done in numerous ways, including DD or microarray analysis after the induction of the gene of interest. Both these methods will allow for further elucidation of its function.

Another area to pursue is that of the role of HSIX1 in vivo. Because HSIX1 is a homeobox gene, its role in normal development may provide us with insight into its function. It is possibly involved in the normal development of the mammary gland through its role in cell cycle control and its dysregulation then leads to tumorigenesis. Knockout models and mammary-specific inducible transgenic models should allow us to address this question.

Identifying the Function of Novel Genes Identified by DD

Excellent examples come from Okamura et al. (45) and Tanaka et al. (65), who used DD to identify targets of p53. Two identified targets, p53-inducible ribonucleotide reductase small subunit 2 homolog (p53R2) and p53-dependent damage-inducible nuclear protein (p53DINP1), were unknown genes that have since been well characterized by the group. P53R2 was identified as a p53-inducible gene that contains a 351 amino acid peptide, with high homology to regions of both human and yeast ribonucleotide reductase (65). While the gene was novel, this homology made the task of deciphering its role somewhat simpler. After demonstrating that p53R2 is a direct target of p53, the authors confirmed that it was induced following DNA damage in a p53-dependent manner. Both by ectopically expressing p53R2 in p53-deficient cells and by inhibiting it with antisense oligonucleotides, they were able to demonstrate that p53R2 encodes a ribonucleotide reductase important in the repair of damaged DNA at the p53-induced checkpoint. Thus, the ability of p53 to induce p53R2 in response to DNA damage, which then aids in DNA repair, is one mechanism by which p53 suppresses tumorigenesis (65).

P53DINP1 was also isolated by DD after p53 induction (45). However, unlike p53R2, it had no identity or homology to known genes. As previously done for p53R2, the authors quickly determined that it was dependent on p53 for induction after genotoxic stress. However, this still did not identify the function of the gene. Therefore, they used an antisense approach to assess the function of

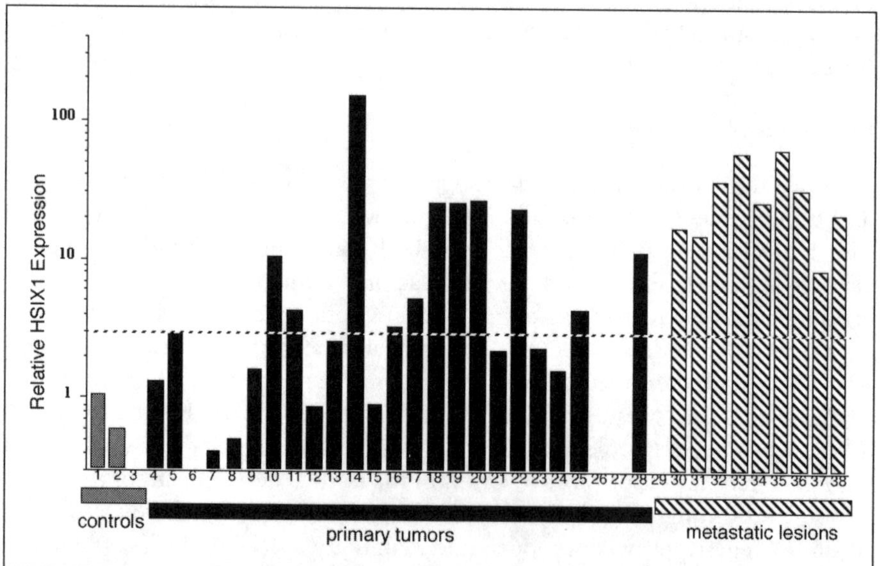

Figure 1. Quantitative representation of Northern blot analysis. Three control tissues Lane 1, adjacent normal breast; lane 2, normal luminal cells; and lane 3, normal myoepithelial cells. We tested twenty-five primary breast tumor biopsy samples (lanes 4–28) and 10 metastatic lesions (lanes 29–38). The blot was stripped and re-probed with 36B4 for normalization, and the relative HSIX1 expression was plotted. A 3-fold increase over adjacent normal breast was considered positive for HSIX1 and is marked by a dashed line. Forty-four percent of the primary and 90% of the metastatic lesions express greater than a 3-fold increase in HSIX1 mRNA over the adjacent normal breast control. It is unknown whether the patients that expressed HSIX1 in the primary tumor developed metastatic disease.

p53DINP1 and found that cell death following DNA damage was inhibited when p53DINP1 protein was down-regulated. This approach further demonstrated that p53DINP1 was involved in the Ser46 phosphorylation of p53 and in the induction of p53AIP1, another recently identified p53 target involved in p53-induced apoptosis (10). This methodology coupled with overexpression analysis led the authors to conclude that p53DINP1 regulates p53-dependent apoptosis by serving as a cofactor for the p53 Ser46 kinase, thus enabling the phosphorylation of p53.

Conclusion

Once a gene is discovered via DD and confirmed to be differentially expressed, many techniques can be employed to determine its function. The methods used will, of course, vary enormously with the type of gene discovered; therefore, we cannot address all possible solutions and outcomes. However, some general principles can be surmised from this review. First, one need not exhaust all possible gene changes when performing DD. Rather, differentially expressed genes can be cloned and sequenced, and those that appear interesting (i.e., through known functions, homologies, and robust expression profiles) can be pursued further. When a new gene is identified (or an already identified gene is found in an unknown context), the first step should always be to compare the gene to available databases (e.g., GenBank® and motif databases). An enormous amount of information can be gained by homologies to other proteins and by conserved functional domains. The second natural step for identifying gene function is to overexpress and/or knock out the gene and then determine whether the phenotype observed is relevant to the original conditions compared (e.g., tumor vs. normal, onset vs. exit of S-phase, and response to DNA damage). Inducible expression systems can then be utilized to examine pathways that may be affected by the gene of interest. This can be examined by again using DD or by another methodology such as microarray analysis or proteomics. Although not described in this review, protein-protein interaction studies can also provide insight into the function of an unknown, newly identified gene. Finally, animal models are important for determining the relevance of the observed gene change in a physiological setting. These are only some of the many analyses that can be performed to assign function to a differentially expressed gene. If an interesting gene expression change is discovered, then it can easily lead to a long period of stimulating experiments.

PROPERTIES AND DETECTION OF CANCERS

An important objective of current cancer research is to understand, at a highly detailed level, the changes in gene expression that take place during tumor formation. The expression of approximately five percent of a cell's genes (approximately 1000 total genes) can be altered in tumor cells compared to normal cells (37,67). Knowledge of the nature of these expression changes can allow the development of powerful new assays to detect and identify cancer cells. In addition, since tumor formation is a multistep process in which sequential changes occur as cells progress from normal to premalignant, carcinogenic, and finally metastatic cancer, understanding of the specific expression changes that correspond to specific stages of tumor cells can lead to an entirely new generation of assays for determining prognosis, predicting optimal treatment, and developing novel anticancer therapies.

In principle, the concept of using expression changes as a cellular "fingerprint" or "bar code" to classify cell type is straightforward. However, in practice, there are numerous hurdles. Cancer cell genetics is a particularly complex subject; numerous genetic changes combine to elicit the disease, and the specific changes can vary considerably even within a given clinical entity. Further, the process of sorting out informative markers from the vastness of the genome is a significant challenge. Several gene expression technologies, including subtractive hybridization DD, cDNA microarrays, and serial analysis of gene expression (SAGE), have recently made significant strides toward the development of clinically useful gene expression assays for cancer care (17,21,25,37,38,49,61). Likewise, protein expression patterns analyzed by using 2-D gel electrophoresis and mass spectrometry represent a useful approach to the development of cancer fingerprinting methods (24).

Our work with DD and cDNA microarrays is summarized here as an example of progress toward the development of clinically useful gene expression assays. We have achieved accomplishments in two specific areas. First, we identified a set of genes whose expression distinguished between different clinical classes of breast cancer biopsies. Second, we identified a set of genes that detected early breast cancer in peripheral blood samples. Our studies, like most similar studies reported to date, focused on relatively small groups of clinical samples (20–50), and hence our results are indicative of the those that will likely follow from full-size studies of very large sample populations.

Method Summary

To identify cancer-related gene expression changes, we developed a two-step screening approach. The first step, DD, was used to search all cellular mRNAs, known and unknown, for candidate "cancer marker genes" whose expression level was altered in comparisons between malignant and normal breast epithelial cells. The principle of DD was outlined earlier. The second step, custom cDNA microarrays, then allowed us to screen these candidate marker genes and to identify those that were the most informative in clinical samples. The microarray procedure made use of nylon membranes that we spotted with an orderly matrix of PCR-amplified cDNA tags for 100–200 different genes. The arrayed genes included our DD-identified candidate cancer marker genes as well as literature-identified markers and control genes. Expression levels of all arrayed genes were measured in individual biological samples (e.g., blood, tissue, or cultured cells) by assessing the amount of cellular cDNA that bound by a hybridization reaction to the membrane-fixed tag.

An important advantage of this approach is its sensitivity, which is approximately 10 000 times greater than glass arrays (34,38). This sensitivity was a key feature in the detection of low numbers of tumors cells disseminated in the peripheral blood.

Gene Expression Changes in Breast Cancer

To date, we have used DD to identify nearly 200 genes differentially expressed between the highly metastatic breast tumor cell line MDA-MB-435 and normal breast epithelia (37), immunomagnetically separating myoepithelial and luminal epithelial cells from the primary cultures of normal breast tissue (14). These represent approximately one-third of all genes aberrantly expressed in our comparisons. Interestingly, a marked contrast was seen in the types of genes that comprised the down-regulated versus the up-regulated categories. These changes strongly reflect our established knowledge of tumorigenesis pathways. Nearly 75% of the genes that decreased in the tumor cells play roles in maintaining the normal cell phenotype, including filamentous, cell surface, and secreted genes important in the adhesion, communication, and maintenance of cell shape. Similarly, a SAGE study that compared normal mammary epithelial cells in situ with invasive and metastatic breast carcinomas showed that many of the genes lost in carcinomas encoded secreted proteins, cytokines, and chemokines, implicating abnormal paracrine and autocrine signaling in the initiation of breast tumorigenesis (50). In contrast, we found that 75% of the genes increased in tumor cells had roles in conferring the tumor cell phenotype (e.g., enzymes involved in metabolism, macromolecular synthesis, and disruption of the extracellular matrix) (37).

cDNA Microarray Results

We used microarrays to study the mRNA expression patterns of 26 different samples, including both cultured cell lines and archived tumor biopsy tissue (37). We found 35 genes whose expression levels distinguished between three different clinical classes in breast cancer: estrogen receptor (ER) status, tumor size, and tumor stage.

These 35 genes represented four different types of expression changes, or gene clusters. Two gene clusters had expression patterns that strongly correlated with ER status but were inversely related to each other. The p53 gene represented a cluster that was overexpressed in ER-positive tumors, while the maspin gene represented a cluster overexpressed in ER-negative tumors. A third cluster represented by HSP-90 was associated with clinical tumor stage. A fourth cluster, including keratin 14, was associated with tumor size. The expression levels of these genes allowed these breast tumors to be classified into categories that currently are of central importance in the treatment of breast cancer. ER status is key in predicting tamoxifen sensitivity, while tumor stage and size are indicative of tumor aggressiveness and patient prognosis.

The two ER status-associated gene clusters were the largest and the most statistically significant of those we identified. Similarly, remarkably distinct ER-related expression clusters have been found in several other studies of breast tumors using genome-wide cDNA microarrays (22,49,70). Microarrays have recently been used to divide ER-positive breast tumors into subgroups that differed in the outcome of the disease (62,66). Work in the direction of identifying and subclassifying breast cancers based on the estrogen pathway is particularly important because as many as 40% of breast cancers currently classified as ER-positive by the conventional ligand-binding assay do not respond to anti-estrogen therapy (2). A likely reason is that these nonresponsive tumors express defects at various stages along the estrogen pathway or other growth regulatory pathways (5). Tumors that express ER do not necessarily have an intact estrogen response pathway, nor do they necessarily require estrogen for their growth. Mutations can render the ER pathway constitutively active, inactive, or irrelevant to a tumor's growth. A new assay that examines numerous genes along key pathways has strong potential to accurately predict anti-estrogen responsiveness by not merely determining "ER status," but rather, the more relevant issue of "ER pathway status" (6). More accurate predictions of nonresponsiveness would allow for a significant number of patients to consider receiving alternative treatments.

Detection of Cancer Markers in Blood Samples

Early detection is an effective means of reducing cancer mortality. Mammography, in particular, has been widely considered to reduce the relative risk of breast cancer death by 15%–20% by improving the options and chances for successful treatment (23), although much confusion has been generated by recent reviews of these data (41,47). However, numerous cancers currently lack any means of early detection (e.g., lung, ovarian, and pancreatic cancer), while others have only partially effective assays (e.g., prostate and colon cancer). A blood-based gene expression assay for the detection of disseminated solid tumor cells presents a powerful tool, with the potential to detect and identify cancers originating from virtually any primary site in the body.

Solid tumors as small as 2 mm in diameter typically display active angiogenesis (68) and are therefore capable of releasing cells into the peripheral blood. Numerous PCR and immunocytochemical studies using individual breast tumor cell markers have reported the detection of disseminated tumor cells in the blood of breast cancer patients, including those with early, localized disease (15,30,51). In the early stages, the disseminated cells are not yet capable of forming metastases. These cells, however, represent an easily sampled source for the detection of cancer.

Numbers of disseminated solid tumor cells present in the blood have been previously quantified using individual and double markers by a variety of immunology-based and PCR-based techniques. Maximum reported levels of tumors cells in the blood of breast cancer patients are approximately 3×10^3 cells/mL (30,51), together with 5×10^6 normal lymphocytic cells/mL. Averages reported for earlier stages of disease include: localized breast cancer, 0.8 cells/mL; invasive breast cancer that has spread to lymph nodes, only 2.4 cells/mL; and metastatic breast cancer, 6 cells/mL (51). Thus, a sensitivity of 1 in a million cells is necessary for measurements of disseminated tumor cells in blood. Standard microarray approaches are not sufficiently sensitive; limits for genome-wide glass microarrays are typically 2000 cells/mL blood (2 in 10^6 cells or 1 in 3×10^5 RNA transcripts) (34). Our cDNA microarray technology detected tumor cells in the blood of early breast cancer patients at 5 cells/mL (1 in 10^6 cells or 1 in 10^8 transcripts) (38).

We identified a set of 12 cancer marker genes that were generally expressed at higher levels in the blood of breast cancer patients than in the blood of healthy individuals (38). In this study, 48 microarray experiments were performed using blood samples from 41 women (15 normal volunteers and 26 breast cancer patients). The 12 genes included previously known markers for breast cancer (e.g., the oncogene mdm-2) and our novel genes. Together the 12 marker genes accurately identified (10 of 13; 77%) of patients with invasive breast cancer (Figure 2). These genes were generally not elevated in patients treated with chemotherapy (29%; 2 of 7) or in patients with local disease (ductal carcinomas in situ [DCIS]) (17%; 1 of 6). Initially, three false positives resulted among the 15 healthy volunteers (19%); however, two follow-up tests of one of these three corrected the result, bringing our false-positive rate to 13% (2 of 15).

We found the identities of the 12 individual genes interesting. Several were previously described as useful markers for tumor cells in the blood. Maspin, in particular, is a breast-specific serine-protease inhibitor (60) described as a useful marker for the PCR-based detection of breast cancer cells in the blood (35,57). Although maspin functions as a breast tumor suppressor that inhibits tumor cell motility and angiogenesis (60,74), its levels are relatively high in early cancer and decline later during tumor progression (53,75). It may be that expressed maspin will not detect advanced breast cancers.

Additional blood marker genes that we found include Mdm-2, a cellular oncoprotein that is overexpressed in 30%–70% of breast tumors (10,63). Levels of mdm-2 in tumor tissue correlated with patient survival (27). CD44 has also been described as useful blood marker for breast cancer (42). Gro-α is overexpressed in colon cancers (16) and induces the migration of breast tumor cells (73). HSIX1 is a homeobox transcription factor that is elevated in breast and other cancers (18,28,32) and was addressed earlier in this review.

The inclusion of a large number of marker genes is essential for a useful cancer assay. Assays that test individual markers, including the highly sensitive PCR and immunology-based studies cited above, were able to detect circulating tumor cells in fewer than 50% of the cancer patients tested (30,51). These low detection rates likely result from the wide genetic diversity of cancer. In particular, microarray studies showed that a significantly larger degree of variability in expression gene patterns exist among breast cancers, even within individual clinical groups, than among normal breast tissue (49). This genetic diversity is best accommodated by assays that incorporate numerous markers.

We envision the expansion of panels with both additional breast cancer markers and markers for numerous cancers from other tissues. In this manner, the assay will detect many types of cancer, the specific marker profile indicating the cancer source. We envision secondary assays that can sub-classify cancers. For instance, a breast cancer panel of markers might determine ER status, prognosis, tumor stage, grade, size, and optimum therapy, all from the gene expression profile from a blood sample. A blood-based assay for the early detection and characterization of cancer has the potential to markedly impact cancer care.

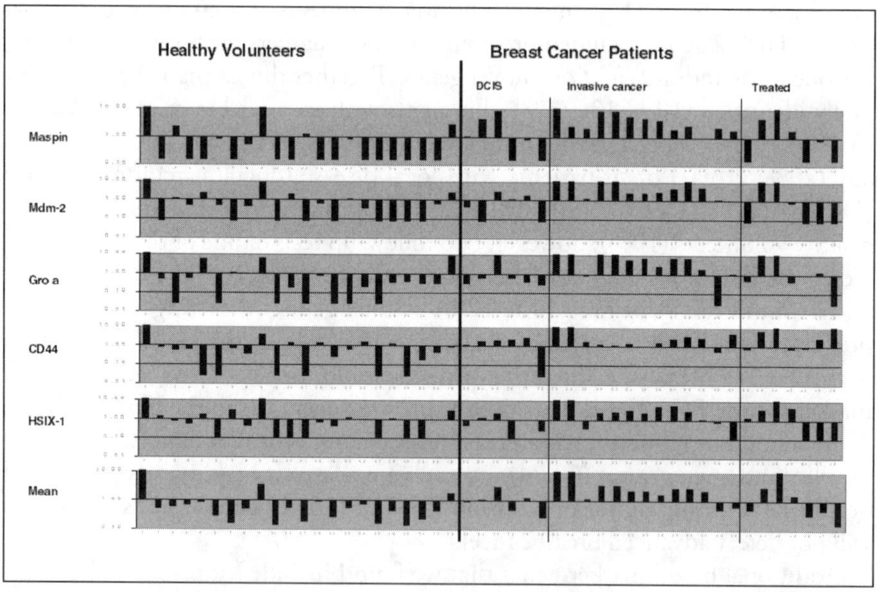

Figure 2. Expression levels of candidate breast cancer marker genes in 48 blood samples. Healthy volunteers included 15 women with no history of breast disease; patients included 26 women with a variety of stages of breast cancer. Cancer patients are ordered by stage of disease at the time of blood collection. Within the category labeled "Invasive cancer," the order is from early (Stage I) to advanced (Stage III) disease, from left to right. "Treated" patients received chemotherapy prior to blood collection. Expression levels were normalized to the mean expression level for each gene. "Mean" data represent the averages of all 12 blood markers reported in the study, including maspin, mdm-2, gro-α, CD44, HSIX-1, srp, myosin light chain, tubulin, laminin α3, N33 gene, and two unknown genes.

TESTING EFFECTIVENESS OF CANCER THERAPY WITH DD MARKERS

Testing therapeutic protocols for cancers is complicated due to the lack of general criteria for determining their effectiveness. This is currently assessed by the eventual disappearance or recurrence of cancer-associated clinical manifestations. But, with few exceptions (prostate-specific antigen [PSA] in prostate cancer), potentially disease-related markers such as those described above have not often been utilized as criteria for assessing the effectiveness of therapy. A significant advance would be made if expression levels of selected genes could be correlated with the regression or progression of other cancers after therapy.

Because disseminated tumor cell markers can be detected in a very small sample of peripheral blood of cancer patients, the determination of expression levels of tumor-specific genes could provide a simple and highly sensitive assay. This technique is feasible for routine examinations because body fluids are easily accessible. We are approaching this goal by using a tumor-bearing mouse model (5,6), and our preliminary data are summarized here.

Individual breast cancers show different responses to therapies. Some (ER+) have receptors for the ovarian hormones estrogen and progesterone, which can enhance their cell proliferation. Therapy for these cancers is mainly based on the application of anti-estrogens such as tamoxifen. However, many cancers are unresponsive to anti-estrogens and are stimulated by alternative cell proliferation signals and pathways. One such pathway is initiated by the interaction of epidermal growth factor (EGF) with overexpressed EGFR family receptors. Improved therapies against these cancers are badly needed.

To perform tests of drug efficacy, we utilized a compound Go6976 that we demonstrated blocks the growth in culture of ER-Her2/neu+ human breast cancer cells and induces their apoptotic death. This drug's mechanism involves the interaction of EGF with its receptor, which activates a series of kinases, including PIK3, protein kinase C, and Iκ kinase, which phosphorylates the inhibitor IκB. This reaction releases IκB from NF-κB, and NF-κB is then translocated into the nucleus where it binds to specific sequences in promoter regions and transactivates responsive genes. The products of some of these genes regulate cell cycle progression through the production of cyclin D and also prevent apoptosis (Figure 3). These properties qualify NF-κB as a potential novel target for cancer therapy.

Go6976 inhibits the catalytic activity of the protein kinase C subspecies alpha and beta and thereby effectively blocks the activation of NF-κB and subsequent downstream anti-apoptotic and pro-growth events in the subclass of ER-Her2/neu+ cells. It is thus a potential agent for the treatment of ER-breast cancer patients, for which available therapies are currently very limited.

Go6976 reverted ER-cancer cells from their highly proliferative and anti-apoptotic state to the reduced proliferation and apoptotic phenotypes of normal cells (5,6). Repeated treatments with Go6976 inhibited the growth of tumors produced in syngeneic mice after the implantation of ER-mouse mammary epithelial cancer cells. Strikingly, these tumors regressed and did not reappear following treatment with Go6976. Very importantly, this compound did not show any toxic side effects and did not cause any damage to the vital organs of the treated animals, such as to the liver, lung, and spleen, even after intravenous administration of the compound for 16 weeks.

To initiate tests of the efficacy of a therapy, we used ER-tumor-bearing mice exposed to Go6976. At intervals, we examined the blood and tumor tissues of untreated and treated mice for levels of some of the differentially expressed genes that were detected in the tumors and blood of breast cancer patients. The expression levels of two genes, MDM2 and Maspin, which were elevated in blood of breast cancer patients (38), were increased in the tumor tissues and blood of animals bearing tumors and were elevated with the progression of the disease, as measured by arrays and also by quantitative PCR (Biswas et al., unpublished data). These levels initially increased sharply after treatment. This effect could be explained by the release of cells into the blood upon the regression of the tumor caused by the anti-tumorigenic agent. We are testing the prediction that an effective treatment will eventually cause the markers to return to their levels in the blood of normal animals.

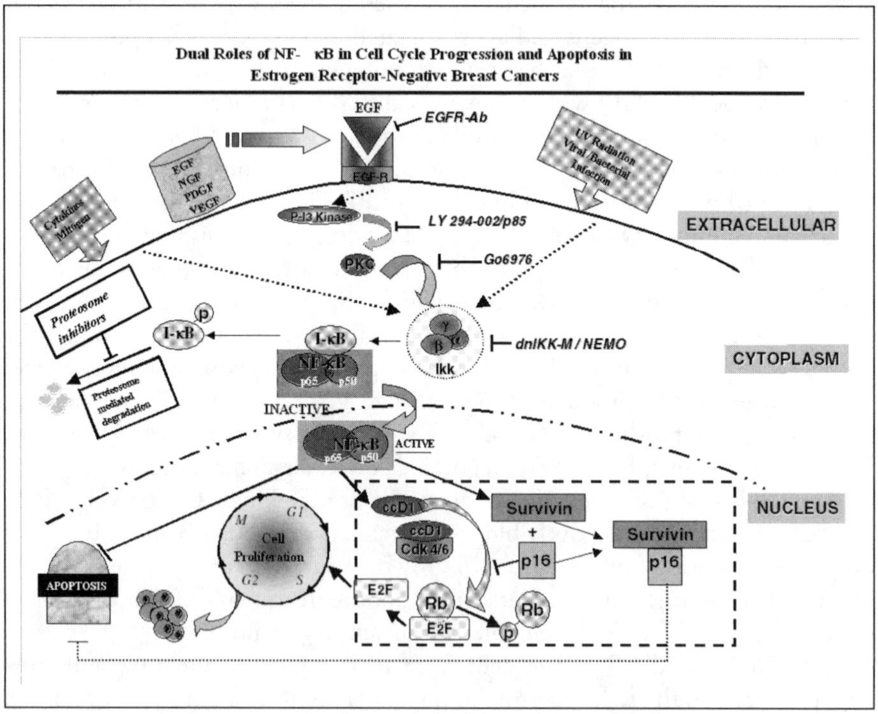

Figure 3. Schematic diagram of action of G06976 in the EGF-stimulated pathway.

These results suggest that gene expression profiles could be applied to monitor therapeutic efficacy in breast cancer patients. This experimental approach can be adopted for the analysis of both biopsy specimens and the more readily available blood samples from patients before and after treatment. These studies will be important for the identification of clusters of genes overexpressed in the tumor tissue or detected in blood that are down-regulated by specific therapies. This procedure is potentially useful for monitoring the efficacy of current and new therapies, including target-directed therapeutic agents such as calmodulin antagonists (4) and inhibitors of activation of NF-κB (5,6).

SUMMARY

The identification of novel differentially expressed genes provides potential targets for molecular-based therapies. Genes that are specifically expressed in cancers can also provide novel detection and diagnostic markers. Current methods of cancer chemotherapy are like trying to put out a fire when the house is already in flames. Mammography, pap smears, and PSA provide early warning systems to permit more effective therapies. We need to invent molecular biologically based "smoke-detector" techniques that provide very early warning of all kinds of cancers. Tumors detected early might be more effectively treated than advanced ones, by current and possibly novel modes that are less stringent. These methods would permit the treatment of fewer cancer cells before their mutations, invasion, distant metastases, and drug resistance.

ACKNOWLEDGMENTS

We thank The Bruce and Giovanna Ames Foundation for financial support.

REFERENCES

1. **Averboukh, L., P. Liang, P.W. Kantoff, and A.B. Pardee.** 1996. Regulation of S100P expression by androgen. Prostate *29*:350-355.
2. **Baum, M.** 1998. Tamoxifen—the treatment of choice. Why look for alternatives? Br. J. Cancer Suppl. *4*:1-4.
3. **Biswas, D.K., L. Averboukh, S. Sheng, K. Martin, D.S. Ewaniuk, T.F. Jawde, F. Wang, and A.B. Pardee.** 1998. Classification of breast cancer cells on the basis of a functional assay for estrogen receptor. Mol. Med. *4*:454-467.
4. **Biswas, D.K., P.V. Reddy, M. Pickard, B. Makkad, N. Pettit, and A.B. Pardee.** 1998. Calmodulin is essential for estrogen receptor interaction with its motif and activation of responsive promoter. J. Biol. Chem. *273*:33817-33824.
5. **Biswas, D.K., A.P. Cruz, E. Gansberger, and A.B. Pardee.** 2000. Epidermal growth factor-induced nuclear factor κB activation: a major pathway of cell-cycle progression in estrogen-receptor negative breast cancer cells. Proc. Natl. Acad. Sci. USA *97*:8542-8547.
6. **Biswas, D.K., S.C. Dai, A. Cruz, B. Weiser, E. Graner, and A.B. Pardee.** 2001. The nuclear factor κB (NF-κB): a potential therapeutic target for estrogen receptor negative breast cancers. Proc. Natl. Acad. Sci. USA *98*:10386-10391.

7. **Biswas, D.K., A. Cruz, N. Pettit, G.L. Mutter, and A.B. Pardee.** 2001. A therapeutic target for hormone-independent estrogen receptor-positive breast cancers. Mol. Med. *7*:59-67.
8. **Boucher, C.A., S.K. King, N. Carey, R. Krahe, C.L. Winchester, S. Rahman, T. Creavin, P. Meghji, et al.** 1995. A novel homeodomain-encoding gene is associated with a large CpG island interrupted by the myotonic dystrophy unstable (CTG)n repeat. Hum. Mol. Genet. *4*:1919-1925.
9. **Boucher, C.A., N. Carey, Y.H. Edwards, M.J. Siciliano, and K.J. Johnson.** 1996. Cloning of the human SIX1 gene and its assignment to chromosome 14. Genomics *33*:140-142.
10. **Bueso-Ramos, C.E., T. Manshouri, M.A. Haidar, Y. Yang, P. McCown, N. Ordonez, A. Glassman, N. Sneige, et al.** 1996. Abnormal expression of MDM-2 in breast carcinomas. Breast Cancer Res. Treat. *37*:179-188.
11. **Cheyette, B.N., P.J. Green, K. Martin, H. Garren, V. Hartenstein, and S.L. Zipursky.** 1994. The Drosophila sine oculis locus encodes a homeodomain-containing protein required for the development of the entire visual system. Neuron *12*:977-996.
12. **Cillo, C.** 1994. HOX genes in human cancers. Invasion Metastasis *14*:38-49.
13. **Cillo, C., A. Faiella, M. Cantile, and E. Boncinelli.** 1999. Homeobox genes and cancer. Exp. Cell Res. *248*:1-9.
14. **Clarke, C., J. Titley, S. Davies, and M.J. O'Hare.** 1994. An immunomagnetic separation method using superparamagnetic (MACS) beads for large-scale purification of human mammary luminal and myoepithelial cells. Epithelial Cell Biol. *3*:38-46.
15. **de Cremoux, P., J.M. Extra, M.G. Denis, J.Y. Pierga, E. Bourstyn, C. Nos, K.B. Clough, E. Boudou, et al.** 2000. Detection of MUC1-expressing mammary carcinoma cells in the peripheral blood of breast cancer patients by real-time polymerase chain reaction. Clin. Cancer Res. *6*:3117-3122.
16. **Cuenca, R.E., R.G. Azizkhan, and S. Haskill.** 1992. Characterization of GRO α, β, and γ expression in human colonic tumors: potential significance of cytokine involvement. Surg. Oncol. *1*:323-329.
17. **Eisen, M.B., P.T. Spellman, P.O. Brown, and D. Botstein.** 1998. Cluster analysis and display of genome-wide expression patterns. Proc. Natl. Acad. Sci. USA *95*:14863-14868.
18. **Ford, H.L.** 1998. Homeobox genes: a link between development, cell cycle, and cancer? Cell. Biol. Int. *22*:397-400.
19. **Ford, H.L., E.N. Kabingu, E.A. Bump, G.L. Mutter, and A.B. Pardee.** 1998. Abrogation of the G2 cell cycle checkpoint associated with overexpression of HSIX1: a possible mechanism of breast carcinogenesis. Proc. Natl. Acad. Sci. USA *95*:12608-12613.
20. **Ford, H.L., E. Landesman-Bollag, C.S. Dacwag, P.T. Stukenberg, A.B. Pardee, and D.C. Seldin.** 2000. Cell cycle-regulated phosphorylation of the human SIX1 homeodomain protein. J. Biol. Chem. *275*:22245-22254.
21. **Golub, T.R., D.K. Slonim, P. Tamayo, C. Huard, M. Gaasenbeek, J.P. Mesirov, H. Coller, M.L. Loh, et al.** 1999. Molecular classification of cancer: class discovery and class prediction by gene expression monitoring. Science *286*:531-537.
22. **Gruvberger, S., M. Ringner, Y. Chen, S. Panavally, L.H. Saal, A. Borg, M. Ferno, C. Peterson, et al.** 2001. Estrogen receptor status in breast cancer is associated with remarkably distinct gene expression patterns. Cancer Res. *61*:5979-5984.
23. **Harris, R.** 1997. Variation of benefits and harms of breast cancer screening with age. J. Natl. Cancer. Inst. Monogr. *22*:139-143.
24. **Harris, R.A., A. Yang, R.C. Stein, K. Lucy, L. Brusten, A. Herath, R. Parekh, M.D. Waterfield, et al.** 2002. Cluster analysis of an extensive human breast cancer cell line protein expression map database. Proteomics *2*:212-223.
25. **Hedenfalk, I., D. Duggan, Y. Chen, M. Radmacher, M. Bittner, R. Simon, P. Meltzer, B. Gusterson, et al.** 2001. Gene-expression profiles in hereditary breast cancer. N. Engl. J. Med. *344*:539-548.
26. **Henderson, B.R.** 2000. Nuclear-cytoplasmic shuttling of APC regulates β-catenin subcellular localization and turnover. Nat. Cell Biol. *2*:653-660.
27. **Jiang, M., Z.M. Shao, J. Wu, J.S. Lu, L.M. Yu, J.D. Yuan, Q.X. Han, Z.Z. Shen, et al.** 1997. p21/waf1/cip1 and mdm-2 expression in breast carcinoma patients as related to prognosis. Int. J. Cancer *74*:529-534.
28. **Khan, J., M.L. Bittner, L.H. Saal, U. Teichmann, D.O. Azorsa, G.C. Gooden, W.J. Pavan, J.M. Trent, et al.** 1999. cDNA microarrays detect activation of a myogenic transcription program by the PAX3-FKHR fusion oncogene. Proc. Natl. Acad. Sci. USA *96*:13264-13269.
29. **Kitagawa, M., S. Hatakeyama, M. Shirane, M. Matsumoto, N. Ishida, K. Hattori, I. Nakamichi, A.**

Kikuchi, et al. 1999. An F-box protein, FWD1, mediates ubiquitin-dependent proteolysis of β-catenin. EMBO J. *18*:2401-2410.

30.**Kraeft, S.K., R. Sutherland, L. Gravelin, G.H. Hu, L.H. Ferland, P. Richardson, A. Elias, and L.B. Chen.** 2000. Detection and analysis of cancer cells in blood and bone marrow using a rare event imaging system. Clin. Cancer Res. *6*:434-442.

31.**Lewis, M.T.** 2000. Homeobox genes in mammary gland development and neoplasia. Breast Cancer Res. *2*:158-169.

32.**Li, C.-M., M. Guo, A. Borczuk, C.A. Powell, M. Wei, H.M. Thaker, R. Friedman, U. Klein, et al.** 2002. Gene expression in Wilms' tumor mimics the earliest committed stage in the metanephric mesencymal-epithelial transition. Am. J. Pathol. *160*:2181-2190.

33.**Liang, P. and A.B. Pardee.** 1992. Differential display of eukaryotic messenger RNA by means of the polymerase chain reaction. Science *257*:967-971.

34.**Lockhart, D.J., H. Dong, M.C. Byrne, M.T. Follettie, M.V. Gallo, M.S. Chee, M. Mittmann, C. Wang, et al.** 1996. Expression monitoring by hybridization to high-density oligonucleotide arrays. Nat. Biotechnol. *14*:1675-1680.

35.**Luppi, M., M. Morselli, E. Bandieri, M. Federico, R. Marasca, P. Barozzi, M.G. Ferrari, M. Savarino, et al.** 1996. Spontaneous loss of Ph chromosome with maintenance of clonal hemopoiesis in an untreated patient with myeloproliferative disease and a long survival. Ann. Oncol. *7*:619-624.

36.**Martin, K.J. and A.B. Pardee.** 1999. Principles of differential display. Methods Enzymol. *393*:234-258

37.**Martin, K.J., B.M. Kritzman, L.M. Price, B. Koh, C.P. Kwan, X. Zhang, A. Mackay, M.J. O'Hare, et al.** 2000. Linking gene expression patterns to therapeutic groups in breast cancer. Cancer Res. *60*:2232-2238.

38.**Martin, K.J., E. Graner, Y. Li, L.M. Price, B.M. Kritzman, M.V. Fournier, E. Rhei, and A.B. Pardee.** 2001. High-sensitivity array analysis of gene expression for early detection of disseminated breast tumor cells in peripheral blood. Proc. Nat. Acad. Sci. USA *98*:2646-2651.

39.**Martin, K.J. and A.B. Pardee.** 2001. Identifying expressed genes. Proc. Nat. Acad. Sci. USA *97*:3789-3791.

40.**McClelland, M. and A.B. Pardee.** 1999. Expression Genetics: Accelerated and High-Throughput Methods. BioTechniques Books, Eaton Publishing, Westborough, MA.

41.**Miettinen, O.S., C.I. Henschke, M.W. Pasmantier, J.P. Smith, D.M. Libby, and D.F. Yankelevitz.** 2002. Mammographic screening: no reliable supporting evidence? Lancet *359*:404-405.

42.**Matsumura, Y. and D. Tarin.** 1992. Significance of CD44 gene products for cancer diagnosis and disease evaluation. Lancet *340*:1053-1058.

43.**Neufeld, K.L., F. Zhang, B.R. Cullen, and R.L. White.** 2000. APC-mediated downregulation of β-catenin activity involves nuclear sequestration and nuclear export. EMBO Rep. *1*:519-523.

44.**Oda, K., H. Arakawa, T. Tanaka, K. Matsuda, C. Tanikawa, T. Mori, H. Nishimori, K. Tamai, et al.** 2000. p53AIP1, a potential mediator of p53-dependent apoptosis, and its regulation by Ser-46-phosphorylated p53. Cell *102*:849-862.

45.**Okamura, S., H. Arakawa, T. Tanaka, H. Nakanishi, C.C. Ng, Y. Taya, M. Monden, and Y. Nakamura.** 2001. p53DINP1, a p53-inducible gene, regulates p53-dependent apoptosis. Mol. Cell *8*:85-94.

46.**Oliver, G., R. Wehr, N.A. Jenkins, N.G. Copeland, B.N. Cheyette, V. Hartenstein, S.L. Zipursky, and P. Gruss.** 1995. Homeobox genes and connective tissue patterning. Development *121*:693-705.

47.**Olsen, O. and P.C. Gotzsche.** 2001. Screening for breast cancer with mammography. Cochrane Database Syst. Rev. *4*:CD001877.

48.**Pardee, A.B. and M. McClelland.** 1999. Expression Genetics: Differential Display. BioTechniques Books, Eaton Publishing, Westborough, MA.

49.**Perou, C.M., T. Sorlie, M.B. Eisen, M. van de Rijn, S.S. Jeffrey, C.A. Rees, J.R. Pollack, D.T. Ross, et al.** 2000. Molecular portraits of human breast tumors. Nature *406*:747-752.

50.**Porter, D.A., I.E. Krop, S. Nasser, D. Sgroi, C.M. Kaelin, J.R. Marks, G. Riggins, and K. Polyak.** 2001. A SAGE (serial analysis of gene expression) view of breast tumor progression. Cancer Res. *61*:5697-5702.

51.**Racila, E., D. Euhus, A.J. Weiss, C. Rao, J. McConnell, L.W. Terstappen, and J.W. Uhr.** 1998. Detection and characterization of carcinoma cells in the blood. Proc. Natl. Acad. Sci. USA *95*:4589-4594

52.**Ranganathan, R., K.P. Lu, T. Hunter, and J.P. Noel.** 1997. Structural and functional analysis of the mitotic rotamase Pin1 suggests substrate recognition is phosphorylation dependent. Cell *89*:875-876.

53.**Reddy, K.B., R. McGowen, L. Schuger, D. Visscher, and S. Sheng.** 2001. Maspin expression inversely correlates with breast tumor progression in MMTV/TGF-α transgenic mouse model. Oncogene *20*:6538-6543.

54. **Rosin-Arbesfeld, R., F. Townsley, and M. Bienz.** 2000. The APC tumor suppressor has a nuclear export function. Nature *406*:1009-1012.

55. **Rubinfeld, B., I. Albert, E. Porfiri, C. Fiol, S. Munemitsu, and P. Polakis.** 1996. Binding of GSK3β to the APC-β-catenin complex and regulation of complex assembly. Science *272*:1023-1026.

56. **Ryo, A., M. Nakamura, G. Wulf, Y.C. Liou, and K.P. Lu.** 2001. Pin1 regulates turnover and subcellular localization of β-catenin by inhibiting its interaction with APC. Nat. Cell Biol. *3*:793-801.

57. **Sabbatini, R., M. Federico, M. Morselli, R. Depenni, K. Cagossi, M. Luppi, G. Torelli, and V. Silingardi.** 2000. Detection of circulating tumor cells by reverse transcriptase polymerase chain reaction of maspin in patients with breast cancer undergoing conventional-dose chemotherapy. J. Clin. Oncol. *18*:1914-1920.

58. **Sager, R.** 1997. Expression genetics: shifting the focus from DNA to RNA. Proc. Natl. Acad. Sci. USA *94*:952-955.

59. **Shen, M., P.T. Stukenberg, M.W. Kirschner, and K.P. Lu.** 1998. The essential mitotic peptidyl-prolyl isomerase Pin1 binds and regulates mitosis-specific phosphoproteins. Genes Dev. *12*:706-720.

60. **Sheng, S., J. Carey, E.A. Seftor, L. Dias, M.J. Hendrix, and R. Sager.** 1996. Maspin acts at the cell membrane to inhibit invasion and motility of mammary and prostatic cancer cells. Proc. Natl. Acad. Sci. USA *93*:11669-11674.

61. **Shridhar, V., A. Sen, J. Chien, J. Staub, R. Avula, S. Kovats, J. Lee, J. Lillie, et al.** 2002. Identification of underexpressed genes in early- and late-stage primary ovarian tumors by suppression subtraction hybridization. Cancer Res. *62*:262-270.

62. **Sorlie, T., C.M. Perou, R. Tibshirani, T. Aas, S. Geisler, H. Johnsen, T. Hastie, M.B. Eisen, et al.** 2001. Gene expression patterns of breast carcinomas distinguish tumor subclasses with clinical implications. Proc. Natl. Acad. Sci. USA *98*:10869-10874.

63. **Soslow, R.A., D.L. Carlson, M.G. Horenstein, and M.P. Osborne.** 2000. A comparison of cell cycle markers in well-differentiated lobular and ductal carcinomas. Breast Cancer Res. Treat. *61*:161-170.

64. **Stukenberg, P.T. and M.W. Kirschner.** 2001. Pin1 acts catalytically to promote a conformational change in Cdc25. Mol. Cell *7*:1071-1083.

65. **Tanaka, H., H. Arakawa, T. Yamaguchi, K. Shiraishi, S. Fukuda, K. Matsui, Y. Takei, and Y. Nakamura.** 2000. A ribonucleotide reductase gene involved in a p53-dependent cell-cycle checkpoint for DNA damage. Nature **404**:42-49.

66. **van't Veer, L.J., H. Dai, M.J. van de Vijver, Y.D. He, A.A. Hart, M. Mao, H.L. Peterse, et al.** 2002. Gene expression profiling predicts clinical outcome of breast cancer. Nature *415*:530-536.

67. **Velculescu, V.E., S.L. Madden, L. Zhang, A.E. Lash, J. Yu, C. Rago, A. Lal, C.J. Wang, et al.** 1999. Analysis of human transcriptomes. Nat. Genet. *23*:387-390.

68. **Weidner, N., P.R. Carroll, J. Flax, W. Blumenfeld, and J. Folkman.** 1993. Tumor angiogenesis correlates with metastasis in invasive prostate carcinoma. Am. J. Pathol. *143*:401-409.

69. **Welsh, J., K. Chada, S.S. Dalai, R. Cheng, D. Ralph, and M. McClelland.** 1992. Arbitrary primed PCR fingerprinting of RNA. Nucleic Acids Res. *20*:4965-4970.

70. **West, M., C. Blanchette, H. Dressman, E. Huang, S. Ishida, R. Spang, H. Zuzan, J.A. Olson, Jr., et al.** 2001. Predicting the clinical status of human breast cancer by using gene expression profiling. Proc. Natl. Acad. Sci. USA *98*:11462-11467.

71. **Wulf, G.M., A. Ryo, G.G. Wulf, S.W. Lee, T. Niu, V. Petkova, and K.P. Lu.** 2001. Pin1 is overexpressed in breast cancer and cooperates with Ras signaling in increasing the transcriptional activity of c-Jun towards cyclin D1. EMBO J. *20*:3459-3472.

72. **Yaffe, M.B., M. Schutkowski, M. Shen, X.Z. Zhou, P.T. Stukenberg, J.U. Rahfeld, J. Xu, J. Kuang, et al.** 1997. Sequence-specific and phosphorylation-dependent proline isomerization: a potential mitotic regulatory mechanism. Science *278*:1957-1960.

73. **Youngs, S.J., S.A. Ali, D.D. Taub, and R.C. Rees.** 1997. Chemokines induce migrational responses in human breast carcinoma cell lines. Int. J. Cancer *71*:257-266.

74. **Zhang, M., O. Volpert, Y.H. Shi, and N. Bouck.** 2000. Maspin is an angiogenesis inhibitor. Nat. Med. *6*:196-199.

75. **Zhang, W. and M. Zhang.** 2002. Tissue microarray analysis of maspin expression and its reverse correlation with mutant p53 in various tumors. Int. J. Oncol. *20*:1145-1150.

2 Identification of Genes in Leukemia Using Differential Display

Tomoko Kohno[1], Shinichi Sugita[1,2], Kazuo Yamamoto[1], and Toshifumi Matsuyama[1]

[1]Division of Cytokine Signaling, Department of Molecular Microbiology and Immunology, Nagasaki University Graduate School of Medical Sciences, and [2]Department of Obstetrics and Gynecology, Nagasaki University School of Medicine, Nagasaki, Japan

INTRODUCTION

To elucidate the molecular mechanisms of haematopoietic malignancy, several analyses in comparative gene expression could be applied. These include differential display (DD) (31), subtractive cloning, serial analysis of gene expression (SAGE) (67), and cDNA microarray analysis (37). Although DD is not as high throughput as SAGE or cDNA microarray, it has the advantage when only small amounts of total RNA is applicable as starting material. SAGE provides a thorough quantitative analysis of total gene expression but less extensive sequence information for each cDNA. Compared to 100–350 base pairs in DD, SAGE typically generates 11 base-pair sequence tags for the further screening (67). DD is based on the reverse transcription of mRNAs with anchored oligo-dT primers and PCR amplification using anchored primers and arbitrary primers. The amplified cDNAs were analyzed by electrophoresis on sequencing gel (Figure 1). DD is a versatile technique that is suitable to examine and identify genes and does not require prior knowledge of the genes at the transcriptional level. It is possible to compare more than two relevant RNA samples at one time. This clearly contrasts from SAGE or cDNA microarray analysis, which are based on the pairwise comparison of the genes. A major drawback of DD is the requirement of labor-intensive work to exclude the false-positive clones by the use of another method such as

Perspectives in Gene Expression
Edited by K. Appasani

Northern blot analysis. DD and related technologies have been reviewed (1,32,33,41), and readers are advised to consult these reviews for the detailed descriptions because this chapter is primarily intended to describe our personal experience using DD in adult T cell leukemia research. Several examples in which DD has been applied successfully to leukemia research are also included.

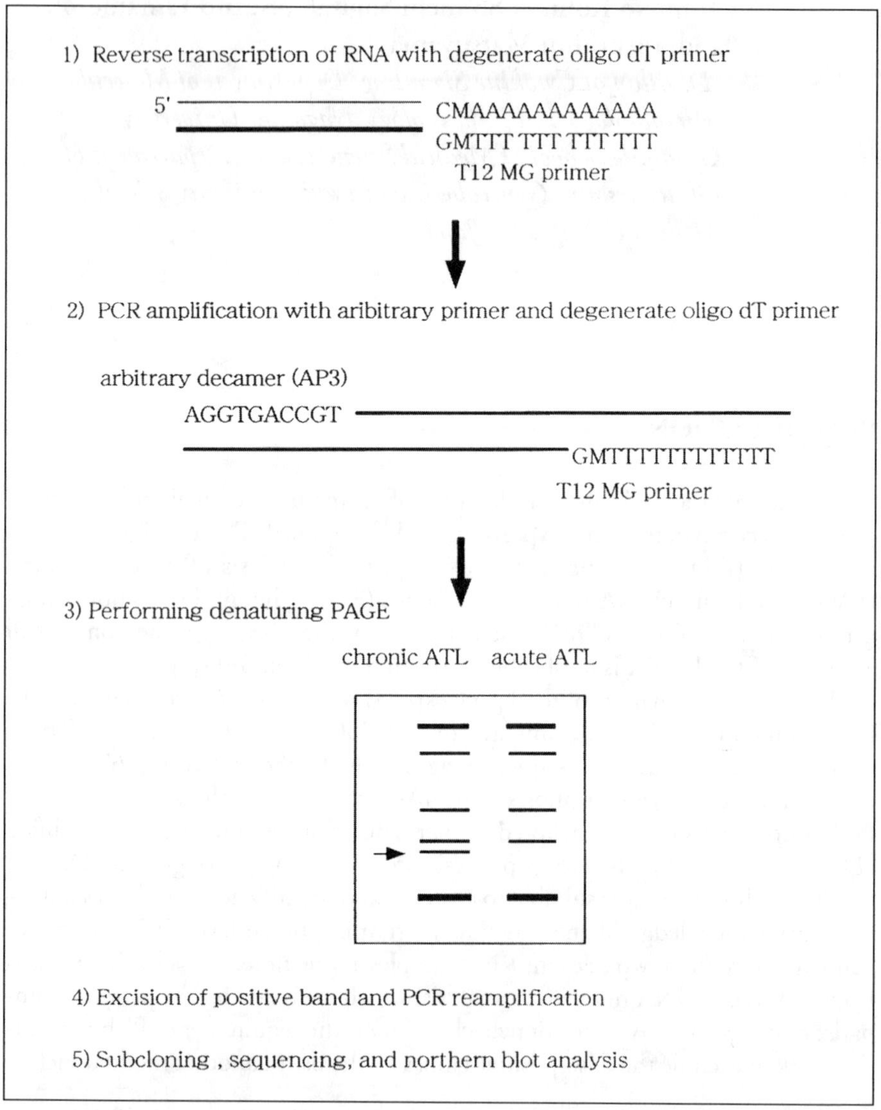

Figure 1. Schematic diagram of DD. To generate cDNA, degenerate oligo-dT primer (in this case, T12MG) is used. An arbitrary primer (in this case, AP3) is then used for PCR amplifications. Using polyacrylamide gel electrophoresis (PAGE), a band that appeared specifically in a chronic ATL sample is schematically shown. M indicates A, C, or G.

METHODOLOGY

Preparation of Starting Materials

It has been advised that obtaining high-quality mRNA is critical to get reproducible results (33). Generally, samples must be quickly frozen or preserved in guanidinium thiocyanate unless subsequent RNA extraction is scheduled. If enough RNA is available, then it is advisable to check its integrity by electrophoresis of an aliquot through a formaldehyde agarose gel with ethidium bromide staining. The RNA bands corresponding to 28S and 18S rRNA should be visible, and that of 28S should be thicker than that of 18S under UV illumination.

Clonal cell populations provide an ideal template for DD because of the homogeneity of the starting materials. This contrasts with tissue specimens from different individuals that render interpretation more difficult because of the multiple cell types combined with individual polymorphisms. For analyzing such non-clonal populations, including peripheral blood mononuclear cells (PBMC), several individuals should be studied so that only changes that are consistently different between the samples are manifested. Similarly, when analyzing a clonal population such as leukemia, one should seek consistent differences between clinical subtypes.

The isolation of RNA that is free of genomic DNA is critical for the success of DD because the contamination of genomic DNA is the major source of false positives or false negatives. Therefore, DNase I treatment to remove cDNA is essential. Some researchers set to verify the exclusion of DNA by performing PCR using primer sets spanning intronic sequences (1). The contamination of genomic DNA is also checked by the use of DD without reverse transcription. With this procedure, possible contamination of genomic DNA was recognized even after DNase I treatment (76).

An additional option is the purification of mRNA from total RNA using oligo-dT columns. Although RNA can be isolated by any method, using oligo-dT primer to isolate poly(A) RNA is sometimes undesirable. It was reported that the contamination of oligo-dT primers, possibly detached from the column, gave a smear on the DNA sequencing gels (33).

To standardize the samples, the quantification of mRNA by the use of the optical absorbance or normalized against a housekeeping transcript such as GAPDH or β-actin by quantitative RT-PCR is frequently applied.

In our case, PBMCs were freshly obtained from patients with acute adult T cell leukemia (ATL), chronic ATL, and normal subjects or cell lines. Total RNA was freshly extracted from cells with ISOGEN™ (Nippongene, Tokyo, Japan) and chloroform and was precipitated in isopropanol. Total cellular RNA (50 μg) was treated with 10 U RNase-free DNase I (Takara Bio, Tokyo, Japan) at 37°C for 30 min in a buffer containing 10 mM Tris-HCl (pH 8.3), 1.5 mM $MgCl_2$, 50 mM KCl, and 10 U RNase inhibitor (RNasin™; Promega, Madison, WI, USA).

RT-PCR

To obtain cDNA, reverse transcription should be performed. The original protocol used nondegenerate two base-anchored oligo-dT primers and then degenerate two base- or one base-anchored oligo-dT primers (31,33). Reverse transcription using short random hexamers and arbitrary primers has also been performed. Logically, this type of primer can anneal any RNA species other than mRNA unless prior purification of poly(A) RNA.

DD was performed using a commercial kit (RNAmap™ kit; GenHunter, Nashville, TN, USA) according to the manufacturer's instructions, with slight modifications. For reverse transcription, we prepared DNase I-treated RNA (0.2 μg each) and four different, degenerate, two base-anchored oligo-dT primer sets (T_{12}MN: T_{12}MG, T_{12}MA, T_{12}MT, and T_{12}MC, where M is a mixture of G, A, and C). Reverse transcription was performed at 37°C for 60 min in a total volume of 20 μL containing 20 μM dNTP, 200 U MMLV-SuperScript™-II, 1× RT buffer (50 mM Tris-HCl, pH 8.3, 75 mM KCl, and 3 mM $MgCl_2$) (Invitrogen, Carlsbad, CA, USA), 10 mM DTT, and 10 U RNase inhibitor (Promega).

When one uses oligo-dT primers for reverse transcription, it is important to realize that almost all the cDNAs are derived from the 3′-untranslated portion of the transcripts. The high frequency of seemingly novel transcripts isolated by DD could be accounted for in part by the lack of complete 3′-untranslated regions for many known genes registered in databases. To avoid this, one could use non-oligo-dT-based 18–20-mers of arbitrary sequences [RNA arbitrarily primed PCR (RAP-PCR)] (72) or rTth DNA polymerase, which possibly amplifies longer cDNA containing their coding regions as well as the 3′-untranslated portions (22). A potential problem with the former solution is that mRNAs are not exclusively selected in the reverse transcription step so that only abundant mRNA or even rRNA and tRNA may be amplified. In fact, a single RAP-PCR typically produced only 10–30 bands per track, and many experiments are required to obtain a collection of differentially expressed genes. The validity of using rTth DNA polymerase for DD should be evaluated by other investigators.

In performing PCR, the original protocol was designed to use two base-anchored oligo-dT primers in combination with arbitrary 10-mer primers to amplify the 3′ termini of mRNA sequences (31). Theoretically, the combinations of the primers for both directions will allow the detection of most of the expressed genes. As a parameter for PCR, the annealing temperature was approximately 40°C, which is lower than that of the protocol, because low annealing temperature allows sufficient annealing for the short primers. On the other hand, the use of this lower annealing temperature is regarded as a parameter of increased false positives because the optimal temperature for *Taq* DNA polymerase (Promega) should be around 60°C. Thus, some researchers prefer to use longer primers to detect reproducible signals at the cost of reduced coverage of the genes in each amplification. Recently, a protocol was proposed to use composite primers, adding 10-base nucleotides at the 5′-ends of the primers used in the original mRNA DD (76). These included a restriction site to aid in cloning. PCR began with one low-stringency cycle (40°C for annealing), followed by 35 high-stringency cycles (60°C for annealing). The modified method seems to improve the reproducibility and sensitivity; however, the coverage of the genes is unclear at present.

After finishing the reverse transcription, inactivating the reverse transcriptase at 95°C for 5 min was performed. For PCR, the mixture (2 μL) was subjected to the reaction in 20 μL PCR buffer (10 mM Tris-HCl, pH 8.4, 50 mM KCl, 1.5 mM $MgCl_2$, and 0.001% gelatin) containing 0.2 μM of one of the arbitrary 10-mers (AP1-AP10) as the 5′-primer, 0.2 μM of $T_{12}MA$, $T_{12}MC$, $T_{12}MG$, or $T_{12}MT$ as the 3′-primer, 2 μM dNTP mixture, 1 U *Taq* DNA polymerase, and 37 kBq [α-^{32}P] dCTP (Amersham Biosciences, Buckinghamshire, UK). Parameters for PCR were as follows: 40 cycles of denaturation at 94°C for 30 s, annealing at 42°C for 2 min, and extension at 72°C for 30 s.

SDS-PAGE and Band Retrieval

The PCR products were size-separated onto denaturing polyacrylamide gel. For the selection of bands that are consistent in DD gels, it is worth considering the use of paired RNA samples from duplicate experiments. Another consideration is about the use of denaturing gel because it is possible that the complementary strands shorter than 150 bp are separated as doublets (33). From this point of view, nondenaturing gel will simplify the band complexity, but any sequence hybridized with some DNA will possibly appear as a different band unless the denaturation step precedes electrophoresis. To date, the routine use of nondenaturing gel after the first PCR amplification has not been reported in DD.

After positive bands were excised from the dried gel, cDNAs were recovered by boiling and usually precipitated with ethanol. The cDNAs were re-amplified by PCR using primer sets and conditions similar to those used in the first PCR. The PCR samples are characterized by subcloning and sequencing.

The radiolabeled PCR products were analyzed by electrophoresis on 6% denaturing sequencing gels. The dried gel was exposed to Kodak® XAR film (Eastman Kodak, Rochester, NY, USA). According to the resulting autoradiogram, the positive bands were excised from the dried gel, boiled in 100 μL of double-distilled water for 15 min, concentrated by Microcon® 30 ultrafiltration (Amicon, Beverly, MA, USA), and suspended in 10 μl of double-distilled water. Four microliters of recovered cDNA were re-amplified by PCR in a 20-μL reaction volume using primer sets and conditions similar to those used in the first PCR, except the dNTP concentration was 20 μM and no radioactive nucleotide was used. The re-amplified samples were cloned into the PCRII vector using the TA Cloning® system (Invitrogen).

Confirming Altered Expressed Genes

There are a number of options for the confirmation of putative differentially regulated sequences that have been derived from an excised band. Among them, Northern blot analysis is the golden standard, using isolated cDNA as a probe. Northern blot analysis is useful for determining the size of the full-length mRNA transcript from which the cDNA is derived and also provides information about alternatively spliced variants. It is recommended that one use mRNAs from the same source to make blots (33); however, it should be kept in mind that the source sometimes does not represent the real source in general. In our application of DD to the study of IRF4 using IRF4 knockout mice (46), we identified a cDNA reproducibly positive by Northern blot analyses. However, it became positive only when we used the same source but not the different sources of mRNA. Therefore, this false positive was derived from the source used from the very beginning. A similar observation was previously reported (36).

Another reason for the isolation of false positives is that the displayed bands are usually composed of more than a single cDNA, representing not only the differentially regulated transcript. A useful procedure to take is the direct sequencing of the cDNA band so that only bands that give an unreadable sequence due to cDNA heterogeneity can be cloned to isolate the differentially expressed sequence. On the other hand, cloning the retrieved cDNAs risks the loss of the real positives. An interesting alternative to the usual cloning of the retrieved cDNAs is the affinity capturing of the cDNA on Northern blots (30). This procedure isolates the differentially regulated transcript as a cDNA fragment that can be cloned directly from the Northern membrane, followed by PCR amplification. Unfortunately, this method does not seem to have been widely used after it was published in 1994. The use of a nondenaturing polyacrylamide gel is another option. Retrieved bands from the denaturing gel could be further separated by nondenaturing gel based on single-stranded conformational polymorphism (SSCP). Numbers of recent reports about successful use of this procedure seem to verify its significance (40,43,53,54,57,68).

To screen large numbers of differentials, reverse Northern blot analysis has been proposed (49). A number of variations of this method have been developed. Basically, the PCR products are dot-blotted on the duplicate membranes and incubated separately with two total cDNA samples for comparison. For standardization, the dots can be normalized against a housekeeping transcript. The advantage of this method is that a smaller amount of mRNA is needed to screen a large number of clones. The drawback of this approach resides in blotting without band separation. Some positive but faint signals will be masked against background noise.

Northern blot analysis. Ten micrograms of total cellular RNA were resolved by electrophoresis on a 1% MOPS-formaldehyde gel and transferred to a Hybond®-N membrane (Amersham Biosciences). The PCR products that had been differentially amplified in the display were used as probes. The cDNA probe and the β-actin probe were labeled with [α-^{32}P]dCTP using the Megaprime DNA Labelling System (Amersham Biosciences) according to the manufacturer's protocol. Hybridization was performed in a buffer containing 50% formamide at 42°C overnight.

The membranes were washed four times at room temperature in 2× SSC containing 0.1% SDS and then three times at 65°C in 0.1× SSC containing 0.5% SDS. The washed membrane was autoradiographed at -80°C overnight.

To amplify the sequences between a defined internal site and unknown sequences at the 5′-end of the mRNA, a 5′-RACE system kit was used (Invitrogen). The first-strand cDNA was synthesized from 1 μg total RNA using the gene-specific primer GSP-1, 5′-CCAGGTAACTTCTTCACTGG-3′. The sample RNA and GSP-1 were incubated at 70°C for 10 min and then added to the 1× PCR buffer (20 mM Tris-HCl, pH 8.4, and 50 mM KCl) with 2.5 mM $MgCl_2$, 1 mM dNTP, 10 mM DTT, and 200 U SuperScript II reverse transcriptase. The mixture was incubated at 42°C for 50 min and then at 70°C for 15 min.

To remove the mRNA template, treatment with an RNase (mixture of RNase H) was performed and a homopolymeric tail was added to the 3′-end of the cDNA. This was achieved by incubation at 37°C for 10 min with terminal deoxynucleotidyl transferase (TdT) in buffer containing 10 mM Tris-HCl, pH 8.4, 25 mM KCl, 1.5 mM $MgCl_2$, and 0.2 mM dCTP.

In the next step, 5′-end primary PCR amplification was performed using a nested gene specific primer, GSP-2 5′-GCCGTTCATCACTCTTTCG-3′, and the 5′-RACE Abridged Anchor Primer 5′-GGCCACGCGTCGACTAGTACGGGIIGGGIIGGGIIG-3′. The PCR parameters were at 94°C for 2 min, followed by 35 cycles at 94°C for 1 min, 55°C for 1 min, and 72°C for 2 min, followed by a final extension step at 72°C for 5 min. Successively nested PCR was performed with the gene-specific primer GSP-3, 5′-GGGACAAGAACATGAACCCA-3′, as the 3′-primer and the Universal Amplification primer, 5′-CUACUACUACUAGGCCACGCGTCGACTAGTAC-3′ as the 5′-primer. The nested PCR conditions were similar to those of the primary PCR, except a diluted primary PCR product and a different primer were used. The resultant product was cloned into the PCRII vector using the TA Cloning system and sequenced.

EXAMPLES OF APPLICATION OF DD IN LEUKEMIA RESEARCH

The application of DD in leukemia research is largely categorized into two subjects. The first one intends to detect a gene in which expression is changed after treatment with some reagents such as differentiation inducers or cytokines. The second one intends to detect a downstream target of a fused gene that is produced by chromosomal translocation.

All-*trans*-retinoic acid (ATLA) acts as a potent differentiation agent for acute promyelocytic leukemia (APL/M3) characterized by a t(15;17) that involves the retinoic acid receptor alpha gene (RARα). Several genes implicated in ATLA-dependent differentiation in cell lines were identified using DD technique. In NB-4 cells derived from a patient with APL, the expression of RIG-E, a human homolog of the murine Ly-6 family and JEM-1, encoding leucine-zipper nuclear factor mRNA were induced by ATLA stimulation (38). In U937 cells treated with ATLA, the expression level of Src-like adapter protein mRNA was increased (50). In HL-60 cells, one of the cloned cDNAs encodes sequences identifying Burkitt's lymphoma receptor 1 (BLR1), a recently described chemokine receptor. Northern blot analysis demonstrates that blr1 mRNA expression increases within 9 h of retinoic acid treatment and peaks at 72 h. In HL-60 cells stably transfected with blr1 cDNA, ectopic expression of blr1 causes an increase in ERK2 MAPK activation and promotes retinoic acid-induced G(1)/G(0) growth arrest and cell differentiation (4). The induction of monocyte chemoattractant protein-1 (MCP-1) was also identified by DD in HL-60 cells (7). On the other hand, in ATLA-stimulated HL-60, the expression of FUS mRNA was down-regulated (45).

Cytokines regulate proliferation, differentiation, and various cellular functions in hematopoietic cells. In the IL-3 dependent OTT-1 cell line, Morikawa et al. (47) identified synaptosomal-associated protein-23 kDa gene (SNAP-23) that was induced by various cytokines. In megakaryoblastic leukemia cell line, MO7e, proliferation-associated SNF-2 like gene (PASG) was down-regulated after the withdrawal of cytokines (29). In erythroleukemia cell lines, several genes implicated with erythropoietin-dependent differentiation were identified (42).

AML1-MTG8 (ETO) fusion transcription factor generated by t(8;21) translocation is thought to dysregulate the proliferation and differentiation of progenitor cells. Shimada et al. (62) identified genes involved in AML1-MTG8-mediated leukemogenesis by DD. Twenty-four genes under the downstream control of AML1-MTG8 were isolated. Nineteen genes were up-regulated, and the other five genes were down-regulated by the AML1-NTG8 gene.

Chronic myeloid leukemia (CML) is characterized by a reciprocal translocation between chromosomes 9 and 22. As a result, a chimeric BCR/ABL gene is formed on the Philadelphia chromosome. BCR/ABL protein is a p210kD tyrosine kinase that plays a crucial role in the pathogenesis of CML. Compelling evidence indicates that malignant transformation by BCR-ABL is critically dependent on its protein tyrosine kinase activity, which activates multiple signaling pathways, and, thus, the activation of such pathways may affect the expression of genes that confer the malignant phenotype. Watari et al. (71) used a DD approach to compare between TF-1 cells and ID-1 cells that had been established from TF-1 cells by the transfection of BCR/ABL. This study demonstrated that PRAME, a melanoma antigen, was markedly expressed in ID-1 cells and primary cells with CML in blast crisis. To define downstream targets of BCR/ABL, Deininger et al. (10) studied alterations in the profile of gene expression in BV173 (derived from a CML lymphoid blast crisis) exposed to ST1571, a specific inhibitor of BCR/ABL protein tyrosine kinase. They identified 12 genes related to the kinase activity, seven of which corresponded to known genes and five of which corresponded to unknown genes. One of known genes is SOCS-2, a suppressor of cytokine signaling, which is markedly expressed in advanced stage of CML. Although refractoriness to STI571 or relapse after the initial response was accompanied by the augmentation of SOCS-2 expression, ectopic overexpression of SOCS-2 in 32Dp210 cells slowed growth, inhibited clonogenicity, and increased their motility and sensitivity to STI571. It is suggested that SOCS-2 is induced by a negative feedback mechanism but cannot reverse its overall growth-promoting effects in blastic transformation (58).

THE USE OF DD IN THE STUDY OF ATL

As described earlier, clonal cell populations are an ideal source for DD because of the homogeneity of the starting material. However, it is also true that clonal cell populations such as cell lines do not represent the original phenotype. From this point of view, clinical samples of leukemic cells meet both requirements to some extent because most of the PBMCs consist of oligoclonally expanded leukemic cells that are easy to enrich. In our experiment, we tried to use PBMC from ATL patients in DD analysis.

ATL is a lymphoproliferative disorder of the helper/inducer T cell origin that is caused by human T lymphotrophic virus type-1 infection (HTLV-1) (20,52,65). HTLV-1 infection mainly occurs early in life via the breast milk of carrier mothers (24), and approximately 4%–5% of HTLV-1 carriers are thought to develop ATL after long latency periods, usually 40–50 years (27). However, patients with ATL exhibit a variety of clinical features, and this disease is therefore clinically subclassified into acute, lymphomatous, chronic, and smoldering types (63). The acute ATL progresses rapidly, and the median survival period after diagnosis is approximately six months. In contrast, the chronic ATL has a less aggressive clinical course and long-term survival even without chemotherapy. Little is known about the underlying molecular mechanisms of the clinical course of ATL. The DD was thus applied to clarify the molecular differences between chronic and acute ATL (26).

After screening the PBMC obtained from three patients with acute ATL and five patients with chronic ATL, we obtained 18 fragments that were differently expressed between the chronic and acute ATL cases. Four of eighteen fragments were confirmed to be differentially expressed in among the samples of ATL patients. However, others were turned out to be false positives (Figure 2 and Table 1). As shown in Figure 3, three products (C316, C416, and G811) were amplified preferentially in samples of PBMC from patients with acute ATL. In contrast, the remaining product (G319) was produced in only chronic ATL samples. PCR of C316 and C416 were performed with AP-3 (5′-AGGTGACCGT-3′), AP-4 (5′-GGTACTCCAC-3′) as the 5′-primer, and $T_{12}MC$ as the 3′-primer, respectively. PCR of G811 and G319 was performed with AP-8 (5′-CTACTAC-

TAG-3′), AP-3 as the 5′-primer, and T_{12}MG as the 3′-primer, respectively.

The sequences of these candidate mRNAs (C316, C416, G319, and G811) were identical or significantly homologous to already published sequences. The sequence of C316 was almost identical (98.1%) to base positions 813-1024 of the MAL protein gene, which is expressed in T cells at intermediate and late

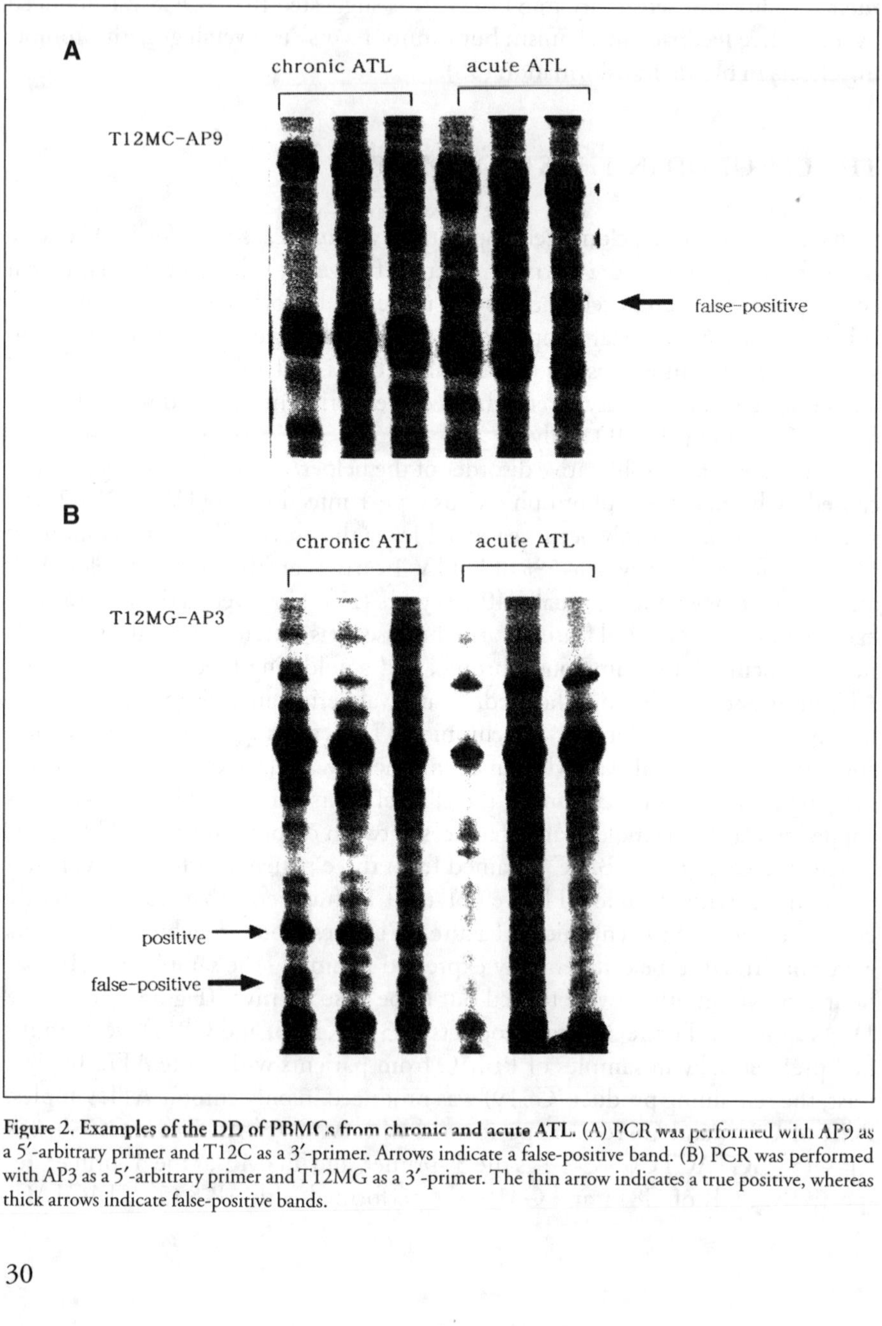

Figure 2. Examples of the DD of PBMCs from chronic and acute ATL. (A) PCR was performed with AP9 as a 5′-arbitrary primer and T12C as a 3′-primer. Arrows indicate a false-positive band. (B) PCR was performed with AP3 as a 5′-arbitrary primer and T12MG as a 3′-primer. The thin arrow indicates a true positive, whereas thick arrows indicate false-positive bands.

Table 1. Summaries after DD Screening

	T12M A	T12M C			T12M G			T12M
	(a)	(a)	(b)	(c)	(a)	(b)	(c)	(a)
AP1	0	1	1	0	0			0
AP2	0	1	1	0	0			0
AP3	0	1	1	1	2	1	1	0
AP4	0	2	2	1	1	0		0
AP5	0	0			0			0
AP6	0	1	1	0	2	0		0
AP7	0	1	1	0	2	0		0
AP8	0	1	1	0	1	1	1	0
AP9	0	1	1	0	1	0		0
AP10	0	1	1	0	1	0		0

PCR was performed in combinations with oligo-dT primers (T12MN) and arbitrary primers (AP1–AP10). Data shown are the numbers of excised bands after the first PCR amplification (a), successfully amplified clones after band retrieval (b), and positive clones verified by Northern blot analysis.

stages of differentiation (2,3). The C416 sequence was completely homologous to base positions 2004–2139 of the published sequences for Burkitt's lymphoma receptor 2 (BLR2)/Epstain-Barr virus-induced gene 1 (EBI-1)/CCR7, a lymphoid-specific member of the G protein-coupled receptor family (59). The 310-bp sequence of G319 was identical to the 3′-untranslated region of the human fibrinogen-like protein (hpT49) (28,55). The G811 sequence (241 bp) showed significant homology to the bovine Cl-MNLL mRNA for the ubiquinone oxidoreductase complex.The level of homology between the two was approximately 80% (69,70).

The differential expression of these genes was reevaluated by Northern blot analysis of RNA samples from the PBMC of ATL patients and normal volunteers, using the amplified PCR products as probes (Figure 4). The sizes of C316 and C416 mRNA were approximately 1.1 and 2.4 kb, respectively. These sizes were consistent with the reported sizes of MAL and EBI-1/CCR7, respectively. The size of G811 mRNA was approximately 0.3 kb, and the G319 probe detected 4.5- and 1.5-kb bands, which was consistent with the previously reported sizes of hpT49 mRNA.

In this study, we found that the MAL, EBI-1/CCR7, and hMNLL genes were up-regulated in acute ATL cells. It was reported that the MAL mRNA is expressed in T cells at intermediate and late stages of differentiation. MAL associates with glycosylphosphatidylinositol (GPI)-anchored proteins and the Src-like tyrosine kinase Lck in T lymphocytes, and cross-linking of GPI-anchored proteins triggers signaling pathways that lead to Lck activation and T cell proliferation (44,61).

Thus, the MAL molecule is closely associated with molecules for T cell activation, and it probably has roles in activation and proliferation. It is possible that MAL is induced during the process of, or as a consequence of, malignant cell transformation in vivo. EBI-1/CCR7 is a lymphoid-specific member of the G protein-coupled receptor family and is expressed exclusively in B and T lymphocyte cell lines and in lymphoid tissues (5). The ligand for EBI-1/CCR7 is ECL, a CC chemokine that is present in the thymus and lymph nodes (74). CCR7 is a major homing receptor and an important regulatory molecule for initiating an antigen-specific immune response (16). The altered expression level in acute ATL cells may reflect the homing/infiltrating ability of ATL cells, rather than malignant cell transformation.

It should be stressed that there is some discrepancy in gene expression between freshly isolated samples and cell lines. The expression of MAL mRNA was significantly high in the PBMCs of patients with acute ATL, while it was

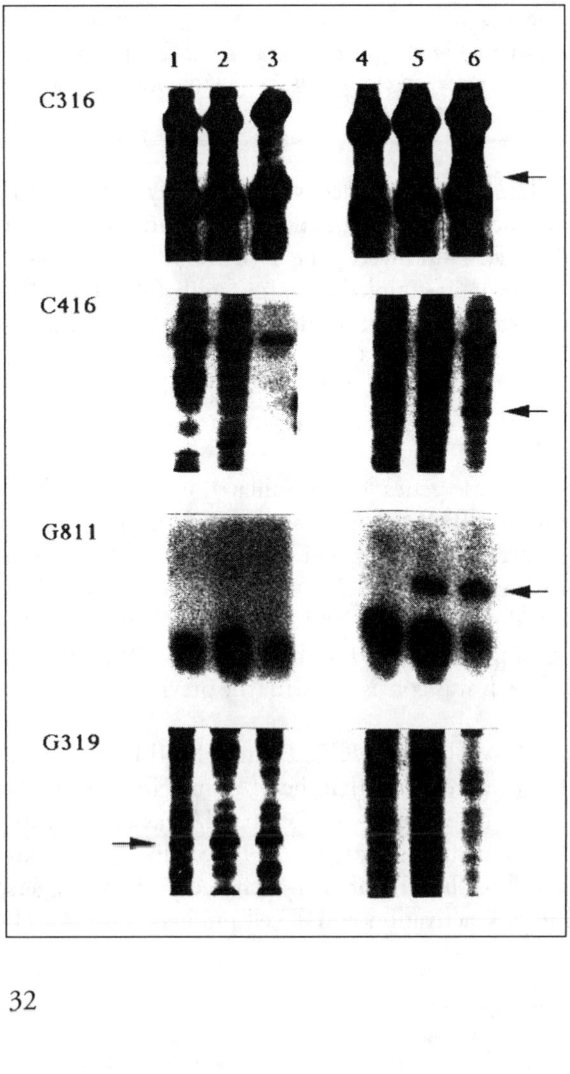

Figure 3. DD clones between chronic and acute ATL. Lanes 1–3 were derived from chronic ATL, and lanes 4–6 were derived from acute ATL samples. The bands (C316, C416, G811, and G319) that were differentially amplified in chronic and acute ATL cells are indicated by arrows. Reproduced from Reference 26, with permission from the Japanese Journal of Cancer Research.

not detected in HTLV-1 infected cell lines. The expression of EBI-1/CCR7 mRNA was down-regulated in chronic ATL cells. Although one of the cell lines was established from a patient with chronic ATL, the expression of EBI-1/CCR7 gene was high. Furthermore, Tax, a potent transactivator protein in many biological settings, is not expressed in clinically isolated ATL cells and becomes expressed several hours after culture (17,25). Similarly, it was reported that the transcripts of MCP-1, a chemokine, were undetectable in freshly isolated neutrophils but readily detectable after incubation in media for 20 h (7). Therefore, it also should be stressed that there is some discrepancy in gene expression between freshly isolated samples and freshly isolated samples that are followed by an incubation step in vitro.

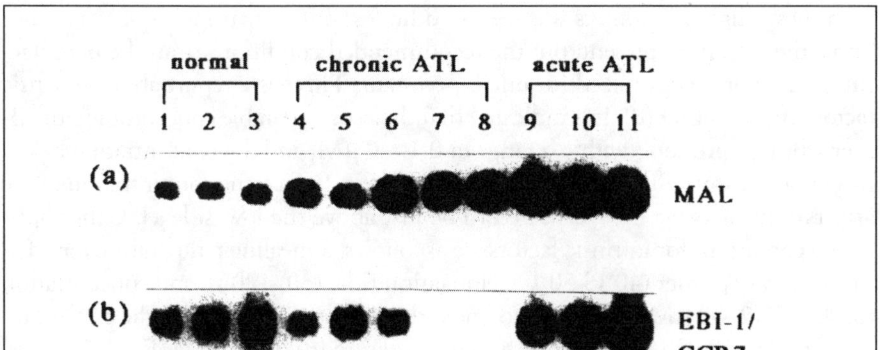

Figure 4 Northern blot analysis for the differentially expressed genes. Panels a–d were probed with DD derived clones, C316, G416, G811, and G391, respectively. Reproduced from Reference 26, with permission from the Japanese Journal of Cancer Research.

REFINEMENTS TO THE DD

Although we successfully identified four genes with aberrantly regulated expression in ATL cells, there are still genes that remain unidentified. For example, we should have isolated genes such as p53 (48,56), cyclin-dependent kinase inhibitors, p15 and p16 (19,73), which have been already reported in acute ATL cells. Recently, we also identified the increased expression of IRF4 (39), which is a member of interferon regulatory family, in acute ATL cells. With associated cell cycle analysis, we proposed the possible involvement of IRF4 in leukemic crisis (21). However, our DD screening failed to detect this difference. The major reason why we lost these signals is that the number of bands that appeared on the gel was much smaller than we had expected. In fact, we could not detect differentially expressed bands when we used oligo-dT primers, T12MA, or T12MT (Table 1). The same problem with these oligo-dT primers was seen in other experiments using DD. We do not know the reasons at present, but the recommended conditions from the manufacturer may not be optimized for our experiment. There is a report about intrinsic factors affecting DD (8). It is indicated that the most favorable concentration of 13-mer arbitrary primer is within a range of 0.1–20 µM, and the concentration below this range greatly reduces the amplification process. In fact, the concentration of the arbitrary primers we used appeared to be just above the lowest level. Other optimized conditions for intrinsic factors are as follows: annealing temperature for a 13-mer arbitrary primer (40°C-50°C), elongation time (30–120 s), and concentration for dNTP (2–50 µM). It is good to know that we can use dNTP at the concentration of up to 50 µM because it is the concentration for standard PCR.

In addition to the refinement of the conditions, there are several alternatives for the original DD. One such method is called molecular or cDNA indexing (23). This technology is based on the finding that class IIS restriction enzymes cleave DNA at a precise location outside their recognition sites and produce overhangs of unknown sequences, by which cDNA fragments were characterized by a library of 64 adaptors for all possible three-base overhangs. This step is achieved with the use of *E. coli* DNA ligase, which discriminates three nucleotides adjacent to the joining site, and selected fragments are amplified by PCR using an adaptor-primer and an anchored oligo-dT primer. This method has several advantages over display techniques based on arbitrarily primed PCR; in particular, genes with low redundancy can be efficiently displayed. This method has successfully applied in several cases (13,60).

Another alternative is called targeted DD, which is based on the conserved sequence information of mRNAs or protein modules (1). For example, if one wants to see the differential expression of cytokines or proto-oncogenes, 3′-primers could be set to amplify AUUUA sequences that regulate early and transient biological responses (11,18,66). For the amplification of conserved protein modules, degenerate oligonucleotides have been designed for small G proteins (34), signal-peptide sequences (15,64), protein kinase (12,51), and nuclear hormone receptors (75). Another type of targeted DD is based on the computational finding that only 30 sets of octanucleotide primers were sufficient to detect approximately 75% of known (and presumably unknown) human protein-coding regions (36). This method almost entails 870 (30 × 29) PCR amplifications to provide greater coverage of the gene. To validate this method, the successful application of specifically targeted primers for the G protein-coupled receptor family was demonstrated. Another example of this method was the statistically designed octamer primer for bHLH transcription factor family. It successfully produced the bHLH family coverage figure of 90%, as estimated by a computer-modeled PCR amplification (6). Based on the similar computational analysis, the use of 96 primers of 12-mer length for the general population of cDNA was proposed (9). Although the coverage of the genes seemed to reduce to 2–3 times lower compared to that of 870 reactions, this method only required 96 reactions.

PERSPECTIVES

With the completion of the human genome draft sequence, along with the expansion of the sequenced mammalian mRNA database, the next revolution should be to assign functions to all the predicted proteins. When homologs of the retrieved clone are found in the database, the next step is to characterize the biological function of the gene by the disruption of the normal expression. This has been a challenging step because several applicable procedures offer different advantages and disadvantages. For example, a gene-targeting experiment is the surest, straightforward way, but it is also time consuming. On the other hand, an antisense oligonucleotide experiment is seemingly easy to apply but often proved ineffective. In this regard, the breakthrough technology is based on RNA interference (RNAi), which is an evolutionarily conserved surveillance mechanism that responds to double-stranded RNA by the sequence-specific silencing of homologous genes. After the evolution of RNAi into a powerful tool for probing gene function in *Caenorhabditis elegans*, it is time to deploy RNAi in mammalian cells (14). We believe that, with the combination of RNAi, DD will allow us to identify genes of interest in large-scale screens, which will eventually unleash the dormant potential of sequence databases in the near future.

ACKNOWLEDGMENTS

We thank Dr. R. Moriuchi for technical advice, Drs. A. Koda, H. Ichinose, and M. Miyazaki for encouragement, and Ms. F. Tujita for secretarial work. This work is supported in part by Grants-in-Aid for Scientific Research from the Ministry of Education, Science, Sports, and Culture, Japan, and by a grant (to T.M.) from NDR Corporation (Gifu, Japan).

REFERENCES

1. **Ali, M., A.F. Markham, and J.D. Isaacs.** 2001. Application of differential display to immunological research. J. Immunol. Methods *250*:29-43.
2. **Alonso, M.A. and S.M. Weissman.** 1987. cDNA cloning and sequence of MAL, a hydrophobic protein associated with human T-cell differentiation. Proc. Natl. Acad. Sci. USA *84*: 1997-2001.
3. **Alonso, M.A., D.E. Barton, and U. Francke.** 1988. Assigment of the T-cell differentiation gene MAL to human chromosome 2, region cen—q13. Immunogenetics *27*: 91-95.
4. **Battle, T.E., R.A. Levine, and A. Yen.** 2000. Retinoic acid-induced blr1 expression promotes ERK2 activation and cell differentiation in HL-60 cells. Exp. Cell Res. *254*:287-298.
5. **Birkenbach, M., K. Josefsen, R. Yalamanchili, G. Lenoir, and E. Kieff.** 1993. Epstein-Barr virus-induced genes: First lymphocyte-specific G protein-coupled peptide receptor. J. Virol. *67*: 2209-2220.
6. **Brown, A.J., C. Hutchings, J.F. Burke, and L.V. Mayne.** 1999. Targeted display: a new technique for the analysis of differential gene expression. Methods Enzymol. *303*:392-408.
7. **Burn, T.C., M.S. Petrovick, S. Hohaus, B.J. Rollins, and D.G. Tenen.** 1994. Monocyte, chemoattractant protein-1 gene is expressed in activated neutrophils and retinoic acid-induced human myeloid cell lines. Blood *84*:2776-2783.
8. **Cho, Y.J., V.R. Prezioso, and P. Liang.** 2002. Systematic analysis of intrinsic factors affecting differential display. Biotechniques *32*:762-766.
9. **Consalez, G.G., A. Cabibbo, A. Corradi, D. Alli, M. Sardella, R. Sitia, and R. Fesce.** 1999. A computer-driven approach to PCR-based differential screening, alternative to differential display. Bioinformatics *15*:93-105.
10. **Deininger, M.W., S. Vieira, R. Mendiola, B. Schultheis, J.M. Goldman, and J.V. Melo.** 2000. BCR-ABL tyrosine kinase activity regulates the expression of multiple genes implicated in the pathogenesis of chronic myeloid leukemia. Cancer Res. *60*:2049-2055.
11. **Dominguez, O., Y. Ashhab, L. Sabater, E. Belloso, P. Caro, and R. Pujol-Borrell.** 1998. Cloning of ARE-containing genes by AU-motif-directed display. Genomics *54*:278-286.
12. **Donohue, P.J., G.F. Alberts, Y. Guo, and J.A. Winkles.** 1995. Identification by targeted differential display of an immediate early gene encoding a putative serine/threonine kinase. J. Biol. Chem. *270*:10351-10357.
13. **Doyu, M., K. Sawada, N. Mitsuma, J. Niwa, M. Yoshimoto, Y. Fujii, G. Sobue, and K. Kato.** 2001. Gene expression profile in Alzheimer's brain screened by molecular indexing. Brain Res. Mol. Brain Res. *87*:1-11.
14. **Elbashir, S.M., J. Harborth, W. Lendeckel, A. Yalcin, K. Weber, and T. Tuschl.** 2001. Duplexes of 21-nucleotide RNAs mediate RNA interference in cultured mammalian cells. Nature *411*:494-498.
15. **Eriksson, A. and K. Nordqvist.** 2001. Protocol for using signal peptide differential display and representational difference analysis to isolate differentially expressed cDNAs from fetal mouse brain. Brain Res. Brain Res. Protoc. *6*:119-128.
16. **Forester, R., A. Schubel, D. Breitfeld, E. Kremmer, I. Renner-Muller, E. Wolf, and M. Lipp.** 1999. CCR7 coordinated the primary immune response by establishing functional microenviroments in secondary lymphoid organs. Cell *99*:23-33.
17. **Furukawa, Y., M. Osame, R. Kubota, M. Tara, and M. Yoshida.** 1995. Human T-cell leukemia virus type-1 (HTLV-1) tax is expressed at the same level in infected cells of HTLV-1-associated myelophathy or tropical spastic paraparesis patients as in asymptomatic carriers but at a lower level in adult T-cell leukemia cells. Blood *85*:1865-1870.
18. **Gonsky, R., J.A. Knauf, R. Elisei, J.W. Wang, S. Su, and J.A. Fagin.** 1997. Identification of rapid turnover transcripts overexpressed in thyroid tumors and thyroid cancer cell lines: use of a targeted differential RNA display method to select for mRNA subsets. Nucleic Acids Res. *25*:3823-3831.
19. **Hatta, Y., T. Hirama, C.W. Miller, Y. Yamada, M. Tomonaga, and H.P. Koeffler.** 1995. Homozygous deletions of the p15 (MTS2) and p16 (CDKN2/MT21) genes in adult T-cell leukemia. Blood *85*: 2699-2704.
20. **Himuna, Y., K. Nagata, M. Hanaoka, M. Nakai, T. Mastumoto, K. Kinoshita, S. Shirakawa, and I. Miyoshi.** 1981. Adult T-cell leukemia: Antigen in an ATL cell line and detection of antibodies to the antigen in human sera. Proc. Natl. Acad. Sci. USA *78*: 6476-6480.
21. **Imaizumi, Y., T. Kohno, Y. Yamada, S. Ikeda, Y. Tanaka, M. Tomonaga and T. Matsuyama.** 2001. Possible involvement of interferon regulatory factor 4 (IRF4) in a clinical subtype of adult T-cell

leukemia. Jpn J Cancer Res. *92*:1284-1292.

22.**Jurecic, R., R.G. Nachtman, S.M. Colicos, and J.W. Belmont.** 1998. Identification and cloning of differentially expressed genes by long-distance differential display. Anal. Biochem. *259*:235-244.

23.**Kato, K.** 1996. RNA fingerprinting by molecular indexing. Nucleic Acids Res. *24*:394-395.

24.**Kinoshita, K., T. Amagasaki, S. Hino, H. Doi, K. Yamanouchi, N. Ban, S. Momita, S. Ikeda, S. Kamihira, M. Ichimaru, S. Katamine, T. Miyamoto, Y. Tsuji, T. Ichimaru, T. Yamabe, M. Ito, S. Kamura, and T. Tsuda.** 1987. Milk-borne transmission of HTLV-1 from carrier mothers to their children. Jpn. J. Cancer Res. *78*: 674-680.

25.**Kinoshita, T., M. Shimoyama, K. Tobinai, M. Ito, S. Ito, S. Ikeda, K. Tajima, K. Shimotohno, and T. Sugimura.** 1989. Detection of mRNA for the tax1/rex1 gene of human T-cell leukemia virus type I in fresh peripheral blood mononuclear cells of adult T-cell leukemia patients and viral carriers by using the polymerase chain reaction. Proc. Natl. Acad. Sci. USA *86*: 5620-5624.

26.**Kohno, T., R. Moriuchi, S. Katamine, Y. Yamada, M. Tomonaga, and T. Matsuyama, T.** 2000. Identification of genes associated with the progression of adult T cell leukemia (ATL). Jpn. J. Cancer Res. *91*:1103-1110.

27.**Kondo, T., H. Kono, H. Nonaka, N. Miyamoto, R. Yoshida, F. Bando, H. Inoue, I. Miyoshi, Y. Himuna, and M. Hanaoka.** 1987. Risk of adult T-cell leukemia/lymphoma in HTLV-1 carriers. J. Lancet *2*:159.

28.**Koyama, T., L.R. Hall, W.G. Haser, S. Tonegawa, and H. Saito.** 1987. Structure of a cytotoxic T-lymphocyte-specific gene shows a strong homology to fibrinogen β and γ chains. Proc. Natl. Acad. Sci. USA *84*:1609-1613.

29.**Lee, D.W., K. Zhang, Z.Q. Ning, E.H. Raabe, S. Tintner, R. Wieland, B.J. Wilkins, J.M. Kim, R.I. Blough, and R.J. Arceci.** 2000. Proliferation-associated SNF2-like gene (PASG): a SNF2 family member altered in leukemia. Cancer Res. *60*:3612-3622.

30.**Li, F., E.S. Barnathan, and K. Kariko.** 1994. Rapid method for screening and cloning cDNAs generated in differential mRNA display: application of northern blot for affinity capturing of cDNAs. Nucleic Acids Res. *22*:1764-1765.

31.**Liang, P. and A.B. Pardee.** 1992. Differential display of eukaryotic messenger RNA by means of the polymerase chain reaction. Science *257*: 967-971.

32.**Liang, P., and A.B. Pardee.** 1994. Differential display of mRNA by PCR. Current protocols in molecular biology. p. 15.8.1-15.8.8. John Wily & Sons. New York.

33.**Liang, P., and Pardee, A.B.** 1995. Recent advances in differential display. Curr. Opin. Immunol. *7*:274-280.

34.**Liu, G., Takano, T. Matsuzuka, F. Higashiyama, T. Kuma, K. and Amino, N.** 1999. Screening of specific changes in mRNAs in thyroid tumors by sequence specific differential display: decreased expression of c-fos mRNA in papillary carcinoma. Endocr. J. *46*:459-466.

35.**Lopez-Nieto, C.E., and Nigam, S.K.** 1996. Selective amplification of protein-coding regions of large sets of genes using statistically designed primer sets. Nat. Biotechnol. *14*:857-861.

36.**Luce, M.J., and Burrows, P.D.** 1998. Minimizing false positives in differential display. Biotechniques *24*:766-768, 770.

37.**Lyer, V.R., M.B. Eisen, D.T. Ross, G. Schuler, T. Moore, J.C.F. Lee, J.M. Trent, L.M. Staudt, J. Hudson, M.S. Boguski, D. Lashkari, D. Shalon, D. Botstein, and P.O. Brown.** 1999. The transcriptional program in the response of human fibroblasts to serum. Science *283*: 83-87.

38.**Mao, M., M. Yu, J.H. Tong, J. Ye, J. Zhu, Q.H. Huang, G. Fu, L. Yu, S.Y. Zhao, S. Waxman, M. Lanotte, Z.Y. Wang, J.Z. Tan, S.J. Chan, and Z. Chen.** 1996. RIG-E, a human homolog of the murine Ly-6 family, is induced by retinoic acid during the differentiation of acute promyelocytic leukemia cell. Proc. Natl. Acad. Sci. USA *93*:5910-5914.

39.**Matsuyama T., A. Grossman, H.W. Mittrucker, D.P. Siderovski, F. Kiefer, T. Kawakami, C.D. Richardson, T. Taniguchi, S.K. Yoshinaga, and T.W. Mak.** 1995. Molecular cloning of LSIRF, a lymphoid-specific member of the interferon regulatory factor family that binds the interferon-stimulated response element (ISRE). Nucleic Acids Res. *23*:2127-2136.

40.**Mathieu-Daude, F., R. Cheng, J. Welsh, and M. McClelland.** 1996. Screening of differentially amplified cDNA products from RNA arbitrarily primed PCR fingerprints using single strand conformation polymorphism (SSCP) gels. Nucleic Acids Res. *24*:1504-1507.

41.**McClelland, M., F. Mathieu-Daude, and J. Welsh.** 1995. RNA fingerprinting and differential display using arbitrarily primed PCR. Trends Genet. *11*:242-246.

42.**McKeveney, P.J., V.M. Hodges, R.N. Mullan, P. Maxwell, D. Simpson, A. Thompson, P.C. Winter, T.R. Lappin, and A.P. Maxwell.** 2001. Characterization and localization of expression of an erythropoietin-induced gene, ERIC-1/TACC3, identified in erythroid precursor cells. Br. J. Haematol. *112*:1016-1024.

43.**Miele, G., L. MacRae, D. McBride, J. Manson, and M. Clinton.** 1998. Elimination of false positives generated through PCR re-amplification of differential display cDNA. Biotechniques *25*:138-144.

44.**Millan, J. and M.A. Alonso.** 1998. MAL, a novel integral membrane protein of human T lymphocytes, associated with glycosylphosphatidylinositol-anchored proteins and Src-like tyrosine kinases. Eur. J. Immunol. *28*:3675-3684.

45.**Mills, K.I., V. Walsh, A.F. Gilkes, M.C. Sweeney, T. Mirza, L.J. Woodgate, G. Brown, and A.K. Burnett.** 2000. High FUS/TLS expression in acute myeloid leukaemia samples. Br. J. Haematol. *108*:316-321.

46.**Mittrucker, H.W., T. Matsuyama, A. Grossman, T.M. Kundig, J. Potter, A. Shahinian, A. Wakeham, B. Patterson, P.S. Ohashi, and T.W. Mak.** 1997. Requirement for the transcription factor LSIRF/IRF4 for mature B and T lymphocyte function. Science *275*:540-543.

47.**Morikawa, Y., H. Nishidam, K. Misawa, T. Nosaka, A. Miyajima, E. Senba, and T. Kitamura.** 1998. Induction of synaptosomal-associated protein-23 kD (SNAP-23) by various cytokines. Blood *92*:129-135.

48.**Nagai, H., T. Kinoshita, J. Imamura, Y. Murakami, K. Hayashi, K. Mukai, S. Ikeda, K. Tobinai, H. Saito, M. Shimoyama, and K. Shimotohno.** 1991. Genetic alternation of p53 in some patients with adult T-cell leukemia. Jpn. J. Cancer Res. *82*:1421-1427.

49.**Nedivi, E., D. Hevroni, D. Naot, D. Israeli, and Y. Citri.** 1993. Numerous candidate plasticity-related genes revealed by differential cDNA cloning. Nature *363*:718-722.

50.**Ohtsuki, T., K. Hatake, M. Ikeda, H. Tomizuka, Y. Terui, M. Uwai, and Y. Miura.** 1997. Expression of Src-like adapter protein mRNA is induced by all-trans retinoic acid. Biochem. Biophys. Res. Commun. *230*:81-84.

51.**Okada, M., S. Yamaga, S. Yasuda, S.M. Weissman, and Y. Yasukochi, Y.** 2002. Differential display of protein tyrosine kinase cDNAs from human fetal and adult brains. Biotechniques *32*:856, 858, 860, 863-865.

52.**Poiez, B.J., F.W. Ruscetti, A.F. Gazdar, P.A. Bunn, J.D. Minna, and R.C. Gallo.** 1980. Detection and isolation of type-C retrovirus particles from fresh and cultured lymphocytes of patient with cutaneous T-cell lymphoma. Proc. Natl. Acad. Sci. USA *77*:7415-7419.

53.**Rios, M., S. Ojeda, L.A. Velasquez, K. Maisey, and H.B. Croxatto.** 2001. A segment and epithelium specific messenger ribonucleic acid fragment up-regulated by estradiol in the rat oviduct. Biol. Res. *34*:15-21.

54.**Roschier, M., E. Kuusisto, and A. Salminen.** 2000. A differential display protocol to identify differentially expressed mRNAs in potassium-deprived cerebellar granule cells. Brain Res. Brain Res. Protoc. *5*:121-131.

55.**Ruegg, C. and R. Pytela.** 1995. Sequence of a human transcript expressed in T- lymphocytes and encoding a fibrinogen-like protein. Gene *160*:257-262.

56.**Sakashita, A., T. Hattori, C.W. Miller, H. Suzushima, N. Asou, K. Takatsuki, and H.P. Koeffler.** 1992. Mutations of the p53 gene in adult T-cell leukemia. Blood *79*: 477-480.

57.**Schneider, O. and U.E. Zelck.** 2001. Differential display analysis of hemocytes from schistosome-resistant and schistosome-susceptible intermediate hosts. Parasitol. Res. *87*:489-491.

58.**Schultheis, B., M. Carapeti-Marootian, A. Hochhaus, A. Weisser, J.M. Goldman, and J.V. Melo.** 2002. Overexpression of SOCS-2 in advanced stages of chronic myeloid leukemia: possible inadequacy of a negative feedback mechanism. Blood *99*:1766-1775.

59.**Schweickart, V.L., C.J. Raport, R. Godiska, M.G. Byers, R.L. Eddy, T.B. Shows, and P.W. Gray.** 1994. Cloning of human and mouse EBI1, a lymphoid-specific G-protein-coupled receptor encoded on human chromosome 17q12-q21.2. Genomics *23*: 643-650.

60.**Shaw-Smith, C.J., A.J. Coffey, E. Huckle, J. Durham, E.A. Campbell, T.C. Freeman, J.R. Walters, and D.R. Bentley.** 2000. Improved method for detecting differentially expressed genes using cDNA indexing. Biotechniques *28*:958-964.

61.**Shenoy-Scaria, A.M., J. Kwong, T. Fujita, M.W. Olszowy, A.S. Shaw, and D.M. Lubin.** 1992. Signal transduction through decay-accelerating factor. Interaction of glycosyl-phosphatidylinositol anchor and protein tyrosine kinases p56lck and p59fyn. J. Immunol. *149*:3535-3541.

62.**Shimada, H., H. Ichikawa, S. Nakamura, R. Katsu, M. Iwasa, I. Kitabayashi, and M. Ohki.** 2000. Analysis of genes under the downstream control of the t(8;21) fusion protein AML1-MTG8: overexpression of the TIS11b (ERF-1, cMG1) gene induces myeloid cell proliferation in response to G-CSF. Blood *96*:655-663.

63.**Shimoyama, M., and members of The Lymphoma Study Group (1984-1987).** 1991. Diagnostic criteria and classification of clinical subtypes of adult T-cell leukemia-lymphoma. Br. J. Haematol. *79*:428-437.

64.**Tohonen, V., C. Osterlund, and K. Nordqvist.** 1998. Testatin: a cystatin-related gene expressed during early testis development. Proc. Natl. Acad. Sci. USA *95*:14208-14213.

65.**Uchiyama, T., J. Yodoi, K. Sagawa, K. Takatsuki, and H. Uchino.** 1977. Adult T- cell leukemia: Clinical and hematologic features of 16 cases. Blood *50*:481-492.

66.**Utans-Schneitz, U., H. Lorez, W.E. Klinkert, J. da Silva, and W. Lesslauer.** 1998. A novel rat CC chemokine, identified by targeted differential display, is upregulated in brain inflammation. J. Neuroimmunol. *92*:179-190.

67.**Velculescu, V.E., L. Zhang, B. Vogelstein, and K.W. Kinzler.** 1995. Serial analysis of gene expression. Science *270*:484-487.

68.**Vila-Ortiz, G.J., M. Radrizzzani, H. Carminatti, V.P. Idoyaga-Vargas, and T.A. Santa-Coloma.** 2001. Single strand mRNA differential display (SSDD) applied to the identification of serine/threonine phosphatases regulated during cerebellar development. J. Neurosci. Methods *105*:87-94.

69.**Walker, J.E., J.M. Arizmendi, A. Dupuis, I.M. Fearnley, M. Finel, S.M. Medd, S.J. Pilkington, M.J. Runswick, and J.M. Skehel.** 1992. Sequences of 20 subunits of NADH: Ubiquinone oxidoreductase from bovine heart mitochondria. Application of a novel strategy for sequencing proteins using the polymerase chain reaction. J. Mol. Biol. *226*:1051-1072.

70.**Walker, J.E., J.M. Skehel, and S.K. Buchana.** 1995. Structural analysis of NADH: ubiquinone oxidoreductase from bovine heart mitochondria. Methods Enzymol. *260*: 14-34.

71.**Watari, K., A. Tojo, T. Nagamura-Inoue, F. Nagamura, A. Takeshita, T. Fukushima, T. Motoji, K. Tani, and S. Asano.** 2000. Identification of a melanoma antigen, PRAME, as a BCR/ABL-inducible gene. FEBS Lett. *466*:367-371.

72.**Welsh, J., K. Chada, S.S. Dalal, R. Cheng, D. Ralph, and M. McClelland.** 1992. Arbitrarily primed PCR fingerprinting of RNA. Nucleic Acids Res. *20*:4965-4970.

73.**Yamada, Y., Y. Hatta, K. Murata, K. Sugamura, S. Ikeda, T. Maeda, Y. Hirakata, S. Kamihira, K. Tsukasaki, S. Ogawa, H. Hirai, H.P. Koeffler, and M. Tomonaga.** 1997. Deletion of p15 and /or p16 genes as a poor prognosis factor in adult T-cell leukemia. J. Clin. Oncol. *15*:1778-1785.

74.**Yoshida, R., T. Imai, K. Hieshima, J. Kusuda, M. Baba, M. Kitaura, M. Nishimura, M. Kakizaki, H. Nomiyama, and O. Yoshie.** 1997. Molecular cloning of a novel human CC chemokine EBI-1ligand chemokine that is a specific functional ligand for EBI-1, CCR7. J. Biol. Chem. *23*:13803-13809.

75.**Yoshikawa, T., G.Q. Xing, and S.D. Detera-Wadleigh.** 1995. Detection, simultaneous display and direct sequencing of multiple nuclear hormone receptor genes using bilaterally targeted RNA fingerprinting. Biochim. Biophys. Acta. *1264*:63-71.

76.**Zhao, S., S.L. Ooi, and A.B. Pardee.** 1996. New primer strategy improves precision of differential display. BioTechniques *18*:842-850.

3 Gene Expression Studies in Lymphoid Cells Using Differential Display

Chanaka Mendis, Rina Das, and Marti Jett
Department of Molecular Pathology, Walter Reed Army Institute of Research, Silver Spring, MD, USA

TECHNOLOGY PLATFORM

We have made tremendous strides as a scientific community in many significant health-related issues that pertain to the improvement of mankind, yet we do not have total dominance in the detection, prevention, and treatment of many devastating illnesses that have been plaguing us for decades. We have laid a solid foundation in each instance by developing and optimizing techniques that were not even thought of, let alone utilized, a few years ago. However, we still cannot pinpoint potential fatal diseases and infections, and even if we are fortunate enough to ascertain such a disastrous situation, we may still be solely dependent on the capacity and the complexity of the situation. In such an event, our ability to predict or detect would depend on a model that combines facts as well as predictions. As any given fraction of human population goes through numerous periodic and abrupt fluctuations, the probability of the success of any such predictive model becomes next to impossible. Since the unraveling of the mysteries of DNA, scientists have focused intensively on genetics and related fields, which is clearly perceived in most recent publications (i.e., books, journal articles, and patents). Lately, scientists have focused on this remarkable avenue of study and have made deeper inroads into the detection, prevention, and therapeutics associated with deadly viral and bacterial toxins that are known to cause concerns in today's society. With this in mind, we utilized an efficient yet highly sensitive differential display reverse transcription PCR (DD RT-PCR) technique to provide insights into potential gene markers and therapeutic targets in human lymphoid cells using staphylococcal enterotoxin B (SEB) as a model stimulant.

Perspectives in Gene Expression
Edited by K. Appasani

Before DD RT-PCR was introduced in 1992 by Liang et al. (15), all accessible methods to screen the alterations of genes were highly inefficient and error-prone. Two frequently utilized techniques to analyze altered gene expression were subtractive hybridization (30) and differential hybridization (26). Both methods identified genes from only one cell type of the respective pair of cells, detected qualitative but not quantitative changes in gene alterations (2), and limited identifications to previously characterized (functional) genes. It was highly desirable to use a method that detects all genes (known and novel) expressed in a particular cell type. The ability to compare patterns of expressed DNA products from two cell types allowed researchers to resolve both qualitative and quantitative changes in gene expression. DD RT-PCR offered an attractive alternative to subtractive hybridization and differential hybridization and has become increasingly popular because of its ease and sensitivity to detect all kinds of differentially expressed genes (21).

The technique of DD RT-PCR has evolved into a powerful method for determining changes in gene expression in several cell lines in response to temporal development, diseases, and mitogenic stimulants. DD RT-PCR has been employed diligently to isolate genes differentially expressed in cancers (15), heart diseases (24,28), diabetes (19), embryogenesis (31), the brain (29), growth factor stimulants (9,10) human tumor mammary epithelical cells (15), and cyclin G as a target of the p53 tumor suppressor protein (20). DD RT-PCR makes feasible the simultaneous detection of induced, inhibited, up-regulated and down-regulated PCR products, and for the first time introduced a technique that was simple yet addressed most of the complex issues twice as efficiently as any other available method. The DD RT-PCR technique also has tremendous potential in many applications such as cell differentiation, cell cycle and life stages, cell activation and signaling, markers and mutations, drug resistance, targets for drugs, nutritional and environmental stress, low abundant samples, microbial pathogenesis, drug and environmental influences, gene expression in vivo, gene expression in the host, cell death and biodefense (27).

It is crucial to apply the current techniques to fulfill the goals and requirements of a project and not let abundantly available savvy techniques dictate the terms and consequently loose the initial objective and purpose of the study. Each and every technique can be used to its fullest potential in some investigations, while for others, the same technique can be highly ineffective. As the strategic design of the technique (Figure 1) and the design of the primers (Figure 2, A and B), there were two key factors to the success of the investigation that we made every effort to design and utilize to our advantage.

Some of the inherent trademark features of DD RT-PCR were a perfect fit to the objectives of our study. These included: the ability to screen simultaneously for multiple samples (cells exposed to multiple dosages of toxin and those exposed for various time intervals); identify novel genes (i.e., genes not well known in the literature and may be specific for the cell type or the condition); systematically detect the whole RNA population (ability to group RNA to multiple sub-populations);

ability to use small quantities of total RNA (as little as 20 ng total RNA/reaction); high sensitivity (ability to detect low expressed RNA that otherwise would have gone undetected); high reproducibility (evident by obtaining similar results for duplicate samples analyzed side by side on long high-resolution gels (Figure 3); minimizing the false-positive results (by implementing stringent PCR conditions); and using long specifically designed anchored (31-mer) and arbitrary (26-mer) primers to identify longer cDNA fragments (thus producing more coding information).

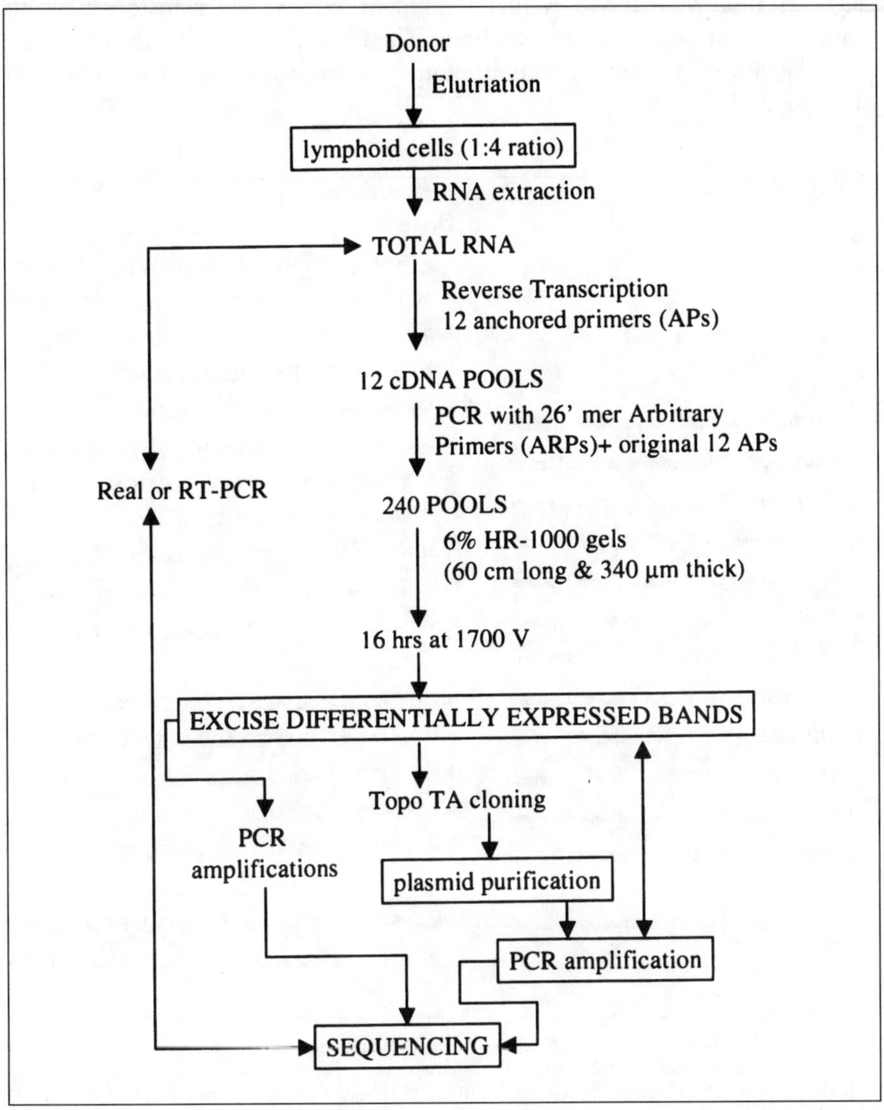

Figure 1. Schematic representation of DD RT-PCR performed on SEB-induced human lymphoid cells.

Because the survival of any organism (including *Homo sapiens*) is consistently challenged by various extracellular stimulants, we preferred to investigate human peripheral blood mononuclear cells (PBMCs), which are part of the most affected system on the intrusion of any foreign particle. For such penetrating foreign particles, the lymphoid cells have the capability to function as foot soldiers and provide the first line of defense, making it a more attractive and interesting investigative model. The ability of the lymphoid system to function as the primary target to most stimulants and toxins makes the study of human lymphoid cells even more valuable to evaluate and identify most recent historical events that have transpired in the human body. The identification of differentially expressed genes in a systematic manner should allow us to develop new avenues for diagnosis and therapy.

A

1. T7(dT$_{12}$) AP1 = T7(dT$_{12}$)GA anchored 3' primer
2. T7(dT$_{12}$)AP2 = T7(dT$_{12}$)GC anchored 3' primer
3. T7(dT$_{12}$)AP3 = T7(dT$_{12}$)GG anchored 3' primer
4. T7(dT$_{12}$)AP4 = T7(dT$_{12}$)GT anchored 3' primer
5. T7(dT$_{12}$)AP5 = T7(dT$_{12}$)CA anchored 3' primer
6. T7(dT$_{12}$)AP6 = T7(dT$_{12}$)CC anchored 3' primer
7. T7(dT$_{12}$)AP7 = T7(dT$_{12}$)CG anchored 3' primer
8. T7(dT$_{12}$)AP8 = T7(dT$_{12}$)AA anchored 3' primer
9. T7(dT$_{12}$)AP9 = T7(dT$_{12}$)AC anchored 3' primer
10. T7(dT$_{12}$)AP10 = T7(dT$_{12}$)AG anchored 3' primer
11. T7(dT$_{12}$)AP11 = T7(dT$_{12}$)AT anchored 3' primer
12. T7(dT$_{12}$)AP12 = T7(dT$_{12}$)CT anchored 3' primer

B

1. M13r-ARP1 = 5' ACAATTTCACACAGGACGACTCCAAG 3'
2. M13r-ARP2 = 5' ACAATTTCACACAGGAGCTAGCATGG 3'
3. M13r-ARP3 = 5' ACAATTTCACACAGGAGACCATTGCA 3'
4. M13r-ARP4 = 5' ACAATTTCACACAGGAGCTAGCAGAC 3'
5. M13r-ARP5 = 5' ACAATTTCACACAGGAATGGTCGTCT 3'
6. M13r-ARP6 = 5' ACAATTTCACACAGGATACAACGAGG 3'
7. M13r-ARP7 = 5' ACAATTTCACACAGGATGGATTGGTC 3'
8. M13r-ARP8 = 5' ACAATTTCACACAGGATGGTAAAGGG 3'
9. M13r-ARP9 = 5' ACAATTTCACACAGGATAAGCCTAGC 3'
10. M13r-ARP10 = 5' ACAATTTCACACAGGACAGTGTGTGA3'
11. M13r-ARP11 = 5' ACAATTTCACACAGGAACGCTAGTGT 3'
12. M13r-ARP12 = 5' ACAATTTCACACAGGAGGTACTAAGG 3'
13. M13r-ARP13 = 5' ACAATTTCACACAGGAGTTGCACCAT 3'
14. M13r-ARP14 = 5' ACAATTTCACACAGGATCCATGACTC 3'
15. M13r-ARP15 = 5' ACAATTTCACACAGGACTTTCTACCC 3'
16. M13r-ARP16 = 5' ACAATTTCACACAGGATCGGTCATAG 3'
17. M13r-ARP17 = 5' ACAATTTCACACAGGACTGCTAGGTA 3'
18. M13r-ARP18 = 5' ACAATTTCACACAGGATGATGCTACC 3'
19. M13r-ARP19 = 5' ACAATTTCACACAGGATTTTGGCTCC 3'
20. M13r-ARP20 = 5' ACAATTTCACACAGGATCGATACAGG 3'

Figure 2. (A) RT-PCR anchored primers (Beckman Coulter). All primers were designed to incorporate two nucleotides immediately upstream of a 12-oligo dT region, followed by a 17-bp T7 RNA polymerase promoter-derived site (5′-ACGACTCACTATAGGGC-3′). (B) Set of arbitrary primers (Beckman Coulter) used in DD RT-PCR. Core annealing sequences of all primers are 10 bases in length and contain a 16-nucleotide segment of the M13 universal reverse (-48) 24-mer priming sequence.

DESIGN OF PRIMERS

A common problem associated with using isolated poly(A) RNA was the non-specific binding of oligo-dT primer to long poly(A) tails (as long as 300 bases), which, in turn, resulted in smearing on DNA sequencing gels. Therefore, we used the more preferable total RNA as our starting material. Smearing was overcome by using oligo(dT)$_{12}$ (with two-base) anchored primers. Although the number of reverse transcription reactions can be dramatically reduced by the introduction of a single-base anchored oligo(dT)$_{12}$ primer, a single-base anchored primer poses serious drawbacks. The use of single-base anchored primers demands an unacceptable number of upstream arbitrary primers to reduce the size of each sub-population. Also, the annealing of a single non-T base may insufficiently stable to ensure the efficient capture of all intended target mRNA in a reproducible manner. Our de-

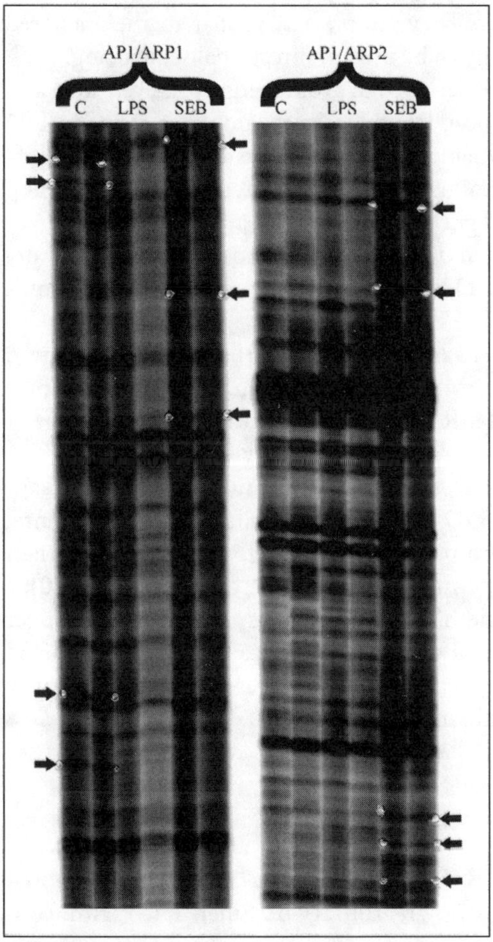

Figure 3. Identification of altered genes through DD RT-PCR. This figure illustrates a portion of a high-resolution DD-PCR gel containing samples with or without SEB. Equal quantities of total RNA from control and SEB and LPS-treated (100 ng/mL) samples, with appropriate housekeeping genes, were first reverse-transcribed, followed by DD RT-PCR (labeling with [^{33}P]-dATP). All samples were analyzed in duplicate on a 6% DD-PCR gel for 16 h at 2700 V. The primer combinations are indicated above the autoradiographs, and the samples are labeled as shown below. C, without treatment of SEB; SEB, SEB-treated (100 ng/mL) samples for 16 h; LPS, LPS-treated (100 ng/mL) samples for 16 h. AP1, anchored primer 1; ARP1 and ARP2, represent arbitrary primers 1 and 2, respectively. All images were quantified using a NIH-image program and done at least twice in duplicate for reproducibility. Bands indicated by arrows were typical examples of altered DNA products that were excised for further evaluation. All standard error (SE) values were less than 10% and are not shown.

signed anchored primers (Figure 2a) contained two nucleotides immediately upstream of a 12-nucleotide oligo-dT region, ensuring the efficient and reproducible capture of target mRNA sub-populations. Assuming an equal statistical distribution of all four bases within these first two non-T sites, each of the 12 anchored primers would generate a different fraction of first-strand cDNA, representing 8% of all mRNA species. While the 12 consecutive oligo-dTs ensured the efficient annealing to the mRNA poly(A) region, the 17-nucleotide T7 RNA polymerase promoter-derived site (5′-ACGACTCACTATAGGGC-3′) allowed direct sequencing from the 3′-end (without the absolute necessity for fragment/vector subcloning) and an antisense "run-off" to be generated (for ribosomal protection assay or Northern blot analysis). To reduce the number of priming events per cDNA, the annealing sequences of the arbitrary 5′-primers were also designed to contain 10 nucleotides. This reduction of priming events diminished the multiple size versions generated for each first-strand cDNA fragment, thus abating the complexity of the DD RT-PCR display. Each of the 26-mer arbitrary primers (Figure 2b) consist of a 16-mer segment of the M13 universal reverse primer, similar to the anchored primer, and has the functional capacity to be used in directional sequencing. Both primers (arbitrary and anchored) were designed to avoid being excessively long and unwieldy because long primers have been known to create ambiguities during DD RT-PCR, resulting in the loss of valuable data. All primers used in the DD RT-PCR study permit for stringent annealing conditions, which prevent mispriming during the amplification process. Designed features of both anchored and arbitrary primers (length) negated the formation of false positives and ensured the high degree of confidence in the analysis of DD RT-PCR patterns by increasing the reproducibility of identical samples.

We strategically designed primers to amplify 0.42% of the mRNA population for any given primer (anchored and arbitrary) combination. In a statistical point of view, this enables us to amplify all mRNA sequences by the systematic usage of all possible primer combinations. Assuming that any given cell contains 15 000–20 000 different mRNA species, each pair of primers (arbitrary and anchored) would generate 100–120 DD RT-PCR products. Our experimental strategy manipulated high-resolution polyacrylamide gels (60 cm; longer than conventional gels) to resolve the above-mentioned number of genes (100–120); a gene per every 0.60–0.50 cm. Some of the tremendous advantages of longer polyacrylamide gels over conventional gels (45 cm) were the minimization of cross-contamination by nearby fragments and the increased ability to identify unique bands due to extended inter-band spacing.

DD RT-PCR METHODOLOGY

The first published DD RT-PCR technique utilized a two-base anchored oligo$(dT)_{12}$ primer in combination with an arbitrary decamer. Theoretical considerations indicated that a high percent of the genes expressed in a cell could be detected systematically by changing the sequences of both the arbitrary decamer and the two-base anchored oligo(dT)$_{12}$ primer. The use of short primers (arbitrary decamer) required a low annealing temperature (40°C). This low stringent condition resulted in the amplification of a few spurious bands, which ultimately resulted in false positives. To avoid the generation of smeared PCR products as a result of non-anchored priming by the carried over primer (14), Watson and Fleming (29) used labeled anchored primers in combination with arbitrary decamers to amplify cDNA. One major drawback was the loss in sensitivity of the detected bands. The use of arbitrary primers longer than 10 bases, together with longer two-base anchored oligo(dT)$_{12}$ primer, enhanced the amplified DNA signal and the reproducibility of the method (15). We performed DD amplifications by applying a few initial cycles of low-stringent annealing temperatures (50°C; to achieve successful annealing of the arbitrary primers), followed by twenty-five cycles of high-stringent annealing temperatures (60°C). This procedure was intended to produce more reproducible bands.

Depending on the specific function of the cell or the tissue (ovalbumin in hen oviducts and globin in reticulocytes), some mRNA can be in as high quantities as 10^5 copies per cell. Our ability not to overexpress any one particular type of mRNA minimized the chances of observing overwhelming or repetitive sequences from the abundant types of mRNA. As a result, we were able to identify rare transcripts (novel and known genes that have never been implicated in SEB infections but somewhat explain the functions of SEB), as shown in Table 1. The single most critical step in many molecular biology experiments (and, in our study, DD analysis) is to obtain high-quality RNA preparations. Consequently, RNase-free DNase treatment of all extracted RNA samples from human lymphoid cells is crucial in obtaining uncontaminated samples (free of contaminating cDNA).

We introduced a few modifications to optimize the adapted the DD amplification model. As shown in the schematic diagram (Figure 1), we first used relatively long (29 bp), two-base anchored oligo(dT)$_{12}$ primers to screen for the whole mRNA population. The use of 3′ anchored, long oligo(dT)$_{12}$ primers in turn would (*i*) increase nonspecific binding, (*ii*) reduce the complexity of cDNA (*iii*) fix the priming site to the beginning of the poly(A) tail, and (*iv*) subdivide the mRNA population into 12 subdivisions.

Our strategy was to amplify short sequences of cDNA obtained from reactions through DD RT-PCR. We used an approach to systematically identify the whole population of RNA by using a set of (12 primers in each set) two-base anchored primers. Each of these 12 different sub-populations of cDNAs were then PCR amplified using a set of arbitrary primers (26-mer) and analyzed on denaturing polyacrylamide gels to identify the presence or difference in the abundance of PCR products. The detection of low abundant mRNA was achieved by incorporating a highly efficient radiolabeling technique using ^{33}P-labeled dATP (Amersham Biosciences, Piscataway, NJ, USA) radioisotope during the amplification of cDNA

Table 1. Comparison of Changes in Gene Expression Levels by DD RT-PCR and RT-PCR

Name	Alteration	Primers	RT-PCR	Accession No.	Function
IL-6	UP / 8	AP1/ARP2	*	M14584	Inflammation
IL-17	UP / 4	AP1/ARP17	*	NM_002190	Inflammation
HIF-1	UP / 9	AP4/ARP3	*	U22431	Respiratory distress
GBP-2	UP / 3	AP1/ARP7	*	M55542	Vadsodilation
PROTEO-GLYCAN	DOWN / 3	AP3/ARP7	*	Embx15998	Low Blood P
HSJ2	UP / 6	AP1/ARP17	*	NM_001539	Heat Shock
FERRETIN HC	DOWN / 4	AP1/ARP6	*	Nm002032	Respiratory
MYOSIN	UP / 3	AP3/ARP3	*	AJ001382	Muscle deprivation
CATHEPSIN L	UP / 20	AP1/ARP18	*	NM_001912	Tissue Degradation
NOVEL	UP / 51	AP1/ARP2	*	N/A	N/A
NOVEL	DOWN / 3	AP3/ARP3	*	N/A	N/A
NOVEL	UP / 13	AP1/ARP2	*	N/A	N/A
NOVEL	DOWN / 5	AP1/ARP18	*	N/A	N/A

A set of genes identified by DD RT-PCR was further examined by RT-PCR to compare their fold-change. After designing specific primers for each gene of interest, RT-PCRs were performed to confirm alterations observed using DD RT-PCR. Identical total RNA samples were used for both RT-PCR and DD RT-PCR, the bands of PCR products were digitized using a Molecular Imager FX after they had been normalized with a housekeeping gene (15S) and quantitated. IL-6, interleukin 6; IL-17, interleukin 17; HIF-1, hypoxia inducible factor 1: GBP-2, guanylate binding protein-2; HSJ2, *Homo sapiens* 40 kDa heat shock protein 2; AP, anchored primers; ARP, arbitrary primers; *, observed similar expression pattern through RT-PCR; N/A, not applicable. A brief functional relevance as well as the accession number are included with each gene.

products. Autoradiographs of analyzed dried gels allowed us to excise the amplified PCR products and subject them to further examination. Once the altered PCR products were eluted from the excised gels, each product was either PCR amplified (Figure 4) or subcloned into a highly efficient pCR®2.1-TOPO vector (Invitrogen, Carlsbad, CA, USA) for further purification (Figure 1). Re-amplification of the excised DD RT-PCR products using the initial pair of primers greatly reduced the probability of undetected or unresolved PCR fragments, which, excised together with the DD RT-PCR amplified products, were further purified through the cloning procedure. Figure 3 illustrates a typical image obtained from a set of statistically possible primer combinations analyzed side by side in duplicate samples. It is also obvious that the number of genes altered would be a small percentage out of the whole gene spectrum, which is clearly illustrated in Figure 3.

Primer efficiency is a combination of annealing and amplification characteristics of the primers. According to our experiments, any possible primer combinations should yield 100–120 PCR products that range from 200 to 2 kb in length. However, it was noticeable that certain primers generate more PCR products than others (Table 2; anchored primer AP1and AP8). This is not surprising because certain cell lines may have a set of mRNA that can recognize a certain specific sequence of the primers. Unless otherwise indicated, all DD RT-PCR products were done in duplicates, and an amplified DNA was categorized as altered (up, down, inhibited, and induced) only if similar alterations were seen in side by side run duplicate samples (Figure 3) and the alteration was at least 3-fold over respective control samples. The autoradiograph corresponding to the high-resolution gel containing various DD amplified products was scanned, and the altered bands were quantified and digitized using a Molecular Imager FX® (Bio-Rad Laboratories, Hercules, CA, USA).

Once they had been analyzed on high-resolution differential display gels (Beckman Coulter, Foster City, CA, USA), the amplified altered genes were excised and eluted by boiling in TE buffer, pH 8.0, at 95°C for 2 min. Excised bands identified through DD analysis (859) were either subjected to another PCR amplification using similar primers as in the initial DD RT-PCR amplification or were subcloned. The analyzed products that showed clear, crisp, single bands were sequenced without further purification (Figure 4, lanes 1–7, 9–11, and 15), but the amplified fragments that were not visible (Figure 4, lanes 12–14) were subjected to PCR with less stringent PCR conditions (lower annealing temperature and higher numbers of PCR cycles). The amplified products that showed contamination (multiple bands of similar intensity) were further purified through subcloning (TOPO-TA cloning; Invitrogen).

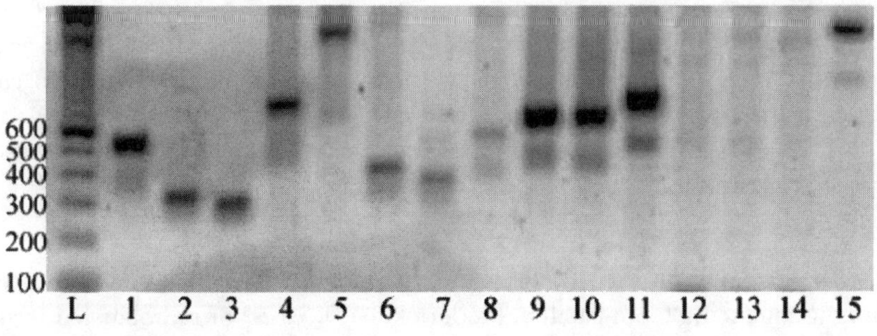

Figure 4. Amplified fragments excised from DD gels. Altered DD amplified samples were excised from gels, eluted, and subjected to PCR, using the same anchored and arbitrary primers used in DD RT-PCR. These samples were analyzed on a 1% agarose gel and scanned using Quantity One Quantitation® Software (Bio-Rad Laboratories). Samples 1–15 were amplified using the following primer combinations: lane 1, AP1/ARP6; lane 2, AP3/ARP7; lane 3, AP3/ARP7; lane 4, AP1/ARP7; lane 5, AP1/ARP11; lane 6, AP1/ARP11; lane 7, AP1/ARP12; lane 8, AP1/ARP12; lane 9, AP1/ARP15; lane 10, AP1/ARP15; lane 11, AP1/ARP15; lane 12, AP1/ARP14; lane 13, AP1/ARP14; lane 14, AP1/ARP14; lane 15, AP1/ARP14; and L, 100-bp ladder.

Table 2. Identification of Induced Genes by SEB Using a Set of Primer Combinations

	AP1	AP2	AP3	AP4	AP5	AP6	AP7	AP8	AP9	AP10	AP11	AP12
ARP1	1	1	1	0	4	7	2	5	4	3	1	0
ARP2	3	0	1	5	1	5	5	6	3	1	1	1
ARP3	0	2	3	2	0	0	5	9	0	7	3	1
ARP4	0	0	1	4	5	6	4	3	0	2	0	2
ARP5	0	6	0	9	0	0	0	0	1	0	6	1
ARP6	2	3	3	3	0	0	0	20	2	0	4	3
ARP7	2	7	6	2	0	9	0	9	1	0	5	2
ARP8	1	5	1	3	0	0	0	6	1	03	1	2
ARP9	13	0	0	0	7	3	9	0	0	16	2	3
ARP10	15	5	0	0	0	4	1	0	1	0	0	4
ARP11	11	1	0	0	1	2	2	0	2	2	3	7
ARP12	14	1	0	0	1	1	1	0	2	0	5	8
ARP13	5	6	7	0	2	0	8	12	0	0	1	0
ARP14	12	6	8	0	5	6	10	20	0	0	12	5
ARP15	13	1	16	0	5	10	12	8	0	0	6	1
ARP16	18	3	11	0	2	5	14	20	0	0	5	9
ARP17	3	3	6	5	0	2	0	0	0	0	3	6
ARP18	6	2	14	5	11	6	0	0	0	0	2	6
ARP19	5	7	12	10	12	5	0	0	0	0	15	0
ARP20	0	8	3	0	0	4	6	0	0	0	8	4

Each DD RT-PCR was carried out using a combination of anchored primers (AP) and arbitrary primers (ARP), totaling 240 possible combinations. Each of the AP (1–12) was used with a set of ARPs (1–20), and the number of genes altered for each combination is shown as a numerical value (0–100). Each of these bands altered were excised, amplified, or cloned, sequenced, and identified using BLASTN algorithm search engine using the NCBI/EMBL databases. Each experiment was performed in duplicate using a pair of primers as illustrated above and conducted using the DD RT-PCR method.

The sequences of purified PCR fragments were deduced either by fluorescent labeling (Sequencing Core Facility, Frederick, MD, USA) or radiolabeled terminator methods [Sequence version 2.0 DNA sequencing kit (USB, Cleveland, OH, USA)]. The nucleotide sequences obtained were compared with available sequences by utilizing EMBL and GenBank® databases with the BLASTN search program of the National Center for Biotechnology Information (NCBI). Determination of the sequences in turn would help lead to functional information about the gene in question.

IDENTIFICATION OF LYMPHOID-RELATED GENES

Out of the 859 differentially expressed products, 650 were cloned, sequenced, and compared against sequence databases using the BLAST algorithms for possible identification. Identified genes were further explored using the NCBI and PubMed databases for specific cellular roles and functions. Some of the identified genes have never been implicated in SEB induction but somewhat explained the symptoms of SEB, while some genes did not match with any available sequences (novel genes) in the GenBank/EMBL databases (Table 2).

Interestingly, SEB is one of the few lethal enterotoxins that has been the focus of many investigations due to its ability to function as a superantigen. Even though a wealth of information is available on the structure, potent regions, binding sites, and symptoms of SEB, little is known about the exact signal transduction pathways or various gene expression patterns. Hence, we took advantage of DD to investigate genes that may be used as potential biomarkers. As a result, we not only had the opportunity to analyze a large number of altered genes (859 in our single study) but also to find functionally important genes previously implicated with the toxin (interleukins associated with inflammation); known genes that have never been implicated in the study of interest but explain some of the symptoms displayed by the stimulant [hypoxia inducible factor-1 (HIF-1), ferreitin heavy chain involved in respiratory distress]; and novel genes that may have significant functional relevance.

Inflammation is one of the earliest conditions associated with SEB. Heat shock proteins, cytokines, and chemokines are characterized for their involvement in inflammation. We were also able to identify a few genes belonging to each group through DD RT-PCR (Table 2) and later confirmed through an alternative method (Table 2). Interleukins (IL-6 and IL-17) that were induced by SEB and cause inflammation (12,25) had similar activation patterns (up-regulation) determined through DD RT-PCR. The alterations were later confirmed semi-quantitatively through RT-PCR (Table 2). We further investigated the time dependence of gene and protein expression of both IL-2 and IL-17 using RT-PCR and ELISA immunoblots, respectively (Reference 17; unpublished data). Both genes showed an induction as early as 2 h and were present over basal levels throughout the investigation (2–24 h). Heat shock proteins are commonly associated with inflammation and inflammation-related events, and we were able to detect a novel 40-kDa heat shock protein (up-regulation) that was recently identified as the homologue of bacterial DnaJ protein. The ability to identify genes that are known to be associated with the toxin and the cell type of interest indicate the efficiency, validity, and reproducibility of the technique.

Tissue damage and cardiac dysfunction are closely associated with inflammation and are two common symptoms that are associated with SEB. Cathepsin L is a lysosomal cysteine protease regulated by cytokines during inflammation and is known to play a crucial role in tissue degradation. We have observed a sharp increase in cathepsin L expression through DD RT-PCR as well as the up-regulation of myosin, which has previously been implicated in significant cardiac dysfunction (6).

Under restrictive oxygen conditions, alternative metabolic pathways that do not require oxygen are activated and genes such as erythropoietin, vascular endothelical growth factors, and glycolytic enzymes are readily observed (5). These restrictive hypoxia-inducible conditions induce a heterodimeric helixprotein complex (consisting of 120 kDa and 90–94 kDa subunits) that is implicated in respiratory distress. The up-regulation of an amplified DNA, which codes for HIF-1, was detected through DD-PCR and later confirmed by RT-PCR. An oxygen-deprived condition may explain the down-regulation of a heme-based PCR product (ferritin heavy chain) detected through DD RT-PCR. The ability to detect HIF-1 (up-regulation) and ferritin heavy chain (down-regulation) through DD RT-PCR may explain the respiratory symptoms associated with SEB. HIF-1 has never been implicated in SEB and may play a role that ultimately leads to lethal shock.

Another gene detected through DD RT-PCR (up-regulated) codes for an interferon-induced cytoplasmic guanylate binding protein-2 (GBP-2) (67 kDa). GBP-2 has a wide variety of basic cellular functions such as protein synthesis, signal transduction, and intracellular protein transcription (4). The ability of GBP-2 to produce high levels of cyclic guanosine monophosphate (cGMP) through guanylate cyclase is responsible for causing vasodilation (22), which is another symptom associated with SEB activation. A gene that is known to code for a chondrin sulphate proteoglycan versican V1, which belongs to the growing family of large aggregating proteoglycans, was identified through DD RT-PCR (down-regulated) (7,8). All proteoglycans (including proteoglycan V1) contain a few chondroitin sulphate chains that protect the endothelium from oxidant injury and direct cytotoxicity (1,18,23). The removal of these sulphate chains from the endothelium in the course of inflammation might lead to profound changes in the physiology of all types of blood vessels, ranging from large caliber aorta to the smallest capillaries (3,11,16). A decrease in proteoglycan (observed through DD RT-PCR and confirmed by RT-PCR) would result in thinner blood vessels, which may contribute to low blood pressure, another frequently occurring symptom in SEB-exposed individuals.

We have also identified a handful of novel genes that were either up- or down-regulated upon SEB treatment that may be potentially crucial for identifying relevant functions and symptoms of SEB. Currently, the expression of these genes are verified by alternative methods, and once confirmed, will be further evaluated (protein expression) to obtain more information. A description of the altered gene profile, pattern, significance, and functional relevance is clearly illustrated in a patent titled "Method of Diagnosis of Exposure to Toxic Agents by Measuring Distinct Pattern in the Levels of Expression of Specific Genes" (R. Das, C. Mendis, and M. Jett, inventors; The United States of America as represented by the Secretary of the Army, assignee. November 13, 2001. US patent no. 6316197.). Conformational studies performed using a different screening method (DNA microarrays) showed activation patterns similar to those observed with DD RT-PCR (17). Out of the functionally relevant 35 altered DD RT-PCR amplified products, 33 showed similar activation patterns through microarray experiments, thus validating DD RT-PCR as a solid gene screening technique. Although the efficiency and reproducibility play a major role in the success of any technique, it is more important that the produced data are confirmed by similar techniques. Our investigation clearly indicates the compatibility of DD RT-PCR to other available techniques.

SUMMARY

It is clear that DD RT-PCR can be utilized to resolve several questions posed at the beginning of the investigation. The results discussed throughout the chapter include the detection of a high number of amplified products (Table 1, 859 PCR amplified products), a high reproducibility due to various strategies and modifications employed in the design of the experimental procedure (Figure 1 and Figure 4), the identity of genes, irrespective of their availability (Table 2; novel genes, genes known to be affiliated with SEB, and genes that were never before implicated with SEB), and, most importantly, the easy confirmation of results using other available screening techniques (Table 2, RT-PCR data). In today's ever-changing world, the systematic analytical power of DD RT-PCR may provide crucial information to many complex diseases and conditions that may otherwise go undetected. The ability of DD RT-PCR to pose vastly different qualities (systematically screen the whole genome yet be sensitive enough to show high reproducibility and comparability) place the technique in a league of its own and difficult to replace.

ACKNOWLEDGMENTS

This work was supported by an award from the United States Army for novel in-house laboratory Research and a grant from Defense Advance Research Project Agency (DARPA). We are thankful to Dr. David Yang for his expert advice and suggestions and to Michael Hartl and Atabak Royaee for their diligent work in various aspects of the DD RT-PCR screening procedure.

REFERENCES

1. **Abrahamsson, T., U. Brandt, S. Marklund, and P. Sjoquist.** 1992. Vascular bound recombinant extracellular superoxide dismutase type C protects against the detrimental effects of superoxide radicals on endothelium-dependent arterial relaxation. Circ. Res. *70*:264-271.
2. **Bauer, D., H. Muller, J. Reich H. Reidel, V. Ahrenkiel, P. Warthoe, and M. Strauss.** 1993. Identifi-

cation of differentially expressed mRNA species by an improved display technique (DDRT-PCR). Nucleic Acids Res. *21*:4272-4280.

3. **Bode-Lesniewska, B., M. Dours-Zimmermann, B. Odermatt, J. Briner, P. Heitz, and D. Zimmermann.** 1996. Distribution of the large aggregating proteoglycan versican in adult human tissues. J. Histochem. Cytochem. *44*:303-312.
4. **Bourne, H.** 1988. Do GTPases direct membrane traffic in secretion? Cell *53*:669-671.
5. **Bunn, H. and R. Poyton.** 1996. Oxygen sensing and molecular adaptation to hypoxia. Physiol. Rev. *76*:839-885.
6. **Colbert, M., D. Hall, T. Kimball, S. Witt, J. Lorenz, M. Kirby, T. Hewett, R. Klevitsky, et al.** 1997. Cardiac compartment-specific overexpression of a modified retinoic acid receptor produces dilated cardiomyopathy and congestive heart failure in transgenic mice. J. Clin. Invest. *100*:1958-1968.
7. **Doege, K., M. Sasaki, E. Horigan, J.R. Hassell, and Y. Yamada.** 1987. Complete primary structure of the rat cartilage proteoglycan core protein deduced from cDNA clones. J. Biol. Chem. *262*:17757-17767.
8. **Doege, K., M. Sasaki, T. Kimura, and Y. Yamada.** 1991. Complete coding sequence and deduced primary structure of the human cartilage large aggregating proteoglycan, aggrecan, human-specific repeats, and additional alternatively spliced forms. J. Biol. Chem. *266*:894-902.
9. **Donohue, P.J., G.F. Alberts, B.S. Hampton, and J.A. Winkles.** 1994. A delayed-early gene activated by fibroblast growth factor-1 encodes a protein related to aldose reductase. J. Biol. Chem. *269*:8606-8609.
10. **Hsu, D.K.W., P.J. Donohue, G.F. Alberts, and J.A. Winkles.** 1993. Fibroblast growth factor-1 induces phosphofructokinase, fatty acid synthase and Ca $^{(2+)}$ -ATPase mRNA expression in NIH 3T3 cells. Biochem. Biophys. Res. Commun. *197*:1483-1491.
11. **Ihrcke, N., W. Parker, K. Reissner, and J. Platt.** 1998. Regulation of platelet heparanase during inflammation: role of pH and proteinases. J. Cell. Physiol. *175*:255-267.
12. **Krakauer, T., L. Pitt, and R.E. Hunt.** 1997. Detection of interleukin-6 and interleukin-2 in serum of rhesus monkeys exposed to a nonlethal dose of staphylococcal enterotoxin B. Mil. Med. *162*:612-615.
13. **Liang, P. and A.B. Pardee.** 1995. Recent advances in differential display. Cur. Opin. Immunol. *7*:274-280.
14. **Liang, P., L. Averboukh, and A.B. Pardee.** 1993. Distribution and cloning of eukaryotic mRNAs by means of differential display: refinements and optimization. Nucleic Acids Res. *21*:3269-3275.
15. **Liang, P., L. Averboukh, K. Keyomarsi, R. Sager, and A.B. Pardee.** 1992. Differential display and cloning of messenger RNAs from human breast cancer versus mammary epithelial cells. Cancer Res. *52*:6966-6968.
16. **McGrath, R.** 1997. Heparan sulphate proteoglycan and wound healing in skin. J. Pathol. *183*:251-252.
17. **Mendis, C., R. Das, R. Hammamieh, A. Royaee, D. Yang, S. Peel, and M. Jett.** 2002. Common and unique host gene expression responses to two shock-inducing toxins provide diagnostic markers. J. Biol. Chem. (In Press.)
18. **Nakazono, K., N. Watanbe, K. Matsuno, J. Sasaki, T. Sato, and M. Inoue.** 1991. Does superoxide underlie the pathogenesis of hypertension? Proc. Natl. Acad. Sci. USA *88*:10045-10048.
19. **Nishio, Y., L. Aiello, and G. King.** 1994. Glucose induced genes in bovine aortic smooth muscle cells identified by mRNA differential display. FASEB J. *8*:103-106.
20. **Okamato, K. and D. Beach.** 1994. Cyclin G is a transcriptional target of the p53 tumor suppressor protein. EMBO J. *13*:4816-4822.
21. **Pardinas, J., N. Combates, S. Prouty, K. Stenn, and S. Parimoo.** 1998. Differential subtraction display: a unified approach for isolation of cDNAs from differentially expressed genes. Anal. Biochem. *257*:161-168.
22. **Qiu, H., D. Henrion, J. Benessiano, C. Heymes, B. Tournier, and B. Levy.** 1998. Decreased flow-induced dilation and increased production of cGMP in spontaneously hypertensive rats. Hypertension *32*:1098-1103.
23. **Redini, F., J. Tixier, M. Petitou, J. Choay, L. Robart, and W. Hornebeck.** 1988. Inhibition of leucocyte elastase by heparin and its derivatives. Biochem. J. *252*:515-519.
24. **Russell, M., U. Utans, A.F. Wallace, P. Liang, R.J. Arceci, M.J. Karnovsky, L.R. Wyner, Y. Yamashita, et al.** 1994. Identification and upregulation of galactose/N-acetylgalactosamine macrophage

lectin in rat cardiac allografts with arteriosclerosis. J. Clin. Invest. *9*:722-730.

25. **Sperber, K., L. Silverstein, C. Brusco, C. Yoou, G.E. Mulliu, and L. Mayer.** 1995. Cytokine secretion induced by superantigens in peripheral blood mononuclear cells, lamina propria lymphocytes, and intraepithelial lymphocytes. Clin. Diagn. Lab. Immunol. *2*:473-477.
26. **St. John, T.P. and R.W. Davis.** 1979. Isolation of galactose-inducible DNA sequences from *Saccharomyces cerevisiae* by differential plaque filter hybridization. Cell *16*:443-452.
27. **Sturtevant, J.** 2000. Applications of differential-display reverse transcription-PCR to molecular pathogenesis and medical mycology. Clin. Microbiol. Rev. *13*:408-427.
28. **Utans, U., P. Liang, L.R. Wyner, M.J. Karnovsky, and M. Russell.** 1994. Chronic cardiac rejection: identification of five upregulated genes in transplanted hearts by differential mRNA display. Proc. Natl. Acad. Sci. USA *91*:6463-6467.
29. **Watson, M. and T. Fleming.** 1994. Isolation of differentially expressed sequence tags from human breast cancer. Cancer Res. *54*:4598-4602.
30. **Zimmermann, C.R., W.C. Orr, R.F. Leclerc, E.C. Barnard, and W.E. Timberlake.** 1980. Molecular cloning and selection of genes regulated in Aspergillus development. Cell *21*:709-715.
31. **Zimmermann, J. and R.M. Schultz.** 1994. Analysis of gene expression in the preimplantation mouse embryo: use of mRNA differential display. Proc. Natl. Acad. Sci. USA *91*:5456-5460.

4 Gene Discovery in Experimental Colitis: mRNA Display Approach

Krishnarao Appasani
Brigham and Women's Hospital, Harvard Medical School, Boston, MA, USA

INTRODUCTION

Inflammatory bowel diseases (IBDs), such as ulcerative colitis (UC) and Crohn's disease (CD), are chronic, relapsing tissue-destructive processes that represent an uncontrolled mucosal immune response to yet unknown immunogens (34,72). While UC is limited to the large bowel, CD can occur anywhere in the small or large intestine. UC affects about 1 in 1000 people in the western world and has characteristic histological features including the loss of globlet cells, crypt elongation, and inflammatory cell infiltrate in the lamina propria (59). Previous clinical, experimental, and transgenic studies suggest that dysregulation of both the epithelial barrier function and the mucosal immune system is a key feature of colitis (18,28,42,54,63,70). Other factors, including several inflammatory mediators and gut constituents (e.g., enteric bacterial flora), have also been implicated in contributing to the chronic and relapsing nature of UC (7,20). Recent mapping studies also confirmed that IBD susceptibility regions exist on chromosomes 1, 3, 6, 7, 12, 14, and 16 (9,29,30,56). However, a comprehensive genetic analysis of gene expression in colitis had not yet been performed when this project was initiated in the mid-1990s.

A series of experimental colitis models have been developed to aid our understanding of IBD (52,57,65). Recently, colleagues at our institution developed an experimental model in which colitis is induced in mice transiently made deficient in high-affinity interleukin-2 receptor (IL-2R)-positive cells (Tx+*DSS* model of colitis) by the injection of a high-affinity IL-2R-targeted immunotoxin (21). Inflammation in colitis involves the increased expression of the cytokines interleukin-1 (IL-1), IL-2, IL-4, IL-6, tumor necrosis factor, and granulocyte macrophage-colony stimulating factor in both the UC and dextran sodium sul-

Perspectives in Gene Expression
Edited by K. Appasani

fate (DSS)-induced colitis animal models (17,23,60). Studies also documented that chemokines such as IL-8 and MCP-1 are primarily responsible for attracting leukocytes (neutrophils, macrophages, and T cells) to the site of colitis (3). In addition, pro-inflammatory cytokines (e.g., IL-1 and IL-2) promote IBD while anti-inflammatory mediators (e.g., IL-1ra) modulate colitis (11). However, none of these mediators are unique to mucosal inflammation.

The evaluation of gene transcripts has become an important tool in the study of normal and pathological processes because the expression of most genes is regulated at the level of transcription. Previous studies used Northern blot analysis or RT-PCR to identify genes expressed in colitis (23,43,60,76), but the scope of these methods was confined to known genes (32). It is yet to be determined whether known factors or rather yet unknown genes play the pivotal role in colitis. No unique genes have been identified for UC or for any form of experimental colitis (reviewed in Reference 18). The major goal of this investigation was to identify IBD-specific diagnostic signatures, to review the entire repertoire of available transcripts, and to elucidate differentially regulated (known and unknown) genes that play a key role in colitis.

To investigate the molecular pathogenesis of bowel inflammation, an elegant comparative genetic analysis of mRNA expression between normal and colitis colon was performed (37) using differential display PCR [DD-PCR (50)]. The mRNA expression pattern in two mouse experimental colitis models (*DSS*-induced and *TxDSS* model of colitis) was analyzed. In contrast to traditional methods, DD-PCR analysis allows: (*i*) the semi-quantitative evaluation of the expression of individual genes between diseased and normal tissues; (*ii*) the simultaneous comparison of a large number of mRNA species between normal and diseased samples/tissues; (*iii*) the cloning of the genes of interest; and (*iv*) the identification of known and unknown genes with equal probability. DD-PCR produces valuable reagents (e.g., cDNA) for a more detailed characterization of gene expression and tissue localization (75). In addition, this technique requires only 1 μg of total RNA for a comparative gene analysis. Thus, differential display has the power to identify a spectrum of molecular factors, known and unknown alike, that are regulated in a disease-related fashion (73). Differential display was used to identify disease-related genes from rat heart transplants undergoing secondary arteriosclerosis (73), from blood vessels after angioplasty (5), and in a model of anastomotic intimal hyperplasia (27). Recently, this technique has been utilized in the discovery of leukemia (35), and malignant pleural mesothelioma (8,26)-related genes. Its high sensitivity made DD-PCR a favorable technique (49) for the identification, cloning, and sequencing of genes from various biological and disease processes (5,8,26,27,35,68,71).

The mouse genome contains an estimated 50 000 to 100 000 genes. The expressed sequence tag (EST) approach has been used extensively to deduce the whole sequence and map of the mouse genome (1,55). It has been demonstrated that cDNA library screening strategies can also be used to identify differentially expressed mouse genes. Recent genome sequencing project results (55) and the availability of microarray techniques provided new molecular tools to understand genome-wide expression profiles in normal and disease states (15,66,67). More importantly, the ability to measure and analyze this type of gene expression data should therefore provide a basis for improved diagnosis, identification of new therapeutic molecules and, subsequently, prognostic information. Recently, DNA microarray technology has been applied in IBD, and distinctive gene expression profiles were detected between UC and CD (16,44). These studies identified some novel susceptibility candidate genes for UC and normal human colonic mucosa.

In this chapter, I summarized the results obtained from differential display analysis of normal and colitis colon RNAs. Using 24 primer combinations, 5000 mRNA species were visualized, and the expression patterns between normal and diseased colon were compared. One hundred and two candidate cDNAs that showed differential expression between normal and colitis colon were cloned and sequenced (38). The cloned cDNAs were used to confirm the disease-related expression of these genes. This chapter details the expression of 17 previously cloned genes (either up-regulated or down-regulated) in experimental colitis. The isolation, cloning, sequencing, and nucleotide sequence comparison of these 17 differentially regulated genes in colitis showed significant sequence homology to previously cloned cDNAs (38). None of these genes, except IL-1ra, have been linked to colitis before. IL-1ra was shown to be expressed in rabbit immune complex colitis (19) and is also known to modulate the disease when exogenous IL-1ra is administered to diseased animals (11). The validation of these differentially expressed genes identified in mouse experimental colitis models will provide information on homologous or similar genes and their functional role in human UC.

In this gene hunting analysis, one of the cDNAs isolated was 94% homologous to the mouse Ly-6A/E gene, a lymphocyte differentiation/activation antigen and T-cell activating protein (14). Although, Ly-6A/E mRNA is present in normal large bowel, its expression enhanced 6-fold early in the course of colitis disease development. Molecular characterization of the Ly-6A/E gene and its involvement in colitis has been established (40). Current studies indicate that DD-PCR is a powerful technique for the rapid identification of the group of genes that are characteristic of IBDs. The characterization of the molecules reported here suggests potential diagnostic markers, which could be useful in the clinical management of IBD-related diseases in healthcare.

METHODOLOGY

Tissue Extraction from Mouse Experimental Colitis Models and RNA Isolation

Experimental colitis was induced in mouse strain C3H/2J by an intraperitoneal injection of IL-2 diphtheria toxin fusion protein and oral administration of DSS as previously described (21,57). These will be described as DSS-induced and TxDSS colitis models throughout this chapter. The histopathology of colon in TxDSS colitis parallels most features of human UC, once the disease is induced by DSS treatment. Thus, these models are more amenable for identifying colitis-related genes (CRGs) from inflamed tissues. Two normal, two DSS-induced, and two TxDSS-induced colitis mice were sacrificed after carbon monoxide poisoning, and their colons (large intestine) were dissected. Immediately, total cellular RNA from whole colon samples was then extracted with Tri-Reagent® in a single-step extraction method (Molecular Research Center, Cincinnati, OH, USA).

mRNA Differential Display

Chromosomal DNA contaminants from the total RNA samples were removed by treatment with DNaseI by using the MessageClean® Kit (GenHunter, Nashville, TN, USA). Three reverse transcription reactions were performed for each RNA sample using 0.2 μg DNA-free total RNA in a reaction volume of 20 μL using GenHunter's RNAimage® Kit. Three one-base anchored oligo-dT primers (51) were used because (*i*) they provide excellent high selectivity; (*ii*) decrease the number of reverse transcription reactions needed for each sample; (*iii*) minimize the redundancy; and (*iv*) minimize the under-representation of certain RNA species due to the degeneracy of the primers. In addition, a *Hin*dIII site was introduced at the 5′-end of both arbitrary and anchored primers, which makes the primers longer and more efficient in cDNA amplification and also allows cDNA to be cloned easily in a plasmid vector (51). PCR was performed with anchor primers. Band recovery, re-amplification, cloning into the pCR-TRAP vector, and colony PCR analysis were followed according to the manufacturer's instructions (GenHunter).

DNA Sequencing and Database Searches

High-quality mini-prep DNA templates were isolated using the Qia-well 8 Plasmid DNA Purification System (Qiagen, Valencia, CA, USA). DNA sequencing of the cDNA fragments that showed specific hybridization patterns was performed manually in both directions using the Aidseq-Kit C (GenHunter) and Sequenase® version 2 kits (USB, Cleveland, OH, USA). Dye terminator chemistry was also adopted using an automated fluorescence DNA sequencer (model 373A; Applied Biosystems, Foster City, CA, USA). The inserts from a minimum of two recombinant plasmids that originated from two separate, different differential gels were sequenced to establish a consensus sequence. Gene database searches were performed at the National Center for Biotechnology Information (NCBI) using the BLAST server (2).

Northern Blot Analysis

Total RNA (30 µg) was fractionated in 2.2 M formaldehyde/1.2% agarose gels, and blots were prepared as previously described (4,41). High-specific probes were generated by labeling re-amplified or cloned cDNA fragments with α-[^{32}P] dCTP (6000 Ci/mmol) and using a random prime labeling kit (NEN® Life Science Products, Boston, MA, USA), except that 1 µL of relevant anchor primer (2 µM) was included in the reaction. mRNA transcript expression levels were quantitated either by exposing the probed filter to a Molecular Devices' cassette or the autoradiograms were further analyzed in a PhosphorImager (Molecular Devices, Sunnyvale, CA, USA).

Fluorescent In Situ Hybridization

To determine the tissue localization of the newly identified CRGs within the colon, frozen tissue sections of colon from both diseased and normal mice were prepared. Samples were fixed in 4% paraformaldehyde for 20 h, treated in 30% sucrose for 16 h at 4°C, and frozen in methyl-butane at -80°C. Six-micron-thin frozen sections were prepared on a cryostat (Carl Zeiss, Thornwood, NY, USA) after imbedding into OCT compound. The cDNA inserts from specific clones were amplified directly from colony lysates using a riboprobe primer set (T7-Lgh and SP6-Rgh primers; GenHunter) in a volume of 20 µL as recommended by the manufacturer. Antisense RNA was prepared for probing, and sense RNA was prepared for specificity and background control (45). The in vitro transcription was performed using fluorescein-12-UTP or rhodamine, and the labeled nucleotides were separated from the unlabeled nucleotides on a Sephadex® G-50 column (Worthington Biochemical, Lakewood, NJ, USA). The processed tissue sections were rehydrated in 70% ethanol and subjected to hybridization. Finally, the slides were analyzed in a Sarastro 2000 laser-based confocal microscope (Molecular Dynamics, Sunnyvale, CA, USA).

RESULTS

Identification and Isolation of a Large Collection of Differentially Expressed Genes

A total of eight arbitrary 13-mer primers and three one-base T11M primers were used in 24 primer combinations, which identified 100–120 discrete bands per primer combination. This allowed for the comparison of the expression of approximately 5000 RNA species between normal and colitis colons. Figure 1 shows the results of such differential display analysis. Significant differential expression spectrum (increase or decrease) has been detected in more than 100 cDNAs in colitis of which 102 were isolated. These cDNA fragments were then cloned, designated as CRGs, and numbered from 1 to 102. Table 1 lists the expression patterns for the representative genes. Although the goal was to isolate genes with either increased or decreased expression in colitis, some genes whose expression was unchanged between normal colon and colitis were also cloned for the purpose of internal controls. Of the 102 cDNAs (120–350 bp), 55 isolated fragments showed an increased expression, while 45 fragments showed decreased levels of expression in colitis. The expression pattern of five fragments was unchanged between the normal and colitis samples. A 90% overall rate of recovery throughout the re-amplification process and 97% success rate was achieved after cloning.

DNA Sequence Comparison with GenBank® Nucleotide Depository

Sequence comparison of the 102 colitis-related cDNAs with the GenBank database using NCBI's Blast server revealed that several of them had no matches in the depository, suggesting that they represented novel, previously uncharacterized mouse genes (data not shown). Twenty percent of the cDNA clones isolated in this study showed significant homology to known sequences at a probability greater than 10^{10} (Table 2). Five cDNAs showed a low level of homology ($<10^{10}$) to known sequences. Further examination of the 20 CRGs homologous to known sequences revealed that 14 of them showed homology with mouse sequences. Of the remaining, two of the CRGs showed homology with human, three with rat, and one with invertebrates (data not shown). The efficacy and

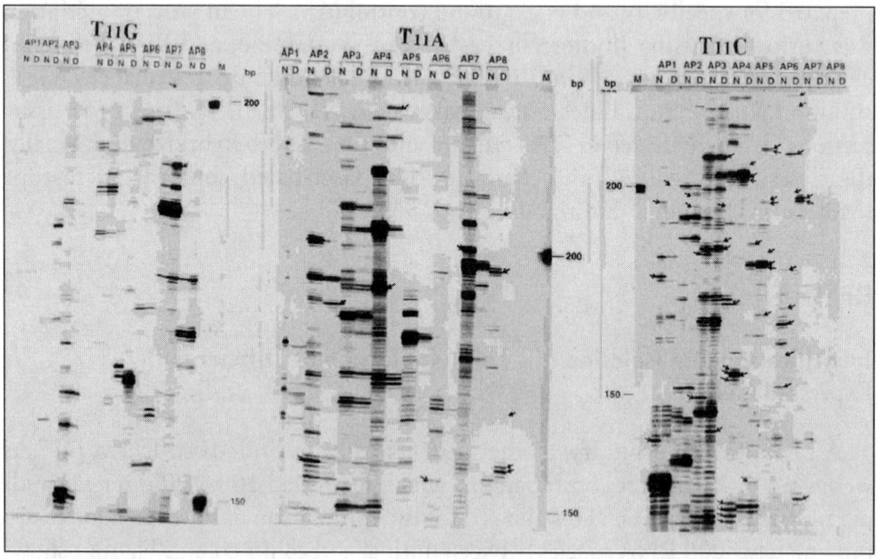

Figure 1. mRNA display analysis of normal and colitis-induced mouse colon RNAs. Three anchored primers and eight arbitrary primers were used in 24 primer combinations for both normal and diseased mouse colon RNAs using the GenHunter's kit. The samples were then analyzed on 6% sequencing gels and dried exposed with an X-ray film. The autoradiogram representing differentially expressed cDNAs is shown. N, normal; D, colitis.

Table 1. Gene Expression Profiles of CRGs

Clone	Sequence Homology	Level of Expression[a] Normal	Level of Expression[a] Colitis	GenBank Accession No.
A. CRGs whose level of expression is increased in colitis:				
CRG15	Mouse clone TA-7 mRNA	+	++	MMU29372
CRG72	Mouse (C57BL/10 X C3H) clone	+	++	MMU13370
CRG95	*Mouse Nicotonic ACh receptor*	+	++++	*MUSACHRRB*
CRG12	*Mouse mRNA for nucleosome ass.*	+	++	*MUSNAP1*
CRG14	*Mouse mitochondrial genes coding..*	+	+++	*MIMM01*
CRG101	*Mouse mitochondrial genes coding..*	+	++++	*MIMM01*
CRG102	*Mouse mitochondrial genes coding ..*	+	+++	*MIMM01*
CRG8	*Mouse IL-1 receptor antagonist mRNA*	+	++++	*MUSYL 1RA*
CRG76	Drosophila Cpo 61.1 gene	+	++++	DMCP0611G
CRG65	*Mouse band 3 anion exchange protein*	+	++	*MUSBAND31*
CRG6	*Mouse lysyl oxidase/ ras recision gene*	+	+++	*MUSRRG*
CRG9	*Mouse lymphocyte diff. antigen (Ly6)*	+	+++	*MUSLY62*
B. CRGs whose level of expression is decreased in colitis:				
CRG91	Rattus sp. cDNA(EST110386)	++++	++	H34995
CRG96	Human STS WI-9333	++	+	G07246
CRG98	Mouse cDNA 3′-end	++++	+	R75129
CRG18	Rattus sp. (EST108865)	+++	+	H33149
CRG2	Human cDNA clone 1..	++	+	R67751

[a]Level of expression was graded as: + = Expressed at low level; ++ = Expressed at moderate level; +++ = Expressed at higher level; ++++ = Expressed at highest level.

yield of the DD-PCR technique used in this study was compared with published results using either random EST (1) or differential cDNA library cloning methods (data not shown). The results suggest that DD-PCR and differential cDNA library cloning methods are superior to the random EST cloning method, especially in the identification of novel genes. Although both the DD-PCR and differential cDNA library cloning methods are equally productive, DD-PCR is a more robust and rapid technique.

Identification of CRGs

The cDNA sequence of CRG8 was highly homologous to the IL-1 receptor antagonist [(MUSYL1RA) (Tables 1 and 2) (53)]. It has been previously demonstrated that IL-1ra levels increased in rabbit immune complex-induced colitis (11). This supported the observation that CRG8 (IL-1ra) mRNA expression was highly increased in two experimental models of mouse colitis [(Figure 2.I, 1 and 2) (39)]. Three cDNA clones out of 102 (designated CRG14, 101, and 102) showed very strong homology with mouse mitochondrial genes and were significantly up-regulated in colitis (Figure 2.II, 1 and 2). These findings were further confirmed by in situ hybridization (Figure 2.I, 3 and 2.II, 3). Using the differential hybridization technique, Fukushima et al. (22,24) also identified five mitochondria-related genes that were highly expressed in IBD. In addition, for the first time, five additional cDNAs (CRG95, 12, 65, 6, and 9), which represent known genes with well-defined biological functions but no previous links to colitis, were identified in this study (described in the next section). Furthermore, eight additional cDNAs (CRG15, 72, 91, 96, 98, 76, 18, and 2) were cloned that had significant homology to previously cloned sequences with unknown biological functions. Two genes (CRG43 and 44), which were cloned as internal controls, showed no expression changes between normal and colitis colon and were highly homologous to repetitive sequences. Differential expression of several of these genes was independently confirmed by Northern blot analyses (data not shown).

Sequence Homology of cDNAs with Unknown Relationships to Colitis

Five genes, originally selected for cloning based on their increased expression in colitis by DD-PCR analysis [CRG95, 12, 65, 6, and 9 (Table 1)], were found to be highly homologous to previously cloned genes. These genes had not been previously linked to colitis. CRG95 is a 252-bp fragment highly expressed in colitis. It is 87% (129/147) homologous to mouse nicotinic acetylcholine receptor β subunit gene (MUSACHRRB) from nucleotides 7627 to 7773 at a probability of 10^{41}. This particular fragment showed 70%–90% homology with 40 other mouse genes. No homology was observed to species other than mouse, indicating that CRG95 was a highly species-specific cDNA. This cDNA was also 85% (122/143) homologous to mouse lymphotoxin gene [MUSLTA (25)] and mouse

Table 2. Sequence Homology of CRGs with Previously Identified Genes

Clone No.	Size	Primers	Sequence Homology	Score	GenBank Accession No.
CRG14	*265*	*T11G +AP8*	*Mouse mitochondrial genes coding for Phe, Val, Leu tRNAs, 12S and 16S r RNAs*	10^{85}	*MIMM01*
CRG15	240	T11G + AP1	Mouse clone TA-7 mRNA	10^{71}	MMU29372
CRG72	216	T11C +AP5	Mouse (C57BL/10 X C3H) clone	10^{43}	MMU13370
CRG95	*252*	*T11C +AP4*	*Mouse Nicotonic ACh receptor*	10^{41}	*MUSACHRRB*
CRG91	210	T11C +AP2	Rattus sp. cDNA(EST110386)	10^{34}	H34995
CRG12	*123*	*T11G +AP7*	*Mouse mRNA for nucleosome ass.*	10^{32}	*MUSNAP1*
CRG96	155	T11C +AP6	Human sts WI-9333	10^{31}	G07246
CRG101	*150*	*T11C +AP8*	*Mouse mitochondrial genes coding..*	10^{31}	*MIMM01*
CRG102	*155*	*T11C +AP8*	*Mouse mitochondrial genes coding ..*	10^{31}	*MIMM01*
CRG8	*167*	*T11G +AP5*	*Mouse IL-1 receptor antagonist mRNA*	10^{28}	*MUSYL 1RA*
CRG98	290	T11C +AP7	Mouse cDNA 3′-end	10^{26}	R75129
CRG76	158	T11C +AP6	Drosophila Cpo 61.1 gene	10^{20}	DMCP0611G
CRG65	*124*	*T11C +AP5*	*Mouse band 3 anion exchange protein*	10^{18}	*MUSBAND31*
CRG6	*169*	*T11G +AP4*	*Mouse lysyl oxidase/ras recision gene*	10^{17}	*MUSRRG*
CRG18	210	T11G +AP4	Rattus sp. (EST108865)	10^{16}	H33149
CRG9	*120*	*T11G +AP6*	*Mouse lymphocyte diff. antigen (Ly6)*	10^{15}	*MUSLY62*
CRG2	113	T11G +AP3	Human cDNA clone 1..	10^{10}	R67751

Genes with defined biological function are printed in *italics*. Genes with sequence homology significant at greater than 10^{10} are listed here.

Anchor primers: T11C, 5′-AAGCTTTTTTTTTTTC-3′; T11G, 5′-AAGCTTTTTTTTTTTG-3′

Arbitary primers: AP1, 5′-AAGCTTGATTGCC-3′; AP2, 5′-AAGCTTCGACTGT-3′; AP3, 5′-AAGCTTTGGTCAG-3′;
AP4, 5'-AAGCTTCTCAACG-3'; AP5, 5'-AAGCTTAGTAGGC-3'; AP6, 5'-AAGCTTGCACCAT-3'
AP7, 5′-AAGCTTAACGAGG-3′; AP8, 5′-AAGCTTTTACCGC-3′

tumor necrosis factor locus (MMTNFAB) at a probability of 10^{37}. CRG65 was identified as a 124-bp cDNA fragment up-regulated in colitis (Table 2), and its cDNA sequence was exclusively homologous to 20 mouse genes. Its highest homology was to mouse band 3 anion exchange protein [MUSBAND31 (36)] at probability of 10^{18}. CRG65 cDNA sequence was also 71% homologous (76/107) to mouse T-cell receptor alpha/delta chain locus [MUSTCRA (64)] between nucleotides 8906 to 9012 at a probability of 10^{15}. CRG6 was identified as a 169-bp fragment up-regulated in diseased tissues but also expressed to a significant level in normal colon (Table 1). The cDNA sequence was 75% homologous to the mouse lysyl oxidase [MUSLYOX (12,13,33)] gene and *ras* recision [MUSRRG (31)] genes at probability of 10^{17} (Table 2). CRG9 sequence was 95% homologous to mouse lymphocyte differentiation antigen [MUSLY62 (58)], mouse T-cell activating protein gene [MUSTAPLY (61,62)], and mouse Ly-6A/E antigens [MUSLY6EA (14)]. Characterization of CRG9 and its implication in colitis disease are described below.

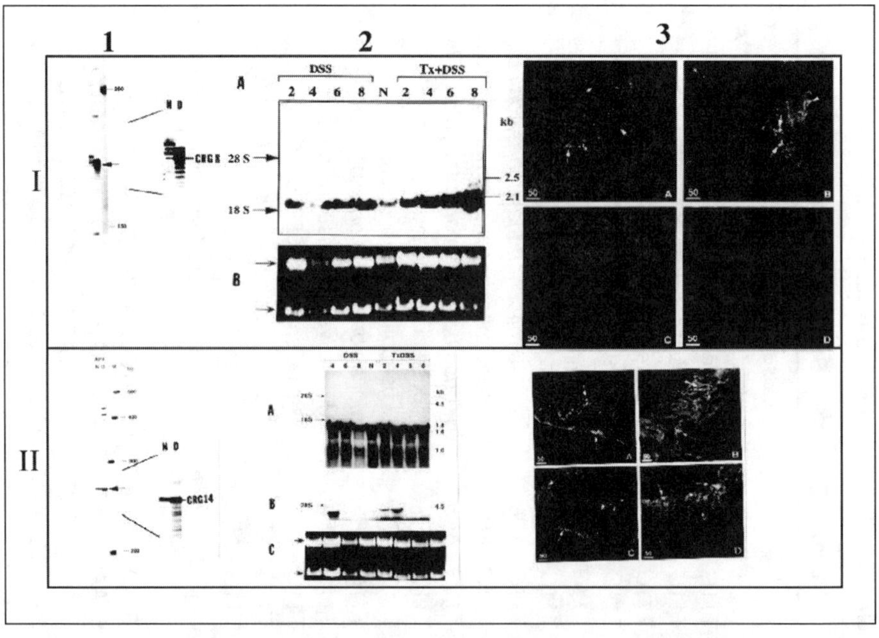

Figure 2. Characterization of two known genes in colitis. (I) IL-1ra. IL1ra was isolated as an up-regulated cDNA (CRG8) in colitis using AP5 and T11A primers. 1. Differential display pattern of CRG8: N, normal; D, colitis. 2. Northern blot analysis of experimental colitis tissue RNAs. 3. In-situ hybridization of IL-1ra message. (A) Normal tissue with antisense probe; (B) colitis tissue with antisense probe; (C) normal tissue with sense probe; and (D) colitis tissue with sense probe. (II) Mitochondrial genes. CRG14 was isolated as an up-regulated cDNA in colitis using AP6 and T11A primers. 1. Differential display pattern of CRG14: N, normal; D, colitis. 2. (A) Northern blot analysis of experimental colitis tissue RNAs; (B) enlarged portion of part of panel A; (C) ethidium bromide gel analysis for checking the RNA samples. 3. In-situ hybridization of mitochondrial nuclear gene. (A) Normal tissue with antisense probe; (B) colitis tissue with antisense probe; (C) normal tissue with sense probe; and (D) colitis tissue with sense probe.

Sequence Homology of Differentially Expressed cDNAs with Unknown Biological Functions

Eight cDNAs cloned in this study (CRG15, 72, 91, 96, 98, 76, 18, and 2) were found to have significant homology to existing sequences in the GenBank database (Table 2). Although the respective full-length cDNA clones have been previously isolated, their biological function is still unknown. CRG2, 18, 91, 96, and 98 showed a decreased expression in colitis (i.e., these were down-regulated in colitis and up-regulated in normal colon tissues). CRG2 and CRG96 showed 84% homology with a variety of human partial cDNA clones isolated by the EST method by the IMAGE consortium of the Washington University-Merck Group. CRG18 is 82% homologous to a differentially expressed EST (H33149) cloned from PC-12 cells (48). CRG91 was 88% homologous, with a probability of 10^{34}, to rat cDNA (H34995), which was originally isolated from adrenal chromaffin PC-12 cells (48). CRG98 was 86% homologous to an unpublished mouse brain cDNA (R75129) at the 3′-end with a probability of 10^{26}. It is noteworthy that all CRGs identified as genes whose expression is decreased in colitis belong to the group of genes with yet unidentified functions, perhaps reflecting the fact that the breadth of cloning efforts are focused on identifying inducible genes. A smaller group of three CRGs with yet undefined biological functions showed elevated expression in colitis. Of these genes, CRG15 is a 231-bp cDNA with increased expression in colitis but with a very low level in normal tissues. Its sequence was 88% (198/224) homologous to mouse clone TA-7 mRNA (MMU29372) between nucleotides 1–224 with a probability of 10^{71}. In addition, clone CRG15 was also homologous to various human cDNA clones, especially to human fibronectin cDNA (T17426). CRG72 was a 216-bp cDNA with 94% homology to mouse kidney cell line protein [MMU13370 (69)]. CRG76 was a 158-bp up-regulated gene in colitis (Table 2), and its cDNA sequence was highly homologous to a variety of *Drosophila* genes, including the couch potato protein [DMCPO611G (6)] and hunchback gene for zinc-finger protein transcription factor (DVHB).

Tissue Localization of mRNA Transcripts

To localize the expression of the novel CRGs within the colon, both normal and diseased colon tissue sections of mice were used for in situ hybridization. As an example, the results for four CRG mRNA transcripts are shown in Figure 3. In most of the cases, an increased expression was detected in diseased colon over normal. It is worth noting that the tissue localization for some genes was different from others in colitis. This means that signals for some genes that were localized into the superficial (epithelial) layer of mucosa in cells seem to be intraepithelial leukocytes, whereas for some genes, transcript signals were distributed throughout the entire diseased mucosa.

ELUCIDATION OF LY-6A/E ANTIGEN FUNCTION IN IBD

Identification and Characterization of Ly-6A/E-Like Gene in Colitis

During this comprehensive genetic analysis, with the help of an anchor primer ($T_{11}G$: 5′-AAGCTTTTTTTTTTTG-3′) and an arbitrary primer (AP6: 5′-AAGCTTGCACCAT-3′), a 120-bp cDNA fragment was identified with an increased expression in colitis (Tables 1 and 2). Figure 4.1 shows the autoradiogram of the DD-PCR analysis. This partial cDNA was designated as CRG9. Northern blot analysis with total RNA collected from the colon of diseased mice on days 2, 4, 6, and 8 from two different models of colitis confirmed the disease-related expression of CRG9 (Figure 4.2). When the expression of CRG9 was cor-

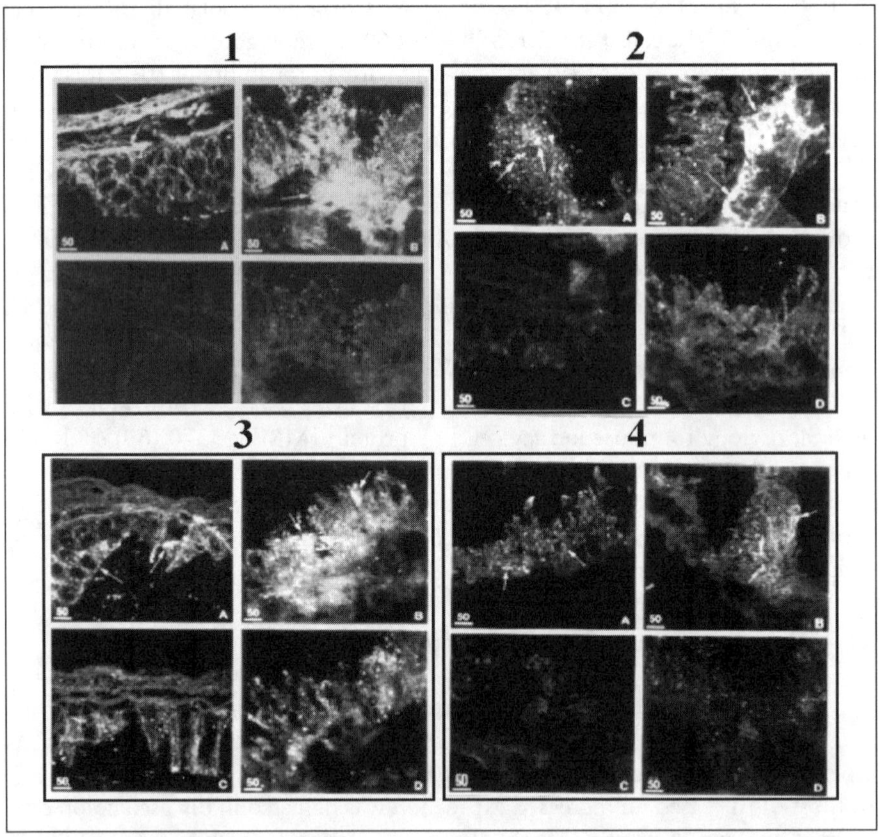

Figure 3. Tissue localization of CRGs by in-situ hybridization. Six micron-thin cryosections were prepared from both normal and diseased colon. Sense and antisense RNA probes were synthesized for four CRGs by incorporating fluorescein-6-UTP into the RNA transcripts. Tissue sections were hybridized with fluorescence-labeled probes as described in the Methodology section. (A) Normal tissue with antisense probe; (B) colitis tissue with antisense probe; (C) normal tissue with sense probe; and (D) colitis tissue with sense probe. The localized signals are shown with arrows.

related with the kinetics of disease development, enhanced messages were detected. The expression levels started increasing from day 2 and reached maximum by day 6 in both colitis models (Figure 4.2B). However, the expression levels were low in normal colon. Densitometry analysis of the Northern blots indicated that CRG9 message levels were increased from 5- to 6-fold over baseline (Figure 4.2A). Total RNA isolated from various organs of normal mouse was used to determine the multiple tissue distribution of CRG9. Although CRG9 mRNA expression was approximately equal in heart, spleen, cecum, and large intestine, it decreased in other organs in the following order: lung > kidney > heart > spleen > muscle > cecum > large intestine > brain > testis. It is noteworthy that the mRNA levels were approximately 9-fold higher in lung and kidney, as shown by densitometry (Figure 4.3).

Ly-6A/E Protein Localization

Protein lysates were isolated from both experimental colitis models during different stages of disease progression; Western blots were prepared and probed with commercially available antibody. The results shown in Figure 5.1 indicated that a

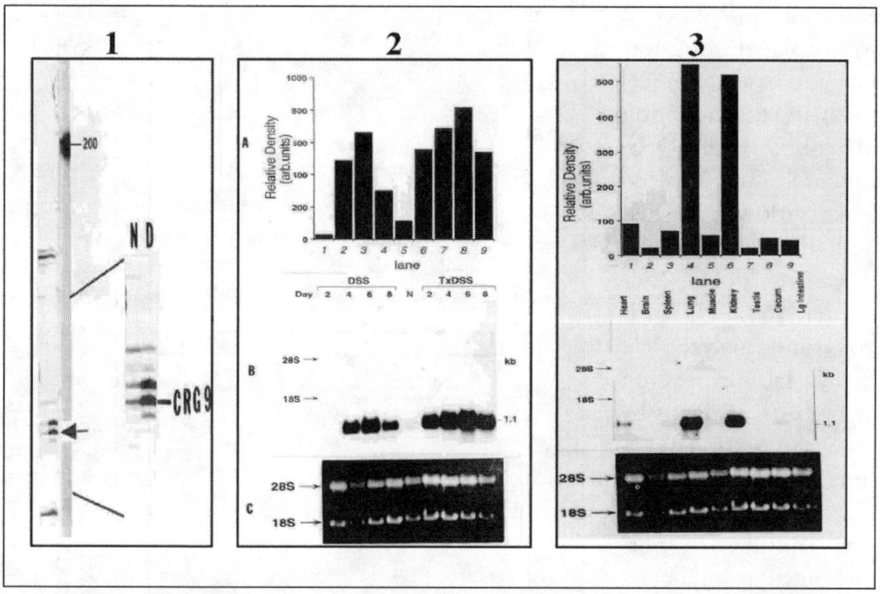

Figure 4. Up-regulation of the CRG9/Ly-6A/E antigen. 1. Differential display pattern. CRG9 was isolated as an up-regulated cDNA in colitis. N, normal; D, colitis. 2. Northern blot analysis of mRNA transcripts in experimental colitis models. Total RNA was obtained from normal (N) mice and both DSS-induced (labeled as DSS) and Tx+DSS (labeled as TxDSS) models of colitis on days 2, 4, 6, and 8 after the development of colitis. RNA blots were prepared, and the blots were hybridized with a ^{32}P-labeled CRG9 probe. 3. Multiple tissue the Northern blot. Total RNA was obtained from various organs of normal mice, and the Northern blot was prepared and hybridized with a ^{32}P-labeled CRG9 probe. (A) Densitometry analysis. (C) Equal amount of RNA was loaded as judged from the intensities of 18S and 28S rRNAs on ethidium bromide-stained gels.

9-kDa Ly-6A/E protein band was detected. CRG9 (Ly-6A/E) mRNA levels were significantly high in the TxDSS-induced model (Figure 4.2, panels A and B). In contrast, CRG9 protein expression levels were lower than the DSS-induced model. This suggests that Ly-6A/E is transcriptionally regulated in the TxDSS model and translationally controlled in the DSS-induced model (40). In addition, both mucosal and spleen leukocytes were isolated from normal and diseased mice, following the induction of colitis with DSS. Mucosal and spleen leukocytes were treated with fluorescein isothiocyanate (FITC)-conjugated Ly-6A/E antibody and analyzed for the surface expression of Ly-6A/E antigens by flow cytometry. The results shown in Figure 5.2 demonstrate that a small number of normal mucosal leukocytes express a detectable level of Ly-6A/E even in normal colon. However, Ly-6A/E levels were significantly increased in the DSS-in-

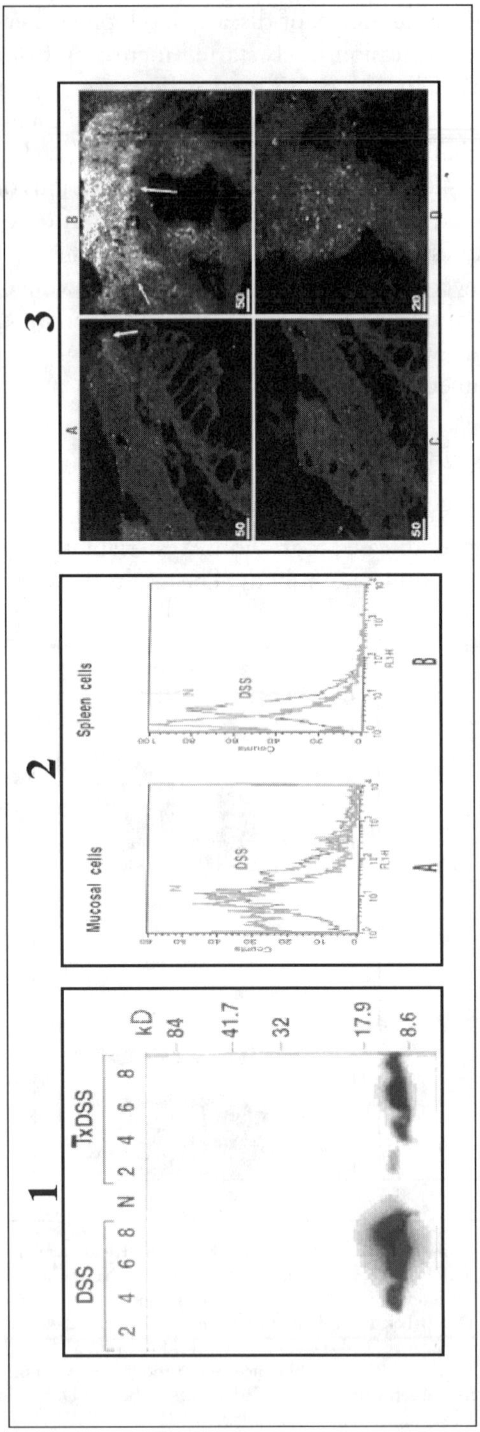

Figure 5. Protein and mRNA localization of the CRG9/Ly-6A/E gene. 1. Western blot analysis. Protein extracts from mice experimental colitis tissue samples were separated on 12% PAGE, blotted to a PVDF membrane, probed with Ly-6A/E antibody, and the blot was developed by chemiluminescent method. 2. FACS® analysis. DSS-induced colitis samples: N, normal; A, mucosal cells; B, spleen cells. 3. Fluorescence in situ hybridization (FISH) analysis. (A) Normal tissue with antisense probe; (B) colitis tissue with antisense probe; (C) normal tissue with sense probe; and (D) colitis tissue with sense probe. The localized signals are shown with arrows. **(See color plate A1.)**

duced model samples. Contrary to mucosal cells, normal spleen cells are negative for Ly-6A/E and seem to develop a Ly-6A/E antigen-positive population as colitis develops in DSS-induced models. The Western blot and flow cytometry results suggest that the partial cDNA clone identified in this study as CRG9 is indeed Ly-6A/E.

Tissue Localization of CRG9 (Ly-6A/E) mRNA in Inflamed Colon Tissues

To localize the expression of CRG9 (Ly6A/E) within the colon, both normal and diseased colon tissue sections of mice were subjected to in situ hybridization. The antisense CRG9 (Ly-6A/E) probe showed a very significant increase of mRNA expression in colitis, with minimal signal in normal colon (Figure 5.3). CRG9 (Ly-6A/E) mRNA signals were distributed throughout the entire diseased mucosa, but its expression was highest in the most superficial layers, where most of the inflammatory cells are crowded. The sensitivity of this assay did not discern whether leukocytes or their neighboring cells expressed the message. These in situ hybridization studies raised the possibility that cells other than leukocytes may also express CRG9 (Ly-6A/E) mRNA in colitis.

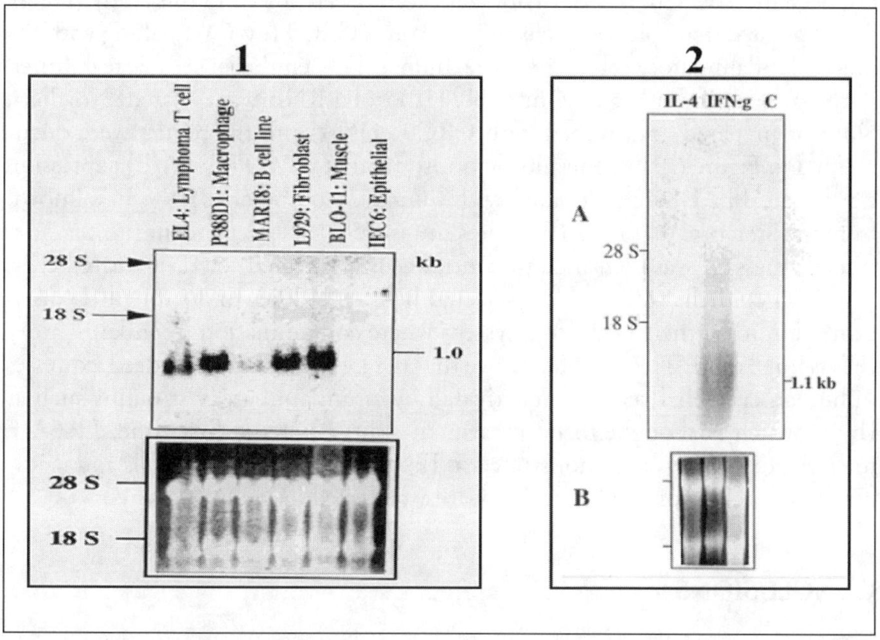

Figure 6. Identification of the CRG9/Ly-6A/E transcript. 1. Multiple cell line Northern. Murine multiple cell lines were grown, total RNA was isolated, and Northern blot was prepared and hybridized with a ^{32}P-labeled CRG9 probe. 2. Cytokine-treated T-84 cell-line Northern. (A) Human epithelial cell line T84 was grown in presence of IL-4 and IFNγ. Total RNA was then isolated, and Northern blot was prepared and hybridized with ^{32}P-labeled CRG9 probe. N, untreated cells. (B) Ethidium bromide-stained gel.

Expression of CRG9 (Ly-6A/E) mRNA in Mouse Multiple Cell Lines

To investigate the presence of the Ly-6A/E message in epithelial cells, initially, various mice cell lines were cultured, RNAs were isolated, and a Northern blot was prepared. Probing the blot with CRG9 detected a 1-kb transcript in lymphoma T cell, macrophage, fibroblast, and muscle (Figure 6.1). The messages were faint in the MAR18-B cell line and undetectable in the intraepithelial cell line (IEC-6). The results suggest that the CRG9 (Ly-6A/E) message is highly present in macrophages, possibly due to inflammation in colitis. Although a 1-kb message was undetected in IEC-6, induction with cytokines such as interferon γ (IFNγ) might help to enhance the mRNA expression. Therefore, this was further investigated in a pure in vitro system, using cultured human colonic epithelial cells as described below.

CRG9 (Ly-6A/E) mRNA Expression in IFNγ-Induced T-84 Human Epithelial Cells

The expression of Ly-6A/E, a murine leukocyte differentiation and activation antigen, has been previously documented in both murine non-leukocyte cells (47) and in human brain cells (14); however, its expression in epithelial cells has not been investigated. In situ hybridization studies reported in this study (Figure 5.3) suggested that epithelial cells might express CRG9 (Ly-6A/E) along with the mucosal inflammatory cells. Therefore, human T-84 epithelial cells were cultured in the presence of either IL-4 or IFNγ (10), total RNA was isolated, Northern blot was prepared and probed with CRG9 cDNA, and the results were documented in Figure 6.2. The results demonstrate that CRG9 (Ly-6A/E) expression was induced in T-84 cells treated with human recombinant IFNγ. It is known from the literature that Ly-6A/E expression is truly IFNγ dependent (46). No detectable message was identified in control cells (untreated) or IL-4-treated cells. The T-84 epithelial cell line was selected here for studies to document Ly6A/E expression in epithelial cells in a system where contamination by inflammatory cells is impossible. These studies prove that the Ly-6A/E antigen indeed expresses in human epithelial cells when activated by an inflammatory stimulus such as IFNγ, which is thought to be present in colitis. For the first time, Ly-6A/E mRNA expression was demonstrated in IFNγ-induced epithelial cells and its expression, localization, and function were established in IBD.

CONCLUSIONS

Utilizing DD-PCR, 5000 mRNA species were visualized, cDNA expression profiles were compared between normal and colitis mice, and 102 fragments were recovered. The reproducibility of the DD-PCR technique was demonstrated by the ability to successfully re-amplify, clone, and sequence over 90% of the cDNAs originally selected for analysis. This is the first report describing the successful utilization of the DD-PCR approach in colitis to identify novel genes from two experimental models. Three partial genes (CRG14, CRG101, and CRG102) that were homologous to mitochondrial genes were identified in the course of the investigation (Table 2). The functional importance of mitochondrial genes in IBD is not well established, but a growing body of evidence suggests that a variety of chronic diseases, including cancer and diabetes, are associated with damage to mtDNA (74). Because mitochondria are constantly exposed to high levels of reactive oxygen species, it is likely that oxidative damage to mtDNA may be responsible for some of these maladies in the tissue injury (colitis) models. It is quite possible that overexpression of mitochondrial messages present in diseased models may protect from epithelial tissue damage and cell injury, which are hallmarks of colitis. Sequence homology of CRG15 with fibronectin (Table 2) raises the possibility that CRG15 may be the gene for fibronectin present in the mucosal epithelial cells that can modulate cell-substrate and cell-cell interactions that occur during the epithelial tissue injury and wound-healing repair mechanism. This study demonstrated that the DD-PCR technique could be efficiently used to isolate novel and unknown genes involved in colitis.

Ly-6A/E is a member of a family of proteins that share either protein or DNA sequence homology but perform various functions in different species, especially in different cell types. Ly-6A/E has multiple functions in adhesion and signaling T-cell activation. Ly-6A/E has been linked to the regulation of T-cell activation because anti-Ly-6A/E monoclonal antibody can either induce or inhibit IL-2 production. This suggests that Ly-6A/E may play an important role in regulating the early phase of the immune response. The mechanism by which Ly-6A might regulate the level of TCR? chain expression is also unclear. Although the role of Ly-6C was demonstrated in spontaneous NOD mouse models, Ly-6A/E has never been implicated in any type of disease. This is the first time Ly-6A/E has been implicated in colitis.

Mouse multiple cell line Northern blot results for CRG9/Ly-6A/E indicated that faint (or no) message was detected in the IEC-6. This raises two possibilities: (*i*) the CRG9/Ly-6A/E message may be present at very low levels (if this is true, then it has to be induced by cytokines) or (*ii*) CRG9/Ly-6A/E may be present in macrophages rather than in epithelial cells. It has been previously shown using various cell lines that the Ly-6 message was enhanced in the presence of IFNγ and IL-1 (46). The expression of the Ly-6A/E can be up-regulated by IFNγ on thymic and bone marrow stromal cells, astroglial cells, and mature T cells. Therefore, to prove the presence of Ly-6A/E in epithelial cells, human colonic epithelial cells T-84 were treated with IFNγ, and the presence of the Ly-6A/E message was indeed demonstrated. It is still unclear how the IFNγ induces the Ly-6A/E transcription and mediates a kind of molecular pathway or whether increased Ly-6A/E expression precedes or follows the production of IFNγ and IL-4 by T cells. It is also unclear whether the conversion of Ly-6A/E low cells to Ly-

6A/E high cells is a prerequisite for, or a consequence of, the Ly6A/E-mediated activation of T cells. These data suggest that Ly-6A/E is likely to play a key role in regulating the function of CD4+ T cells and may be one key element in the development of tissue inflammation such as IBD.

In summary, the differential detection of changes at the mRNA level by differential display not only allowed identification of novel genes in colitis but also identified known gene(s) that play key role(s) in colitis. Therefore, I conclude that this is an innovative gene discovery and hunting technique for the gene analysis of multiple cell types present in the gut. This gene discovery approach applied to IBD-related studies has opened a major avenue to a wealth of questions posed by molecular medicine and basic biology in the field of gastroenterology.

ACKNOWLEDGMENTS

This work was carried out in the Anesthesia Research Laboratories of Brigham and Women's Hospital at Harvard Medical School, Boston, and supported by grants from the Crohn's and Colitis Foundation of America and Brigham and Women's Hospital's Anesthesia Foundation to Dr. G. Frendl. I am thankful to Dr. Simon Gelman of Brigham and Women's Hospital for his continuous support and laboratory facilities during the course of this investigation. I also thank Dr. Arthur B. Pardee of the Dana Farber Cancer Institute for his meticulous advice. Help of Dr. Rick A. Rogers (Harvard School of Public Health) in the confocal image analysis is highly acknowledged. I also thank Dr. S. Colgan of Brigham and Women's Hospital for providing the human epithelial T-84 cell line. Results of this study were presented on the occasion of outstanding investigator award (to K.A.) at the annual meeting of the Eastern section of American Federation of Medical Research (1996), Boston, MA. In addition, part of this work was recognized by the Ranbaxy Science Foundation, New Delhi, India, in the award of Ranbaxy Research Medal (to K.A.) in the field of Applied Medical Sciences.

REFERENCES

1. **Adams., M.D., J.M. Kelley, J.D. Gocayne, M. Dubnick, M.H. Polymeropoulos, H. Xiao, C.R. Merril, et al.** 1991. Science *252*:1651-1656.
2. **Altschul, S.F., W. Gish, W. Miller, E.W. Myers, and D.J. Lipman.** 1990. Basic local alignment search tool. J. Mol. Biol. *215*:403-410.
3. **Arai, F., T. Takahashi, K. Furakawa, K. Matsushima, and H. Asakura.** 1998. Mucosal expression of interleukin-6 and interleukin-8 messenger RNA in ulcerative colitis and Crohn's diseases. Dig. Dis. Sci. *43*:2071-2079.
4. **Ausubel, F.M. et al.** 1994. Current Protocols in Molecular Biology. John Wiley & Sons, New York.
5. **Autieri, M.V., G.Z. Feuerstein, T.L. Yue, E.H. Ohlstein, and S.A. Douglas.** 1995. Use of differential display to identify differentially expressed mRNAs induced by rat carotid artery balloon angioplasty. Lab. Invest. *72*:656-661.
6. **Bellen, H.J., S. Kooyer, D. D'Evelyn, and J. Pearlman.** 1992. The *Drosophila* couch potato protein is expressed in nuclei of peripheral neuronal precursors and shows homology to RNA-binding proteins.

Genes Dev. *6*:2125-2136.

7.**Brandtzaeg, P., T. Halstensen, K. Kett, P. Krajci, D. Kvale, T. Rognum, H. Scott, et al.** 1989. Immunobiolgy and immunopathology of the human gut mucosa: humoral immunity and intraepithelial lymphocytes. Gastroenterology *97*:1562-1584.

8.**Bueno, R., K. Appasani, H. Mercer, S. Lester, and D.J. Sugarbaker.** 2001. The α folate receptor is highly activated in malignant pleural mesothelioma. J. Thorac. Cardiovasc. Surg. *121*:225-233.

9.**Cho, J., D. Nicolae, R. Ramos, C. Fields, K. Rabenau, S. Coradino, S. Brant, et al.** 2000. Linkage and linkage disequilibrium in chromosome band 1p36 in American chaldeans with inflammatory bowel disease. Hum. Mol. Genet. *9*:1425-1432.

10.**Colgan, S.P., M.B. Resnick, C.A. Parkos, C. Delp, A.E. Bacarra, D. McGuirk, P.W. Weller, and J.L. Madara.** 1994. Interleukin-4 directly modulates function of a model intestinal epithelia. J. Immunol. *153*:2122-2129.

11.**Cominelli, F., K. Bortolami, T.T. Pizarro, L. Monasacchi, M. Ferretti, M.T. Brewer, S.P. Eisenberg, and R.K. Ng.** 1994. Rabbit interleukin-1 receptor antagonist. Cloning, expression, functional characterization, and regulation during intestinal inflammation. J. Biol. Chem. *269*:6962-6971.

12.**Contente, S., K. Csiszar, K. Kenyon, and R.M. Friedman.** 1993. Structure of mouse lysyl oxidase gene. Genomics *16*:395-400.

13.**Contente, S., K. Kenyon, D. Rimoldi, and R.I. Friedman.** 1990. Expression of gene rrg is associated with reversion of NIH 3T3 transformed by LTR-c-H-ras. Science *249*:796-798.

14.**Cray, C., R.W. Keane, T.R. Malek, and R.B. Levy.** 1990. Regulation and selective expression of Ly-6A/E, a lymphocyte activation molecule, in the central nervous system. Mol. Brain Res. *8*:9-15.

15.**DeRisi, J., L. Penland, P.O. Brown, M.L. Bittner, P.S. Meltzner, M. Ray, Y. Chen, Y.A. Su, et al.** 1996. Use of a cDNA microarray to analyze gene expression patterns in human cancer. Nat. Genet. *14*:457-460.

16.**Dieckgraefe, B.K., W.F. Stetson, J.R. Korzenik, P.E. Swanson, and C.A. Harrington.** 2000. Analysis of mucosal gene expression in inflammatory bowel disease by parallel oligonucleotide arrays. Physiol. Genomics. *4*:1-11.

17.**Dinarello, C.A.** 1991. Interleukin-1 and interleukin-1 antagonism. Blood *77*:1627-1652.

18.**Elson, C.O., R.B. Sartor, G.S. Tennyson, and R.H. Riddell.** 1995. Experimental models of inflammatory bowel disease. Gastroenterology *109*:1344-1367.

19.**Ferretti, M., V. Casini-Raggi, T.T. Pizarro, S.P. Eisenberg, C.C. Nast, and F. Cominelli.** 1994. Neutralization of endogenous IL-1 receptor antagonist exacerbates and prolongs inflammation in rabbit immune colitis. J. Clin. Invest. *94*:449-453.

20.**Fiocchi, C.** 2002. Inflammatory bowel disease: dogmas and heresies. Dig. Liver Dis. *34*:306-311.

21.**Frendl, G., S. Nicholasen, J.R. Murphy, and J.T. LaMont.** 1994. Gasroenterology *106*:A682.

22.**Fukushima, K., D. Walker, and C. Fiocchi.** 1994. Detection of altered genes in epithelial cells from inflammatory bowel disease by differential screening of cDNA libraries. Gastroenterology *106*:A683.

23.**Fukushima, K., G. West, and C. Fiocchi.** 1995. Adequacy of mucosal biopsies for evaluation of intestinal cytokine-specific mRNA. Comparative study of RT-PCR in biopsies and isolated cells from normal and inflamed intestine. Dig. Dis. Sci. *40*:1498-1505.

24.**Fukushima, K., S. Yuan, and C. Fiocchi.** 1994. Differential display of epithelial cell mRNA by PCR: a novel method to detect inflammatory bowel disease-associated genes. Gastroenterology *106*:A683.

25.**Gardner, S.M., B.A. Mock, J. Hiljers, K.E. Huppi, and W.D. Roeder.** 1987. Mouse lymphotoxin and tumor necrosis factor: structural analysis of the cloned genes, physical linkage, and chromosomal position. J. Immunol. *139*:476-483.

26.**Gordon, G.J., K. Appasani, J.P. Parcells, N.K. Mukhopadhyay, M.T. Jacklitch, W.G. Richards, D.J. Sugarbaker, and R. Bueno.** 2002. Inhibitor of apoptosis protein-1 promotes tumor cell survival in mesothelioma. Carcinogenesis *23*:1017-1024.

27.**Hamdan, A.D., L.P. Aiello, W.C. Quist, C.K. Ozaki, M.A. Contreras, M.D. Phaneuf, C. Ruiz, et al.** 1995. Isolation of genes differentially expressed at the downstream anastomosis of prosthetic arterial grafts with use of mRNA differential display. J. Vasc. Surg. *21*:228-234.

28.**Hammer, R.E., S.D. Maika, J.A. Richardson, J.P. Tang, and J.D. Taurog.** 1990. Spontaneous inflammatory disease in transgenic rats expressing HLA-B27 and human β-2-microglobulin. HLA-B27-associated human disorders. Cell *63*:1099-1112.

29.**Hampe, J., S. Shaw, R. Saiz, N. Leysens, A. Lantermann, S. Mascheretti, N. Lynch, et al.** 1999. Linkage of inflammatory bowel disease to human chromosome 6p. Am. J. Hum. Genet. *65*:1647-1655.

30. **Hugot, J., P. Laurent-Puig, C. Gower-Rousseau, J. Olson, J. Lee, L. Beaugerie, I. Naom, J. Dupas, et al.** 1996. Mapping of susceptibility locus for Crohn's disease on chromosome 16. Nature *379*:821-823.
31. **Kagan, H.M. and P.C. Trackman.** 1991. Properties and function of lysyl oxidase. Am. J. Respir. Cell Mol. Biol. *5*:206-210.
32. **Kam, L., T.T. Pizarro, and F. Cominelli.** 1995. Cytokines and chemokines in inflammatory bowel disease. Curr. Opin. Gastroenterol. *11*:305-309.
33. **Kenyon, K., S. Contente, P.C. Trackman, J. Tang, H.M. Kagan, and R.M. Friedman.** 1991. Lysyl oxidase and rrg messenger RNA. Science *253*:802-806.
34. **Kirsner, J. and R. Shorter.** 1999. Inflammatory Bowel Disease, 5th ed. Lea & Febiger, Philadelphia, PA.
35. **Kohno, T., R. Moriuchi, S. Katamine, Y. Yamada, M. Tomonaga, and T. Matsuyama.** 2000. Identification of genes associated with the progression of adult T cell leukemia. Jpn. J. Cancer. Res. *91*:1103-1110.
36. **Kopito, R.R., M.A. Andersson, and H.F. Lodish.** 1987. Multiple tissue-specific sites of transcriptional initiation of the mouse anion antiport gene in erythroid and renal cells. Proc. Natl. Acad. Sci. USA *84*:7149-7153.
37. **Krishnarao, A.S.M.** 1997. Novel gene discovery in experimental colitis: mRNA display approach. Gastroenterology *112*:1019.
38. **Krishnarao, A.S.M. and G. Frendl.** 1996. Differential display polymerase chain reaction based isolation of colitis related genes from mice with experimental colitis. Gut *38*:A636.
39. **Krishnarao, A.S.M. and G. Frendl.** 1996. Elevated levels of IL-1ra mRNA in DSS and DAB389Il-2 fusion toxin induced colitis mice models by DD-PCR. J. Investig. Med. *44*:A410.
40. **Krishnarao, A.S.M. and G. Frendl.** 1997. Transcriptional regulation of T-cell activating protein Ly-6A/E antigen in colitis. Gastroenterology *112*:A1019.
41. **Krishnarao, A.S.M. and R.E. Hausman.** 1993. cDNA for R-cognin: homology with a multifunctional protein. Proc. Natl. Acad. Sci. USA *90*:2950-2954.
42. **Kuhn, R., J. Lohler, D. Rennick, K. Rajewsky, and W. Muller.** 1993. Interleukin-10-deficient mice develop chronic entercolitis. Cell *75*:263-274.
43. **Lauritsen, K., L. Laursen, K. Bukhave, and J. Rask-Madsen.** 1988. In vivo profiles of eicosanoids in ulcerative colitis, Crohn's colitis and clostridium difficile colitis. Gastroenterology *95*:11-17.
44. **Lawrence, I.C., C. Fiocchi, and S. Chakravarti.** 2001. Ulcerative colitis and Crohn's disease: distinctive gene expression profiles and novel susceptibility candidate genes. Hum. Mol. Genet. *10*:445-456.
45. **Lawrence, J.B., R.H. Singer, C.A. Villnave, J.L. Stein, and G.S. Stein.** 1988. Intracellular distribution of histone mRNAs in human fibroblasts studied by in situ hybridization. Proc. Natl. Acad. Sci. USA *85*:463-467.
46. **LeClair, K.P., M.M. Bridgett, F.J. Dumont, R.G.E. Palfree, U. Hammerling, and A.L.M. Bothwell.** 1989. Kinetic analysis of Ly-6 gene induction in a T lymphoma by IFNs and IL1, and demonstration in diverse cell types. Eur. J. Immunol. *19*:1233-1239.
47. **LeClair, K.P., R.G.E. Palfree, P.M. Flood, U. Hammerling, and A. Bothwell.** 1986. Isolation of a murine Ly-6 cDNA reveals a new multigene family. EMBO J. *5*:3227-3234.
48. **Lee, N.H., K.G. Weinstock, E.F. Kirkness, J.A. Earle-Hughes, R.M. Fuldner, S. Marmaras, A. Glodek, et al.** 1995. Comparative expressed-sequence-tag analysis of differential gene expression profiles in PC-12 cells before and after nerve growth factor treatment. Proc. Natl. Acad. Sci. USA *92*:8303-8307.
49. **Liang, P.** 2002. A decade of differential display. BioTechniques *33*:338-346.
50. **Liang, P. and A.B. Pardee.** 1992. Differential display of eukaryotic mRNA by means of the PCR. Science *257*:967-970.
51. **Liang, P., W. Zhu, X. Zhang, Z. Guo, R.P. O'Connell, L. Averboukh, F. Wang, and A.B. Pardee.** 1995. Differential display using one-base anchored oligo-dT primers. Nucleic Acids Res. *22*:5763-5764.
52. **Madara, J.L., D.K. Podolsky, N.W. King, P.K. Sehgal, P.K.R. Moore, and H.S. Winter.** 1985. Characterization of spontaneous colitis in cotton-top tamarins (Saguinus oedipus) and its response to sulphasalazine. Gastroenterlogy *88*:13-19.
53. **Matsushime, H., M.F Roussel, K. Matsushima, A. Hishinuma, and C.J. Sherr.** 1991. Cloning and expression of murine interleukin-1 receptor antagonist in macrophages stimulated by colony-stimulating factor 1. Blood *78*:616-623.

54. **Mombaerts, P., E. Mizoguchi, M.J. Grusby, L.H. Glimcher, A.K. Bhan, and S. Tonegawa.** 1993. Spontaneous development of inflammatory bowel disease in T cell receptor mutant mice. Cell *75*:275-282.
55. **Mural, R.J., M.D. Adams, E.W. Myers, H.O. Smith, G.L. Miklos, R. Wides, A. Halpern, et al.** 2002. A comparison of whole-genome shotgun-derived mouse chromosome 16 and the human genome. Science *296*:1661-1671.
56. **Ohmen, J., H. Yang, K. Yamamoto, H. Zhao, Y. Ma, L. Bentley, Z. Huang, et al.** 1996. Susceptibility locus for inflammatory bowel disease on chromosome 16 has a role in Crohn's disease, but not in ulcerative colitis. Hum. Mol. Genet. *5*:1679-1683.
57. **Okayasu, I., S. Hatakeyama, M. Yamada, T. Ohkusa, Y. Inagaki, and R. Nakaya.** 1990. A novel method in the induction of reliable experimental acute and chronic ulcerative colitis in mice. Gastroenterology *98*:694-702.
58. **Palfree, R.G.E., K.P. LeClair, A.L.M. Bothwell, and U.M. Haemmerling.** 1987. cDNA characterization of an Ly-6.2 gene expressed in BW5147 tumor cells. Immunogenetics *26*:389-391.
59. **Podolsky D.K.** 2002. Inflammatory bowel disease. N. Engl. J. Med. *347*:417-429.
60. **Reinecker, H.C. and D.K. Podolsky.** 1995. Human intestinal epithelial cells express functional cytokine receptors sharing the common gamma c chain of the interleukin 2 receptor. Proc. Natl. Acad. Sci. USA *92*:8353-8357.
61. **Reiser, H., H. Oettgen, E.T.H. Yeh, C. Terhorst, M. Low, B. Benacerraf, and K.L. Rock.** 1986. Structural characterization of the TAP molecule: a phosphatidylinositol-linked glycoprotein distinct from the T cell receptor/T3 complex and Thy-1. Cell *47*:365-370.
62. **Reisner, H., J. Coligan, E. Palme, B. Benceraf, and K.L. Rock.** 1988. Cloning and expression of a cDNA for the T-cell activating protein TAP. Proc. Natl. Acad. Sci. USA *85*:2255-2259.
63. **Sadlack, B., H. Merz, H. Schorle, A. Schimpl, A.C. Feller, and I. Horak.** 1993. Ulcerative colitis-like disease in mice with a disrupted interleukin 2 gene. Cell *75*:253-261.
64. **Saito, H., D.M. Kranz, Y. Takagaki, A.C. Hayday, H.N. Eisen, and S. Tonegawa.** 1984. Complete primary structure of a heterodimeric T-cell receptor deduced from cDNA sequences. Nature *309*:312:36-39.
65. **Sartor R.B.** 1994. Cytokines in intestinal inflammation: pathology and clinical considerations. Gastroenterology *106*:533-539.
66. **Schena, M., D. Shalon, R.W. Davis, and P.O. Brown.** 1995. Quantitative monitoring of the gene expression patterns with a complementary DNA microarray. Science *270*:467-470.
67. **Schena, M., D. Shalon, R. Heller, A. Chai, P.O. Brown, and R.W. Davis.** 1996. Parallel human genome analysis:microarray-based expression monitoring of 1000 genes. Proc. Natl. Acad. Sci. USA *93*:10614-10619.
68. **Schneider, O. and U.E. Zelck.** 2001. Differential display analysis of hemocytes from schistosome-resistant and schistosome-susceptible intermediate hosts. Parasitol. Res. *87*:489-491.
69. **Simon, H.U., G.B. Millis, M. Kozlowski, D. Hogg, D. Branch, Y. Ishimi, and K.A. Siminovitch.** 1994. Molecular characterization of hNRP, a cDNA encoding a human nucleosome-assembly-protein-1-related gene product involved in the induction of cell proliferation. Biochem. J. *297*:389-397.
70. **Strober, W. and R.O. Ehrhardt.** 1993. Chronic intestinal inflammation: an unexpected outcome in cytokine or T cell receptor mutant mice. Cell *75*:203-205.
71. **Sturtevant, J.** 2000. Applications of differential display reverse transcription-PCR to molecular pathogenesis and medical mycology. Clin. Microbiol. Rev. *13*:408-427.
72. **Talbot, I.C. and A.B. Price.** 1987. p. 117-147. *In* A.M. Neville, F. Walker, and L.S. Gottleib (Eds.), The Biopsy Pathology in Colorectal Disease. Chapman and Hall, London.
73. **Utans, U., P. Liang, L.R. Wyner, M.J. Karnovsky, and M.E. Russell.** 1994. Chronic cardiac rejection: identification of five upregulated genes in transplanted hearts by differential mRNA display. Proc. Natl. Acad. Sci. USA *91*:6463-6467.
74. **Van Etten, R.A., M.W. Walberg, and D.A. Clayton.** 1980. Precise localization and nucleotide sequence of the two mouse mitochondrial rRNA genes and three immediately adjacent novel tRNA genes. Cell *22*:157-170.
75. **Wang, F.L., Y. Wang, W.K. Wong, Y. Liu, F.J. Addivinola, P. Liang, L.B. Chen, et al.** 1996. Two differentially expressed genes in normal human prostate tissue and in carcinoma. Cancer Res. *56*:3634-3637.
76. **Zuo, L., C.K. Ogle, J.E. Fisher, and M.S. Nussbaum.** 1997. mRNA differential display of colonic mucosa cells in ulcerative colitis. J. Surg. Res. *69*:119-127.

Section II

SAGE and Validation

5 Genome-Wide Analysis of Gene Expression in Human CD34+ Stem/Progenitor Cells

Guolin Zhou[1], Jianjun Chen[1], Sanggyu Lee[1], Terry Clark[2], Janet D. Rowley[1], and San Ming Wang[1]
Departments of [1]Medicine and [2]Computer Science, University of Chicago Medical Center, Chicago, IL, USA

INTRODUCTION

Hematopoietic stem cells are self-renewable and can differentiate into different hematopoietic lineages, including myelomonocytic, megakaryocytic, lymphoid, and erythroid cells (12,15). Hematopoietic stem cells have been widely used in the treatment of hematopoietic disorders, such as leukemia and immunodeficiency diseases (6,21). Recent data also suggest that hematopoietic stem cells are highly plastic. Under certain conditions, they can differentiate into nonhematopoietic cells such as brain, liver, and cardiac cells (1,14,16). These features suggest that hematopoietic stem cells may potentially be used for the treatment of nonhematopoietic disorders such as neural and cardiac diseases (17).

Considering the biological importance of stem cells, many attempts have been made to understand the gene expression in these cells using various methods (11,23). Although much knowledge about hematopoietic stem cells has been gained, not much is known about the genetic mechanisms determining their development. We performed a genome-wide analysis of gene expression in human CD34+ stem/progenitor cells to understand the genetic information controlling hematopoietic differentiation. We used serial analysis of gene expression (SAGE) as the major tool in our study. In SAGE, a short tag is released from a transcript, and multiple tags from many transcripts are concatenated for cloning and sequencing (19). SAGE has several unique advantages over other methods, such as collecting tags for nearly every expressed transcript regardless of the expression

Perspectives in Gene Expression
Edited by K. Appasani

levels and providing quantitative information for each detected tag. However, there are two limitations associated with using SAGE for gene identification. One is that multiple genes may share the same tag sequence, resulting in difficulty in identifying the correct gene from the SAGE tag, and the other is that a tag may not have a match to an existing gene as the genes contributing to the SAGE tag may have not been identified. The short length of a SAGE tag prevents the further analysis of its corresponding gene. To overcome these limitations, we developed the generation of longer DNA fragments from SAGE tags for gene identification (GLGI) technique to convert the SAGE tags into the corresponding 3′ cDNA (10). Such extended 3′ cDNAs are 100 bases long. As a result, these 3′ cDNAs provide high specificity for identifying the correct genes from multiple genes sharing the same tag and provide longer sequences for further analysis of the potential novel genes.

MATERIALS AND METHODS

Cell Purification

The human CD34+ cells (Poietics, Gaithersburg, MD, USA) were isolated from mononuclear cells of human bone marrow through positive immunomagnetic selection (CD34 Progenitor Cell Isolation kit, Miltenyi Biotec, Auburn, CA, USA). The CD34+ cells from three donors were pooled for the analysis. The purity of the CD34+ cells was 96.4%, 98.7%, and 97.3% in the three samples, as measured by fluorescence-activated cell sorter analysis (FACS®; BD Biosciences, San Jose, CA, USA).

SAGE Performance

We used a modified SAGE protocol for our study. The main modifications included (*i*) the use of a set of anchored oligo(dT) primers for reverse transcription. The purpose of using the anchored oligo(dT) primers instead of the oligo(dT) primer for DNA synthesis is to generate poly(dA)/(dT) (-) 3′ cDNA to provide fixed-length 3′ cDNA templates. This treatment is important for 3′ cDNA amplification for the SAGE process and for providing templates for GLGI amplification. This also diminishes the generation of truncated cDNAs caused by internal poly(A) priming (13,20). (*ii*) The second modification was to introduce low-cycle PCR amplification of the 3′ DNA templates to generate sufficient 3′ cDNAs from a small quantity of mRNA for SAGE analysis (9). Normal bone marrow contains only 1%–2% CD34+ cells (5,8). This makes it difficult to perform SAGE analysis using the standard protocol. Our modification largely decreased the requirement for starting mRNA. In fact, we completed the SAGE analysis using 2.5 μg total RNA isolated from 2.5×10^6 CD34+ cells. (*iii*) Purifying the *Bsm*FI-released tag fragments before the ditag formation. This modification provides higher quality of ditags for concatermerization (9).

GLGI Performance

In the GLGI process, a SAGE tag sequence is used as the sense primer, and a universal antisense primer located at the 3′-end of cDNA is used as the antisense primer to amplify the original cDNA template from which the SAGE tag was derived. We used a high-throughput GLGI procedure to identify the correct gene for the multi-matched SAGE tags with higher than nine copies and to convert novel SAGE tag sequences into 3′ cDNAs for SAGE tags higher than four copies (3).

RESULTS AND DISCUSSION

To provide the information for the pattern of gene expression in the primary CD34+ cells, we used the cells directly isolated from human bone marrow without any in vitro culture. The qualitative identification and quantitative measure of SAGE tags for each known gene, expressed sequence tag (EST), and novel SAGE tag provide a basis for studying normal gene expression in hematopoietic stem/progenitor cells and abnormal gene expression in hematopoietic diseases.

Distribution of SAGE Tags in CD34+ Cells

We collected a total of 106 021 SAGE tags from 2.5×10^6 CD34+ cells. From these tags, we identified 42 399 unique SAGE tags (Figure 1). All the unique SAGE tags were matched to the SAGE reference database for gene identification. The results show that in CD34+ cells, a few genes are expressed at high levels, and most genes are expressed at low levels. For example, only 91 (0.2%) SAGE tags were present in more than 100 copies, whereas 32 453 (76.5%) SAGE tags were present as single copy. This distribution is consistent with those observed in other matured somatic cell types (18), indicating that this is a universal pattern of gene expression in different cell types. Of the unique SAGE tags, 21 546 (50.9%) matched existing known expressed sequences, and 20 854 (49.1%) were novel SAGE tags without matches (Figure 2). The distribution of matched tags and novel SAGE tags shows a reciprocal relationship with their copy numbers; the matched tags tended to be the ones at high copies, whereas the novel SAGE tags tended to be at lower copies (Figure 3). This observation suggests that many novel transcripts expressed in the human genome may not be identified yet (10), particularly for the genes expressed at low levels in CD34+ cells. We also observed that 34% of the matched SAGE tags have multiple matches to sequences in different UniGene clusters. The distribution of these multi-matched SAGE tags was also parallel with their copy numbers.

Identification of the Correct Genes for SAGE Tags with Multiple Matches

Because the SAGE tags with multiple matches tend to be the ones at higher copies, we used the GLGI method to identify the correct genes for the 440 multi-matched SAGE tags higher than nine copies. We identified the correct gene for 385 (88%) of these 440 SAGE tags. Analysis of these confirmed genes

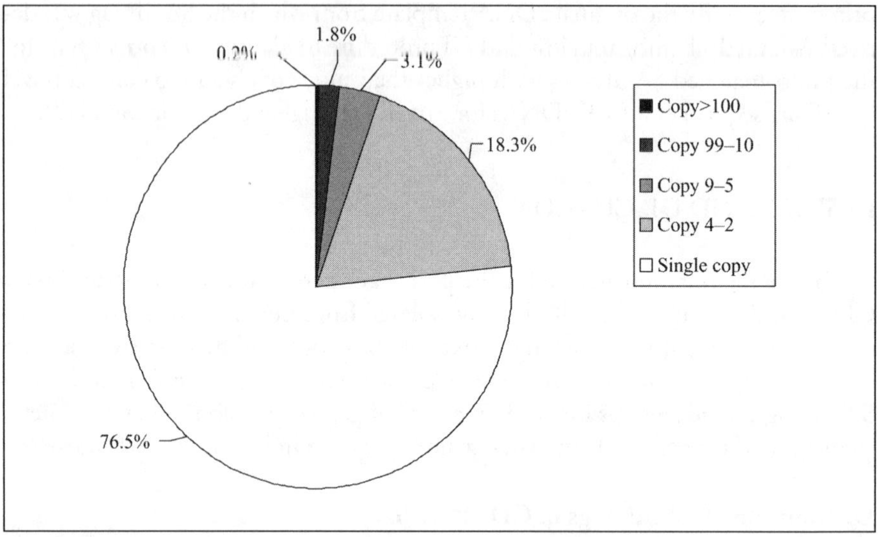

Figure 1. Distribution of 42 399 unique tags in CD34⁺ cells. We identified 42 399 unique tags from total 106 021 SAGE tags. Most tags are at low levels, and a few tags are at high levels.

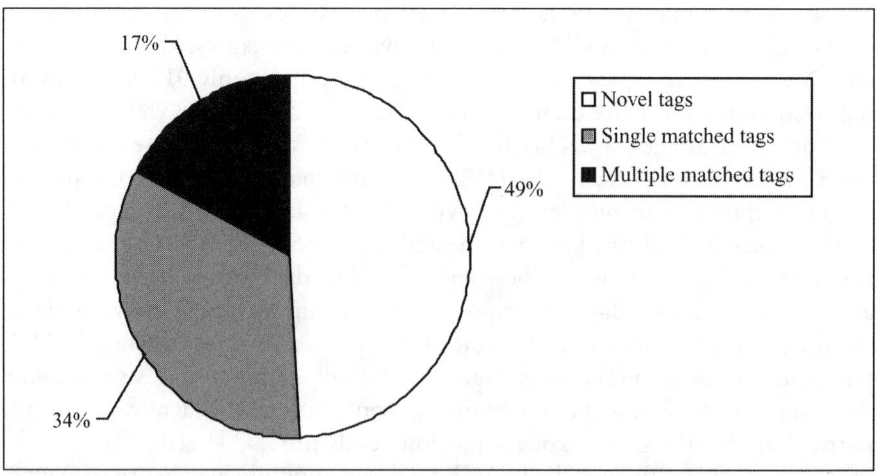

Figure 2. Distribution of single, multi-matched tags and novel tags. From a total of 42 399 unique tags, we identified 14 212 single matched tags, 7 334 multiple matched tags, and 20 853 as novel tags without matches to expressed sequences.

shows that many housekeeping genes are among the highly expressed genes in CD34+ cells, including 55 ribosomal proteins. However, we also observed that many genes with specified function are also expressed at high levels in CD34+ cells, such as v-fos and tumor necrosis factor (TNF). Interestingly, many novel genes are also among these highly expressed genes, including 77 ESTs, 13 hypothetical genes, 10 KIAA protein genes, and 198 novel 3′ cDNAs converted from novel SAGE tags (Table 1). The high-level expression of these genes suggests their functional importance in the development of hematopoietic stem cells.

Conversion of Novel SAGE Tags into Novel 3′ ESTs

The novel SAGE tags account for half of the unique SAGE tags detected in this analysis. Using the GLGI method, we tried to convert 219 novel SAGE tags present in more than four copies into 3′ ESTs. A total of 198 sequences generated from 138 novel SAGE tags were confirmed to be novel 3′ cDNAs (Figure 4). Sequences generated from 32 novel tags matched known expressed sequences, and 49 reactions from 49 novel tags did not generate qualified sequences. These 3′ cDNAs range in length from 8 to 454 bp, with a mean of 152 bp. This result suggests that the novel SAGE tags represent a large number of novel transcripts.

Known Genes Identified in CD34+ Cells

Based on the UniGene clusters matched by SAGE tags, we identified 3687 known genes expressed in CD34+ cells. This is the largest number of known

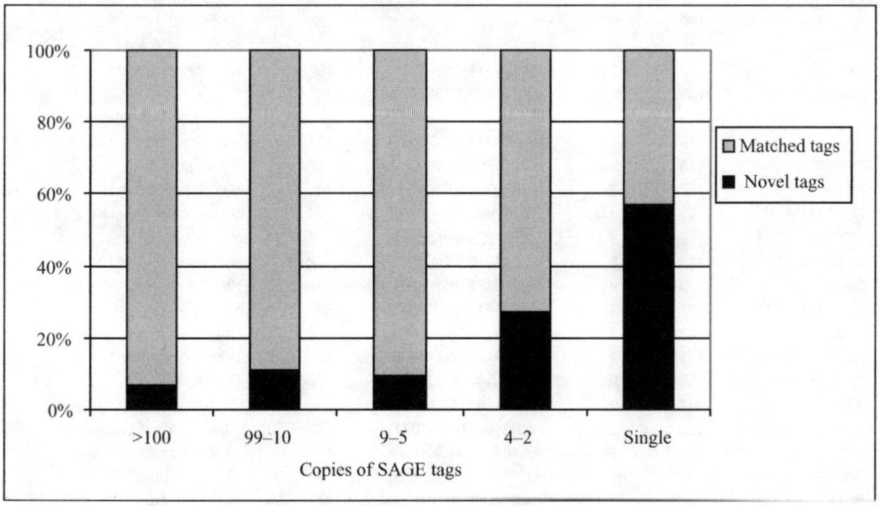

Figure 3. Distribution of matched tags and novel tags. The matched SAGE tags and novel SAGE tags exist in a reciprocal relationship with their copy numbers. Most matched tags are in high copies, but most novel tags are in low copies.

Table 1. Top 100 Genes Expressed in CD34⁺ Cells

Tag	Tag Copy	Matched Clusters	GLGI Confirmation	GLGI Length (bp)	Gene
TGTGTTGAGA	1711	6	Hs.181165 (AK026650)	149	Eukaryotic translation factor
ATCAAGGGTG	874	2	Hs.157850 (BF689002)	318	Ribosomal protein L9
GTGAAGGCAG	783	2	Hs.77039 (AI862222)	144	Ribosomal protein S3A
ATGGCTGGTA	775	2	Hs.182426 (BG272391)	269	Ribosomal protein S2 (RPS2)
CCTGTAATCC	516	430	Hs.106004 (AK025503)	131	Hypothetical protein FLJ22347
GCCGTGTCCG	482	2	Hs.241507 (AV649570)	157	Ribosomal protein S6
GTGAAACCCC	471	325	Hs.288917 (AK026323)	23	Hypothetical protein FLJ22670
TCACCCACAC	470	3	Hs.234518 (BF476398)	192	Ribosomal protein L23
AGCTCTCCCT	442	2	Hs.82202 (BF448541)	170	Ribosomal protein L17
TGGGCAAAGC	377	2	Hs.2186 (BG222874)	118	Eukaryotic translation factor
GGACCACTGA	347	3	Hs.119598 (BG655756)	100	Ribosomal protein L3
CCACTGCACT	331	175	Hs.201593 (AK027225)	49	FLJ23572 fis
TTGGTCCTCT	311	4	Hs.108124 (BG271707)	116	"Ribosomal protein S4, X-linked"
AAGACAGTGG	289	3	Hs.184109 (AI187181)	159	Ribosomal protein L37a
GGGCTGGGGT	275	3	Hs.183698 (BG759031)	51	Ribosomal protein L29
ATTATTTTTC	273	5	Hs.153 (BG272465)	54	Ribosomal protein L7
ATAATTCTTT	270	6	Hs.539 (AI339736)	43	Ribosomal protein S29
GCCTCAGTTC	256	3	Hs.5508 (AF000381)	22	Nonfunctional folate binding protein
AATGGATGAA	253	3	Hs.5508 (AW864746)	88	Nonfunctional folate binding protein
AGCCCTACAA	253	2	Hs.180532 (AI064973)	118	Glucose phosphate isomerase
CTGGGTTAAT	246	9	Hs.298262 (BG491251)	29	Ribosomal protein S19
CGCTGGTTCC	242	4	Hs.179943 (AW196616)	102	Ribosomal protein L11
GTGAAACCCT	239	155	Hs.289053 (AK022536)	23	Hypothetical protein FLJ14733
ACATCATCGA	236	2	Hs.182979 (BG272569)	97	Ribosomal protein L12
CCCGTCCGGA	234	8	Hs.180842 (BG370441)	206	Ribosomal protein L13
TCAGATCTTT	229	6	Hs.108124 (AL569555)	38	"Ribosomal protein S4, X-linked"
GGATTTGGCC	211	5	Hs.119500 (BG272444)	65	"Ribosomal protein, large P2"
TTATGGGATC	203	2	Hs.241507 (BG745160)	280	Ribosomal protein S6
TAATAAAGGT	199	9	Hs.151604 (BG232142)	34	Ribosomal protein S8
TTCAATAAAA	183	5	Hs.177592 (BE676868)	26	"Ribosomal protein, large, P1"
AAGGAGATGG	182	2	Hs.184014 (BG656345)	257	Ribosomal protein L31
CCAGAGAACT	165	2	Hs.6975 (AW582859)	30	PRO1073 protein
TTTTTGATAA	156	2	Hs.181165 (BG770933)	363	Eukaryotic translation factor
CAATAAATGT	155	4	Hs.179779 (AI073855)	34	Ribosomal protein L37
TAGGTTGTCT	148	5	Hs.279860 (AW167109)	37	Tumor protein
TGCACGTTTT	147	7	Hs.169793 (BF437755)	45	Ribosomal protein L32
TTCATACACC	147	3	Hs.75760 (AW193452)	71	Sterol carrier protein 2
AGGGCTTCCA	146	2	Hs.29797 (AA629546)	67	Ribosomal protein L10
ATTCTCCAGT	142	2	Hs.234518 (BF476398)	22	Ribosomal protein L23
CACCTAATTG	140	2	Hs.282283 (AW243783)	170	Proteasome subunit
TGATTTCACT	138	4	Hs.240443 (AW583759)	286	FLJ23538 fis
GCAAGCCAAC	132	2	Hs.80562 (BE695848)	42	"Casein kinase 1, epsilon"
AGGCTACGGA	131	2	Hs.119122 (BG744857)	116	Ribosomal protein L13a
CATTTGTAAT	131	2	Hs.240443 (AI133529)	289	FLJ23538 fis
ATTGTTTATG	127	2	Hs.181163 (BF593317)	47	High-mobility group protein 17
CTCATAAGGA	126	3	Hs.80562 (AV741032)	18	"Casein kinase 1, epsilon"
GCTCCCCTTT	126	2	Hs.1817 (AV736453)	140	Myeloperoxidase
AGGTCAGGAG	122	47	Hs.23413 (AB033099)	57	KIAA1273 protein
ACCCTTGGCC	122	3	Hs.279009 (AL134959)	371	Matrix Gla protein
AAGGTGGAGG	118	3	Hs.163593 (AV742869)	205	Ribosomal protein L18a
TGTAATCAAT	117	3	Hs.249495 (BG655713)	33	Nuclear ribonucleoprotein A1

Table 1. Continued

Tag	Tag Copy	Matched Clusters	GLGI Confirmation	GLGI Length (bp)	Gene
GAAAAATGGT	115	6	Hs.151604 (BG232142)	34	Ribosomal protein S8
TTGGCCAGGC	109	96	Hs.193361 (AK022676)	133	Hypothetical protein FLJ12614
TCACAAGCAA	109	3	Hs.146763 (NM_005594)	201	Nascent-polypeptide-associated
AATAGGTCCA	108	2	Hs.113029 (AW090450)	31	Ribosomal protein S25
TCGTCTTTAT	106	2	Hs.75538 (M77233)	47	Ribosomal protein S7
GGGCATCTCT	102	2	Hs.76807 (AU157203)	91	HLA-DRA
CCTGTAGTCC	100	116	Hs.240136 (AL049983)	63	DKFZp564D042 cDNA
TACCCTAAAA	100	49	Hs.165662 (AI824585)	24	KIAA0675 gene product
GTGTTAACCA	98	2	Hs.74267 (AF283772)	411	Ribosomal protein L15
GTTCGTGCCA	96	3	Hs.179666 (BG236685)	95	Hypothalamus protein
AAGAAGATAG	92	2	Hs.278406 (BG831765)	268	Ribosomal protein S2
CCATTGCACT	91	59	Hs.200113 (AK021441)	49	FLJ11379 fis cDNA
AGGAAAGCTG	88	2	Hs.76437 (AF077043)	73	Ribosomal protein L36
CTCATAGCAG	84	2	Hs.279860 (BG654607)	116	Tumor protein
TTCTTGTGGC	84	2	Hs.182740 (BG492065)	342	Ribosomal protein S11
CTCCTCACCT	79	4	Hs.119122 (BG655779)	48	Ribosomal protein L13a
CCCCAGCCAG	79	3	Hs.252259 (AW246658)	107	Ribosomal protein S3
GCCGAGGAAG	78	2	Hs.285405 (BF942220)	305	Ribosomal protein S12
ACTTTTTCAA	76	46	BG099326	16	EST
AAAAGAAACT	76	4	Hs.172182(BG744897)	26	Poly(A)-binding protein
GCTTTATTTG	75	2	Hs.288061(AA554747)	31	"Actin, beta"
GCCTTCCAAT	74	3	Hs.76053(BF941985)	33	DEAD/H box polypeptide 5
GGCAAGCCCC	72	6	Hs.252574(BG656357)	58	Ribosomal protein L10a
TACCATCAAT	72	6	Hs.169476(BG370213)	34	Phosphate dehydrogenase
ACTCCAAAAA	72	5	BG236559	15	EST
GCAAAACCCC	70	44	Hs.283546(AK023770)	31	FLJ13708 fis cDNA
GCGAAACCCC	68	39	Hs.269899(AI079278)	189	ESTs
GCATTTAAAT	64	4	Hs.275959(BG655489)	30	Eukaryotic translation factor
GTCTGGGGCT	63	3	Hs.75725 (BF591438)	61	Transgelin 2
GCTTTTAAGG	61	2	Hs.8102 (AI423086)	400	Ribosomal protein S20
ATTTGTCCCA	61	2	Hs.139800 (AU160425)	79	High-mobility group protein isoforms
CAGCTCACTG	59	2	Hs.738 (BG775811)	197	Ribosomal protein L14
TCTGCTAAAG	58	4	Hs.274472 (BF434300)	62	High-mobility group protein 1
TGTACCTGTA	58	3	Hs.278242 (BG222897)	54	Hypothetical protein MGC12992
AACCCGGGAG	56	19	AA428792	90	EST
AACTAACAAA	56	2	Hs.3297 (BG272886)	16	Ribosomal protein S27a
GTTCCCTGGC	55	3	Hs.177415 (BG271519)	224	Murine sarcoma virus
CCTGTAATCT	53	44	Hs.35088 (BF515942)	102	ESTs
GGCTTTACCC	51	2	Hs.119140 (BF432256)	83	Eukaryotic translation factor 5A
CCTAGCTGGA	50	4	Hs.182937 (BG655492)	57	Peptidylprolyl isomerase A
TTGGCTTTTC	50	2	Hs.41569 (BF475411)	65	Phosphatidic acid phosphatase
CCTATAATCC	49	73	Hs.25328 (AL048825)	124	EST
GATGCTGCCA	49	3	Hs.129914 (L21756)	178	Runt-related transcription factor 1
ATTTGAGAAG	48	5	Hs.169921 (BE621880)	42	General transcription factor II
TAGAAAGGCA	48	2	Hs.78909 (W37407)	267	Butyrate response factor 2
TGTGTTAAGA	47	4	Hs.288036 (AI804500)	15	Isopentenylpyrophosphate transferase
GTGGCTCACA	46	44	Hs.120769 (AI982685)	113	Homo sapiens cDNA FLJ20463 fis
TTGGTCAGGC	45	34	Hs.12094 (AA650333)	119	Hypothetical protein
GTGCACTGAG	45	5	Hs.181244 (D32129)	51	Major histocompatibility complex

genes identified in human hematopoietic CD34+ cells (4,7,11,22,23) (Figure 5). Considering the identification of 42 399 unique SAGE tags and the matches of 21 546 unique SAGE tags to known expressed sequences, the 3687 known genes account for only 9% of the total unique SAGE tags and 17.6% of the matched SAGE tags. This indicates that the functions of many genes expressed in CD34+ cells are unknown.

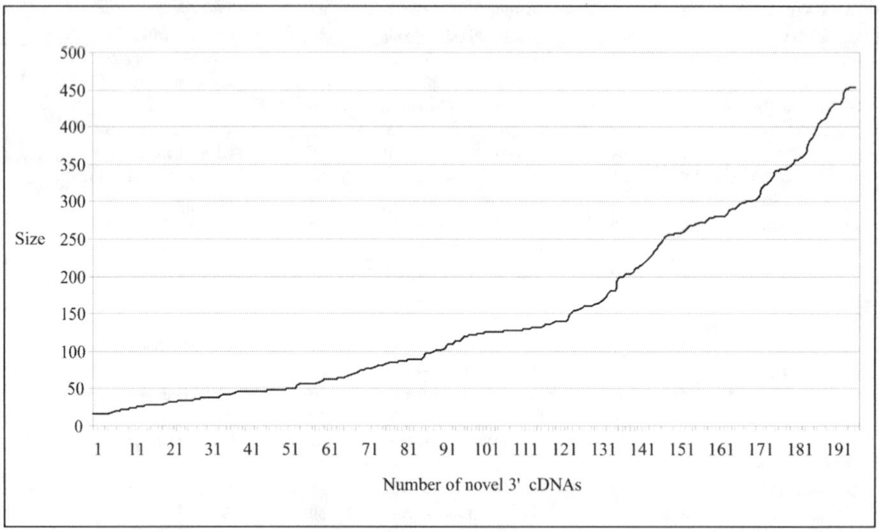

Figure 4. Size distribution of novel 3′ cDNAs. A total of 198 novel 3′ cDNAs were confirmed from 138 novel SAGE tags, ranging from 8 to 454 bp, with an average mean of 152 bp.

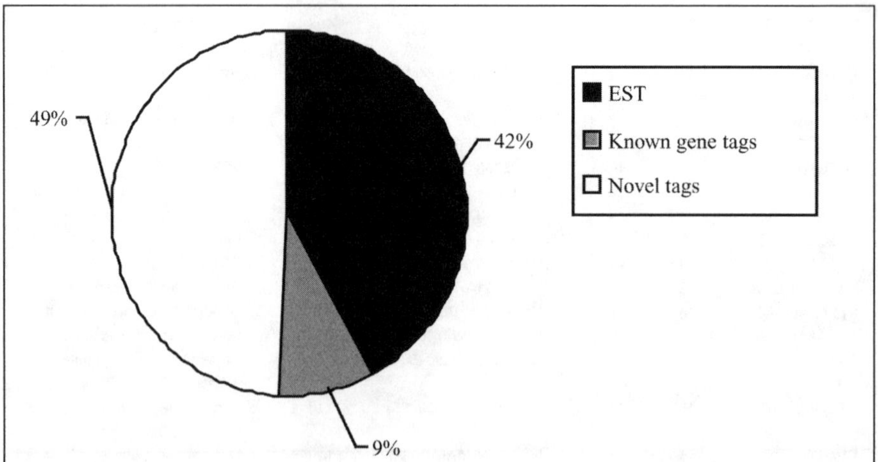

Figure 5. Matches of SAGE tags identified in CD34+ cells. From 42 399 unique SAGE tags, 3687 match to known genes (9%), 17 859 to ESTS (42%), and 20 854 do not have matches (49%).

Genes Important for Hematopoietic Differentiation

We analyzed the genes known to be important for hematopoiesis. We have confirmed many of these genes in this study. For example, seven HOX genes in HOX-A, B, and C clusters were identified, including A3, A5, A9, B2, B7, B13, and C9 (2). All of these HOX genes were expressed at low levels.

In summary, our study provides a general view for the pattern of gene expression in normal human CD34 stem/progenitor cells. The data generated from this study will provide a basis for further studies of stem cell biology and hematopoiesis.

ACKNOWLEDGMENTS

Supported by National Institutes of Health grant no. CA78862-01 to J.D.R. and S.M.W. and by the G. Harold and Lelia Y. Mathers Foundation to S.M.W.

REFERENCES

1. **Brazelton, T.R., F.M. Rossi, G.I. Keshet, and H.M. Blau.** 2000. From marrow to brain: expression of neuronal phenotypes in adult mice. Science *290*:1775-1779.
2. **Buske, C. and R.K. Humphries.** 2000. Homeobox genes in leukemogenesis. Int. J. Hematol. *7*:301-308.
3. **Chen, J., S. Lee, G. Zhou, and S.M. Wang.** 2002. High-throughput GLGI procedure for converting a large number of serial analysis of gene expression tag sequences into 3′ complementary DNAs. Genes Chromosomes Cancer *33*:252-261.
4. **Claudio, J.O., C.C. Liew, A.A. Dempsey, E. Cukerman, A.K. Stewart, and E. Na.** 1998. Identification of sequence-tagged transcripts differentially expressed within the human hematopoietic hierarchy. Genomics *50*:44-52.
5. **D'Arena, G., P. Musto, N. Cascavilla, G. Di Giorgio, F. Zendoli, and M. Carotenuto.** 1996. Human umbilical cord blood: immunophenotypic heterogeneity of CD34+ hematopoietic progenitor cells. Haematologica *81*:404-409.
6. **Espanol, T.** 2001. Substitution therapy with hematopoietic progenitors in the primary immunodeficiencies. Allergol. Immunopathol. *29*:118-121.
7. **Gu, J., Q.H. Zhang, Q.H. Huang, S.X. Ren, X.Y. Wu, M. Ye, C.H. Huang, G. Fu, et al.** 2000. Gene expression in CD34(+) cells from normal bone marrow and leukemic origins. Hematol. J. *1*:206-217.
8. **Krause, D.S., M.J. Fackler, C.I. Civin, and W.S. May.** 1996. CD34: structure, biology, and clinical utility. Blood *87*:1-13.
9. **Lee, S., J. Chen, G. Zhou, and S.M. Wang.** 2001. Generation of high-quantity and quality tag/ditag cDNAs for SAGE analysis. BioTechniques *31*:348-354.
10. **Lee, S., G. Zhou, T. Clark, J. Chen, J.D. Rowley, and S.M. Wang.** 2001. The pattern of gene expression in human CD15+ myeloid progenitor cells. Proc. Natl. Acad. Sci. USA *98*:3340-3345.
11. **Mao, M., G. Fu, J.S. Wu, Q.H. Zhang, J. Zhou, L.X. Kan, Q.H. Huang, K.L. He, et al.** 1998. Identification of genes expressed in human CD34(+) hematopoietic stem/progenitor cells by expressed sequence tags and efficient full-length cDNA cloning. Proc. Natl. Acad. Sci. USA *95*:8175-8180.
12. **Morrison, S.J., D.E. Wright, S.H. Cheshier, and I.L. Weissman.** 1997. Hematopoietic stem cells: challenges to expectations. Curr. Opin. Immunol. *9*:216-221.
13. **Nam, D.K., S. Lee, G. Zhou, X. Cao, C. Wang, T. Clark, J. Chen, J.D. Rowley, et al.** 2002. Oligo

(dT) primer generates a high frequency of truncated cDNAs through internal poly(A) priming during reverse transcription. Proc. Natl. Acad. Sci. USA *99*:6152-6156.

14. **Orlic, D., J. Kajstura, S. Chimenti, I. Jakoniuk, S.M. Anderson, B. Li, J. Pickel, R. McKay, et al.** 2001. Bone marrow cells regenerate infarcted myocardium. Nature *410*:701-705.

15. **Peschle, C., R. Botta, R. Muller, M. Valtieri, and B.L. Ziegler.** 2001. Purification and functional assay of pluripotent hematopoietic stem cells. Rev. Clin. Exp. Hematol. *5*:3-14.

16. **Punzel, M. and A.D. Ho.** 2001. Divisional history and pluripotency of human hematopoietic stem cells. Ann. NY Acad. Sci. *938*:72-81.

17. **Steidl, U., R. Kronenwett, U.P. Rohr, R. Fenk, S. Kliszewski, C. Maercker, P. Neubert, M. Aivado, et al.** 2002. Gene expression profiling identifies significant differences between the molecular phenotypes of bone marrow-derived and circulating human CD34+ hematopoietic stem cells. Blood *99*:2037-2044.

18. **Velculescu, V.E., S.L. Madden, L. Zhang, A.E. Lash, J. Yu, C. Rago, A. Lal, C. Wang, et al.** 1999. Analysis of human transcriptomes. Nat. Genet. *23*:387-388.

19. **Velculescu, V.E., L. Zhang, B. Vogelstein, and K.W. Kinzler.** 1995. Serial analysis of gene expression. Science *270*:484-487.

20. **Wang, S.M., S.C. Fears, L. Zhang, J. Chen, and J.D. Rowley.** 2000. Screening poly (dA/dT)-cDNAs for gene identification. Proc. Natl. Acad. Sci. USA *97*:4162-4167.

21. **Weissman, I.L.** 2000. Translating stem and progenitor cell biology to the clinic: barriers and opportunities. Science *287*:1442-1446.

22. **Yang, Y., K.R. Peterson, G. Stamatoyannopoulos, and T. Papayannopoulou.** 1996. Human CD34+ cell EST database: single-pass sequencing of 402 clones from a directional cDNA library. Exp. Hematol. *24*:605-612.

23. **Zhang, Q.H., M. Ye, X.Y. Wu, S.X. Ren, M. Zhao, C.J. Zhao, G. Fu, Y. Shen, et al.** 2000. Cloning and functional analysis of cDNAs with open reading frames for 300 previously undefined genes expressed in CD34+ hematopoietic stem/progenitor cells. Genome Res. *10*:1546-1560.

6 Expression of Novel Genes in Normal and Diseased Human Tissues

Shelby A. Martensen and Jay C. Strum
Quantitative Expression Group, Department of Cellular Genomics, GlaxoSmithKline Pharmaceuticals, Research Triangle Park, NC, USA

INTRODUCTION

The elucidation of the 3.2-gigabase human genomic sequence (7) has had a tremendous impact on the pharmaceutical industry and its quest for the identification of tractable targets. While this tremendous scientific achievement excited researchers and pharmaceutical companies across the world with its potential, it also overwhelmed us all with its sheer magnitude. Traditionally, biotechnology and pharmaceutical companies had devised and implemented strategies to mine sequence databases in the hope of finding new and better targets. Sequences were analyzed, often by experienced bioinformaticians, and grouped into designated target classes. These target classes include G-protein-coupled receptors (GPCRs), kinases, proteases, ion channels, and nuclear hormone receptors. It has been reported that there are approximately 500 molecular drug targets (1,3) that form the basis of today's medicines, with approximately 45% of those being GPCRs. Subsequent to the publication of the human genome sequence, it has been postulated that there are actually 5000–10 000 molecular drug targets (2), and therein lies the challenge that pharmaceutical companies face today.

The key question is how to make efficient and cost-effective use of this wealth of information. To meet this challenge, we devised a strategy that consists of a combined effort between bioinformatics and genomics. To begin the task of target identification, novel genes are first identified using various sophisticated in silico tactics and software programs to identify and classify novel sequences. (6,8). It is not enough, however, to generate a long list of genes. The unique

Perspectives in Gene Expression
Edited by K. Appasani

tissue distribution of a novel gene is a key step for establishing the potential of a novel sequence to be involved in the progression of human disease. One way to accomplish this is through the analysis of protein expression either by Western blot analysis or immunohistochemistry. However, protein expression analysis is often a timely, costly, and labor-intensive process that requires the procurement of various fixed or paraffin-embedded human tissues, the generation and characterization of a specific antibody, and the acquisition of reagents. The expression of a protein can be, instead, inferred by the presence of its mRNA in a given tissue. However, conventional mRNA analysis by Northern blot analysis, RNase protection assay (RPA), or in situ hybridization is every bit as costly and labor intensive as protein analysis by Western blot analysis or immunohistochemistry. Unfortunately, none of these traditional methods allow us to quantitate gene expression in a high-throughput manner with a minimum of time and cost.

The advent of PCR offers several advantages to these traditional techniques (4). The two most immediate advantages are the decrease in the amount of necessary starting material and time needed to carry out the assay and obtain a useful result. In spite of these advantages, PCR still does not allow the accurate quantitation of mRNA in a given tissue. Various techniques have been introduced, such as competitive PCR, which allow the quantitation of mRNA. However, these techniques are labor intensive and cannot be easily adapted for high-throughput gene expression analysis. The introduction of real-time quantitative PCR resolves all these issues and forms the second facet of our strategy for target identification and validation.

There are two major chemistries that are used in real-time quantitative PCR. The first chemistry is referred to as TaqMan® (Applied Biosystems). This uses a fluorescence resonance energy transfer (FRET) probe in a 5′ nuclease assay. In the 5′ nuclease assay, two traditional PCR primers are used in conjunction with a fluorogenic probe. The probe is a nonextendable oligonucleotide labeled with a 5-carboxyfluorescein (FAM) fluorescent dye at its 5′-end and a 6-carboxytetramethylrhodamine (TAMRA) quencher dye at its 3′-end. The 5′ nuclease activity of the *Taq* DNA polymerase results in the hydrolysis and release of the fluorescent FAM label. The level of fluorescence is monitored and recorded at each PCR cycle, and the increase in fluorescence is proportional to the amount of starting template. The second type of chemistry uses the DNA intercalating dye, SYBR® Green (Molecular Probes). The difference between the TaqMan and SYBR Green chemistry is that the SYBR Green dye simply binds to all double-stranded products, thereby allowing the products to be quantitated at each PCR cycle. Regardless of the chemistry chosen, the assays themselves can be completed in just over 2 h. Additionally, the technology built around the real-time quantitative PCR methodology allows for both 96- and 384-well formats. The tremendous potential throughput of this methodology makes real-time quantitative PCR an attractive new alternative to traditional mRNA expression analysis and is the optimal method of choice for the tissue expression profiling of novel genes.

The first step to successfully performing real-time quantitative PCR is to design specific primers for the gene of interest. It is necessary to determine if the cDNA sequence of interest has any closely related family members. This can be accomplished by using the BLAST program in GenBank®. The entire cDNA sequence is blasted against the GenBank nonredundant nucleotide database to identify any homologous sequences. If closely related homologues are found, then a nucleotide alignment program such as Megalign (Lasergene) must be employed to determine the areas of sequence that can be used to design specific probes and primers. Specific rules are applied to the primer design to obtain optimal results. These rules include choosing: primers with no more than three Cs or Gs on the 3′-end of the primers and with a T_m of approximately 60; probes with more Cs than Gs and a T_m of approximately 70; and products that are 50–150 bases. As a final measure of specificity, the entire product is then blasted against GenBank to ensure there are no known sequences that could be detected with the probe and primer set.

PROCUREMENT OF HUMAN TISSUE SPECIMENS

An important step for discerning the mRNA expression of novel genes in normal and diseased samples is the procurement of human tissue specimens. Access to a wide variety of high-quality normal and diseased human tissue specimens has been limited by the lack of organizations established to supply these samples and by their tremendous demand. Before the establishment of government-sponsored tissue banks, it was common for individual researchers to collaborate with surgeons or pathologists to receive extra tissue that was discarded following diagnosis. This was useful for accessing a small, limited number of samples of variable quality but is not sufficient to support a large expression profiling effort and localization follow-up. Government-sponsored tissue banks and transplant programs have been important for providing a variety of specimens for academic researchers and, to some extent, researchers in biotechnology and pharmaceutical companies. There are now several commercial ventures that have formed to specifically provide clinically relevant tissue specimens to researchers. Most of these ventures were started by scientists who had experienced the difficulties of procuring human tissues for their own research purposes.

There are many issues to consider regarding the procurement of tissues. One must first decide which tissues will best answer the question that is being asked. For the investigation of novel genes in which function is entirely unknown, it is best to look at a large array of normal tissues and a sampling of certain diseases of interest. To most accurately assess the possible function of the novel genes of interest, we constructed a normal plate and a disease plate. Our goal for the normal plate was to gather as many tissues as possible from each physiological system. Many tissues were gathered that represent the gastrointestinal tract, including the esophagus, stomach, duodenum, jejunum, ileum, colon, and rectum. Other tissues such as adenoids, tonsils, and spleen were used to assess immune cell expression. Additional tissues included whole brain, cerebellum, hypothalamus, liver, skin, and testis, with a total of 45 different tissues represented. The disease plate consists of a total of 96 samples, representing eight different diseases of interest, including breast, lung, and colon cancer, rheumatoid arthritis, osteoarthritis, Alzheimer's disease, and cells infected by the human immunodeficiency virus, herpes simplex virus, and human papillomavirus.

Whether the samples of interest are meant to represent normal physiological systems or various diseases of interest, there are several issues to consider. The first issue is how the tissue is collected. Human tissues can be collected either by autopsy or by biopsy or resection during a surgical procedure. There are pros and cons to both types of tissue collection. If the tissues are to be collected by autopsy, then it is important for the researcher and the tissue bank to establish an acceptable period of time between the death of the donor and the collection of the tissue. The greater the length of time between the death and tissue collection, the more likely it is that the valuable RNA will be degraded and thereby irreparably compromised. It is also important for the researcher to establish appropriate and inappropriate causes of death. For example, a death caused by septic shock would deem all tissues in the body useless because all of the tissues would be greatly altered. If there is no massive loss of blood, then the most optimal collection is from an individual that has suffered sudden death (i.e., car wreck), where the patient was otherwise healthy with no known medical problems. Alternatively, if the tissues are to be collected by biopsy, then in addition to the factors mentioned above, the researcher must also have the full disclosure of the medical history of the donor. It is important to remember that biopsies are only performed in the suspicion of illness and, therefore, seemingly normal tissues may in fact be somehow altered. There are still more caveats to the collection of matched normal and diseased pairs. This type of collection is most often performed in cancer-related surgical procedures in which a surgeon will resect malignancies and normal-appearing margins surrounding the malignancies. This begs the question of whether tissue surrounding a malignancy, while appearing morphologically and histologically normal, is also genetically normal. Because there is so much patient-to-patient variation with respect to levels of gene expression, there is still considerable utility in using normal and diseased tissue from the same patient. Still, there are many diseases of interest in which it is simply not possible to obtain normal and diseased tissue from the same patient, and that also comes with its caveats. Many of the neurological or neurodegenerative diseases fall into this category. As one can see, it is extremely important for both the researcher and the tissue provider to be aware of the many issues surrounding the acquisition of tissues so that the best decision can be made with respect the question being asked. It is only with this type of truly collaborative effort that the resultant data will be properly analyzed and interpreted and have the most value to the research effort and to the ultimate goal of target validation.

SAMPLE PREPARATION

The process through which tissues progress to achieve the final goal of gene expression information can be highly standardized for ultimate efficiency. The tissues obtained are subjected to one of two commercial RNA isolation protocols: RNeasy® or Tri Reagent® (both from Qiagen, Valencia, CA, USA). Following the elution of RNA, the $A_{260/280}$ ratios of the samples are determined using a suitable spectrophotometer so that a good-quality preparation should have a ratio of 1.7:2.0. Additionally, the amount of total RNA is quantitated based on the absorbance readings (A of 1.0 = 40 μg RNA). For further quality control, 0.5 μg total RNA are resolved on a formaldehyde gel, and the 18S and 28S bands are visually inspected for intact RNA. If there is not significant degradation of the RNA, as indicated by a smearing effect below the 28S band, then the total RNA is ready to be DNase treated.

The total RNA preparation is treated with RNase-free DNase (Ambion, Austin, TX, USA) to remove contaminating genomic DNA. To accomplish this, all samples are subjected to the protocol published with the DNA-free kit (Ambion). The success of this procedure is determined by subjecting the expectedly DNA-free RNA samples to amplification in the absence of reverse transcriptase. This is an assay in which a negative result or lack of amplification is the desired result and, therefore, a positive DNA control must be used. Following the real-time quantitative PCR assay, the data are analyzed using the SDS system software (Applied Biosystems, Foster City, CA, USA). The software should be allowed to set the threshold and use the default baseline of 3–15 cycles unless some of the samples have cycle thresholds (C_ts) (i.e., the point at which the fluorescent signal crosses the threshold) less than 15 (Figure 1). If the DNase is complete, then all the samples will have C_ts equal to 40, and we can assume that the sample is sufficiently clean for further use. If both duplicates are less than 35, then the DNase treatment must be repeated. If there is no amplification of the house keeping gene of interest in this quality-control amplification assay, then the samples are ready to proceed to the next step.

To obtain the best quantitative results for the majority of probe and primer sets, it is necessary to convert the total RNA to cDNA. Before the conversion to cDNA, the DNA-free total RNA samples are quantitated using the RNA-specific dye, RiboGreen® (Molecular Probes, Eugene, OR, USA). The samples are prepared according to the manufacturer's instructions. Briefly, the unknowns are diluted 1:100, and the RNA standards are prepared by diluting 100 μg/mL rRNA standard in a ratio of 1:50 to make a 2 μg/mL stock. Five 2-fold serial dilutions of the 2 μg/mL stock are then made for the generation of a wide-range standard curve. The standards, blanks, and unknowns are each added to individual wells of a microplate, and 100 μL of the RiboGreen Dye Reagent are added to each sample. The assay is then incubated at room temperature for 4 min in the dark. The results of the assay can be read on the ABI PRISM® 7700 (Applied Biosystems).

The RNA concentration is calculated by the interpolation to the standard curve.

Once quantitated, the samples can be converted to cDNA using random hexamers and murine leukemia virus (MuLV) reverse transcriptase. A typical reaction is as follows: approximately 1 µg total RNA is added to an Eppendorf® tube with 5 mM $MgCl_2$, 1× PCR buffer II, 1 mM dNTPs, 0.5 U/µL RNase inhibitor, 2.5 U/µL MuLV reverse transcriptase, and 2.5 µM random hexamers for a total reaction volume of 100 µL. The cDNA conversion reaction is incubated at 42°C for 30 min. All samples are assessed for the quality of the conversion by the analysis of 18S rRNA, GAPDH, cyclophilin, and β-actin expression.

REAL-TIME QUANTITATIVE PCR ASSAYS

Following the conversion of the RNA to cDNA, the samples are then ready for the intended experiments. The expression of the genes of interest can now be analyzed in the appropriate normal and diseased samples as assessed by real-time quantitative PCR. The data are analyzed by setting an arbitrary reference point to which all samples can be compared. The no template control (NTC) control C_t value of 40 is used for this purpose. Additionally, the efficiency of the PCR is assumed to be 100%, which means that there is a 2-fold increase in the products with each

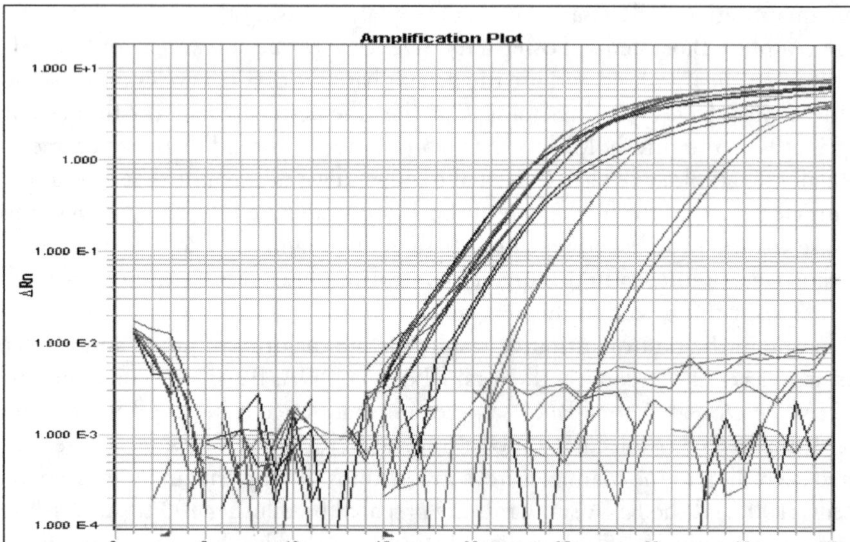

Figure 1. Amplification plot generated by a real-time quantitative PCR assay using the ABI PRISM 7900. The change in fluorescence (ΔRn) is calculated by dividing both the background emission intensity and the experimental emission intensity by the passive reference emission intensity and then subtracting the background value from the experimental value. The result is a logarithmic plot that depicts the amount of product at each PCR cycle. **(See color plate A2.)**

PCR cycle. All C_t values of the samples of interest are then subtracted from 40. With these assumptions, we used $2^{\Delta C_t}$ to determine the relative abundance.

To compare certain samples such as treated cell lines to untreated cell lines or tumor samples to normal samples, we calculated the expression level of a gene of interest as a fold change relative to normal. To perform this calculation, the appropriate baseline must be chosen. Because a higher C_t correlates with lower gene expression, it is necessary to subtract the C_t of the experimental samples from the C_t for the chosen baseline (described as the ΔC_t). When the calculation is done in this manner, a positive fold change indicates an up-regulation of gene expression, and a negative fold change indicates a down-regulation of gene expression. For example, when comparing matched normal and tumor pairs, the C_t for the tumor would be subtracted from the C_t for the matched normal. We can then use $2^{\Delta C_t}$ (when the ΔC_t is positive) and $-[2^{ABS\Delta CT}]$ (when the ΔC_t is negative). Figure 2, panels a and b, shows the expression profiles of a novel gene in normal and diseased human tissues. Upon the initial interpretation of the normal expression profile for this gene of interest, one may conclude that this gene may be involved in diseases associated with the gastrointestinal tract and those that entail immune cell infiltration. The normal data and the preliminary conclusions drawn from it are further supported by the profile of this gene of interest in human diseased tissues. In this panel of tissues, the highest expression is seen in both the colon and lung matched normal/tumor pairs and the fibroblast growth factor (FGF) endothelial cells (proliferation indicator), the dendritic cells, and those cells infected with virus. One potential implication of these results is that this gene may be involved in proliferation and, therefore, may be associated with various cancers. The follow-up expression analysis that would be performed on this gene is to determine its expression profile on a larger number of colon- and lung-matched normal/tumor pairs. For this purpose, we generated disease-specific plates that contained 19 matched pairs of colon tumors and 19 matched pairs of lung tumors. Figure 3, panels a and b, shows the expression profiles of this novel gene in colon and lung tumors. These results show this gene to be overexpressed in 11 of 19 colon tumors and 11 of 19 lung tumors compared to adjacent normal tissue. Further analysis of the pathological and clinical data provided with the samples often offer important insight as to why a particular gene may be up-regulated in some tumors and not in others, rather than relying solely on the patient-to-patient variability reasoning. Thorough analysis of the clinical data associated with sample groups may segregate the data into particular tumor types or patient treatments. In the case of the lung tumor data, we have seen genes that are up-regulated only in non-small cell lung cancer. Additionally, we have seen both colon and lung tumor data that show the up-regulation of genes only in patients that have never been treated with chemotherapy, while those genes remain unchanged or down-regulated in patients that have been treated with particular chemotherapeutic agents. This type of result provides important information regarding certain pathways in which the gene of interest may be involved. It is only when we combine gene expression data with its clinical and pathological data that we may obtain the most powerful information and make the most accurate predictions regarding the tractability of a target and its future as an effective drug in today's market.

CONCLUSIONS

It is easy to see that real-time quantitative PCR is quickly becoming a mainstream technology with a tremendous number of applications. A quick keyword search for real-time quantitative PCR in any one of today's public databases will easily retrieve more than 1100 articles, which is a impressive number consider-

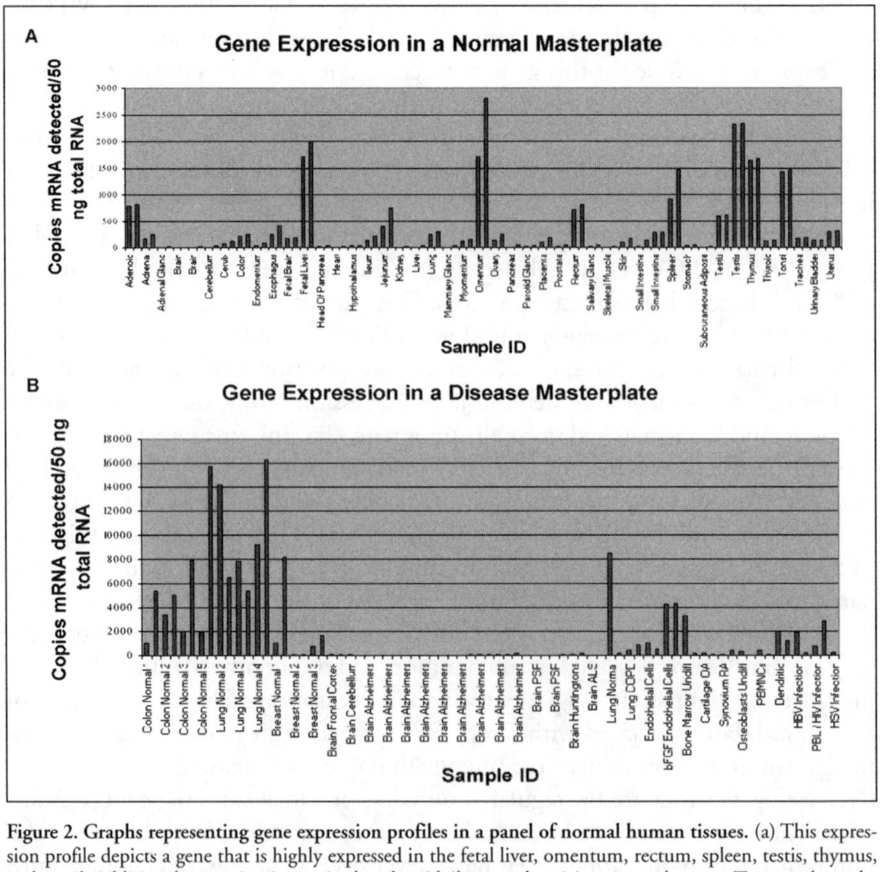

Figure 2. Graphs representing gene expression profiles in a panel of normal human tissues. (a) This expression profile depicts a gene that is highly expressed in the fetal liver, omentum, rectum, spleen, testis, thymus, and tonsil. Additional expression is seen in the adenoid, ileum, colon, jejunum, and uterus. Two trends can be identified in this expression profile, as evidenced by the expression in many of the tissues that represent the gastrointestinal tract, immune system, and the male and female reproductive organs. (b) This expression profile depicts a gene that is highly expressed in both the colon and lung normal/tumor samples, bone marrow cells, and virally infected cells.

ing that the primitive beginnings of this technology only date back to 1993 and more than half of the articles have been published within the past year. We have found real-time quantitative PCR to be robust and reproducible. It has allowed us to rapidly quantitate the expression of thousands of novel genes and predicted sequences in hundreds of samples. With the assistance of bioinformatics, we can quickly and successfully identify novel genes and then, after careful choices regarding the selection of appropriate tissues, we can employ real-time quantitative PCR to determine a tissue distribution profile. After careful analysis of the normal and diseased tissue expression, we can then make educated decisions regarding additional expression analysis with a more focused set of diseased samples. This way, the tissue expression information originally discerned becomes increasingly specific. This information can then be used to direct further research

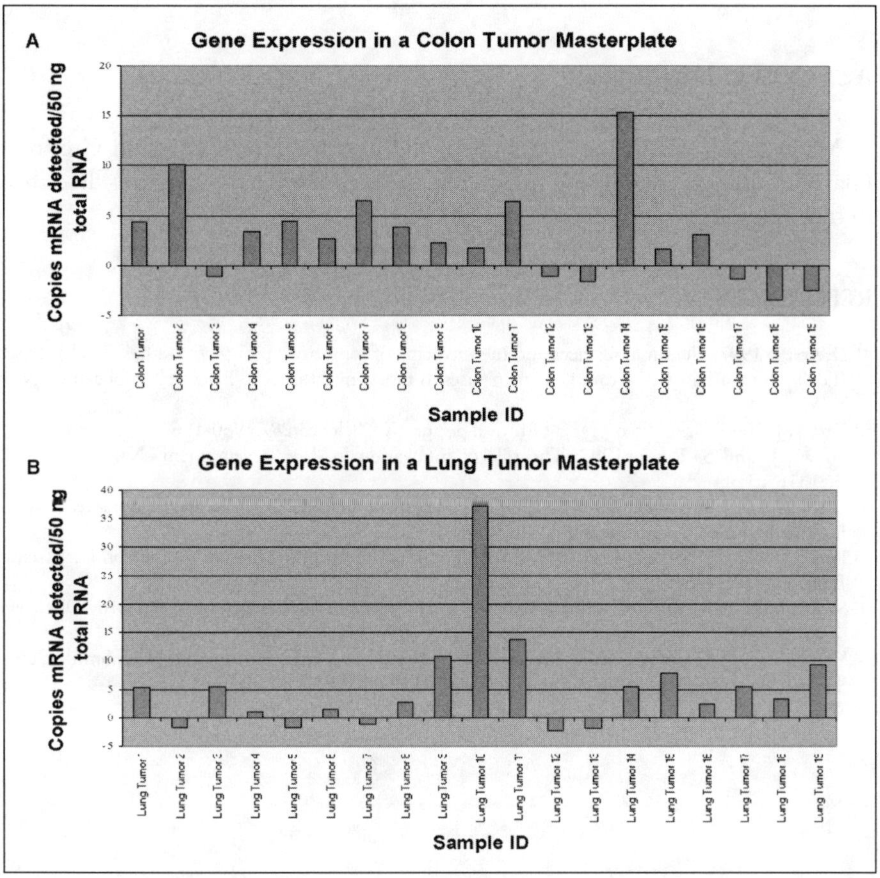

Figure 3. Graphs representing gene expression profiles in a panel of colon tumors relative to their matched normal margins. (a) This graph shows that the gene of interest is up-regulated greater than 2-fold in 11 of 19 colon tumors. (b) This graph shows that the gene of interest is up-regulated greater than 2-fold in 11 of 19 lung tumors.

to localize expression to a particular cell type and determine whether the protein expression correlates with the presence of mRNA.

Through an easily facilitated network of individuals including bioinformaticians, scientists, pathologists, and physicians, and the appropriate laboratory technology, it is possible to gain immensely detailed information regarding particular genes of interest in a tremendously high-throughput and cost-effective manner. This sophisticated approach has also allowed us to make quick and accurate decisions regarding potential disease implications, thereby saving additional time and money by decreasing the time a target spends in the target validation part of the drug discovery process. These new approaches to drug discovery are revolutionizing the drug discovery process (5), which will more than likely serve to deliver an explosion of better medicines focused on curing the disease rather than treating the symptoms.

ACKNOWLEDGMENTS

Many thanks to all the members of the Quantitative Expression Group at GlaxoSmithKline for helping to develop and standardize the methods described in this chapter.

REFERENCES

1. **Drews, J.** 1997. Genomic sciences and the medicine of tomorrow, p. 5-9. *In* J. Drews and J. Ryser (Eds.), Human Disease—From Genetic Causes to Biochemical Effects. Blackwell Publication, New York.
2. **Drews, J.** 2000. Drug discovery: a historical perspective. Science *287*:1960-1964.
3. **Drews, J. and S. Ryser.** 1997. The role of innovation in drug development. Nat. Biotechnol. *15*:1318-1319.
4. **Ginzinger, D.** 2002. Gene quantification using real-time quantitative PCR: an emerging technology hits the mainstream. Exp. Hematol. *30*:503-512.
5. **Reiss, T.** 2001. Drug discovery of the future: the implications of the human genome project. Trends Biotechnol. *19*:496-499.
6. **Sanseau, P.** 2001. Impact of human genome sequencing for in silico target discovery. Drug Discovery Today *6*:316-323.
7. **Venter, J.C., M.D. Adams, E.W. Myers, P.W. Li, R.J. Mural, G.G. Sutton, and H.O. Smith.** 2001. The sequence of the human genome. Science *291*:1304-1351.
8. **Ward, S.J.** 2001. Impact of genomics in drug discovery. BioTechniques *31*:626-634.

Section III

cDNA Arrays

7 Study of Mouse Metabolic Pathways with RIKEN Full-Length cDNA Microarrays

Hidemasa Bono and Yasushi Okazaki
Laboratory for Genome Exploration Research Group, RIKEN Genomic Sciences Center (GSC), RIKEN Yokohama Institute, Yokohama, Kanagawa, Japan

INTRODUCTION

Microarray technology enables scientists to monitor multiple whole transcriptomes simultaneously (20). The first definitive example of this approach was published by DeRisi et al. (9), who assessed the metabolic reprogramming that occurs during metabolic shifts in yeast. Another important feature of this particular publication was that all the data presented were accessible through the Internet as supplementary data, thereby facilitating subsequent, more advanced analyses. The main features of transcriptome data are that they are dynamic and time-dependent, whereas genome sequence data are static and robust. This implies that the notion of comparison is a more basic feature of the transcriptome data. The assignment of an Enzyme Commission (EC) number to each gene is the first step toward this type of analysis. Although the EC numbers of yeast genes have been assigned well after the completion of the yeast (*Saccharomyces cerevisiae*) genome sequence (12), mouse EC number assignment has been slower. The RIKEN Genomic Sciences Center, Genome Exploration Research Group (RIKEN Yokohama Institute, Yokohama, Kanagawa, Japan) has made the largest full-length mouse cDNA collections (RIKEN full-length cDNAs) to date (22). The RIKEN cDNA set includes many novel genes that need EC number assignment. The assignment of an EC number to each gene is laborious, requiring human curation, which was done at mouse annotation jamborees [e.g., the Functional Annotation of RIKEN Mouse cDNA Clones (FANTOM)1 and FANTOM2]. We show the strategy we used to assign EC numbers to the RIKEN full-length cDNAs.

Perspectives in Gene Expression
Edited by K. Appasani

In this review, we provide an overview of the microarray, introducing RIKEN full-length cDNAs as a resource for microarray analyses and additional functional analyses. This is followed by a more specific section on the expression profile of genes involved in metabolic pathways. We describe our experiments, current strategy, and tools for the analyses of metabolic pathways using gene expression profiles.

TWO TYPES OF DNA ARRAYS: OLIGO CHIPS AND cDNA MICROARRAYS

Oligo chips are microscopic arrays of large sets of short DNA sequences (oligomers) immobilized on solid substrates. They are valuable in areas of research that require the identification or quantification of many specific DNA sequences in complex nucleic acid samples. Oligo chips have a variety of applications, including genetic mapping studies, mutational analyses, and genome-wide monitoring of gene expression, and will likely become standard tools in research and clinical applications. Affymetrix (Santa Clara, CA, USA) and Agilent Technologies (Palo Alto, CA, USA) are the main providers of this type of oligo chip. These commercially available oligo chips are made by the systemic synthesis of the oligomers on the chip. The procedures used in experiments and data analyses are well established; the only hindrance is expense. An alternative procedure is to print synthesized oligomers on a chip using custom arrayers, which can be easily done in research laboratories. Because the cost for synthesizing thousands of oligomers is very high, there are companies who provide sets of oligomers at reasonable costs [e.g., Qiagen (Valencia, CA, USA) and Operon Technologies (Alameda, CA, USA)].

cDNA microarrays [developed by Eisen and Brown (11)] are being more widely used in the laboratory because customization is easy and the cost is minimal. cDNA clones are amplified and printed on glass, and these microarrays are

Figure 1. RIKEN full-length mouse cDNA 19K array. The clones on the array are 18 816 different clones.

used in analyses. We use the cDNA microarray system in our laboratory for two main reasons. (*i*) The cost of setting up cDNA microarrays is lower than that of collecting thousands of oligomers and printing them. (*ii*) We had been engaged in the collection of full-length mouse cDNAs (RIKEN Mouse Encyclopedia Project) and, thus, already had access to novel cDNA clones. Figure 1 shows a RIKEN cDNA microarray. The array design for printing is a 21 × 21 grid using 16 pins (TeleChem International, Sunnyvale, CA, USA); hence, each block contains 7056 (21 × 21 × 16) clones. We printed positive (housekeeping genes) and negative (plant cDNAs) controls in each 21 × 21 grid. The original 19K arrays (17) were composed of three meta-blocks, which means we had 21 × 21 × 16 × 3 = 21 168 clones. The spot diameter is approximately 100 μm. We are now using 48 pins; therefore, we can produce high-density arrays of 21 168 (21 × 21 × 48) clones on one slide. The arrayer we use was developed in-house (Figure 2) and has two heads, each carried on an arm that is moved independently by means of a linear guide motor that operates very fast and quietly using magnetic levitation.

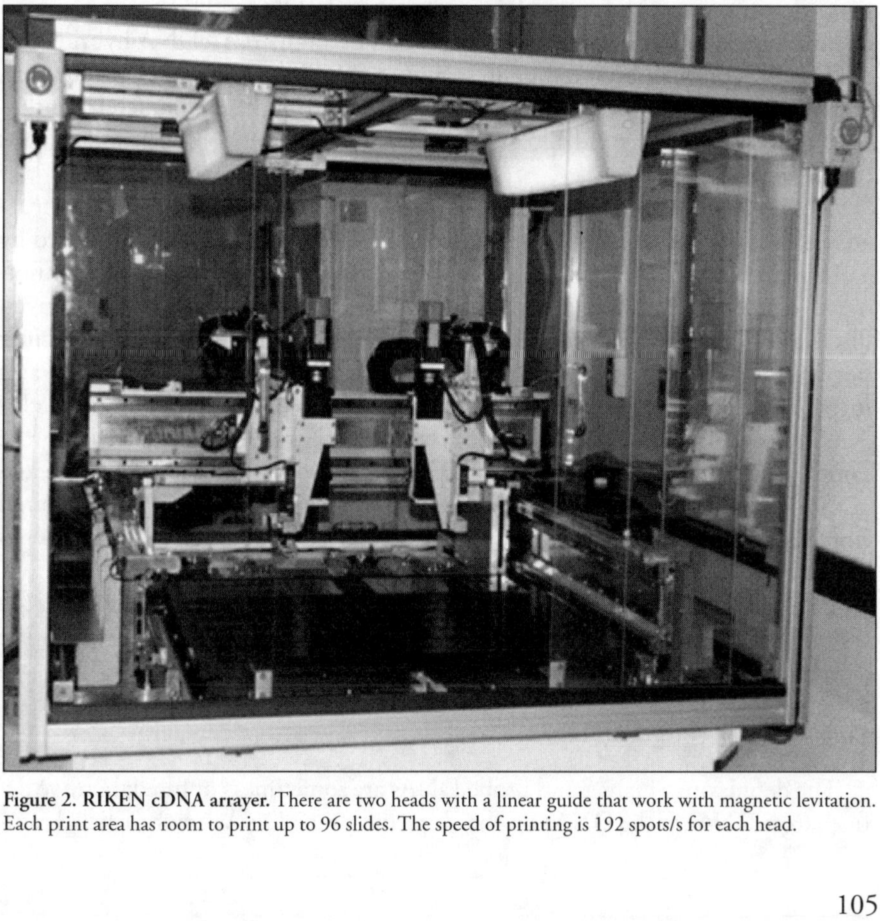

Figure 2. RIKEN cDNA arrayer. There are two heads with a linear guide that work with magnetic levitation. Each print area has room to print up to 96 slides. The speed of printing is 192 spots/s for each head.

METHODOLOGY

Procedure for Preparing cDNA Microarrays and Construction of RIKEN Full-Length cDNAs

Three major features are important in the construction of the RIKEN full-length cDNAs. The first is the use of the disaccharide trehalose to thermoactivate the reverse transcriptase. The second is the use of a CAP-trapper to efficiently capture full-length cDNAs. The third is the introduction of a normalization and subtraction method for constructing cDNA libraries. Since the details of these procedures are described elsewhere, we explain them only briefly here. Trehalose, which confers thermoresistance to yeast cells (8), stabilizes several enzymes, including reverse transcriptase. A mixture of trehalose and sorbitol is used because these compounds work cooperatively to stimulate reverse transcriptase. Commercially available buffers further stimulate this reaction. These conditions allowed us to prepare very long first-strand cDNA, which is essential for constructing full-length cDNA libraries.

The biotinylated CAP-trapper method is used to separate full-length cDNAs from the mixture with incompletely synthesized cDNAs. The cap is the biochemical feature that is specific for the 5′-end of an mRNA/cDNA hybrid. The addition of a biotin group to the cap structure of full-length cDNA with the concomitant physical removal of the biotinylated cap from the incompletely synthesized cDNA/mRNA hybrid allows for the selective recovery of full-length cDNA (Figure 3) (6).

To increase the efficiency of one-pass sequencing, abundantly and intermediately expressed cDNAs should be reduced in frequency. This is accomplished by a cDNA normalization technique based on reassociation hybridization technology. Normalization, or equalization, allows all the species of cDNAs in a given library to be expressed at similar frequencies. In large-scale cDNA sequencing, normalization is convenient not only for reducing the frequency of highly and intermediately expressed cDNAs but also for possibly eliminating unwanted cDNAs from the newly constructed library. This is achieved by normalizing the samples by using an excess of self RNA and by subtracting already-collected cDNA clones from the new library. Subtracting drivers are therefore prepared from known sets of cDNAs or from tissues that need to be specifically subtracted as follows. Known sets of cDNA clones are plated at a high density and amplified, and plasmid DNA is extracted in bulk. From the plasmid DNA, RNAs are transcribed in vitro with T3 or T7 RNA polymerase promoters present on the cloning vector or PCR primers and are used as subtracting drivers.

Preparation of Target DNAs

The definitions of target and probe DNAs are sometimes confused; here we define target DNAs as the DNAs on the chip, and probe DNAs are the labeled sample DNA. Our target DNAs were collected from RIKEN mouse cDNA libraries (22) that were constructed using the CAP-trapper method to enrich them for full-length inserts. The cDNAs were amplified using M13 forward and reverse primers in a 100-µL PCR mixture with a 0.2-µM each final concentration of forward (F1224; 5′-CGCCAGGGTTTTCCCAGTCACGA-3′) and reverse (R1233; 5′-AGCGGATAACAATTTCACACAGGA-3′) primers, 250 µM dNTPs, and 1.25 U Ex *Taq* in 1× Ex *Taq* buffer (TAKARA BIO, Otsu, Japan). The PCR product was precipitated with isopropanol and resuspended in 15 µL 3× SSC. The DNA solution was spotted onto poly-L-lysine-coated slides with a DNA arrayer (http://cmgm.stanford.edu/pbrown/mguide/index.html) with 16 tips (SMP3; TeleChem International). The diameter of the spots was approximately 100 µm. Mouse β-actin and G3PDH cDNAs were used as positive controls, and *Arabidopsis* cDNAs were used as negative controls (GenBank® accession nos. X98108, X13611, X90769, Z99707, AF004393, Z49777, Q03943, and U58284). We have released the expression profiles of the 49 adult and embryonic mouse tissues that we obtained with the RIKEN 19K array (http://pga.tigr.org/tigr-scripts/magic/r1.pl) (17). Tissues from which the cDNA libraries were made and the

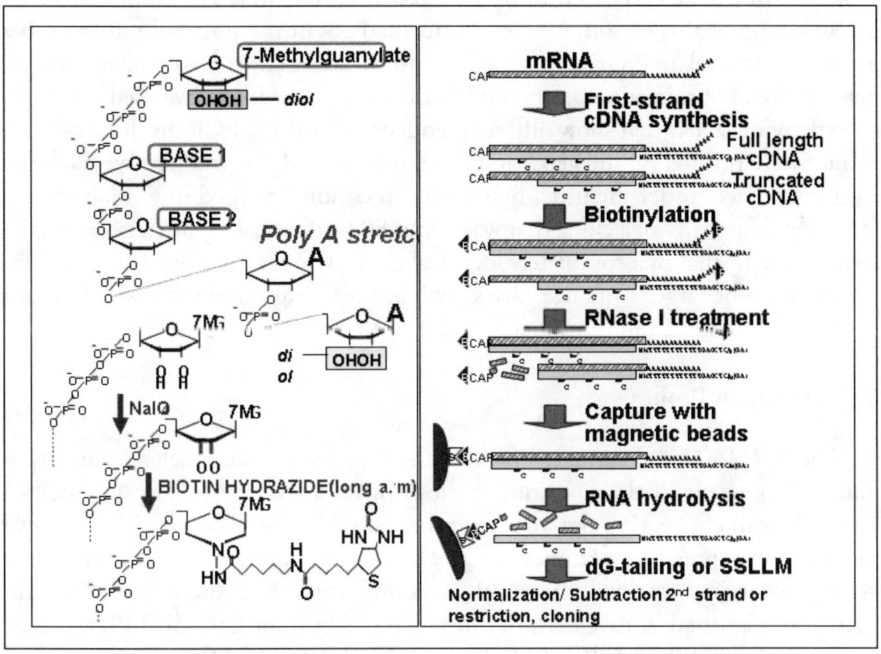

Figure 3. Overview of the CAP-trapping method. The detailed procedure is described in Carninci and Hayashizaki (6). Left panel: diol groups are present only on the extremities of the mRNA (CAP and 3′-end). Diol groups are biotinylated in a two-step reaction. Right panel: use of the CAP-trapping technology to select full-length cDNA. In the first step, diol groups are oxidized with a strong oxidizer, $NaIO_4$, producing a dialdehyde group that is covalently bound to the biotin. Only the biotinylated CAP structure is then trapped by the magnetic streptavidin beads. SSLM, single-strand linker ligation method.

number of clones used (shown in parentheses) from each cDNA library for this 19K array were: kidney (338); brain (35); spleen (17); heart (397); 18 day embryo (1771); lung (514); cerebellum (1200); placenta (437); testis (1591); pancreas (1816); small intestine (1178); stomach (1517); tongue (3110); embryonic stem cell (995); 13 day embryo liver (782); 10 day embryo (976); 11 day embryo (426); 10 day and 11 day embryos (1010); hippocampus (197); E12 embryo (168); 13 day embryo head (158); 14 day embryo head and 17 day embryo head (92); and 17 day embryo head (91). The data are available from the RIKEN Expression Array Database (READ) (http://READ.gsc.riken.go.jp/). The clone correspondence among other publicly or commercially available clones and the RIKEN 19K array was cited in the RESOURCERER (23).

Performance of RIKEN Microarrays

Three different RNAs transcribed in vitro from *Arabidopsis* full-length cDNAs were serially diluted and mixed with 1 μg mRNAs from mouse brain. The *Arabidopsis* mRNA signal was detectable when only 0.33 pg was added to 1 μg brain mRNA, corresponding to a sensitivity of 1 to 3 copies of mRNA per cell (Figure 4).

Many of the cDNAs used to prepare target DNAs were full length and relatively long. For this reason, we were concerned about the possibility that probes from multiple related transcripts might bind to single targets. To evaluate the extent of cross-hybridization under our hybridization conditions, we used four pairs of different cDNAs that show different sequence identities. PCR products of these clones were printed on the glass, and the transcribed RNA products were labeled and hybridized under the usual hybridization conditions used in our laboratory. The signal intensity of a clone that was about 80% identical to the target sequence was only 0.1× that of a completely identical clone. Clones that were less than 80% identical to the target sequence gave signals at the background level, which means that cross-hybridization is negligible under our conditions (Figure 5).

Preparation of Probes

The mRNA (1 μg) extracted from each of 49 tissues was labeled by incorporating fluorescent label Cy3 during random-primed reverse transcription. cDNA derived from entire 17.5 day mouse embryos, which we labeled with Cy5, was used as the expression reference for all tissues. The labeling was carried out at 42°C for 1 h in a total volume of 30 μL containing 400 U SuperScript® II (Invitrogen, Carlsbad, CA, USA); 0.1 mM Cy3-dUTP (or Cy5-dUTP); 0.5 mM each dATP, dCTP, and dGTP; 0.2 mM dTTP, 10 mM DTT, 6 μL 5× first-strand buffer; and 6 μg random primers. To remove unincorporated nucleotides, labeled cDNA was mixed with 500 μL binding buffer (5 M guanidine-SCN, 10 mM Tris, pH 7.0, 0.1 mM EDTA, 0.03% gelatin, and 2 ng/μL tRNA) and 50 μL silica matrix buffer (10% matrix, 3.5 M guanidine chloride, 20% glycerol,

0.1 mM EDTA, and 200 mM sodium acetate, pH 4.8–5.0, transferred to a GFX column (Amersham Biosciences, Piscataway, NJ, USA), and centrifuged at 15 000 rpm for 30 s. The flow-through was discarded, and the column was washed with 500 μL wash buffer. The adsorbed probe was eluted into a final volume of 17 μL distilled water. This labeled probe was mixed with blocking solution containing 3 μL 10 μg/μL oligo-dA, 3 μL 20 μg/μL yeast tRNA, 1 μL 20 μg/μL mouse *Cot1* DNA, 5.1 μL 20× SSC, and 0.9 μL 10% SDS.

Array Hybridization and Data Analysis

The RIKEN full-length mouse cDNA that constituted the target was hybridized in a final volume of 30 μL (the entire array consists of three meta-blocks, and each meta-block requires 10 μL hybridization solution). Before hybridization, probe aliquots were heated at 95°C for 1 min and cooled to room temperature. The slides were covered with coverslips and hybridized overnight at 65°C in a hybricasette (obtained from www.ArrayIt.com). After hybridization, the slides were washed in 2× SSC, 0.1% SDS until the coverslips dropped off. The slides were then transferred into 1× SSC, shaken gently for 2 min, and rinsed with 0.1× SSC for 2 min. After being washed, the slides were spun at 800 rpm in a Sorvall® centrifuge (RC-3B plus; H6000A/HBB6 rotor; NEN® Life Science Products,

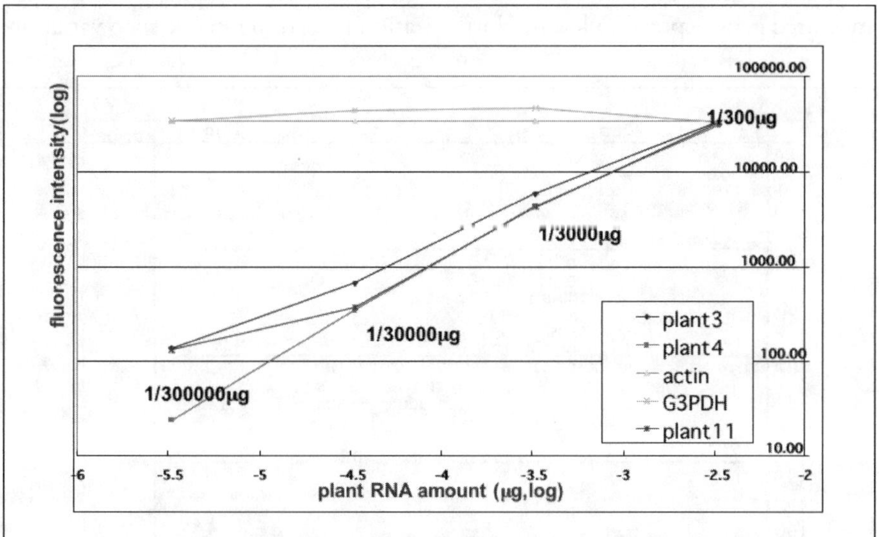

Figure 4. Experiment to determine the minimal detectivity of microarray analyses. RNA was transcribed in vitro from *Arabidopsis* plant cDNA by using T7 RNA polymerase. The RNA was serially diluted and added to 1 μg mouse brain mRNA. This mixture was labeled with fluorescent dye (Cy3 and Cy5) and hybridized to the slides. The y-axis shows the log-transformed fluorescence intensity, and the x-axis shows the log-transformed amount of plant RNA added. Because the background signal was only 50, a significant signal was obtained even when only 0.33 pg plant RNA was added.

Boston, MA, USA). These slides were scanned on a ScanArray® 5000 confocal laser scanner (PerkinElmer Life Sciences, Gaithersburg, MD, USA), and the images were analyzed by using ImaGene™ (BioDiscovery, Marina Del Rey, CA, USA). Representative images of cDNA microarrays are shown in Figure 6. Whole-body DNA from 17.5-day-old mouse embryos (E17.5) was used as a reference. Tissue-specific genes are represented as red spots, and E17.5-specific genes are represented as green spots. In the brain, more genes seem to be expressed than in other tissues, which reflects the high complexity of brain tissue.

When cDNA microarray data are released to the public and compared among laboratories, the use of different reference DNAs can hinder comparisons. Therefore, we strongly recommend the use of whole-body E17.5 DNA as a reference sample, for several reasons. (*i*) The 17.5-day-old embryo is just before full term and thus is recovered by Caesarian section, which means that the extracted RNA is not affected by infection or feeding. (*ii*) The proportions of various tissues are nearly the same among individual whole E17.5 embryos.

METHODS OF ANALYZING cDNA MICROARRAY DATA

Normalization

Many sources of systematic variation in microarray experiments affect the measured gene expression levels. Normalization is used to remove such variation.

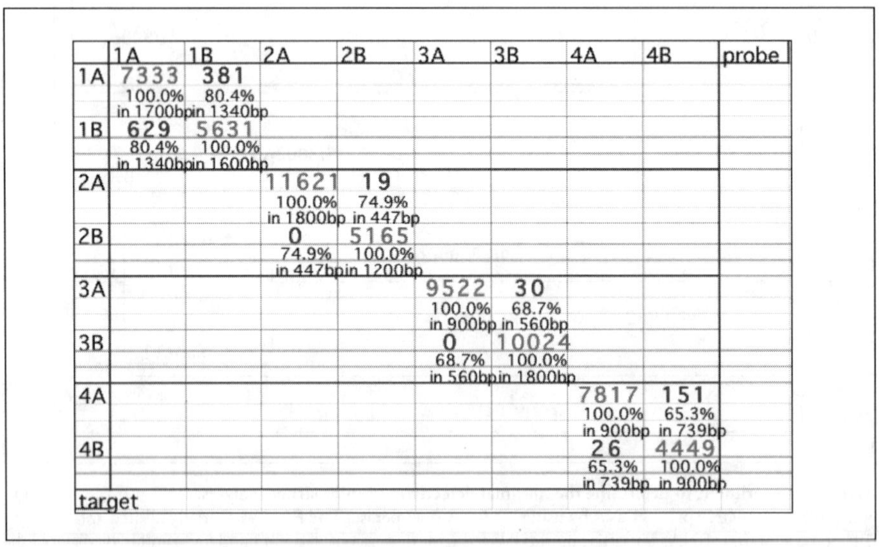

Figure 5. Cross-hybridization experiment. Clones 1A and 1B have 80.4% sequence similarity; 2A and 2B have 74.9%; 3A and 3B have 68.7%; and 4A and 4B have 65.3%. The length of each clone is shown in each box. The signal intensity (red) shows the intensity of self-hybridization. (**See color plate A2.**)

We will review practical efforts to enhance the reproducibility of microarray experiments in the Normalization and Filtering Procedure in RIKEN section.

Microarray Gene Expression Data Group Recommendation

The Microarray Gene Expression Data (MGED) Group is a grass-roots movement whose goals are to facilitate the adoption of standards for annotation and data representation in DNA array experiments and to reach a consensus concerning standard experimental controls and data-normalization methods (5). The MGED Normalization Working Group's meeting records and other information are available at http://www.dnachip.org/mged/normalization.html.

Normalization and Filtering Procedure in RIKEN

For the automated data processing of high-throughput cDNA microarray data, we devised an efficient method called Preprocessing Implementation for Microarray (PRIM) (14).

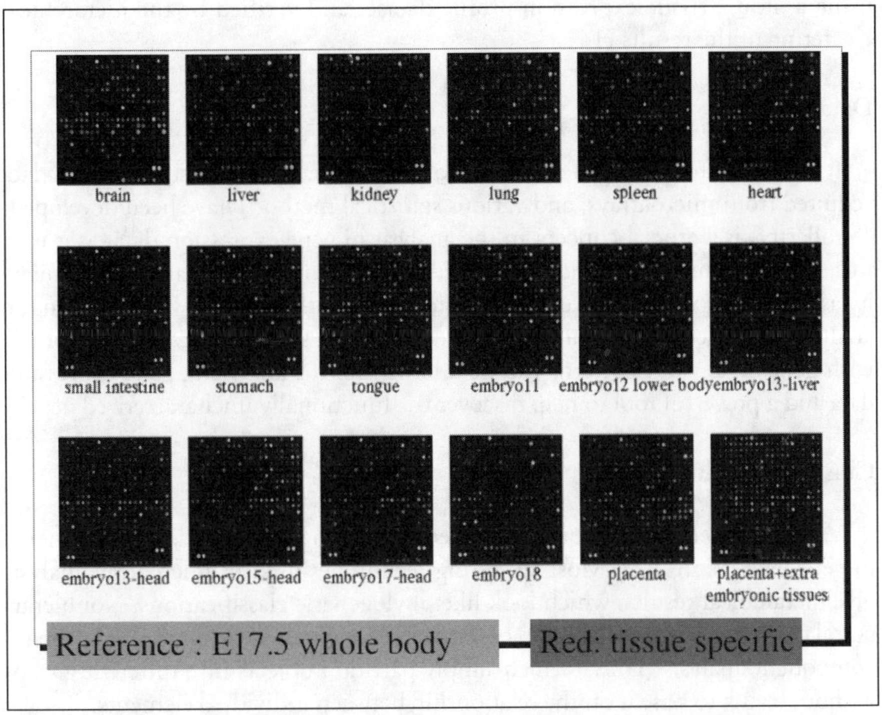

Figure 6. False-color overlaid images of various tissues using RIKEN 19K mouse cDNA microarrays. The reference RNA for each hybridization was from a 17.5-day-old embryo. Only a part of the microarray (21 × 21 = 441 spots) is shown. A red spot shows that the genes are highly expressed in each tissue, and a green spot shows that the gene was expressed less in that tissue. **(See color plate A3.)**

There is some controversy with regards to the reproducibility of cDNA microarray results, which can be affected, for example, by the quality of PCR amplification, spot typing, probe labeling, slide processing, hybridization conditions, and image analysis. Several ways to minimize the variance of the data have been reported [LOWESS Normalization (25)], and the best way of obtaining reproducible data is to perform the experiment at least twice under identical conditions. The PRIM method automatically produces a reproducible dataset from a series of duplicate experiments by filtering the log-transformed ratio data in three steps: (*i*) the removal of spots that were flagged during visual inspections of the microarray images; (*ii*) the removal of spots that have a signal intensity below a threshold value; and (*iii*) the removal of spots that are beyond a threshold distance from the least-mean-square line through a plot of the data obtained in one duplicate experiment against the data obtained in the other duplicate experiment. The product of *N* and *R* is used to set the thresholds in the second and third steps, where *N* is the ratio of the number of spots that passed the filtering program to the initial number of spots (just after the removal of the flagged spots), and *R* is the correlation coefficient of the expression ratio in the first and second experiments (Figure 7). The feasibility of this filtering method was tested using a mouse tissue expression profile dataset and verified by the hierarchical clustering of the results (14).

Data Mining

It is difficult to give a general answer of how to analyze the vast amount of data acquired from microarrays, and various statistical methods have been developed. "Similarity" is a crucial concept in the analysis of gene expression data, as it is in the analysis of nucleotide and amino acid sequences (10). By treating the similarity as a distance, we can make use of clustering algorithms, as has been done in the analyses of nucleotide and amino acid sequences. A similarity search tool for use with gene expression patterns can be a basic tool for analyzing gene expression data and a powerful tool to help discover the functionally uncharacterized genes.

Clustering Techniques

Various clustering techniques have been applied to the identification of patterns in gene expression data. Most clustering techniques used in microarray analysis give hierarchical results, which look like phylogenetic classifications. Nonhierarchical clustering methods can also be used, such as *k*-means clustering and primary component analysis (PCA), which simply partition objects into different groups without trying to base it on the relationship between individual elements.

Clustering techniques are useful because we can group genes of similar expression patterns and easily find patterns in large amounts of gene expression data. However, one must exercise great care in interpreting the output. Cluster and TreeView (both written by Michael Eisen, the Lawrence Berkeley National Lab and The University of California at Berkeley) are an integrated pair of programs for analyzing and visualizing the results of complex microarray experiments. Both programs can be downloaded free of charge by academic and other nonprofit researchers from http://rana.lbl.gov/EisenSoftware.htm.

Cluster performs a variety of types of cluster analysis and other types of processing on large microarray data sets. Although it runs only on Microsoft® Windows® platforms, it includes hierarchical clustering and nonhierarchical clustering [self-organizing maps (SOMs), *k*-means clustering, and PCA]. It implements two-way clustering, which is clustering in both gene and array (experiment) directions at the same time. It produces files for tree information that have gtr (gene tree) and atr (array tree) filename extensions. Other information, such as names of genes, their functional annotations, and numerical values for monitored gene expression, is contained in a file whose extension is cdt (complete data table).

The XCluster program (developed by Gavin Sherlock) is an alternative to Cluster and also implements the above clustering techniques. It runs on several platforms and is also free for academic and nonprofit use. A detailed description of XCluster, including instructions for obtaining and using the program is available at http://genome-www.stanford.edu/~sherlock/cluster.html. Especially for large datasets such as the RIKEN mouse 19K microarray, it is generally more practical to run XCluster on a UNIX platform because clustering programs need large amounts of computer memory. The classification output from XCluster and Cluster can be visualized with the TreeView program, which allows users to

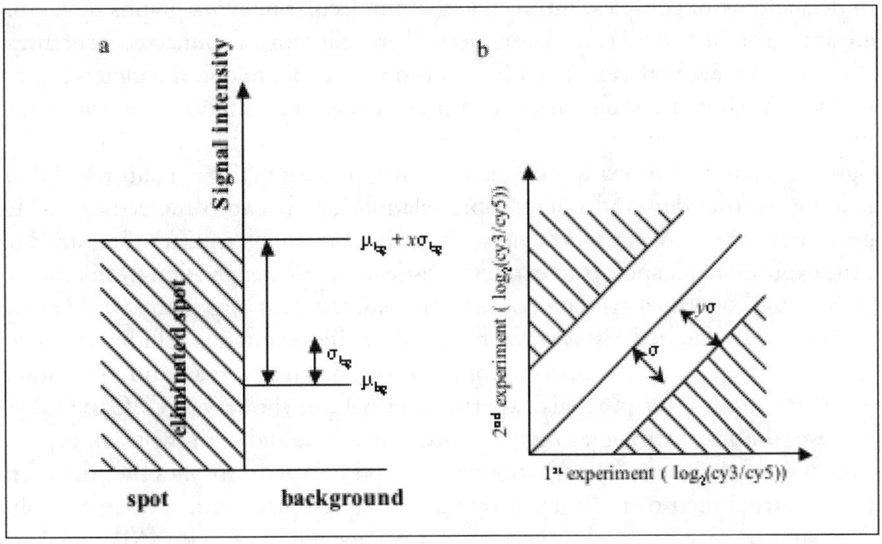

Figure 7. Schema of data processing in PRIM: (a) second filtration and (b) third filtration (see Filtering procedure). Spots in areas shaded with oblique lines are eliminated from subsequent analysis. The first filtration is not shown because it is just the removal of the flagged data.

graphically view the results of clustering and other analyses. It supports the tree-based and image-based browsing of hierarchical trees on-screen and produces output in multiple formats to create images for publication.

Other cdt file-compatible gene expression viewers have been developed by other groups. One is the Multiple Experiment Viewer (from the Institute for Genomic Research [TIGR]) (19), which implements many types of clustering algorithms, including hierarchical clustering (e.g., single-linkage, complete-linkage, and average-linkage clustering) and nonhierarchical clustering, and it allows the comparison of multiple clustering results.

Integration with Genetic Network Data: GSCope

Some projects for integrating clustering results with genetic network data are underway. Although the usefulness of integrating microarray data with the Kyoto Encyclopedia of Genes and Genomes (KEGG) pathways has been pointed out, there are currently no academic projects for integrating metabolic pathway information obtained from microarray data with cellular signaling cascade data. Recently, the Gene MicroArray Pathway Profiler (GENMAPP) Group (http://www.genmapp.org/) developed a tool for viewing and analyzing microarray (gene expression) data on maps representing biological pathways (7). Several projects for this type of integration are ongoing, and in this review, we introduce an example that is underway at RIKEN, called GSCope (http://gscope.gsc.riken.go.jp/) (Figure 8).

GSCope is an integrated viewer that consists of software tools that facilitate the understanding of complex, multi-linked biomolecular network graphs by taking advantage of the fisheye conversion method and clipping of connected structures. Biomolecular networks extracted from various sorts of biological data tend to be highly complicated and difficult to comprehend because connections among molecules are far from simple and cross many other connection lines. GSCope is designed to analyze gene expression data and map it onto biological pathways, thus helping us to understand microscopic relationships between molecules and to grasp an overview of the whole network graph. GSCope can filter thousands of gene expression data points to extract statistically similar expression profiles.

Keeping experimental data confidential is important, especially while using central servers for analysis. Although several public resources, including EcoCyc (16) and KEGG (15), perform computations on database servers and, thus, force users to submit their precious experimental data to the server, GSCope safely achieves this task by integrating the pathway database and with the users' expression data on their own (client) computers (Figure 9). This implies that the users can construct their own diagrams by modifying template pathways on a public Web site. Because the pathways used in GSCope are written in XML, they are not hard to interpret. Another tool, Knowledge Editor, is available for making network graphs and is intended mainly for wet-bench pathway specialists. A key-

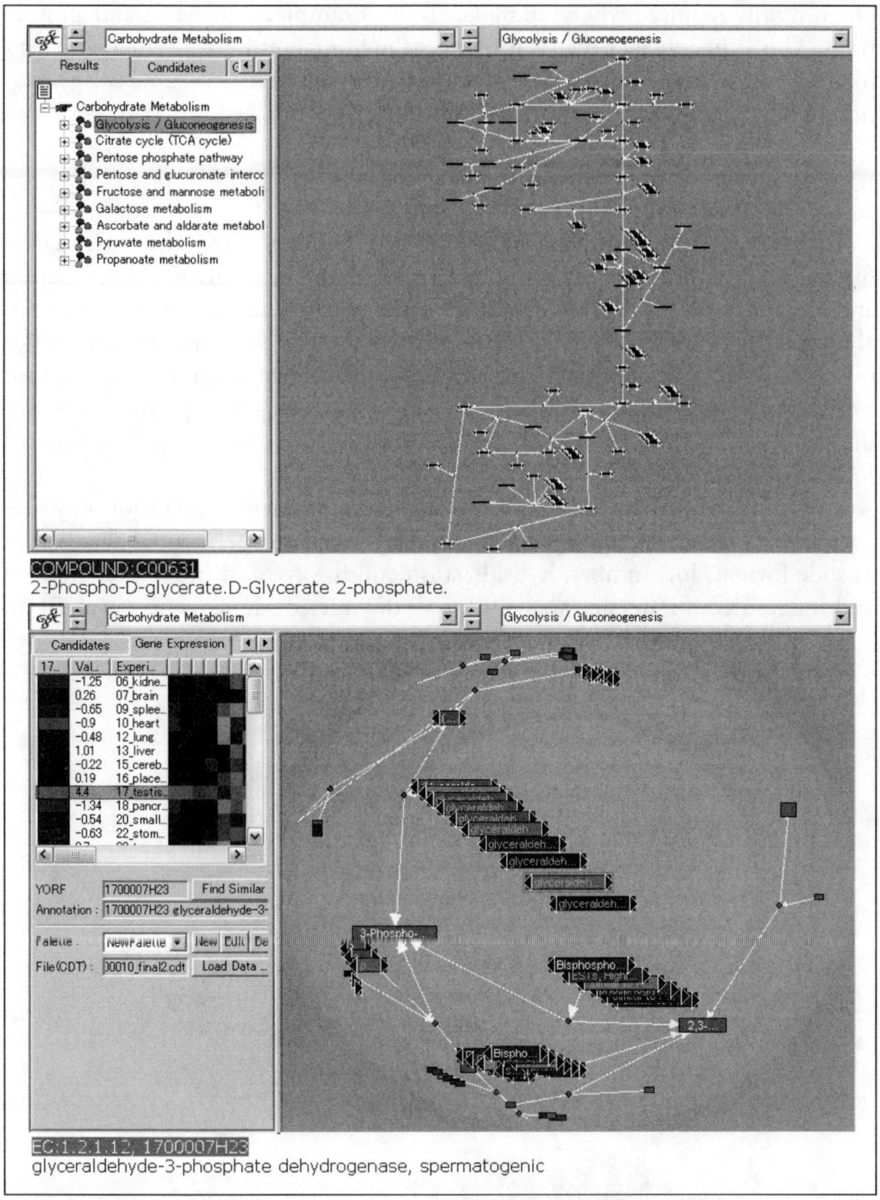

Figure 8. Screen shots of GSCope interface. Upper panel is a bird's-eye view of GSCope. In this view, pathways are listed in the left column, and the metabolic cascade from glucose (glycolysis and TCA cycle) is shown on the right. Lower panel is a wide-angle view of GSCope. A part of the upper metabolic pathway is magnified, and boxes representing enzymes are colored. In the right window, each node represents an enzyme, and a testis-specific G3PDH is colored and surrounded with an orange frame. The corresponding expression profile in each tissue is shown in the left window. The second column shows the base-2 log ratio relative to the reference (E17.5 whole-embryo RNA). The first column shows the color code based on this log ratio. The third column shows the tissue used for the experiment. The other genes are represented in the fourth and subsequent columns. **(See color plate A4.)**

board is only required when typing labels, for example, gene name and annotations. This is beneficial because it reduces the human error associated with typing. We are convinced that interactive use of this tool will accelerate the generation of new models for the regulation of gene expression.

READ

READ is a database for the expression profile of the RIKEN mouse cDNA microarray (4). It contains the microarray experimental data and information and provides a World Wide Web interface that researchers can use to retrieve, analyze, and display their data. The goals of READ are to serve as a storage site for microarray data from ongoing research in the RIKEN Mouse Encyclopedia Project and to provide useful links and tools for deciphering biologically important information. The information on genes is primarily based on the fully annotated FANTOM database.

Currently, READ includes information from microarray data (not only log-transformed ratios of gene expression intensity but also experimental details such as slide format, lot number, hybridization conditions, and probe preparation) and the results of sequence analyses for all the nucleotide sequences arrayed on

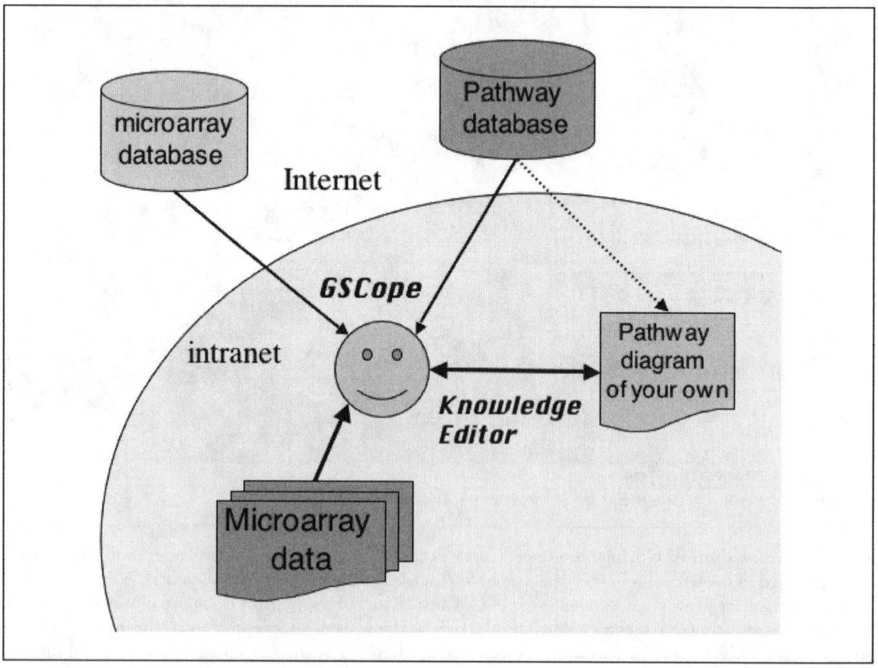

Figure 9. Schema of database integration using GSCope software. Microarray data and pathway data can be retrieved from the Internet or an institutional intranet. The pathway data can be modified and even made from scratch using the Knowledge Editor application.

the RIKEN microarray chips. READ can be queried by the RIKEN cDNA clone identifier, cloneid, and by keywords used in the database descriptions (Figure 10). It also can be searched by the log-transformed ratio of the gene expression intensity in all 49 embryonic and adult mouse tissues in the collection. BLAST is available and can also be used for searching a set of sequences on the chip, called chipBLAST. The chipBLAST query enables users to search against all available sequences of the clones on the RIKEN chip, with their sequence of interest as the query term. Since we have three sets of slides, each containing approximately 20 000 clones (RIKEN 60K arrays), this search tool is very useful to determine if the clone of interest is included on any of the chips.

Figure 11 shows the results obtained with a search tool called the READ Integrates Gene Expression Neighbor (RINGENE). The special feature of this search tool is that similarities in the gene expression patterns of specified tissues can be calculated dynamically. Values of Pearson's correlation coefficient, for correlations between the data of a specified clone identifier and those of all other clones on the chip, are dynamically calculated to identify pairs showing high correlation. Because the correlation coefficient is used to measure gene expression similarity, genes that show an inverse correlation can also be queried in a RINGENE search. Inversely correlated genes may be gene sets that are regulated to control metabolic reactions in opposite directions. Figure 11 also shows the expression profiling of a selected set of genes. A similarity search for gene expression patterns in specified tissues enables us to predict the function of genes, which is difficult to infer from just their sequence similarity.

METABOLIC PATHWAY ANALYSIS WITH cDNA MICROARRAY DATA

The first analysis of metabolic pathways using DNA microarrays was done by Joseph DeRisi and colleagues (9), who investigated the temporal program of gene expression accompanying the metabolic shift from fermentation to respiration ("diauxic shift") in *S. cerevisiae*. The expression profiles observed for genes with known metabolic functions pointed toward features of the metabolic reprogramming that occur during the diauxic shift, and the expression patterns of many previously uncharacterized genes provided clues to their possible functions.

Eisen et al. (11) published the results of an analysis of gene expression profiles taken from 79 experiments with *S. cerevisiae*, including the diauxic-shift data of DeRisi et al. (9). Although the gene expression profiles in this dataset contained those of 2469 well-annotated ORFs, many bioinformatists were prompted to analyze this first large, publicly available gene expression dataset. Some of these researchers devised clustering algorithms specifically designed for microarray data, and others tested the feasibility of predicting gene function from gene expression data by adopting the same clustering algorithms used to infer the function of ORFs according to sequence similarity.

Such a large set of microarray data was helpful in calibrating the results of the cluster analysis of gene expression data from microarray experiments. Simple hierarchical clustering algorithms [i.e., single-linkage and complete-linkage clustering and unweighted pair group method with arithmetic mean (UPGMA)] were initially tested, and UPGMA seemed to be best for finding functional coupling within these clusters because this method yields gene clusters of the appropriate size. The largest cluster resulting from a UPGMA clustering analysis of the yeast data by Eisen et al. (10) comprised ribosomal protein, and the second-largest cluster corresponded to ORFs that encode enzymes of the glycolysis pathway (Figure 12). The assignment of EC numbers to the ORFs in yeast was taken

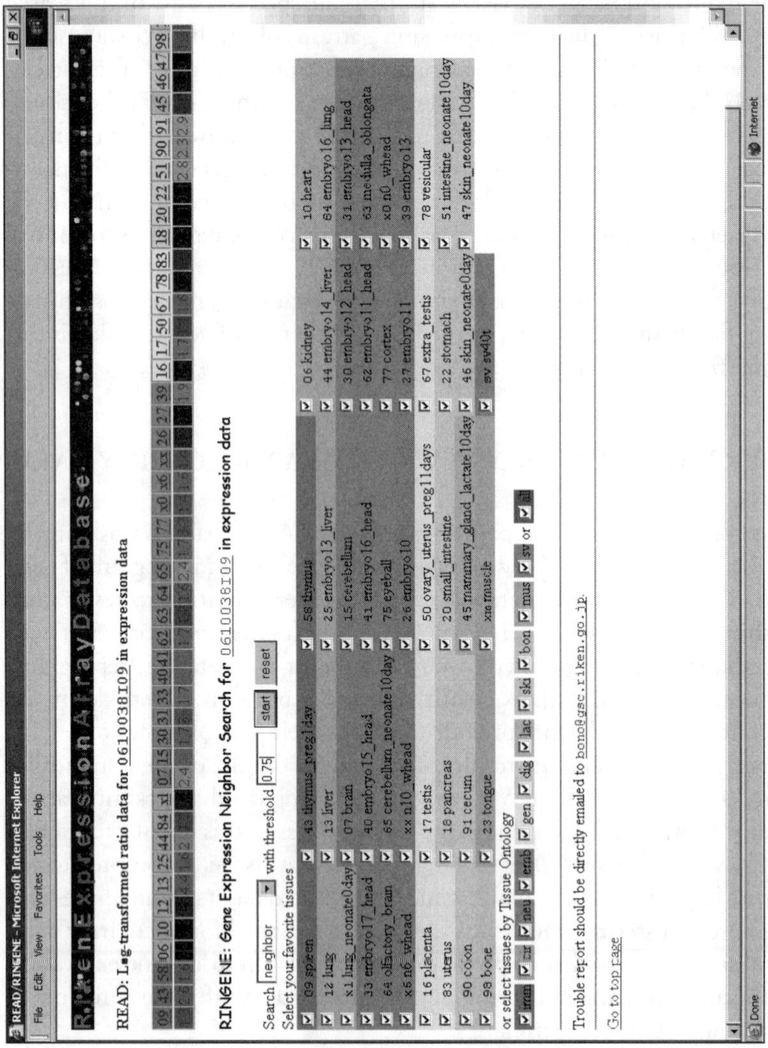

Figure 10. A screenshot of ID search results from READ. The gene expression profiles for 49 adult and embryonic tissues are listed at the top of the figure. Anatomically similar tissues are clustered and colored the same (e.g., digestive organs such as the pancreas, stomach, etc., are blue and on the right side of the table). The lower part of the figure is a Web form for searching RINGENE. Users can check boxes to indicate which tissues they want to search for a similar expression pattern. **(See color plate A5.)**

from the KEGG database (15). This database is a key resource for enzymes because it describes their functions and enables the mapping of genes coding for enzymes to their respective metabolic pathways. The ORFs in the glycolysis pathway that have corresponding EC numbers were summarized and color-coded with respect to the threshold for the correlation coefficient value. The gene expression patterns of the ORFs in the orthologous gene table from KEGG and those in the gene expression cluster were extracted from the entire dataset to display the functional features of this pathway (Figure 12). Significant pattern differences can be seen for the ORFs YOL056W and YDL021W, both of which encode phosphoglycerate mutase (EC 5.4.2.1). Their gene names are *GPM3* and

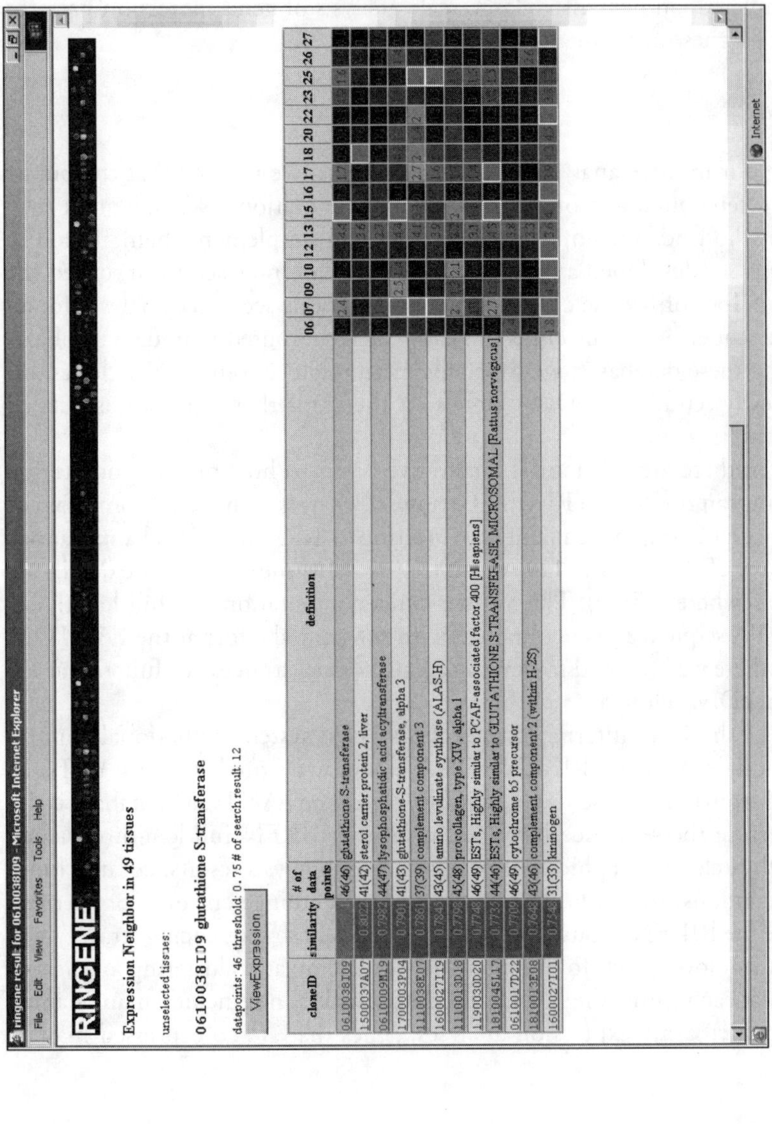

Figure 11. A screenshot of RINGENE search results. Users can enter an arbitrary number (0.75 in this example) as a threshold and select the tissues they would like to search for as described in Figure 9. A group of clones that shows a correlation coefficient higher than the chosen threshold can be extracted and displayed. The expression profile for each clone is presented by clicking its cloneid, and users can repeatedly search for the neighbor of that gene.

GPM2, respectively. These ORFs have been described as nonfunctional homologs of *GPM1* (YKL152C), even though they have conserved key amino acid residues that are considered to be involved in catalysis (13). We can easily perform similar analyses by comparing transcriptome data.

Although genes lacking functional descriptions were not reported in the analysis by Eisen et al. (11), the inclusion of functionally unassigned gene data in studies designed to find co-expression clusters will help researchers to identify or infer the functions of these unassigned genes. Because EC numbers have been assigned to all metabolic enzymes in yeast, ORFs can be mapped to appropriate positions in the metabolic pathway (2). However, for higher organisms, such as humans, mice, and rats, the assignment of EC numbers to cDNA clones has yet to be completed, and extensive large-scale analysis of gene expression patterns must wait for these assignments.

FANTOM

During our massive analyses of microarray data, we noticed that the public databases often contain incomplete functional annotations. We hope that each researcher who finds incomplete annotations will supplement them. To aid in this process, we developed a database system and an interface for it, called the Annotation Tool of Mouse cDNAs (ATOM). Various sequence analyses for all nucleotide sequences on the chips can easily be recomputed to update databases daily. Using these databases, we were able to annotate all our cDNA clones, including newly sequenced cDNA clones on the chips that were not associated with any public sequences.

However, there are substantial qualitative limits to how much annotation a single group can do. So the RIKEN Genomic Sciences Center, Genome Exploration Research Group held a jamboree meeting to assign functional annotations to our mouse cDNA clones. This was the FANTOM meeting, whose details are reported elsewhere (15,19). The human-curated annotations for full-length enriched cDNA sequences were obtained with substantial effort at the FANTOM meeting, where we intended to develop an authorized strategy for functional annotation of cDNA clones.

Figure 13 shows an interface for the FANTOM system. Functional annotations for fully sequenced RIKEN cDNA clones are stored in the FANTOM database (http://fantom.gsc.riken.go.jp/db/) (3). The FANTOM database is designed to store the sequence information of the RIKEN full-length enriched mouse cDNA clones, graphical views of sequence analysis results, curated functional annotations, and additional descriptions, including gene ontology terms.

Because the RIKEN Mouse Encyclopedia full-length sequencing strategy was biased toward novel genes found after the sequencing and clustering analysis of the 3′-end of each clone, the FANTOM database does not include many named genes. For microarray expression analyses, genes that are co-expressed may be

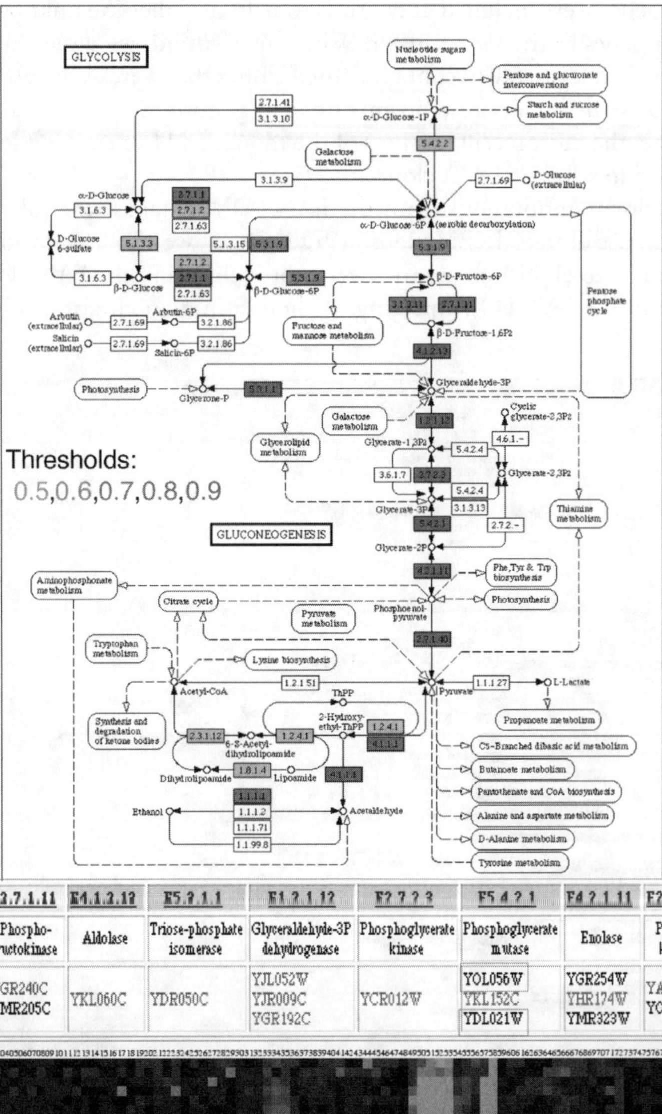

Organism	E2.7.1.11 Phospho-fructokinase	E4.1.2.13 Aldolase	E5.3.1.1 Triose-phosphate isomerase	E1.2.1.12 Glyceraldehyde-3P dehydrogenase	E2.7.2.3 Phosphoglycerate kinase	E5.4.2.1 Phosphoglycerate mutase	E4.2.1.11 Enolase	E2.7.1.40 Pyruvate kinase
sce [P | G | T]	YGR240C YMR205C	YKL060C	YDR050C	YJL052W YJR009C YGR192C	YCR012W	YOL056W YKL152C YDL021W	YGR254W YHR174W YMR323W	YAL038W YOR347C

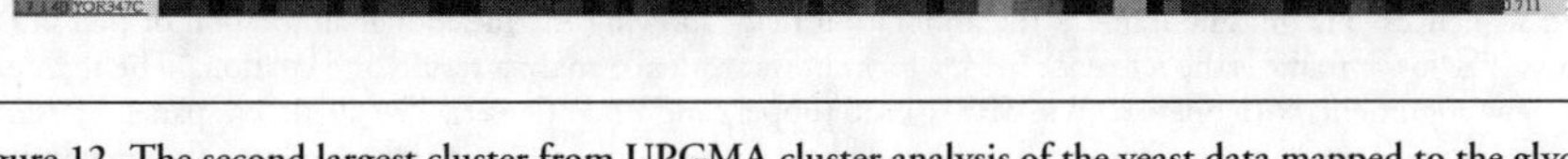

Figure 12. The second largest cluster from UPGMA cluster analysis of the yeast data mapped to the glycolysis pathway (top). At the top of the figure and in the middle table, the relatedness in the gene expression pattern is represented in color (red for closely related genes). The gene expression pattern at the bottom (red for up-regulated and green for down-regulated) was drawn using a tool in the EPTable software (http://eptable.sourceforge.net/). **(See color plate A6.)**

considered functionally related in some way. Therefore, it would be better if named genes were included as controls to help decipher the unknown functions of other genes in the same cluster. With this in mind, we designed the RIKEN mouse 19K microarray to contain named clones that were excluded from further sequence analyses. Because no mammalian genome sequence was available at that time, the most recent versions of the mouse and human UniGene databases were used to select a cDNA clone.

Functional annotations from the FANTOM database are valuable, but, to make full use of the RIKEN mouse 19K array set, we also have to add functional annotations to cDNA clones that were not included in the FANTOM clone set curated in the FANTOM meeting. Although human curation is important to

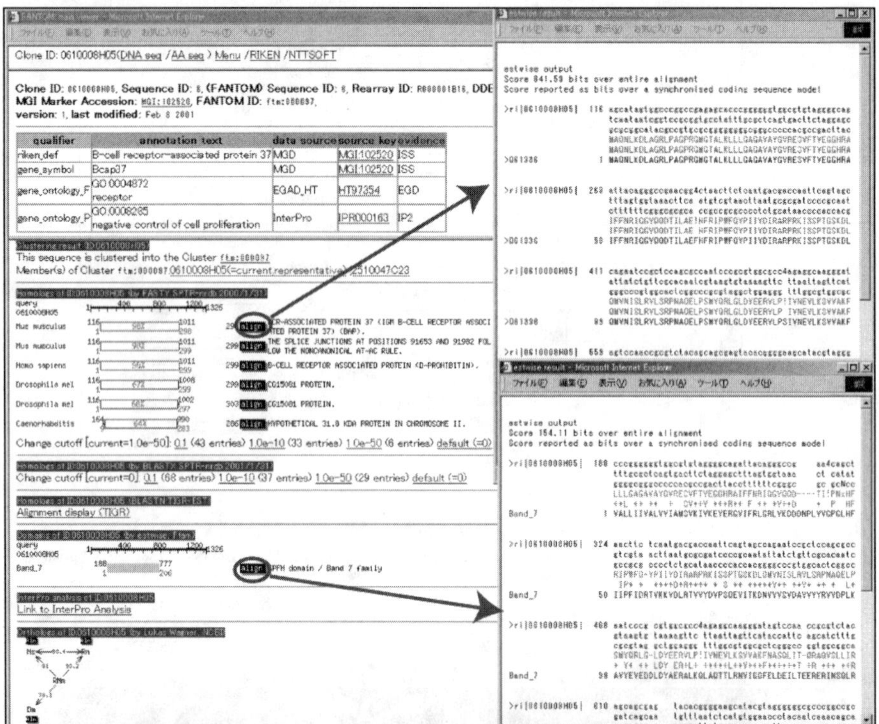

Figure 13. The viewer for the FANTOM database. The left panel shows a search result page for one cDNA clone. This panel contains three frames. The top frame is the summary field, with links to nucleotide and amino acid sequences. The middle frame is the annotation field, showing the functional annotation of that cDNA clone. The lower frame is the reference field, which shows sequence analysis results and citations. The right panels show alignments with one record in SPTR-nrdb (upper) and Pfam (lower). The alignment panels appear after the "align" buttons in the reference frame are clicked. Links in the reference and annotation frames take users to various biological databases including MGED, InterPro, LocusLink, GenPept, EGAT_HT, TIGR-Mouse Gene Index (TIGR-MGI), DDBJ/EMBL/Genbank, NCBI-nr, Protein Information Resources (PIR), UniGene, and the Untranslated Region Database (UTRdb). The reference field also has links to Mapping (http://www.gsc.riken.go.jp/e/FANTOM/map/) and Superfamily (http://stash.mrc-lmb.cam.ac.uk/SUPERFAMILY/).

functional annotation, it is laborious to update the database as often as the genome data are updated. Therefore, we developed a fully computational functional annotation pipeline called the Functional Inference Descriptor (FIND) and evaluated the results by comparison with the human-curated database. The FIND outputs were quite consistent with the human-curated results, and we concluded that the FIND results can feasibly be used as functional annotation information. Further improvement of the automated annotation pipeline was developed at the FANTOM2 meeting held at the RIKEN Genomic Sciences Center, Yokohama, in May, 2002 (17a).

Gene Ontology as a Computerized Description of Gene Function

Gene ontology (GO) is the result of a collaborative effort to produce a dynamic, controlled vocabulary that can be applied to all eukaryotes, even as our knowledge of gene and protein roles in cells accumulates and changes. To this end, three independent ontologies were defined by the GO Consortium, which cover biological processes, molecular functions, and cellular components (1). This terminology has become the de facto standard for the functional annotation of genes. During the first FANTOM meeting (FANTOM1), GO was actively used for annotating the mouse transcriptome (18).

The proper assignment of GO terms to functionally known genes represented on microarray slides can be helpful to infer the roles of functionally unknown genes when they are found in the same co-expression clusters as known genes. A simple way of making this association is to use as a reference the well-curated gene association tables maintained by the GO Consortium or a GO-supporting database for model organisms, which are listed on the GO Web page (http://www.geneontology.org/). Figure 14 shows how such gene associations for clones on the RIKEN mouse 19K cDNA microarray are made. Here, the FANTOM database was primarily used to assign GO terms. GO terms were also assigned directly, in light of information on protein sequences in SWISS-PROT and on protein motifs and domains in InterPro, both of which are GO-supporting databases.

Once GO terms are assigned to all the ORFs in the genome, we can employ functional analysis using the GO hierarchy. Another hurdle is the exploration of this hierarchy. By using GO as a common language for gene function, a comparison of genes from multiple species using GO terms becomes feasible. But how can we explore the hierarchical structure of the ontologies? No common methods exist, but a new type of GO viewer using XML technology can make it easier to analyze the GO hierarchy in conjunction with gene expression profiles. The GO viewer, named Gene Around, makes use of scalable vector graphics to visualize GO hierarchical ontologies and is available from http://db.aist-nara.ac.jp/generaound/ (21).

EC numbers can also be assigned from the GO-to-EC-number conversion table (GO2EC). The assignment allows us to study all the metabolic pathways (metabolome) by matching the EC numbers to pathways. Using this information, we can analyze metabolomes with the gene expression profiles obtained by cDNA microarray (Figure 15).

Mapping Gene Expression Profiles to Metabolic Pathways

As we have described, one can study all metabolic pathways by using EC numbers to match enzymes to pathways in the KEGG database. Parts of the mammalian metabolic pathways are the same as those in yeast and bacteria, including the glycolysis and tricarboxylic acid (TCA) cycle pathways. However, some pathways are specific to higher organisms. In yeast and bacteria, the synthetic pathways for 20 different amino acids have been reconstructed from predicted ORFs (2). Similar, but different, degradation pathways for these amino acids are found in mammalian cells. Because we cannot assume that data obtained from the yeast genome are completely applicable to mammals, it is important to analyze metabolic pathways with gene expression data from mammals.

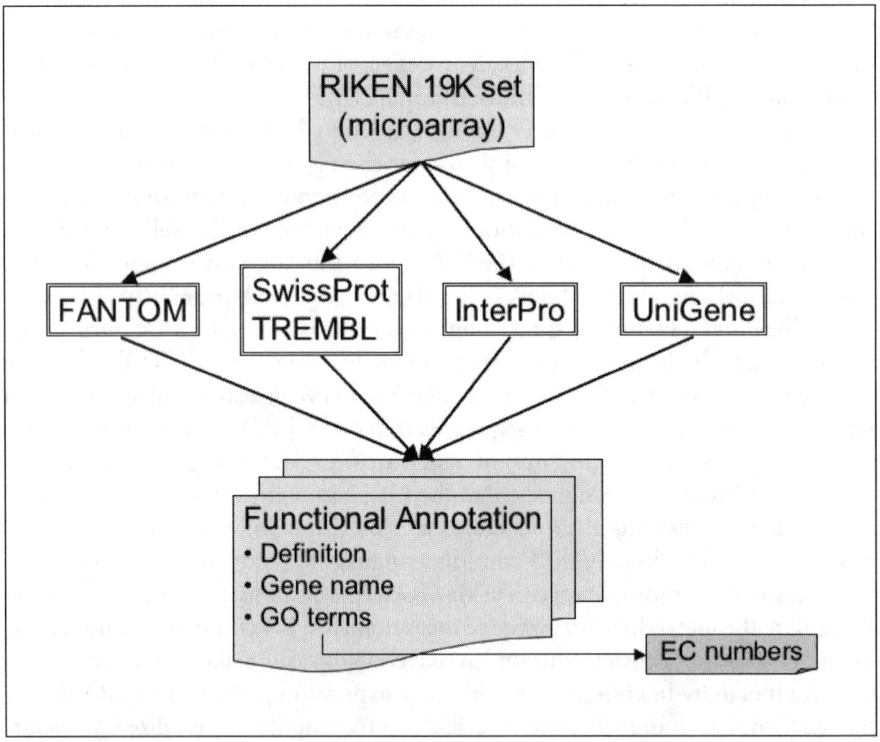

Figure 14. A schematic diagram of the functional annotation process of the RIKEN 19K microarray clone set. Functional annotations are primarily assigned from the FANTOM database. For those clones that are not included in the FANTOM set, the SWISS-PROT/TREMBL and InterPro databases are used to assign functional annotations to clones. The correspondence table between the RIKEN cDNA clone and the UniGene Cluster database is also maintained.

Enzymes that belong to all classes of metabolic pathway are represented in the RIKEN 19K set. We classified these enzymes into 78 synthetic and degradative metabolic pathways and determined the relatedness of gene expression patterns during embryonic development and in adult tissues. The similarities in expression patterns among genes coding for enzymes in the same metabolic pathway were striking (Figure 16). Although expression patterns for individual enzymes have been examined in various tissues of the whole animal, leading to hypotheses regarding the whole-body regulation of metabolism, it has not previously been possible to examine simultaneously the coordinated regulation of most metabolic pathways. Using microarray expression analysis, we have provided strong evidence that metabolic pathways are coordinately regulated throughout the body of an organism during development and in the adult. Intriguingly, the genes were clustered into two major groups in all the metabolic pathways (Figure 16). One group was a ubiquitously expressed gene set, and the other was a tissue-specific expressed gene set. Genes expressed in a tissue-specific pattern were mainly clustered into muscle-specific and liver- or kidney-specific gene sets. For example, tissue-specific genes involved in amino acid metabolism were primarily found in the liver and kidney, whereas those involved in glycolysis were in muscle (Figure 16).

We examined the expression patterns of genes encoding enzymes in the glycolysis pathway in more detail. We mapped various genes in the glycolysis pathway and color-coded each group created by hierarchical clustering according to their expression level (Figure 17). Genes that were closely clustered are shown in the same color. Genes colored yellow have a correlation coefficient of 0.55, those in blue have a value of 0.44, and those in green have a value of 0.57. The genes shown in pink were categorized in the same group as those in yellow but with a correlation coefficient of 0.44 (Figure 17). Surprising differences were observed when the expression patterns of genes involved in glycolysis were examined at this level of detail. The genes in yellow and pink show the muscle-specific expression pattern, the genes shown in green are testis-specific, and those in blue are liver- and kidney-specific. As shown in the muscle-specific gene cluster, genes closely clustered in expression pattern (Figure 17) were more closely linked in the glycolysis pathway (yellow), and genes weakly linked for co-expression (pink) were more distant in the pathway. Interestingly, even though some enzymes may have the same EC number (and presumably arise from a single gene), genes for those enzymes that were clustered in different groups differ in their library of origin and their tissue-specific pattern of expression. [Figure 17; the first two digits in the clone label (20) correspond to the library from which they arose]. This finding provides support for the existence of isoforms that may be specifically expressed and functioning in the testis (24). These genes are hypothesized to play an important role in regulating the switch between various pathways of energy production during spermatogenesis and in the spermatozoon. Similarly, tissue-specific differences in gene products were found for glycolytic enzymes derived from muscle (Figure 17; yellow and pink) and from liver and kidney (Figure 17;

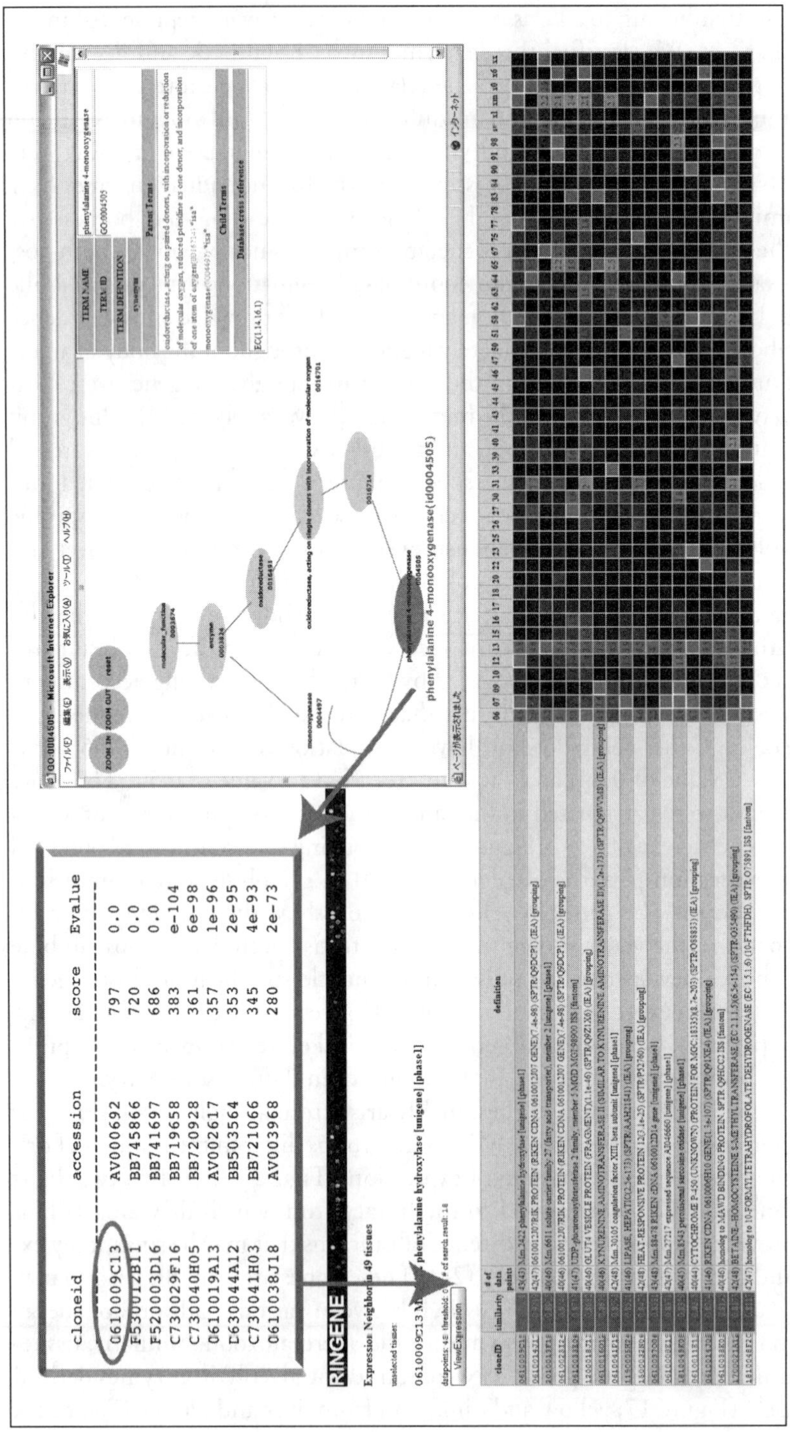

Figure 15. GO viewer (Gene Around, upper right panel) and the process of in silico functional analyses with GO. The Gene Around GO viewer can show a schematic view of complex ancestry relationships in which one term has multiple parents (upper right panel). When one GO term is selected, clones related by sequence similarity can easily be listed by dcing a BLAST search (upper left panel).

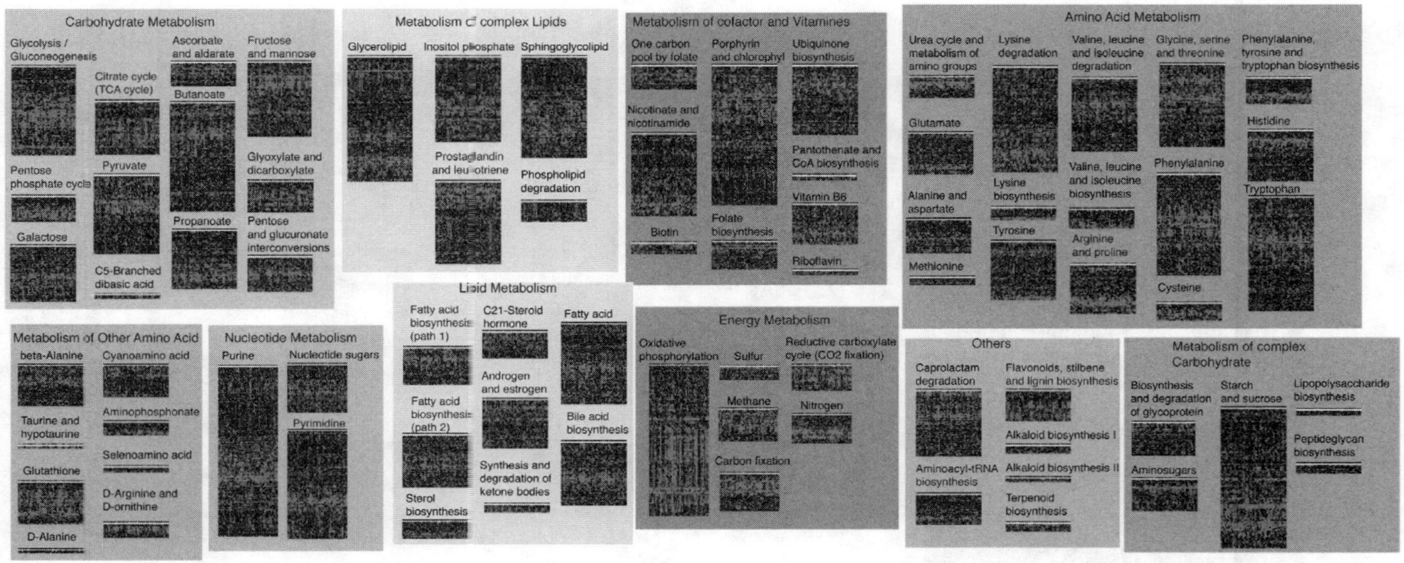

Figure 16. Thumbnail images of expression profiles (in hierarchical tree view) of the clustered genes for all metabolic pathways (metabolome). The y-axis of each thumbnail represents clones, and the x-axis of each figure shows 49 tissues. Groups of functionally similar metabolic pathways are shaded with the same colors (metabolism of carbohydrates, red; amino acids, blue; and lipids, yellow). All the genes included in each pathway are listed, together with links to gene expression information, at (http://genome.gsc.riken.go.jp/READ/supplementary/allpathway/). (**See color plate A7.**)

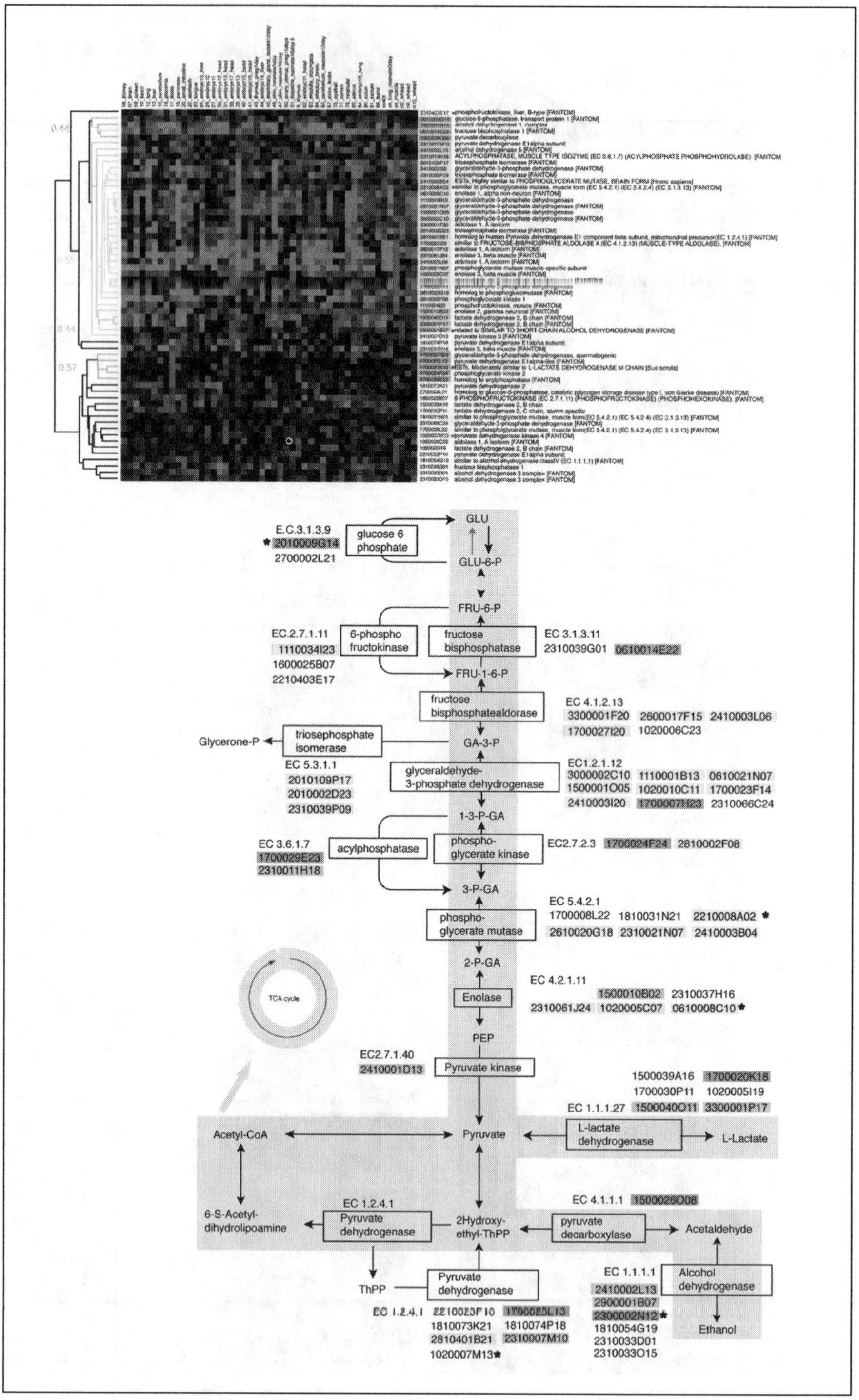

E.C.3.1.3.9
2010009G14
2700002L21
glucose 6 phosphate
GLU
GLU-6-P
FRU-6-P
EC.2.7.1.11
1110034I23
1600025B07
2210403E17
6-phospho fructokinase
fructose bisphosphatase
EC 3.1.3.11
2310039G01 0610014E22
FRU-1-6-P
fructose bisphosphatealdorase
EC 4.1.2.13
3300001F20 2600017F15 2410003L06
1700027I20 1020006C23
Glycerone-P
triosephosphate isomerase
GA-3-P
EC 5.3.1.1
2010109P17
2010002D23
2310039P09
glyceraldehyde-3-phosphate dehydrogenase
EC1.2.1.12
3000002C10 1110001B13 0610021N07
1500001O05 1020010C11 1700023F14
2410003I20 1700007H23 2310066C24
1-3-P-GA
EC 3.6.1.7
1700029E23
2310011H18
acylphosphatase
phospho-glycerate kinase
EC2.7.2.3 1700024F24 2810002F08
3-P-GA
EC 5.4.2.1
1700008L22 1810031N21 2210008A02
2610020G18 2310021N07 2410003B04
phospho-glycerate mutase
2-P-GA
TCA cycle
EC 4.2.1.11
1500010B02 2310037H16
2310061J24 1020005C07 0610008C10
Enolase
PEP
EC2.7.1.40
2410001D13
Pyruvate kinase
1500039A16 1700020K18
1700030P11 1020005I19
EC 1.1.1.27 1500040O11 3300001P17
Acetyl-CoA
Pyruvate
L-lactate dehydrogenase
L-Lactate
EC 4.1.1.1 1500026O08
6-S-Acetyl-dihydrolipoamine
EC 1.2.4.1
Pyruvate dehydrogenase
2Hydroxy-ethyl-ThPP
pyruvate decarboxylase
Acetaldehyde
ThPP
Pyruvate dehydrogenase
EC 1.1.1.1
Alcohol dehydrogenase
2410002L13
2900001B07
2300002N12
1810054G19
2310033D01
2310033O15
Ethanol
EC 1.2.4.1 2210023F10 1700023L13
1810073K21 1810074P18
2810401B21 2310007M10
1020007M13

blue). Our results strongly suggest that clustering efficiently distinguishes genes or perhaps isoforms of those genes that function in a tissue-specific manner.

CONCLUSIONS

Metabolome analysis of 49 mouse tissues using the RIKEN mouse 19K cDNA microarray yields interesting insights into understanding the coordinated regulation of gene expression in mammalian tissues. Although there may be some false-positive predictions of the functional assignment of the genes, tissue-specific and ubiquitously expressed genes can be seen in many pathways, and in the glycolysis pathway, muscle-specific and testis-specific clusters of genes can be observed.

By using the GO terms as the common language for assigning gene function, comparative analyses become more feasible in many ways. By integrating a variety of data, such as transcriptome and interactome, the functional assignment of a gene reflects not only the comparative analysis of nucleotide and/or peptide sequences but also gene and protein expression data. However, it might be more difficult to interpret these results from the biological point of view. There are, of course, some hurdles to overcome to achieve more sophisticated analyses of the entire gene network. These include, for example, increasing the sensitivity and reproducibility of the experiments; the use of a common reference in cDNA microarray experiments; and establishing a method for normalizing the data. These points are actively being discussed by the society in this field (5). Recently, the activities of proteomic analyses have been increasing, and the methodologies for the massive analyses of protein have been developing. A more precise description of the gene network could be achieved by obtaining multi-point time course data and comparing the transcriptome and proteome data. The bridging of the transcriptome and proteome data will be the next challenge to the post-sequence functional genomics era.

ACKNOWLEDGMENTS

We thank Drs. Yoshihide Hayashizaki and Piero Carninci for their technical comments on the preparation of the manuscript. We also thank Tetsuro Toyoda in the RIKEN Genomic Sciences Center for giving us their software in development before publication, RIKEN Genome Exploration Research group members for

---

Figure 17. Hierarchical tree view of the clustered genes involved in the glycolysis pathway (upper panel). The shaded colors in cloneids correspond to the color labels in the bottom pathway map and correspond to genes that have high correlations with each other. Genes colored yellow, blue, and green have correlation coefficient values of 0.55, 0.44, and 0.57, respectively. The yellow genes (high correlation coefficient) were mapped close to each other in the pathway. The pink genes clustered with a lower coefficient of correlation and mapped outside the core pathway are shown in yellow. The asterisk indicates that the functional annotations for these genes were inherited from the FANTOM database. **(See color plate A8.)**

cDNA clone preparation, and chip team members for production of the expression profiling data. This study was supported in part by a research grant to the RIKEN Genome Exploration Research Project from the Ministry of Education, Culture, Sports, Science and Technology of the Japanese Government and by ACT-JST (Research and Development for Applying Advanced Computational Science and Technology) of the Japan Science and Technology Corporation (JST).

REFERENCES

1. **Ashburner, M., C.A. Ball, J.A. Blake, D. Botstein, H. Butler, J.M. Cherry, A.P. Davis, K. Dolinski, et al.** 2000. Gene ontology: tool for the unification of biology. The Gene Ontology Consortium. Nat. Genet. *25*:25-29.
2. **Bono, H., H. Ogata, S. Goto, and M. Kanehisa.** 1998. Reconstruction of amino acid biosynthesis pathways from the complete genome sequence. Genome Res. *8*:203-210.
3. **Bono, H., T. Kasukawa, M. Furuno, Y. Hayashizaki, and Y. Okazaki.** 2002. FANTOM DB: database of functional annotation of RIKEN mouse cDNA clones. Nucleic Acids Res. *30*:116-118.
4. **Bono, H., T. Kasukawa, Y. Hayashizaki, and Y. Okazaki.** 2002. READ: RIKEN expression array database. Nucleic Acids Res. *30*:211-213.
5. **Brazma, A., P. Hingamp, J. Quackenbush, G. Sherlock, P. Spellman, C. Stoeckert, J. Aach, W. Ansorge, et al.** 2001. Minimum information about a microarray experiment (MIAME)—toward standards for microarray data. Nat. Genet. *29*:365-371.
6. **Carninci, P. and Y. Hayashizaki.** 1999. High-efficiency full-length cDNA cloning. Methods Enzymol. *303*:19-44.
7. **Dahlquist, K.D., N. Salomonis, K. Vranizan, S.C. Lawlor, and B.R. Conklin.** 2002. GenMAPP, a new tool for viewing and analyzing microarray data on biological pathways. Nat. Genet. *31*:19-20.
8. **De Virgilio, C., T. Hottiger, J. Dominguez, T. Boller, and A. Wiemken.** 1994. The role of trehalose synthesis for the acquisition of thermotolerance in yeast. I. Genetic evidence that trehalose is a thermoprotectant. Eur. J. Biochem. *219*:179-186.
9. **DeRisi, J.L., V.R. Iyer, and P.O. Brown.** 1997. Exploring the metabolic and genetic control of gene expression on a genomic scale. Science *278*:680-686.
10. **Eisen, M.B., P.T. Spellman, P.O. Brown, and D. Botstein.** 1998. Cluster analysis and display of genome-wide expression patterns. Proc. Natl. Acad. Sci. USA *95*:14863-14868.
11. **Eisen, M.B. and P.O. Brown.** 1999. DNA arrays for analysis of gene expression. Methods Enzymol. *303*:179-205.
12. **Goffeau, A., B.G. Barrell, H. Bussey, R.W. Davis, B. Dujon, H. Feldmann, F. Galibert, J.D. Hoheisel, et al.** 1996. Life with 6000 genes. Science *274*:546, 563-567.
13. **Heinisch, J.J., S. Muller, E. Schluter, J. Jacoby, and R. Rodicio.** 1998. Investigation of two yeast genes encoding putative isoenzymes of phosphoglycerate mutase. Yeast *14*:203-213.
14. **Kadota, K., R. Miki, H. Bono, K. Shimizu, Y. Okazaki, and Y. Hayashizaki.** 2001. Preprocessing implementation for microarray (PRIM): an efficient method for processing cDNA microarray data. Physiol. Genomics *4*:183-188.
15. **Kanehisa, M., S. Goto, S. Kawashima, and A. Nakaya.** 2002. The KEGG databases at GenomeNet. Nucleic Acids Res. *30*:42-46.
16. **Karp, P.D., M. Riley, M. Saier, I.T. Paulsen, J. Collado-Vides, S.M. Paley, A. Pellegrini-Toole, C. Bonavides, et al.** 2002. The EcoCyc Database. Nucleic Acids Res. *30*:56-58.
17. **Miki, R., K. Kadota, H. Bono, Y. Mizuno, Y. Tomaru, P. Carninci, M. Itoh, K. Shibata, et al.** 2001. Delineating developmental and metabolic pathways in vivo by expression profiling using the RIKEN set of 18 816 full-length enriched mouse cDNA arrays. Proc. Natl. Acad. Sci. USA *98*:2199-2204.

17a. **The FANTOM Consortium and the RIKEN Genome Exploration Research Group Phase I & II Team.** 2002. Analysis of the mouse transcriptome based on functional annotation of 60,770 full-length cDNAs. Nature *420*:563-573.

18. **Quackenbush, J.** 2000. Viva la revolution! A report from the FANTOM meeting. Nat. Genet. *26*:255-256.

19. **Quackenbush, J.** 2001. Computational analysis of microarray data. Nat. Rev. Genet. *2*:418-427.
20. **Schena, M., D. Shalon, R.W. Davis, and P.O. Brown.** 1995. Quantitative monitoring of gene expression patterns with a complementary DNA microarray. Science *270*:467-470.
21. **Tanoue, J., M. Yoshikawa, and S. Uemura.** 2002. The GeneAround GO viewer. Bioinformatics *18*:1705-1706.
22. **The RIKEN Genome Exploration Research Group Phase II Team and the FANTOM Consortium.** 2001. Functional annotation of a full-length mouse cDNA collection. Nature *409*:685-690.
23. **Tsai, J., R. Sultana, Y. Lee, G. Pertea, S. Karamycheva, V. Antonescu, J. Cho, B. Parvizi, et al.** 2001. RESOURCERER: a database for annotating and linking microarray resources within and across species. Genome Biol. *2*:software0002.0001-0002.0004.
24. **Welch, J.E., E.C. Schatte, D.A. O'Brien, and E.M. Eddy.** 1992. Expression of a glyceraldehyde 3-phosphate dehydrogenase gene specific to mouse spermatogenic cells. Biol. Reprod. *46*:869-878.
25. **Yang, Y.H., S. Dudoit, P. Luu, D.M. Lin, V. Peng, J. Ngai, and T.P. Speed.** 2002. Normalization for cDNA microarray data: a robust composite method addressing single and multiple slide systematic variation. Nucleic Acids Res. *30*:e15.

8 Molecular Studies in Ovarian Cancer: A Tailored cDNA Arrays Approach

Ann-Marie Martoglio
Department of Pathology, University of Cambridge, Cambridge, UK

INTRODUCTION

Ovarian cancer is the third most common malignancy in European women, after breast cancer and colorectal cancer (22). It accounts for more than half of all gynecological tumor deaths among American women (90), and its occurrence is 30–50/100 000, with the lifetime incidence of 1 in 70 women. Although the cause of ovarian cancer is unknown, several risk factors have been identified, including age, personal history of endometrial, breast, or colon cancer, family history of ovarian cancer, ethnicity, nulliparity, lifestyle, environment, and use of fertility drugs and hormones. Genetic predisposition to ovarian cancer accounts for a relatively low proportion of cases. Although a family history of first-degree relatives with ovarian cancer increases the risk of developing the disease 3.6 times the general population rate (125), these familial cases account for only 5%–10% of ovarian cancer cases.

The poor prognosis and large number of deaths from this disease is largely due to its asymptomatic nature in the early stages of development and late diagnosis. The lack of effective therapeutic interventions in late-stage disease further impairs the prognosis for ovarian cancer cases. Thus, developments in both diagnostic and therapeutic approaches are needed to decrease the mortality rate of the disease.

To improve the screening, prevention, and management of ovarian cancer, two main areas remain central tasks of investigation: (*i*) better understanding of the underlying disease processes that lead to the development and progression of the disease, thus potentially leading to better therapeutic developments and (*ii*) the development of better screening tools to implement proper diagnosis and prognosis on the manifestation of the malignant lesions.

Perspectives in Gene Expression
Edited by K. Appasani

OBJECTIVES OF THE STUDY

The main focus of the study presented in this chapter has been the design, production, and application of tailored cDNA arrays to study the molecular biology underlying the disease process of ovarian cancer. Although the construction of generic cDNA arrays is becoming increasingly achievable with the complete sequencing of the human genome, the interpretation of the resulting data is limited to the known function of only a proportion of these genes. Considering the available resources, an alternative, tailored approach was pursued in this study and genes with known function or potential mediation in ovarian cancer and related processes, such as angiogenesis, were selected for inclusion onto the arrays. This proved an effective approach for focused insights and interpretation of the resulting data.

Because of sensitivity, sample quantity requirements, and additional technical issues at the time of this work, it was decided to use nylon membranes as the support surface for the construction of the arrays. Since then, the field has evolved, with increasing optimization of support surfaces and sample labeling methods. The technical, biological, and analytical principles presented here can be broadly applied to the field, especially in terms of the experimental design, biological significance, and interpretation of the resulting high-throughput data.

PATHOGENESIS OF OVARIAN CANCER

The Origin of Ovarian Cancer

There are several theories of the origins of ovarian cancer. One of the most widely discussed theories is that malignant transformation occurs in the surface epithelial lining, or mesothelium. The argument is centered on the predisposition of this cell layer to malignant transformation because of its repeated disruption and repair in response to the expulsion of the oocyte at ovulation. Epidemiologic studies showing the correlation between ovulatory events and incidence of ovarian cancer have been of major support to this theory (30,45,89,158) and, in general, not much challenge has been presented against it. The initial event in the pathogenesis of ovarian cancer is accordingly thought to be the formation of a cortical inclusion cyst. These are thought to arise from entrapped surface epithelium, following the surface repair of a ruptured follicle or sequestration of epithelial cells in deep invaginations within the aging ovary (45,89). Sporadic transformation and proliferation on the ovarian surface has also been proposed, and the ovarian hormonal milieu, including fluctuations in progesterone and androgen receptors, is among the favored hypothesis for this neoplastic induction (44,78,114).

Considering the rich heterogeneity of cells that populate the ovary and their dynamic potential to differentiate (e.g., granulosa cells; theca interna cells), proliferate, and change in response to appropriate stimuli, an array of alternative hypotheses to the origin of ovarian cancer could be considered, including the possibility of "seed" or undifferentiated/pluripotent cells among the ovarian tissue mass proper. Because the study presented here relies indirectly on the traditional classification of ovarian tumors by their "epithelial" nature, and because the aims of the research focus on diagnostic and therapeutic applications, rather than addressing the origins of the tumor mass directly, this topic will not be considered further. However, it is noteworthy that this issue remains a controversial matter in the field, and that further investigation is needed to better understand the onset and development of ovarian tumorigenesis.

Markers of Ovarian Cancer

The past decades of research on ovarian cancer have seen the elucidation of several markers, stimulatory/inhibitory factors, and potential pathologic pathways in the initiation and sustainment of ovarian malignancy. Both in vivo and in vitro studies have reported on the potential role of growth factors (13,14), gonadotrophic hormones (21,88,134), cytokines (17,150), activation of proto-oncogenes (46,76), and loss of tumor-supressor genes (103,123) in the development of the disease. The most prominent and resonant factors described in ovarian tumorigenesis are as follows: (*i*) poor prognosis associated with the overexpression of epidermal growth factor receptor, fms, and c-erbB-2 (HER-2/neu), which occurs in approximately 30% of the cases (14,118); (*ii*) overexpression of the p53 protein in approximately 50% of the advanced ovarian cancers, with serous adenocarcinomas showing significantly higher incidence than mucinous and endometrioid type tumors (68); (*iii*) overexpression of serine protease hepsin gene (142); (*iv*) expression of basic fibroblast growth factor (bFGF) and its FGF-R1 receptor subtype (36); (*v*) amplification of the c-myc oncogene in approximately one third of specimens (15,143); (*vi*) mutation of k-ras in 2%–12% of the cases, a potentially important and relatively early event in mucinous ovarian tumors (24,55,60,148,156,164); (*vii*) loss of heterozygosity in many chromosomes, with important emphasis on 17q, the location of the predisposing tumor-supressor BRCA1 gene (54,104,133); (*viii*) loss of growth inhibition by transforming growth factor-β (TGF-β) (16); (*ix*) expression of the c-fms proto-oncogene in 56% of cases, with potential autocrine stimulation of the ovarian cancer cells by macrophage colony-stimulating factor (M-CSF or CSF-1) (10,31,77,86); (*x*) decreased progesterone receptor (PR) levels and increased expression of insulin-growth factor-1 (IGF-1) and estrogen receptor (ER) (19,161); (*xi*) positive correlation between elevated serum levels of cancer antigen 125 (CA125) and insulin-like growth factor binding protein-2 (IGF-BP2) (49); (*xii*) expression of platelet-derived growth factor (PDGF) in greater than 70% of ovarian cancer tissue or cell lines and the less frequent (<30%) expression of its α receptor (PDGF-Rα) (57,62,151); (*xiii*) a greater than 17-fold higher production of urinary-type plasminogen activator (uPA) by ovarian carcinoma cells compared to their normal counterparts (108); (*xiv*) autocrine growth stimulation of ovarian cancer cell lines by TGF-α (135); (*xv*) and paracrine stimulation by macrophage-derived interleukin-1 (IL-1), IL-6, and tumor necrosis factor (TNF) (7,112,154,159).

Although the vast amount of past research has identified the aberrant expression of specific genes in a variety of ovarian cancer types, little has been achieved in defining an ideal cohort of tumor markers for appropriate diagnostic or therapeutic application. Furthermore, there are currently no markers of sufficient predictive value to identify patients who are likely to be more responsive to one treatment option instead of another. Because it is a combination of numerous factors that eventually lead to the tumorigenic outcome, it is necessary to apply more comprehensive methods to study the molecular biology of ovarian cancer. The use of high-throughput comparative technologies, such as cDNA arrays, may prove useful in understanding the interdependent roles and potential pathologic pathways that underlie the disease process, leading to better diagnostic, prognostic, and therapeutic developments.

Angiogenesis in Tumor Development

Angiogenesis is defined as new vessel growth from pre-existing vessels. It occurs rarely in normal adult physiology, except in the ovary (e.g., corpus luteum formation), endometrium, and placenta (48). In pathology, angiogenesis is reported to occur in conditions such as rheumatoid arthritis, psoriasis, liver cirrhosis, diabetic retinopathy, endometriosis, and tumor growth (72,79,110,128, 137,153). Tumor growth is thought to be dependent on adequate blood supply, and the major cause of a poor clinical prognosis is the invasion and metastasis of tumor mass cells at distant sites, which likely occurs via these newly formed vessels (26,50–52).

Although much research has been invested in factors that stimulate or suppress tumor angiogenesis, the precise synchrony of the molecular events that take place is not yet well understood. A recent report by Maniotis et al. (99) further suggests that there may be cases in which aggressive or metastatic tumors (e.g., melanomas) acquire the ability to form intratumoral vascular channels lacking an endothelial lining proper. Thus, considering the ongoing research to understand angiogenesis, careful attention is needed in the design and application of "anti-angiogenic" therapies (29) and in the interpretation of single-marker experimental results. Concurrent studies on the many genes involved in setting and maintaining the angiogenic state in the ovary, such as by the application of cDNA arrays, could lead to a better understanding of the genetic progression of ovarian tumor development, dissemination, and growth. Such studies may lead to the development of more precise diagnostic measurements and improved treatment strategies through multi-targeted approaches.

DIAGNOSIS AND PROGNOSIS OF OVARIAN CANCER

Current Methods and Trends

Recent evidence suggests that the frequency of some gynecological cancers, such as invasive cervical cancer and ovarian cancer, is either decreasing or stable and that the survival of patients has improved in recent years (11,73,129). These observations correlate with the increased implementation of routine clinical screening, which identifies premalignant lesions and can anticipate the diagnosis of malignant lesions that respond more effectively to therapy (56,105). There are no optimal screening schemes currently available for ovarian cancer.

There are three main clinical examinations currently accepted as screening measures: (*i*) the bimanual rectovaginal pelvic examination, an important part of routine gynecological care yet inadequately sensitive or specific for ovarian cancer; (*ii*) CA125, an antigenic determinant elevated in the serum of 80% of women with epithelial ovarian cancers yet present in only half of the patients in the early stages of the disease and elevated in a significant proportion of healthy women or women with benign disease; and (*iii*) transvaginal ultrasonography (TVS), which, although preferred for its noninvasive nature, is inadequate for use as a single screening tool because of its lack of specificity for ovarian cancer. The combination of bimanual rectovaginal pelvic examination, CA125, and TVS significantly improves the specificity of screening and has reduced the number of women requiring unnecessary surgical intervention (109). However, there is currently no evidence that these screening modalities can effectively reduce mortality from ovarian cancer. Again, this is mainly due to the lack of sensitivity and specificity of the available screening methods and is compounded by both observer variability and the heterogeneity of test results or response to treatment for patients that are otherwise considered from the same pathological group.

Therefore, although increased implementation of screening routines has lead to the better management of ovarian cancer cases, the availability of screening tools and treatment protocols is far from optimization. New strategies that apply the vast amount of information obtained from past studies of ovarian cancer markers and other prospective indicators of disease may provide a sufficiently robust platform to reform the approach to clinical screening. The simultaneous assessment of multiple molecular markers by cDNA arrays is one such possibility, offering potential as a much-needed diagnostic and prognostic tool of increased sensitivity and specificity for ovarian cancer.

Classification and Staging

Histologic cell type, tumor grade, and tumor stage are the three main prognostic or stratification factors currently used in the clinical setting for the evaluation of ovarian cancer specimens. Initial observation of samples allows for the classification by cell type, location on or within the ovary, and general architecture (e.g., papillary, cystic, and adenomatous) of the tumor mass. Based on standardized typing, approximately 60% of ovarian cancers are classified as epithelial or derived from tissues that come from the coelomic epithelium, and malignant epithelial tumors account for approximately 90% of all ovarian cancers (157). Of these, 40%–50% are defined as serous type, resembling the glandular epithelium lining the fallopian tube; 15%–30% mucinous type, resembling the endocervical gland; and 5%–10% endometrioid type, resembling the endometrial lining. Five to twenty percent of ovarian epithelial carcinomas are defined as poorly differentiated or undifferentiated (53), while clear cell (i.e., resembling renal cell carcinoma or recently considered of the endometriod variety) and Brenner (i.e., resembling transitional cell) carcinomas account for less than 1% of the remaining cases of epithelial ovarian cancers. Tumors of low potential malignancy or "borderline" tumors are lesions that tend to remain confined to the ovary for long periods of time and are associated with a rather good prognosis (98). Peritoneal carcinoma or primary peritoneal papillary serous carcinoma result from a primary malignant transformation of the peritoneum and mimic ovarian cancer clinically, therefore leading to the false diagnosis of "ovarian cancer" (33,144). The work presented here is limited to the discussion of ovarian tumor samples of the epithelial type (by classical definition), therefore other subtypes will not be further considered.

Histologic grading, which classifies samples based on the architecture and nuclear appearance of the tumor, is an important indicator of tumor malignancy and is closely related to prognosis (85). Tumors consisting of a predominantly solid sheet of malignant cells and less than 25% glandular area are referred to as grade 3 or poorly-differentiated, while those consisting of more than 75% glandular area are considered to be grade 1 or well-differentiated.

Staging, or classification based on the extent of tumor spread, is the most important prognostic factor in ovarian epithelial malignancies to date. Tumor stage is assigned in accordance with the International Federation of Gynecology and Obstetrics (FIGO) system and is based on findings at surgery, such as during exploratory laparotomy. Surgical staging improves the decision of appropriate treatment routines, thus increasing the survival rate of ovarian cancer patients.

Based on histologic observation, ovarian epithelial tumors are relatively easily distinguished from other primary types of tumor. However, subdivision into serous, mucinous, endometrioid, or other subtypes of the epithelial group (described earlier) is often a challenging task (9). Several studies have reported on the difficulty of applying standardized histologic typing of ovarian tumor samples, especially in the case of endometrioid/mucinous and endometrioid/serous borderlines, mixed epithelial, and undifferentiated carcinomas (20,35,126,131). The variability in histologic classification of ovarian cancer specimens comes, in part, from the use of different assignment criteria [e.g., World Health Organization (WHO) and FIGO classification schemes]. However, inter- and intra-observer variability are leading factors in the lack of reproducibility in the classification of some cases of ovarian cancer. This is due to the subjective nature of the approaches currently employed and raises the need to search for alternative methods of evaluating tumor samples. More robust and observer-independent methods of evaluating representative samples could eventually lead to more precise diagnosis, prognosis, and treatment strategies for ovarian cancer.

TAILORED cDNA ARRAYS

Design and Production

The design and construction of the tailored cDNA arrays used in this study have been presented in detail in Martoglio et al. (100) and are summarized in Figure 1. Only specific genes of known function or potential role in the disease process were chosen for deposition onto the arrays. This route was first pursued for ease of interpretation of the results, which currently tends to be limited to genes of known function in more generic array studies (81,149) and to focus the surveying process to process-specific markers or a family of markers of interest. Three main areas of interest were principally searched: ovarian cancer tumorigenesis; angiogenesis; and further potentially influential genes or genes observed to play a role in other tissue tumors or relevant biological pathways, such as chemoresistance factors and markers of immune response. The compiled list of selected genes then consisted of growth factors, cytokines, hormones, receptors, angiogenic factors, cell adhesion molecules, cytoskeletal components, intracellular signaling genes, transcription factors, nuclear receptors, and enzymes, among others.

A collaboration with the Medical Research Council Human Genome Mapping Program Resource Centre (MRC HGMP-RC), based in Hinxton, UK, was established for access to the Integrated Molecular Analysis of Genomes and their Expression (IMAGE) cDNA clone bank and use of on site facilities from picking to gridding procedures.

A total of 332 purified PCR products were spotted onto 7.8 cm × 11.9 cm nylon membranes (Hybond® N+; Amersham Biosciences, Little Chalfont, UK), as previously described (100). Following hybridizations, intra- and inter-membrane reproducibility tests were performed, and a final set of 175 genes was selected for analysis (100). The genes represented are listed in Table 1.

Samples and Hybridizations

The sample set consisted of 17 total RNA extracts from normal ovary ($n = 5$); serous papillary adenocarcinoma ($n = 5$); poorly differentiated serous papillary adenocarcinoma ($n = 4$); benign serous cystadenoma ($n = 1$); and benign mucinous cystadenoma ($n = 2$) from ovarian biopsy samples collected during salpingo-oophorectomies or tumor debulking, following Ethical Committee approval. The samples were numbered 1–17, respectively.

Samples were labeled with [α-^{32}P] and hybridized to the cDNA gridded membrane arrays described above. Intra- and inter-membrane reproducibility were then assessed before considering differences in the resulting gene expression profiles. Additional details on sample preparation, labeling, hybridization to cDNA arrays, and reproducibility controls are described in Martoglio et al. (100).

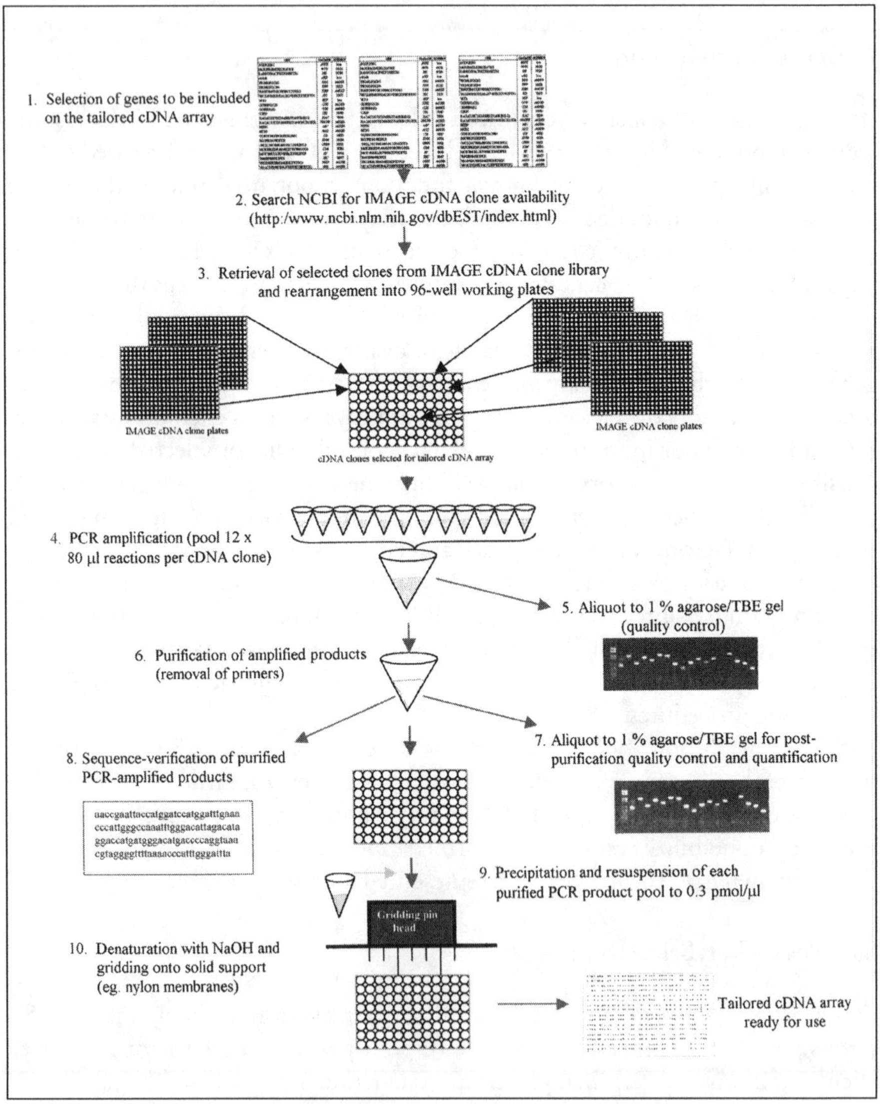

Figure 1. Steps involved in the production of tailored cDNA arrays. These steps briefly outline the process followed for the production of the arrays, as fully described in Martoglio et al. (101).

Table 1. Genes in the Tailored cDNA Array Used for Data Analysis of Ovarian Cancer Samples[a]

Gene name
Actin, α-
Actin, γ-
Alcohol dehydrogenase β
Aldose reductase
Alkaline phosphatase
Alzheimer's disease amyloid A4 protein
AMP deaminase 3
Amphiregulin
ANG1 (angiopoietin-1)
ANG2 (angiopoietin-2)
Angiopoietin-1
Antimullerian hormone type II receptor
Arabidopsis thaliana cytochrome c554 (control)
Asialoglycoprotein receptor 1
ATP synthase lipid-binding protein P2
B219/OB receptor isoform HUB219.3
BAD protein
β-Platelet-derived growth factor receptor
Burkitt's lymphoma receptor (C3A-R)
Cadherin-2 (EPITH-)
Cadherin-3 (PL-)
Cadherin-6 (K-)
Cadherin-11 (OB-)
Calgranulin B
CD44 antigen, epithelial form
Choriogonadotropin β chain
Ovarian carcinoma antigen CA125 (1A1-3B)
Chromosome 21q22.1, containing part of GLUR5
Cofilin
Complement receptor type 2
Cyclin-dependent kinase inhibitor P27
Cytochrome B-245 heavy chain (GP91-PHOX)
Cytochrome C
Dematin
Early response protein NAK1
Elongation factor 1-α 1
Endothelin-1 receptor
Epidermal growth factor
Epidermal growth factor receptor

Table 1. Continued

ERBB-2 receptor protein-tyrosine kinase
ERK5-extracellular signal-regulated kinase 5
Erythrocyte cytosolic protein (TIP49)
Oviductin
Estrogen sulfotransferase
Fibroblast growth factor 3 (FGF-3)
Fibroblast growth factor receptor 3
Fibronectin receptor α
Follistatin 1 and 2
G452519 mRNA, ORF X
γ-Aminobutyric-acid receptor α-2
Glucocorticoid receptor, α
Glucose transporter type 1 (GLUT1)
Glucose transporter type 5
Glutamate
Glutathione reductase
Glutathione S-transferase muscle
Glutathione S-transferase P (GSTP1-1)
Glutathione synthetase
Gonadotropin-releasing hormone receptor (RAT)
Granulocyte colony-stimulating factor (CSF-1)
GM-Colony-stimulating factor receptor
Growth hormone receptor (GHR)
Growth hormone variant II
Haptoglobin-2
Heat shock cognate 71-kDa protein (HSP71)
Hemoglobin γ-A and γ-G chains
Hepatocyte growth factor receptor (HGFR)
Heregulin-α
Hepatocyte growth factor (HGF)
High density lipoprotein-binding protein (HDL-BP)
Histone deacetylase HD1
HLA class I antigen (B-12)
HLA class II antigen, DR-1 β
5-Hydroxytryptamine 2B receptor
Inhibin βA chain
Insulin-like growth factor IB, (IGF-I)
Insulin-like growth factor I receptor (IGF-1 R)
Integrin α-6
Integrin β-4 subunit
Integrin VLA-4 α subunit (VLA-4)

Table 1. Continued

Intercellular adhesion molecule-2 (ICAM-2)
Interferon-γ receptor (α chain)
Interferon-γ receptor (β chain)
Interleukin-1 β convertase
Interleukin-13 receptor (IL-13Ra)
Interleukin-8
Interstitial collagenase (MMP-1)
KDR
Kininogen
Macrophage colony stimulating factor I receptor (CSF-1 R)
Macrophage inflammatory protein 1-β (MIP-1 b)
Macrophage metalloelastase (MMP-12)
Mast/stem cell growth factor receptor (SCF R)
Matrilysin (MMP-7)
Membrane glycoprotein GP130
MET
Metalloproteinase inhibitor 1 (TIMP-1)
Metalloproteinase inhibitor 3 (TIMP-3)
Mevalonate kinase
Moesin
Macrophage stimulating protein (MSP)
Mucin 1
Mullerian inhibiting substance
MYB-related protein B (B-myb)
NADH-ubiquinone dehydrogenase 24-kDa subunit
NEU differentiation factor
Neuron-restrictive silencer factor (REST)
Neuropilin-2
Neurotensin
Neutrophil cytosol factor 1 (P47-PHOX)
Nuclear factor NF-κ-B P105
Oxytocin receptor
PAC 308I13 on chromosome 1p35-1p36.3; 22q11.2; 17
Placental growth factor
Placental thrombin inhibitor
Plasminogen activator inhibitor-1 (PAI-1)
Platelet factor 4 (oncostatin A)
Placenta growth factor (PLGF)
Presenilin I-463
Prolactin receptor type 2
Prostaglandin D synthase

Table 1. Continued

40-kDa Protein kinase,(CDK activating) kinase 7
QM protein
RING3 protein
Serine protease hepsin
Soluble VEGF receptor 1
Site-1 protease (S1P)
Sodium-dependent dopamine transporter
SPARC (mouse/human homologue)
Steroid hormone receptor ERR2
Steroidogenic acute regulatory protein, CAB1
Stromelysin-1 (MMP-3)
Stromelysin-2, (MMP-10)
Stromelysin-3 (MMP-11)
Superoxide dismutase, Cu/Zn-
Superoxide dismutase, Mn-
TATA box binding protein-related factor 2
TEK (TIE 2)
Thrombospondin 1
Thrombospondin 4
Thyrotropin receptor
TIE1
Tissue factor
Titin
Transferrin receptor
Transforming growth factor β 1 (TGF-β 1)
Transforming growth factor β 3 (TGF-β 3)
Transforming protein P21/H-RAS-1 (p21)
Tyrosine-protein kinase receptor FLT4
Tyrosine-protein kinase receptor RON
Tyrosine-protein kinase receptor TEK (TIE-2)
Urokinase plasminogen activator surface receptor (uPA R)
Vascular endothelial growth factor (VEGF) receptor 1
Vasoactive intestinal polypeptide receptor 2 (VIP2 receptor)
Vasopressin V1A receptor
VEGF A
VEGF B
VEGF C
von Willebrand factor (vWF)
WNT-2 protein

[a]Detailed in Martoglio et al. (100,101).

ANALYSIS OF VARIATION IN GENE TRANSCRIPT ABUNDANCE IN OVARIAN CANCER

cDNA arrays generate a large volume of data that represent relative gene transcript abundance in the samples that had been surveyed. The analysis and interpretation of such data depends on two main factors: (*i*) the experimental design (e.g., details of the biological samples tested) and (*ii*) the hypothesis or set of questions to be tested. Both factors are paramount to the choice of analytical tools to be employed, as the use of inappropriate algorithms can result in misleading conclusions from the experimental data. cDNA array studies can be designed to investigate the difference in gene expression levels for a variety of sample sets, including normal and pathologic samples, treated and untreated systems (e.g., cell cultures and animal models), or between different time courses in a process or pathway of interest. Depending on the immobilized targets contained on the array, the signal intensity data can be assessed for correlations in gene expression based on similarities in function or regulation, or for causative differences in specific disease states or treatment regimens.

With a clear understanding of the experimental design and the questions for which the data will be assessed, analysis remains complex due to technical and biological variables inherent in high-density array expression profiling. Of these, once technical parameters such as reproducibility are accounted for, one of the most important factor to consider is the nature of the biological samples that are queried. In the first instance, gene expression profiles arising from as less heterogeneous samples as possible would be preferred to facilitate interpretation of the data. These samples can be produced from specific cell lines, flow- or magnetically sorted cells, or laser microdissection of specific cell populations from whole tissue samples (102). However, these methods are time-consuming and laborious, and results from such isolated samples may miss or mask important changes in gene expression from the surrounding tissue environment in vivo. Thus, in a clinical setting, it would be preferable to design cDNA array-based gene expression studies to encompass complex or generic specimens, such as biopsy samples. The challenge then becomes rooted in the development and application of appropriate mathematical models and bioinformatics tools for the interpretation of the data.

A First Look at Tailored cDNA Array Ovarian Cancer Data

There are two main factors contributing to the challenge of analyzing and interpreting high-density cDNA array data: (*i*) the mass amount of data generated and (*ii*) the limited knowledge currently available on the meaning of co-expressed genes or genes observed to alter in different combinations under different conditions tested (e.g., normal versus pathological samples). A further complexity is presented in large-scale arrays encompassing targets of unknown function [e.g., expressed sequence tags (ESTs)] for which a "fishing expedition" for novel genes tends to be a central focus of the analysis. In any case, a starting point for the biological interpretation of cDNA array results can be based on knowledge of gene families, receptor-ligand interactions, or the known regulatory function of various factors in specific biological processes (e.g., tumorigenesis, angiogenesis, apoptosis, cell cycle, and differentiation) described in the literature.

The use of tailored cDNA arrays eliminated many of the challenges in the interpretation of the data due to the smaller number of genes surveyed and the a priori selection of targets based on knowledge of their function and potential roles from previous independent studies. The smaller scale of the tailored cDNA arrays also allowed for thorough analysis of data reproducibility and the application of data clean-up procedures (100), providing a high level of confidence in the comparison of the hybridization signal profiles.

The optimal tool could come with technology for large-scale, highly reproducible arrays, increased the knowledge of gene function and gene co-expression patterns, and integrative software to help in the data interpretation process.

In the current study, data were first analyzed by gene-specific comparisons between the tissue classes ($n \geq 4$) using Student's *t* tests. This allowed for a focused observation of single-gene and process-specific alterations in transcript abundance but is not recommended for larger data sets because repeated *t* tests have the potential to define false positives, especially as the sample size increases. The results of this exploratory first look at the data revealed some specific differences in single genes between the normal ovary ($n = 5$) and poorly differentiated serous papillary adenocarcinoma (pd-spa) ($n = 4$) tissue types (Table 2), as discussed in Martoglio et al. (100).

Exploring cDNA Expression Data: Pattern Recognition for Molecular Classification of Ovarian Cancer

Second to the fishing expedition type of analysis, a more encompassing method was employed to assess the whole data set (i.e., all tissue classes). The current methods proposed for the analysis or assessment of gene expression array data include hierarchical clustering (42,116), two-way clustering (5), quality cluster algorithm (63), Pearson's correlation (130), *k*-means clustering (145), principal components analysis (64), self-organizing maps (141), and support vector machine analysis (28). Although these clustering methods have proven valuable in describing the respective data sets that they were tested on, they are restricted by lack of robustness, a necessary dependency on predefining the number of expected clusters or cluster behavior (e.g., points that are grouped based on local decisions such as cluster size and expected cluster boundaries) and an inherent deterministic nature that can lock genes into fixed cluster features (107,141).

Alternatively, an approach based on independent component analysis (ICA) was employed, the concepts of which are described by D.J.C. MacKay (96) and further detailed at http://www.inference.phy.cam.ac.uk/mackay/abstracts/icagenes.html and in Miskin (106) and Martoglio et al. (101). The independent

Table 2. Single Genes Found to Differ in Transcript Abundance Using *t* Test between Normal and pd-spa Ovarian Samples (100)

Gene Name
Angiopoietin-1
BAD protein
Cadherin-11 (OB-)
*Cadherin-6 (K-)
Choriogonadotropin β chain
Cofilin
Cytochrome C
*Endothelin-1 receptor
Glucose transporter type 1
Glutathione S-transferase p
GM-colony-stimulating factor receptor
HLA class I antigen (B-12)
HLA class II antigen, DR-1 β
Macrophage inflammatory protein 1-β
Macrophage metallowelastase
Mast/stem cell growth factor receptor
Moesin
Mucin 1
MYB-related protein B (B-myb)
NADH-ubiquinone dehydrogenase
Neuron-restrictive silencer factor
Neutrophil cytosol factor 1 (P47-PHOX)
Plasminogen activator inhibitor-1
Soluble VEGF receptor 1
Superoxide dismutase (Mn-)
Thrombospondin 1
Thrombospondin 4
Transferrin receptor
Transforming growth factor β 3
Tyrosine-protein kinase receptor RON
Vasoactive intestinal polypeptide receptor 2
VEGF A
VEGF receptor 1

All genes were more abundant in the pd-spa samples, compared to the normal samples, except those marked with an asterisk (*), in which the inverse was the case.

component analysis-based method was a more robust way of analyzing high-throughput data and revealed patterns of gene co-expression ("gene signatures") based on all genes on the cDNA array (Figure 2). The ovarian cancer data revealed several gene signatures characteristic to a specific tissue-type or physiopathological process, based on the leading genes defining a given gene co-expression pattern. Due to the limited number of samples used in the current study, it is not fruitful to attempt extensive interpretation or conclusive definitions of the various gene signatures that emerged, but several important observations are noteworthy and could be considered in future experimental designs, as presented in Martoglio et al. (101), as follows.

(*i*) **Pd-spa gene signature.** This was the most striking tissue-type discriminant gene co-expression pattern. Of all ovarian tissue samples surveyed, all pd-spa samples were consistently seen to have the most prominent abundance for this gene signature (gene signature 12) (Figure 3). The leading genes were indicative of angiogenesis [Tie1 (83,113), placental growth factor (41,162)], increased immune priming [HLA I (38,84), HLA-DR (94,163)], changes in tissue matrix integrity [cadherin-3 and cadherin-11 (27,61,111)], proliferative regulation [gp130 (65,95,124)] and chemoresistant traits [cofilin (120,121,127,132)] (Figure 4). The consistency of this gene signature, specifically among pd-spa samples indicates that the gene signature may be an effective molecular signature upon which to classify biopsy samples "on blind" and that the histopathological method was unequivocal in defining the disease type by observation of phenotypic traits. Considering the latter, it is important to note that, although histopathologic classification is supported by the type-specific gene signature, heterogeneities among the pd-spa samples were captured by other gene signatures, such as gene signatures 1, 8, and 14 (Figure 5). The pathophysiological traits represented by these gene signatures are not yet defined. Nevertheless, they indicate that phenotypic tissue-type classification is not enough to capture important differences inherent among similar tissue samples. The use of a combination of tissue-type- and pathophysiol-

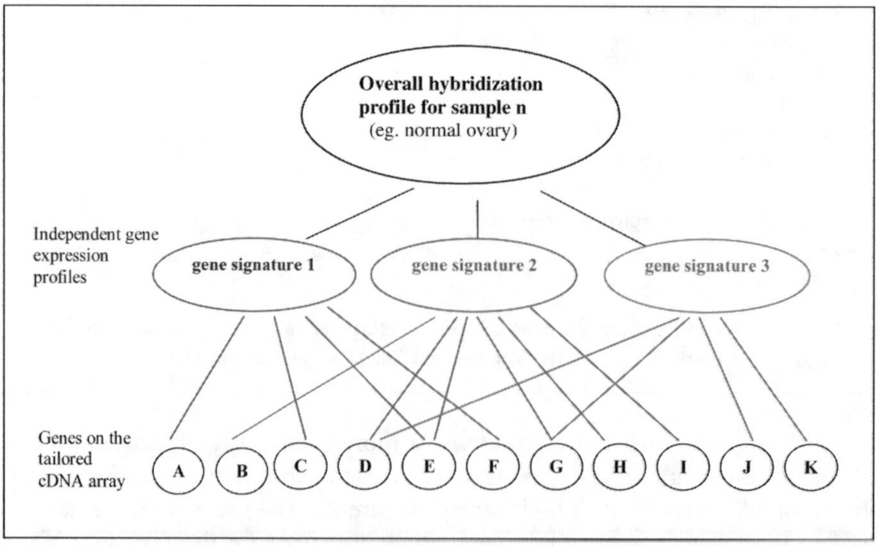

Figure 2. ICA: decomposition of gene expression patterns from all genes on the array. The ica model considers the overall gene expression profile from hybridized cDNA arrays to define patterns of gene expression networks (gene signatures) composed of independent combinations of the genes surveyed (101).

ogy-specific gene signatures holds promising application for the sub-classification of patients beyond current inspection (e,g., histopathology) methods. This could lead to improved diagnostic and prognostic interpretations and the development of a targeted selection of treatment strategies.

(*ii*) **Benign mucinous cystadenoma (bmc) gene signature.** This signature (gene signature 3) (Figure 3) was defined by leading genes metalloproteinase inhibitor 1 (TIMP-1), 5-hydroxytryptamine 2B receptor (5-HT2B), AMP deaminase 3, calgranulin B, and mucin-1, was seen to be most prominent in the bmc samples, among all the ovarian tissue samples surveyed. Normal samples demonstrated a consistently subtracted profile for this gene signature; interestingly, tissue 4, the "normal" contralateral ovary from the same patient as bmc-ridden tissue

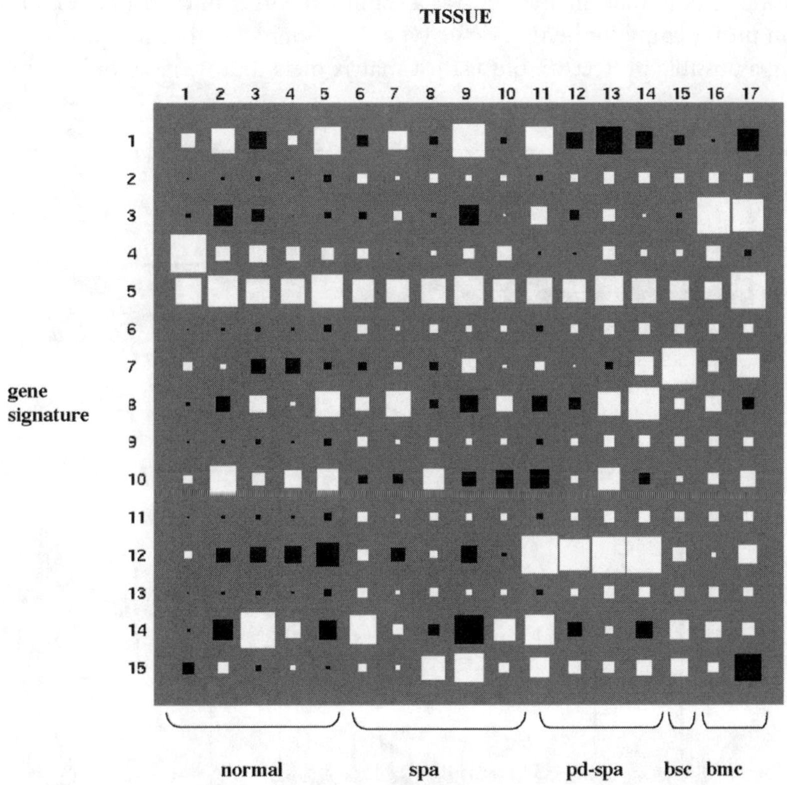

Figure 3. ICA of tailored cDNA array data: unveiling gene expression patterns in ovarian hybridization profiles. Tissues 1–5 correspond to normal ovary; tissues 6–10 to spa; tissues 11–14 to pd-spa; tissue 15 to bsc; and tissues 16 and 17 correspond to samples of ovarian bmc. Several gene expression profiles (gene signatures) were unveiled by the ica run and are represented at each row. Gene signatures 3, 7, and 12 are seen to be distinctively prominent in bmc, bsc, and pd-spa samples, respectively. Intra-tissue type differences could then be assessed, based on a number of further defined gene signatures (see Figure 5) (101). **(See color plate A9.)**

16, demonstrated the least subtracted profile among all the normal samples. This could indicate that the phenotypically normal contralateral ovary might have been in the process of undergoing a benign transformation, although larger scale studies would be needed to confirm such comparative observations. There are no previous reports on the co-expression of the leading genes listed. Increased TIMP-1, 5-HT2B, and mucin-1 transcript abundance levels agree with previous reports in ovarian cancer and tumor progression (40,67,82,91,140,146). It is interesting to note that gene signature 7, a potential discriminator for benign serous cystadenoma (bsc) samples (most prominent in tissue 15, the only bsc sample of all 17 ovarian tissues surveyed), also shared TIMP-1 among its top three leading genes. TIMP-1, a matrix metalloproteinase inhibitor shown to be involved in tumor development (34,75), is also considered to be an endogenous inhibitor of angiogenesis and has been reported to be significantly elevated in ovarian cancer (67). It is interesting to note that the marker was a leading member of discriminant co-expression profiles only for benign tissue types (e.g., bmc, bsc, but not pd-spa), indicating a possible protective role against matrix metalloproteinases only in non-

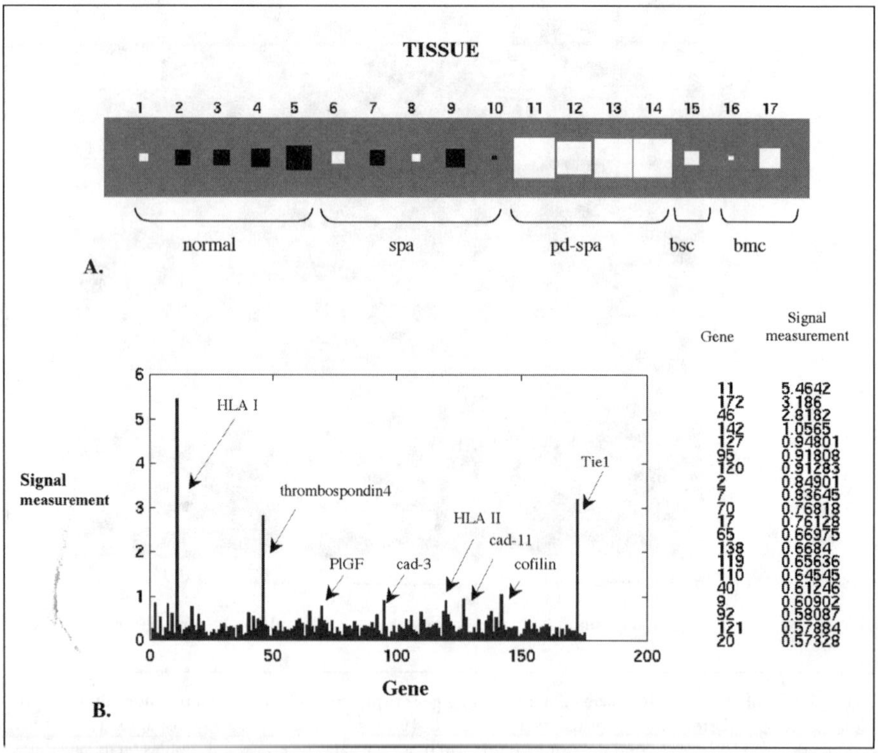

Figure 4. Pd-spa gene signature. (A) ICA of ovarian tailored cDNA array data; excerpt from Hinton diagram (101) showing gene signature 12, which was shown to be characteristic of pd-spa ovarian samples. (B) Diagram of the gene expression profile and the leading genes defining the gene signature.

malignant transformation in the ovary. Calgranulin B is a calcium-binding myeloid-associated protein highly expressed in neutrophils and monocytes (32). The increased levels of transcript for this protein indicate a possible inflammatory environment in the bmc ovaries (136) and correlates with similar findings in colorectal cancer (74). The dominant prevalence of this gene signature in the bmc ovarian samples indicates their potential application for gene transcript abundance-based distinction of this cancer type from other ovarian biopsy samples.

(*iii*) **Premenopause gene signature.** Gene signature 4 (Figure 3), defined by leading genes endothelin-1 receptor (EDNRA), cadherin-6, and Cu/Zn-superoxide dismutase, was distinctively prominent in tissue 1, a member of the normal

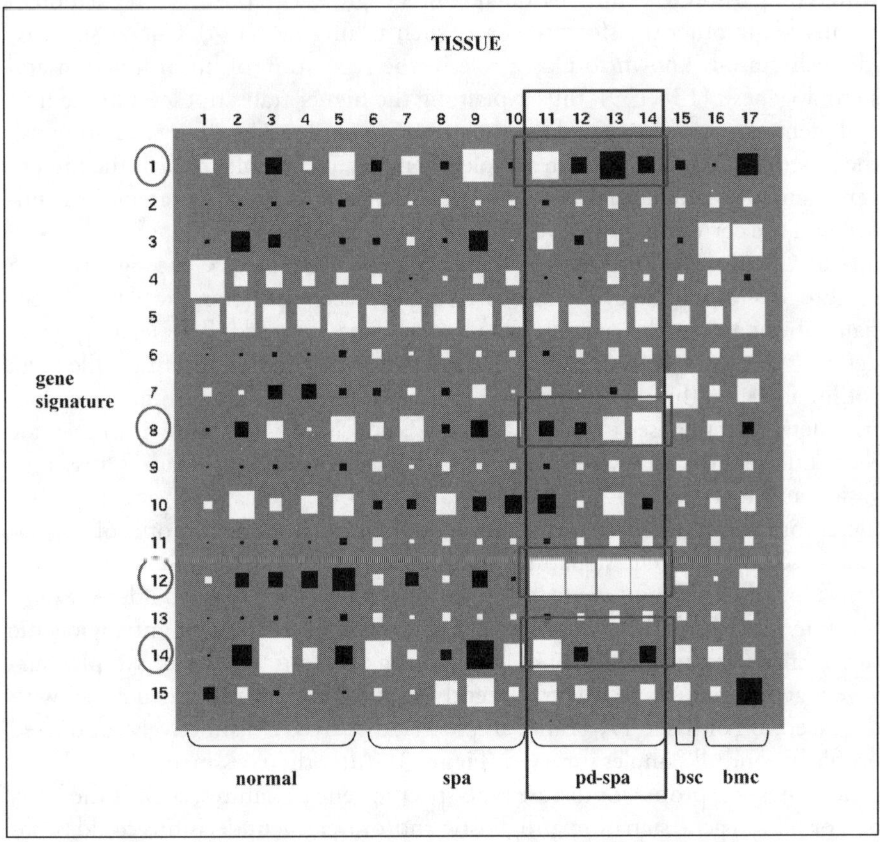

Figure 5. Assessing the significance of gene signatures: tissue-type-specific profiles versus other gene expression patterns. ICA of ovarian tailored cDNA array results revealed tissue-type-specific gene signatures and tissue type-independent gene signatures present in varying amounts among the ovarian samples surveyed (red circles). For example, gene signature 12 can be observed consistently predominant in the pd-spa sample group (delineated in blue), while inherent variations within the histologically similar samples (i.e., pd-spa) can be observed at gene signatures 1, 8, and 14. Molecular analysis of tissue samples in similar multi-parameter schemes may help to classify tissues based on several underlying disease processes. This could help to identify patients who would benefit from different treatment strategies (101). **(See color plate A10.)**

ovaries group, but originating from a premenopausal patient. This observation showed the sensitivity of the ICA method in discriminating functionally distinct samples that are not detected by conventional intra-class comparative methods such as *t* test analyses. It is interesting that EDNRA and cadherin-6, which are noted to be higher in all normal ovaries when compared to pd-spa and spa ovarian samples (Table 2), were further prominent in the functionally active normal ovary above other normal (postmenopausal) samples. Endothelin-1 receptor mediates the mitogenic effect of endothelin-1, a powerful vasoconstrictor expressed in ovarian luteal cells and reported to be involved in the regulation of ovarian steroid synthesis (6). Its predominance in the normal premenopausal ovary agrees with the observed abundance of endothelin-1 receptor in ovulatory follicles, while comparison with tumor-ridden samples agrees with the down-regulation of the marker in other transformed tissues such as lung cancer (4). Cu/Zn-superoxide dismutase is known to play a role in the regulation of luteal function and steroidogenesis (138,139), thus explaining the higher transcript level in the normal premenopausal compared to other ovarian samples. The ability to distinguish the functional status of ovarian samples by molecular profiles such as the present gene signature could lead to better understanding of the overall ovarian tissue milieu and the corresponding development of targeted therapies.

(*iv*) **Chemoresistance/matrix integrity gene signature.** Gene signature 15 (Figure 3) showed one of the most striking gene co-expression profiles differentially abundant in the ovarian tissues that were surveyed. The leading genes defining this gene signature were characteristic of a chemoresistant profile (e.g., cofilin and glutathione S-transferase P). In addition, the predominance of matrix metalloproteinases (MMP-3, MMP-7, and MMP-10), known to degrade the cellular matrix in several pathophysiological conditions including vessel invasion and metastasis (12,58,71,119,122,152,160), indicate a measuring parameter for tissue integrity. Thus, an inspection of the molecular profile of a specimen based on this gene signature could indicate the invasive state of the tissue. This type of checkpoint could aid in treatment decisions that would, for example, interfere with TIMPs or other related molecules, as in recent antiangiogenic treatment strategies (23,69,117). Comparing the various ovarian samples surveyed, gene signature 15 was consistently least prominent in normal ovary, with the exception of tissue 17, a bmc sample that displayed a distinctively subtracted profile beyond all samples surveyed (Figure 3). After the assessment of the cDNA gene expression profile for disease type-specific gene signatures, such as the bmc, bsc, or pd-spa gene signature, prognostic and/or treatment decisions could be facilitated by the subsequent inspection of gene signatures such as the chemoresistant/matrix integrity and others discussed above.

The results here show promising application to future molecular diagnostic, prognostic, and therapeutic applications. Additional large-scale studies would be needed to confirm and assign the combined relevance of the observed gene signatures or others that would result from the use of high-throughput arrays.

GENES, GENE CO-EXPRESSION PATTERNS, AND DATA INTERPRETATION

The Context of Data Interpretation

To understand the individual components that comprise a specific gene profile, it is important to keep in mind the heterogeneity of knowledge or observations from past reports. Single genes and proteins have been reported to have distinct functions in vivo and in vitro, in animal models, and human studies. Therefore, the context in which the observations are drawn is an important parameter to consider. A single gene may have different effects in different tissues due to tissue-specific effectors and negative regulators, or overall tissue environment. At the same time, specific biochemical pathways can induce unique, overlapping, or redundant alterations in the physiological and/or pathological behavior of a tissue.

Although there are many mathematical models that can be applied and tested to analyze gene expression data, there is a current lack in the quantitative definition of probabilities for gene expression patterns. These are lacking not only on a gene-to-gene or gene co-expression pattern basis but also in a system-specific environment. That is, the analysis and interpretation of gene expression data could be facilitated once the quantitative probability of gene co-expression profiles is defined in tissue-, disease-, and pathophysiology-specific contexts. These questions may start to be answered once data sets from large enough sample groups are tested.

Fine-Tuned Classification of Ovarian Samples

Assessment of the ICA results revealed several gene signatures that represent potential inherent pathophysiological processes in the various ovarian tissue samples, as described earlier. Some gene signatures were highly prominent in tissue-specific samples, such as the pd-spa and bmc gene signatures. Other gene signatures, such as 1, 8, 14, and 15, were present at varying levels among the ovarian tissue samples, with no consistent patterns within a given tissue type.

Gene signatures defining a given physiopathological pathway or disease state are expected to be present at similar levels among the representative samples of a given tissue type. This is true if one assumes that histopathological classification identifies tissues that are undergoing a finite regime of molecular changes that lead to the disease phenotype, which may not be the case. The mixed representation of some gene signatures among the ovarian samples may therefore be due to one of two reasons: (*i*) gene signatures may be identifying inherent pathophysiological processes in the ovary that vary independently of the normal or tumorigenic state of the tissue or (*ii*) some tissues may have been incorrectly classified by traditional histopathological methods. Should (*i*) hold true, larger studies and further comparative parameters are necessary to define the molecular trait defined by the specific gene signatures present at non-tissue-type-specific levels. The second explanation proposed could be due to observer-dependent variance in tissue classification, which would support the lack of emergence of a gene signature able to define the spa tissue sample group. To test this hypothesis, an overall tissue-type probability test can be performed on the ICA results. Another explanation for the mixed expression levels of some gene signatures and differential clustering of the samples from the same histopathological group (i.e., spa) may be the monoclonal nature of ovarian cancer (2,87,155). Ovarian cancer has been proposed to arise from the malignant transformation of a single cell, which then undergoes uncontrolled proliferation, leading to the formation of a tumor mass, and the eventual loss of controlled differentiation in the advanced or more malignant stages of the disease (1,87,92). Thus, a group of samples from the same broad histopathological classification (e.g., spa) may host specimens undergoing different stages of molecular progression toward malignancy (e.g., pd-spa), demonstrating an overall similar phenotype.

The application of concepts such as gene signatures has potential for the development of improved diagnostic, prognostic, and therapeutic screening regimes. With such goals in mind, it is paramount that larger scale studies be performed in order to assign proper traits to emerging gene (or protein) co-expression profiles and set informed standards for the architecture of observer-independent analyses.

SUMMARY

Tailored cDNA Arrays, Ovarian Cancer, and Diagnostic Strategies

The poor prognosis and high mortality rate of ovarian cancer are largely due to a lack of proper screening schemes and to the inability to detect the disease in its early stages of development. Current tests are based on gross anatomical examination or limited marker-specific screening tests (e.g., transvaginal ultrasonography and CA125), which lack the specificity and sensitivity for the disease in its various phases. Histopathological typing, grading, and staging can greatly facilitate the classification of tissue samples and prognosis of the patient, but these observer-dependent methods alone lack robustness. Because of the lack of detailed diagnostic measures, there are no tailored treatments available for individual ovarian cancer cases, and a general nonspecific approach is usually followed, resulting in mixed responses from patients who suffer from the same type of disease. Current diagnoses based on broad phenotypic or limited marker-specific ovarian cancer traits may be missing important sub-classification of patients into appropriate treatment groups. An integrative format for testing multiple pathological determinants is thus proposed in the attempt to reveal a range of molecular changes underlying ovarian cancer.

The tailored cDNA array results revealed some marker specific changes in tran script abundance common to all ovarian cancer samples when compared to healthy counterparts by t test analysis; for example, c-kit and cadherin-6. This is in line with other markers known to be broadly altered in ovarian cancer, such as Her-2/neu (8,39,115) and c-myc (3,18,37). On the other hand, pathway-specific marker profiles (e.g., angiogenesis-related factors) were quite different in distinct ovarian cancer types (pd-spa and spa), indicating that a general approach to therapeutic strategies may be inappropriate for the broad range of ovarian disease states (100). Together, these results suggest two distinct strategies for future therapeutic development: (*i*) a general approach, targeting therapy to signaling molecules that are overexpressed in all ovarian cancer cases (e.g., c-kit-specific immunotherapy) and (*ii*) a specific approach, targeting therapy to signaling molecules that are overexpressed in a specific subset of tumor type [e.g., pd-spa angiogenesis (VEGF-A and ANG-1) and spa angiogenesis (Tie2 and ANG-1) (100). Further studies incorporating larger sample groups will aid in the assessment of the most promising possibilities for diagnostic and therapeutic applications.

ICA-based analysis of the cDNA array ovarian cancer data proved to be a powerful tool for unveiling a variety of gene expression profiles (gene signatures) in the tissue samples. Many of the gene signatures were ovarian cancer type-specific (e.g., profiles prominent in ovarian pd-spa, bmc, or bsc, compared to other ovarian samples), yet a number of gene signatures varied independently of cancer type (e.g., chemoresistance/matrix integrity gene signature). These cancer type-independent variations suggest that there is heterogeneity within the patient population and raise a hopeful method to sub-classify candidates who need different treatment regimes. To develop this method of examining diseases and the underlying processes driving the various states of disease, larger scale studies are required. The definition of proper gene (or protein) signatures to test for would be dependent on the disease or pathophysiological process under investigation and choice of the high-throughput screening tool employed.

Perspectives on High-Throughput Studies for Biomedical Applications

There is currently a rapid-evolving wave of new technologies for biomedical research applications, with a growing emphasis on high-throughput tools. This chapter has presented practical issues on the management and analysis of cDNA array data; similar issues at the level of high-throughput proteomic studies are equally important. Experimental design and the quality of the surveying and analytical tools employed are important factors in the outcome of applying these high-throughput technologies.

Parameters such as gene signatures or other gene co-expression pattern descriptors could prove an efficient method to query inherent underlying conditions in a given sample and assign a molecular definition of specimens accordingly. Similarly, variations in protein content, or "protein signatures" in samples could be defined using protein arrays. Thus, diagnostic capabilities could be greatly improved by querying the gene- or protein-signature content of a given sample under investigation. With improved methods of describing disease states and improved multi-targeted therapeutic developments may follow.

One of the main aims of biomedical research is the improvement of diagnosis and treatment strategies. Mixed patient response, drug resistance, and lack of targeted drug delivery (i.e., systemic versus tissue-specific treatment) are common hurdles in the development and administration of therapies (47). In terms of treatment, several relevant issues could be considered: (*i*) there is currently no comprehensive parameter to measure or assess the heterogeneous response to treatment [e.g., capacity for drug detoxification; susceptibility to drug resistance (59,97,147)] among patients otherwise grouped solely on the diagnosed pathology, and (*ii*) the development of diagnostic tools able to define diseases based on multiple processes may allow for the design of multi-targeted therapies tailored to different pathophysiological processes that are inherent in similar disease phenotypes. Bioinformatics is playing a substantial role in achieving these goals, working with data emerging from a range of high-throughput studies, be it from genomic, proteomic, or other related fields (43,66,80,93).

The interpretation of results depends on knowledge available from past and ongoing studies of specific markers, disease states, and relevant model systems. The importance of experimental design is therefore again emphasized. Animal model systems and in vitro studies do not always correlate with the in vivo situation in humans (25,70). Thus, it is important to consider the context of data interpretation and the validity of comparable systems in the analytical process.

High-throughput arrays and further technological advances continue to evolve. With increased accessibility, some level of standardization may be reached and several of these tools may become part of routine screening protocols in the future. In parallel, diagnostic and therapeutic regimens based on gene-, protein-, or other multi-marker co-expression patterns may develop, and advanced treatment strategies for ovarian cancer and other diseases may emerge.

ACKNOWLEDGMENTS

I would like to thank the many people at the Department of Obstetrics and Gynaecology and Department of Pathology at the University of Cambridge, the MRC UK HGMP Resource Centre, and the Cavendish Laboratories at the University of Cambridge that have contributed to the development and analysis of the work presented in this chapter. I especially thank Stephen K. Smith for the many fruitful discussions and support throughout the study; Anthony N. Corps for insight and feedback on experimental design; Cristin Print, Kate Day, Amanda Evans, Heidi Sowter, Diana Licence, Andrew Sharkey, and Stephen Charnock-Jones for their assistance with technical and administrative issues and many fruitful discussions; Mike Starkey for experienced advice on production of the arrays; and Brian D.M. Tom for advice and collaboration on the statistical analysis of the data. I am very grateful to David J.C. MacKay and James W. Miskin of the Cavendish

Laboratories for invaluable discussions and collaborative work on the pattern recognition and ICA-based analysis of the data. Finally, my gratitude to the Cambridge Commonwealth Trust for funding throughout the period of the study.

REFERENCES

1. **Abeln, E.C., N.J. Kuipers-Dijkshoorn, E.M. Berns, S.C. Henzen-Logmans, G.J. Fleuren, and C.J. Cornelisse.** 1995. Molecular genetic evidence for unifocal origin of advanced epithelial ovarian cancer and for minor clonal divergence. Br. J. Cancer *72*:1330-1336.
2. **Abeln, E.C., V.T. Smit, J.W. Wessels, W.J. de Leeuw, C.J. Cornelisse, and G.J. Fleuren.** 1997. Molecular genetic evidence for the conversion hypothesis of the origin of malignant mixed mullerian tumours. J. Pathol. *183*:424-431.
3. **Abeysinghe, H.R., E. Cedrone, T. Tyan, J. Xu, and N. Wang.** 1999. Amplification of C-MYC as the origin of the homogeneous staining region in ovarian carcinoma detected by micro-FISH. Cancer Genet. Cytogenet. *114*:136-143.
4. **Ahmed, S.I., J. Thompson, J.M. Coulson, and P.J. Woll.** 2000. Studies on the expression of endothelin, its receptor subtypes, and converting enzymes in lung cancer and in human bronchial epithelium. Am. J. Respir. Cell Mol. Biol. *22*:422-431.
5. **Alon, U., N. Barkai, D.A. Notterman, K. Gish, S. Ybarra, D. Mack, and A.J. Levine.** 1999. Broad patterns of gene expression revealed by clustering analysis of tumor and normal colon tissues probed by oligonucleotide arrays. Proc. Natl. Acad. Sci. USA *96*:6745-6750.
6. **Apa, R., F. Miceli, D. de Feo, E. Pierro, G. Ayala, S. Mancuso, M. Napolitano, and A. Lanzone.** 1998. Endothelin-1: expression and role in human corpus luteum. Am. J. Reprod. Immunol. *40*:370-376.
7. **Asschert, J.G., E. Vellenga, H. Hollema, A.G. van der Zee, and E.G. de Vries.** 1997. Expression of macrophage colony-stimulating factor (M-CSF), interleukin-6, (IL-6), interleukin-1 β (IL-1 β), interleukin-11 (IL-11) and tumour necrosis factor-α (TNF-α) in p53-characterised human ovarian carcinomas. Eur. J. Cancer *33*:2246-2251.
8. **Aunoble, B., R. Sanches, E. Didier, and Y.J. Bignon.** 2000. Major oncogenes and tumor suppressor genes involved in epithelial ovarian cancer. Int. J. Oncol. *16*:567-576.
9. **Baak, J.P., F.A. Langley, A. Talerman, and J.F. Delemarre.** 1986. Interpathologist and intrapathologist disagreement in ovarian tumor grading and typing. Anal. Quant. Cytol. Histol. *8*:354-357.
10. **Baker, V.V., M.P. Borst, D. Dixon, K.D. Hatch, H.M. Shingleton, and D. Miller.** 1990. c-myc amplification in ovarian cancer. Gynecol. Oncol. *38*:340-342.
11. **Balvert-Locht, H.R., J.W. Coebergh, W.C. Hop, H.A. Brolmann, M. Crommelin, D.J. van Wijck, and M.T. Verhagen-Teulings.** 1991. Improved prognosis of ovarian cancer in The Netherlands during the period 1975–1985: a registry-based study. Gynecol. Oncol. *42*:3-8.
12. **Belaaouaj, A., J.M. Shipley, D.K. Kobayashi, D.B. Zimonjic, N. Popescu, G.A. Silverman, and S.D. Shapiro.** 1995. Human macrophage metalloelastase. Genomic organization, chromosomal location, gene linkage, and tissue-specific expression. J. Biol. Chem. *270*:14568-14575.
13. **Berchuck, A., G.J. Olt, L. Everitt, A.P. Soisson, R.C. Bast, Jr., and C.M. Boyer.** 1990. The role of peptide growth factors in epithelial ovarian cancer. Obstet. Gynecol. *75*:255-262.
14. **Berchuck, A., G.C. Rodriguez, A. Kamel, R.K. Dodge, J.T. Soper, D.L. Clarke-Pearson, and R.C. Bast, Jr.** 1991. Epidermal growth factor receptor expression in normal ovarian epithelium and ovarian cancer. I. Correlation of receptor expression with prognostic factors in patients with ovarian cancer. Am. J. Obstet. Gynecol. *164*:669-674.
15. **Berchuck, A., M.F. Kohler, and R.C. Bast, Jr.** 1992A. Oncogenes in ovarian cancer. Hematol. Oncol. Clin. North Am. *6*:813-827.
16. **Berchuck, A., G. Rodriguez, G. Olt, R. Whitaker, M.P. Boente, B.A. Arrick, D.L. Clarke-Pearson, and R.C. Bast, Jr.** 1992B. Regulation of growth of normal ovarian epithelial cells and ovarian cancer cell lines by transforming growth factor-β. Am. J. Obstet. Gynecol. *166*:676-684.
17. **Berchuck, A., M.F. Kohler, M.P. Boente, G.C. Rodriguez, R.S. Whitaker, and R.C. Bast, Jr.** 1993. Growth regulation and transformation of ovarian epithelium. Cancer *71*:545-551.
18. **Berchuck, A., A. Elbendary, L. Havrilesky, G.C. Rodriguez, and R.C. Bast, Jr.** 1994. Pathogenesis of ovarian cancers. J. Soc. Gynecol. Investig. *1*:181-190.

19. **Berns, E.M., J.G. Klijn, S.C. Henzen-Logmans, C.J. Rodenburg, M.E. van der Burg, and J.A. Foekens.** 1992. Receptors for hormones and growth factors and (onco)-gene amplification in human ovarian cancer. Int. J. Cancer *52*:218-224.
20. **Bertelsen, K., B. Holund, and E. Andersen.** 1993. Reproducibility and prognostic value of histologic type and grade in early epithelial ovarian cancer. Int. J. Gynecol. Cancer *3*:72-79.
21. **Blaakaer, J., M. Baeksted, S. Micic, P. Albrecksen, J. Rygaard, and J. Bock.** 1995. Gonadotrophin-releasing hormone agonist suppression of ovarian tumorigenesis in mice of the Wx/Wv genotype. Biol. Reprod. *53*:775-776.
22. **Black, R.J., F. Bray, J. Ferlay, and D.M. Parkin.** 1997. Cancer incidence and mortality in the European Union: cancer registry data and estimates of national incidence for 1990 [see comments] [published erratum appears in Eur. J. Cancer 1997. *33*:2440]. Eur. J. Cancer *33*:1075-1107.
23. **Blavier, L., P. Henriet, S. Imren, and Y.A. Declerck.** 1999. Tissue inhibitors of matrix metalloproteinases in cancer. Ann. NY Acad. Sci. *878*:108-119.
24. **Boltz, E.M., R.F. Kefford, J.A. Leary, C.R. Houghton, and M.L. Friedlander.** 1989. Amplification of c-ras-Ki oncogene in human ovarian tumors. Int. J. Cancer *43*:428-430.
25. **Bostwick, D.G., D. Ramnani, and J. Qian.** 2000. Prostatic intraepithelial neoplasia: animal models. Prostate *43*:286-294.
26. **Bouck, N.** 1996. p53 and angiogenesis. Biochim. Biophys. Acta *1287*:63-66.
27. **Braungart, E., C. Schumacher, E. Hartmann, H. Nekarda, K.F. Becker, H. Hofler, and M.J. Atkinson.** 1999. Functional loss of E-cadherin and cadherin-11 alleles on chromosome 16q22 in colonic cancer. J. Pathol. *187*:530-534.
28. **Brown, M.P., W.N. Grundy, D. Lin, N. Cristianini, C.W. Sugnet, T.S. Furey, M. Ares, Jr., and D. Haussler.** 2000. Knowledge-based analysis of microarray gene expression data by using support vector machines. Proc. Natl. Acad. Sci. USA *97*:262-267.
29. **Carter, S.K.** 2000. Clinical strategy for the development of angiogenesis inhibitors. Oncologist *5*:51-54.
30. **Casagrande, J.T., E.W. Louie, M.C. Pike, S. Roy, R.K. Ross, and B.E. Henderson.** 1979. "Incessant ovulation" and ovarian cancer. Lancet *2*:170-173.
31. **Chambers, S.K., B.M. Kacinski, C.M. Ivins, and M.L. Carcangiu.** 1997. Overexpression of epithelial macrophage colony-stimulating factor (CSF-1) and CSF-1 receptor: a poor prognostic factor in epithelial ovarian cancer, contrasted with a protective effect of stromal CSF-1. Clin. Cancer Res. *3*:999-1007.
32. **Chilosi, M., A. Mombello, L. Montagna, A. Benedetti, M. Lestani, G. Semenzato, and F. Menestrina.** 1990. Multimarker immunohistochemical staining of calgranulins, chloroacetate esterase, and S100 for simultaneous demonstration of inflammatory cells on paraffin sections. J. Histochem. Cytochem. *38*:1669-1675.
33. **Chu, C.S., A.W. Menzin, D.G. Leonard, S.C. Rubin, and J.E. Wheeler.** 1999. Primary peritoneal carcinoma: a review of the literature. Obstet. Gynecol. Surv. *54*:323-335.
34. **Contasta, I., A.M. Berghella, P. Pellegrini, T. Del Beato, C.A. Casciani, and D. Adorno.** 1999. Relationships between the activity of MMP1/TIMP1 enzymes and the TH1/TH2 cytokine network. Cancer Biother. Radiopharm. *14*:465-475.
35. **Cramer, S.F., L.M. Roth, T.M. Ulbright, M.T. Mazur, C.A. Nunez, D.J. Gersell, S.E. Mills, and F.T. Kraus.** 1987. Evaluation of the reproducibility of the World Health Organization clasification of common ovarian cancers. With emphasis on methodology. Arch. Pathol. Lab. Med. *111*:819-829.
36. **Crickard, K., J.L. Gross, U. Crickard, M. Yoonessi, S. Lele, W.F. Herblin, and K. Eidsvoog.** 1994. Basic fibroblast growth factor and receptor expression in human ovarian cancer. Gynecol. Oncol. *55*:277-284.
37. **Diebold, J., B. Suchy, G.B. Baretton, S. Blasenbreu, W. Meier, M. Schmidt, H. Rabes, and U. Lohrs.** 1996. DNA ploidy and MYC DNA amplification in ovarian carcinomas. Correlation with p53 and bcl-2 expression, proliferative activity and prognosis. Virchows Arch. *429*:221-227.
38. **Dietl, J., H.P. Horny, P. Ruck, and E. Kaiserling.** 1993. Dysgerminoma of the ovary. An immunohistochemical study of tumor-infiltrating lymphoreticular cells and tumor cells. Cancer *71*:2562-2568.
39. **Doherty, J.K., C.T. Bond, W. Hua, J.P. Adelman, and G.M. Clinton.** 1999. An alternative HER-2/neu transcript of 8 kb has an extended 3′UTR and displayes increased stability in SKOV-3 ovarian carcinoma cells. Gynecol. Oncol. *74*:408-415.
40. **Dong, Y., M.D. Walsh, M.C. Cummings, R.G. Wright, S.K. Khoo, P.G. Parsons, and M.A.**

McGuckin. 1997. Expression of MUC1 and MUC2 mucins in epithelial ovarian tumors. J. Pathol. *183*:311-317.

41.**Donnini, S., M.R. Machein, K.H. Plate, and H.A. Weich.** 1999. Expression and localization of placenta growth factor and PlGF receptors in human meningiomas. J. Pathol. *189*:66-71.

42.**Eisen, M.B., P.T. Spellman, P.O. Brown, and D. Botstein.** 1998. Cluster analysis and display of genome-wide expression patterns. Proc. Natl. Acad. Sci. USA *95*:14863-14868.

43.**Eisenberg, D., E.M. Marcotte, I. Xenarios, and T.O. Yeates.** 2000. Protein function in the post-genomic era. Nature *405*:823-826.

44.**Emons, G. and J.J. Kavanagh.** 1999. Hormonal interactions in ovarian cancer. Hematol. Oncol. Clin. North Am. *13*:145-161.

45.**Fathalla, M.F.** 1972. Factors in the causation and incidence of ovarian cancer. Gynecol. Surv. *27*:751-777.

46.**Feig, L.A., R.C. Bast, Jr., R.C. Knapp, and G.M. Cooper.** 1984. Somatic activation of rasK gene in a human ovarian carcinoma. Science *223*:698-701.

47.**Feng, Y., X. Zhang, and H. Sun.** 1996. Effect of carboplatin based combination chemotherapy on ovarian cancer. Chin. Med. J. (Engl). *109*:349-352.

48.**Fidler, I.J. and L.M. Ellis.** 1994. The implications of angiogenesis for the biology and therapy of cancer metastasis. Cell *79*:185-188.

49.**Flyvbjerg, A., O. Mogensen, B. Mogensen, and O.S. Nielsen.** 1997. Elevated serum insulin-like growth factor-binding protein 2 (IGFBP-2) and decreased IGFBP-3 in epithelial ovarian cancer: correlation with cancer antigen 125 and tumor-associated trypsin inhibitor. J. Clin. Endocrinol. Metab. *82*:2308-2313.

50.**Folkman, J., K. Watson, D. Ingber, and D. Hanahan.** 1989. Induction of angiogenesis during the transition from hyperplasia to neoplasia. Nature *339*:58-61.

51.**Folkman, J.** 1990A. What is the evidence that tumors are angiogenesis dependent? J. Natl. Cancer Inst. *82*:4-6.

52.**Folkman, J.** 1990B. Endothelial cells and angiogenic growth factors in cancer growth and metastasis. Introduction. Cancer Metastasis Rev. *9*:171-174.

53.**Fox, H.** 1985. Pathology of surface epithelial tumors, p. 23-32. *In* Ovarian Cancer. Oxford University Press, New York.

54.**Friedman, L.S., E.A. Ostermeyer, E.D. Lynch, C.I. Szabo, L.A. Anderson, P. Dowd, M.K. Lee, S.E. Rowell, et al.** 1994. The search for BRCA1. Cancer Res. *54*:6374-6382.

55.**Fukumoto, M., R.D. Estensen, L. Sha, G.J. Oakley, L.B. Twiggs, L.L. Adcock, L.F. Carson, and I.B. Roninson.** 1989. Association of Ki-ras with amplified DNA sequences, detected in human ovarian carcinomas by a modified in-gel renaturation assay. Cancer Res. *49*:1693-1697.

56.**Gatta, G., M.B. Lasota, and A. Verdecchia.** 1998. Survival of European women with gynaecological tumours, during the period 1978-1989. Eur. J. Cancer *34*:2218-2225.

57.**Golombick, T., D. Dajee, and W.R. Bezwoda.** 1995. Extracellular matrix interactions. 2: Extracellular matrix structure is important for growth factor localization and function. In Vitro Cell. Dev. Biol. Anim. *31*:396-403.

58.**Grant, G.M., J.K. Cobb, B. Castillo, and R.J. Klebe.** 1996. Regulation of matrix metalloproteinases following cellular transformation. J. Cell Physiol. *167*:177-183.

59.**Gregory, R.K., M.E. Hill, J. Moore, R.P. A'Hern, S.R. Johnston, P. Blake, J. Shephard, D. Barton, et al.** 2000. Combining platinum, paclitaxel and anthracycline in patients with advanced gynaecological malignancy. Eur. J. Cancer *36*:503-507.

60.**Haas, M., J. Isakov, and S.B. Howell.** 1987. Evidence against ras activation in human ovarian carcinomas. Mol. Biol. Med. *4*:265-275.

61.**Han, A.C., A.P. Soler, K.A. Knudsen, and H. Salazar.** 1999. Distinct cadherin profiles in special variant carcinomas and other tumors of the breast. Hum. Pathol. *30*:1035-1039.

62.**Henriksen, R., K. Funa, E. Wilander, T. Backstrom, M. Ridderheim, and K. Oberg.** 1993. Expression and prognostic significance of platelet-derived growth factor and its receptors in epithelial ovarian neoplasms. Cancer Res. *53*:4550-4554.

63.**Heyer, L.J., S. Kruglyak, and S. Yooseph.** 1999. Exploring expression data: identification and analysis of coexpressed genes. Genome Res. *9*:1106-1115.

64.**Hilsenbeck, S.G., W.E. Friedrichs, R. Schiff, P. O'Connell, R.K. Hansen, C.K. Osborne, and S.A. Fuqua.** 1999. Statistical analysis of array expression data as applied to the problem of tamoxifen resistance. J. Natl. Cancer Inst. *91*:453-459.

65.**Hirano, T., K. Ishihara, and M. Hibi.** 2000. Roles of STAT3 in mediating the cell growth, differentiation and survival signals relayed through the IL-6 family of cytokine receptors. Oncogene *19*:2548-2556.

66.**Hsiao, L.L., R.L. Stears, R.L. Hong, and S.R. Gullans.** 2000. Prospective use of DNA microarrays for evaluating renal funcion and disease. Curr. Opin. Nephrol. Hypertens. *9*:253-258.

67.**Huang, L.W., A.P. Garrett, D.A. Bell, W.R. Welch, R.S. Berkowitz, and S.C. Mok.** 2000. Differential expression of matrix metalloproteinase-9 and tissue inhibitor of metalloproteinase-1 protein and mRNA in epithelial ovarian tumors. Gynecol. Oncol. *77*:369-376.

68.**Hutson, R., J. Ramsdale, and M. Wells.** 1995. p53 protein expression in putative precursor lesions of epithelial ovarian cancer. Histopathology *27*:367-371.

69.**Indraccolo, S., S. Minuzzo, E. Gola, W. Habeler, F. Carrozzino, D. Noonan, A. Albini, L. Santi, et al.** 1999. Generation of expression plasmids for angiostatin, endostatin and TIMP-2 for cancer gene therapy. Int. J. Biol. Markers *14*:251-256.

70.**Ip, C.** 1996. Mammary tumorigenesis and chemoprevention studies in carcinogen-treated rats. J. Mammary Gland Biol. Neoplasia *1*:37-47.

71.**Ishii, T. and N. Asuwa.** 2000. Collagen and elastin degradation by matrix metalloproteinases and tissue inhibitors of matrix metalloproteinase in aortic dissection. Hum. Pathol. *31*:640-646.

72.**Jackson, J.R., M.P. Seed, C.H. Kircher, D.A. Willoughby, and J.D. Winkler.** 1997. The codependence of angiogenesis and chronic inflammation. FASEB J. *11*:457-465.

73.**Jayant, K., R.S. Rao, B.M. Nene, P.S. Dale, and A. Nandakumar.** 1996. Improved survival in cervical cancer cases in a rural Indian population. Br. J. Cancer *74*:285-287.

74.**Jungblut, P.R., U. Zimny-Arndt, E. Zeindl-Eberhart, J. Stulik, K. Koupilova, K.P. Pleissner, A. Otto, E.C. Muller, et al.** 1999. Proteomics in human disease: cancer, heart and infectious diseases. Electrophoresis *20*:2100-2110.

75.**Kachra, Z., E. Beaulieu, L. Delbecchi, N. Mousseau, F. Berthelet, R. Moumdjian, R. Del Maestro, and R. Beliveau.** 1999. Expression of matrix metalloproteinases and their inhibitors in human brain tumors. Clin. Exp. Metastasis *17*:555-566.

76.**Kacinski, B.M., D. Carter, K. Mittal, L.D. Yee, K.A. Scata, L. Donofrio, S.K. Chambers, K.I. Wang, et al.** 1990. Ovarian adenocarcinomas express fms-complementary transcripts and fms antigen, often with coexpression of CSF-1. Am. J. Pathol. *137*:135-147.

77.**Kacinski, B.M.** 1995. CSF-1 and its receptor in ovarian, endometrial and breast cancer. Ann. Med. *27*:79-85.

78.**Kang, S.K., K.C. Choi, K.W. Cheung, P.S. Nathwani, N. Auersperg, and P.C. Leung.** 2000. Role of gonadotropin-releasing hormone as an autocrine growth factor in human ovarian surface epithelium. Endocrinology *141*:72-80.

79.**Karasek, M.A.** 1999. Progress in our understanding of the biology of psoriasis. Cutis *64*:319-322.

80.**Karp, P.D.** 2000. An ontology for biological function based on molecular interactions. Bioinformatics *16*:269-285.

81.**Khan, J., L.H. Saal, M.L. Bittner, Y. Chen, J.M. Trent, and P.S. Meltzer.** 1999. Expression profiling in cancer using cDNA microarrays. Electrophoresis *20*:223-229.

82.**Kikkawa, F., K. Tamakoshi, A. Nawa, K. Shibata, S. Yamagata, T. Yamagata, and N. Suganuma.** 1997. Positive correlation between inhibitors of matrix metalloproteinase 1 and matrix metalloproteinases in malignant ovarian tumor tissues. Cancer Lett. *120*:109-115.

83.**Kim, I., H.G. Kim, S.O. Moon, S.W. Chae, J.N. So, K.N. Koh, B.C. Ahn, and G.Y. Koh.** 2000. Angiopoietin-1 induces endothelial cell sprouting through the activation of focal adhesion kinase and plasmin secretion. Circ. Res. *86*:952-959.

84.**Kooi, S., H.Z. Zhang, R. Patenia, C.L. Edwards, C.D. Platsoucas, and R.S. Freedman.** 1996. HLA class I expression on human ovarian carcinoma cells correlates with T-cell infiltration in vivo and T-cell expansion in vitro in low concentrations of recombinant interleukin-2. Cell. Immunol. *174*:116-128.

85.**Krag, K.J., G.P. Canellos, C.T. Griffiths, R.C. Knapp, L.M. Parker, W.R. Welch, M. Klatt, and J. Andersen.** 1989. Predictive factors for long-term survival in patients with advanced ovarian cancer. Gynecol. Oncol. *34*:88-93.

86.**Krupitza, G., R. Fritsche, E. Dittrich, H. Harant, H. Huber, T. Grunt, and C. Dittrich.** 1995. Macrophage colony-stimulating factor is expressed by an ovarian carcinoma subline and stimulates the c-myc proto-oncogene. Br. J. Cancer *72*:35-40.

87.**Kupryjanczyk, J., A.D. Thor, R. Beauchamp, C. Poremba, R.E. Scully, and D.W. Yandell.** 1996. Ovarian, peritoneal, and endometrial serous carcinoma: clonal origin of multifocal disease. Mod. Pathol. *9*:166-173.

88.**Kuroda, H., I. Konishi, M. Mandai, K. Nanbu, C.V. Rao, and T. Mori.** 1997. Ovarian cancer in infertile women during or after ovulation induction therapy: expression of LH/hCG receptors and sex steroid receptors. Int. J. Gynecol. Cancer *7*:451-457.

89.**Kvale, G., I. Heuch, S. Nilssen, and V. Beral.** 1988. Reproductive factors and risk of ovarian cancer: a prospective study. Int. J. Cancer *42*:246-251.

90.**Landis, S.H., T. Murray, S. Bolden, and P.A. Wingo.** 1999. Cancer statistics, 1999. CA Cancer J. Clin. *49*:8-31, 1.

91.**Launay, J.M., G. Birraux, D. Bondoux, J. Callebert, D.S. Choi, S. Loric, and L. Maroteaux.** 1996. Ras involvement in signal transduction by the serotonin 5-HT2B receptor. J. Biol. Chem. *271*:3141-3147.

92.**Li, S., H. Han, E. Resnik, M.L. Carcangiu, P.E. Schwartz, and T.L. Yang-Feng.** 1993. Advanced ovarian carcinoma: molecular evidence of unifocal origin. Gynecol. Oncol. *51*:21-25.

93.**Liam, H.A. and T.V. Venkatesh.** 2000. Bioinformatics in the pre- and post-genomic eras. Trends Biotechnol. *18*:133-135.

94.**Loercher, A.E., M.A. Nash, J.J. Kavanagh, C.D. Platsoucas, and R.S. Freedman.** 1999. Identification of an IL-10-producing HLA-DR-negative monocyte subset in the malignant ascites of patients with ovarian carcinoma that inhibits cytokine protein expression and proliferation of autologous T cells. J. Immunol. *163*:6251-6260.

95.**Lou, W., Z. Ni, K. Dyer, D.J. Tweardy, and A.C. Gao.** 2000. Interleukin-6 induces prostate cancer cell growth accompanied by activation of stat3 signaling pathway. Prostate *42*:239-242.

96.**MacKay, D.J.C.** 1995. Developments in probabilistic modelling with neural networks—ensemble learning, p. 191-198. *In* Neural networks: Artificial Intelligence and Industrial Applications. Proceedings of the 3rd Annual Symposium on Neural Networks. Springer-Verlag, New York.

97.**Makar, A.P.** 2000. Hormone therapy in epithelial ovarian cancer. Endocr. Relat. Cancer *7*:85-93.

98.**Manek, S. and M. Wells.** 1999. Pathology of borderline ovarian tumours. Clin. Oncol. (R. Coll. Radiol.) *11*:73-77.

99.**Maniotis, A.J., R. Folberg, A. Hess, E.A. Seftor, L.M. Gardner, J. Pe'er, J.M. Trent, P.S. Meltzer, et al.** 1999. Vascular channel formation by human melanoma cells in vivo and in vitro: vasculogenic mimicry. Am. J. Pathol. *155*:739-752.

100.**Martoglio, A.M., B.D. Tom, M. Starkey, A.N. Corps, D.S. Charnock-Jones, and S.K. Smith.** 2000. Changes in tumorigenesis- and angiogenesis-related gene transcrpt abundance profiles in ovarian cancer detected by tailored high density cDNA arrays. Mol. Med. *6*:750-765.

101.**Martoglio, A.M., J.W. Miskin, S.K. Smith, D.J.C. MacKay.** 2002. A decomposition model to track gene expression signatures: preview on observer-independent classification of ovarian cancer. Bioinformatics *18*:1617-1624.

102.**Masters, J.R. and S.R. Lakhani.** 2000. How diagnosis with microarrays can help cancer patients. Nature *404*:921.

103.**Mazars, R., P. Pujol, T. Maudelone, P. Jeanteur, and C. Theillet.** 1991. p53 mutations in ovarian cancer: a late event? Cancer Res. *6*:1685-1690.

104.**Miki, Y., J. Swensen, D. Shattuck-Eidens, P.A. Futreal, K. Harshman, S. Tavtigian, Q. Liu, C. Cochran, et al.** 1994. A strong candidate for the breast and ovarian cancer susceptibility gene BRCA1. Science *266*:66-71.

105.**Miller, A.B.** 1995. An epidemiological perspective on cancer screening. Clin. Biochem. *28*:41-48.

106.**Miskin, J.W.** 2001. Ensamble learning for independent component analysis. PhD Thesis. Department of Physics, University of Cambridge, UK.

107.**Morgan, B.J.T.** 1992. Analysis of Quantal Response Data. Chapman and Hall, London.

108.**Moser, T.L., T.N. Young, G.C. Rodriguez, S.V. Pizzo, R.C. Bast, Jr., and M.S. Stack.** 1994. Secretion of extracellular matrix-degrading proteinases is increased in epithelial ovarian carcinoma. Int. J. Cancer *56*:552-559.

109.**National Institutes of Health.** 1994. Ovarian cancer: screening, treatment, and followup. National Institutes of Health Consensus Statement. National Institutes of Health, Bethesda, MD. *12*:5-6.

110.**Norrby, K.** 1997. Angiogenesis: new aspects relating to its initiation and control. APMIS *105*:417-437.

111.**Nose, A., A. Nagafuchi, and M. Takeichi.** 1987. Isolation of placental cadherin cDNA: identification of a novel gene family of cell-cell adhesion molecules. EMBO J. *6*:3655-3661.
112.**Offner, F.A., P. Obrist, S. Stadlmann, H. Feichtinger, P. Klingler, M. Herold, H. Zwierzina, A. Hittmair, et al.** 1995. IL-6 secretion by human peritoneal mesothelial and ovarian cancer cells. Cytokine *7*:542-547.
113.**Patan, S.** 1998. TIE1 and TIE2 receptor tyrosine kinases inversely regulate embryonic angiogenesis by the mechanism of intussusceptive microvascular growth. Microvasc. Res. *56*:1-21.
114.**Riman, T., I. Persson, and S. Nilsson.** 1998. Hormonal aspects of epithelial ovarian cancer: review of epidemiological evidence. Clin. Endocrinol. (Oxf). *49*:695-707.
115.**Ross, J.S., F. Yang, B.V. Kallakwy, C.E. Sheehan, R.A. Ambros, and P.J. Muraca.** 1999. HER-2/neu oncogene amplification by fluorescence in situ hybridization in epithelial tumors of the ovary. Am. J. Clin. Pathol. *111*:311-316.
116.**Ross, D.T., U. Scherf, M.B. Eisen, C.M. Perou, C. Rees, P. Spellman, V. Iyer, S.S. Jeffrey, et al.** 2000. Systematic variation in gene expression patterns in human cancer cell lines. Nat. Genet. *24*:227-235.
117.**Rowinsky, E.K., R. Humphrey, L.A. Hammond, C. Aylesworth, L. Smetzer, M. Hidalgo, M. Morrow, L. Smith, et al.** 2000. Phase I and pharmacologic study of the specific matrix metalloproteinase inhibitor BAY 12-9566 on a protracted oral daily dosing schedule in patients with solid malignancies. J. Clin. Oncol. *18*:178-186.
118.**Rubin, S.C., C.L. Finstad, M.G. Federici, L. Scheiner, K.O. Lloyd, and W.J. Hoskins.** 1994. Prevalence and significance of HER-2/neu expression in early epithelial ovarian cancer. Cancer *73*:1456-1459.
119.**Saarialho-Kere, U.K.** 1998. Patterns of matrix metalloproteinase and TIMP expression in chronic ulcers. Arch. Dermatol. Res. *290(Suppl)*:S47-54.
120.**Samstag, Y., C. Eckerskorn, S. Wesselborg, S. Henning, R. Wallich, and S.C. Meuer.** 1994. Costimulatory signals for human T-cell activation induce nuclear translocation of pp19/cofilin. Proc. Natl. Acad. Sci. USA *91*:4494-4498.
121.**Samstag, Y., E.M. Dreizler, A. Ambach, G. Sczakiel, and S.C. Meuer.** 1996. Inhibition of constitutive serine phosphatase activity in T lymphoma cells results in phosphorylation of pp19/cofilin and induces apoptosis. J. Immunol. *156*:4167-4173.
122.**Sang, Q.X.** 1998. Complex role of matrix metalloproteinases in angiogenesis. Cell Res. *8*:171-177.
123.**Sato, T., H. Saito, R. Morita, S. Koi, J.H. Lee, and Y. Nakamura.** 1991. Allelotype of human ovarian cancer. Cancer Res. *51*:5118-5122.
124.**Schaefer, L.K., D.G. Menter, and T.S. Schaefer.** 2000. Activation of sta3 and sta1 DNA binding and transcriptional activity in human brain tumour cell lines by gp130 cytokines. Cell Signal. *12*:143-151.
125.**Schildkraut, J.M. and W. D. Thompson.** 1988. Familial ovarian cancer: a population-based case-control study. Am. J. Epidemiol. *128*:456-466.
126.**Scully, R.E.** 1979. Treatment of ovarian teratoma. Am. J. Obstet. Gynecol. *135*:161-162.
127.**Sinha, P., G. Hutter, E. Kottgen, M. Dietel, D. Schadendorf, and H. Lage.** 1999. Increased expression of epidermal fatty acid binding protein, cofilin, and 14-3-3-sigma (stratifin) detected by two-dimensional gel electrophoresis, mass spectrometry and microsequencing of drug-resistant human adenocarcinoma of the pancreas. Electrophoresis *20*:2952-2960.
128.**Smith, S.K.** 1998. Angiogenesis, vascular endothelial growth factor and the endometrium. Hum. Reprod. Update *4*:509-519.
129.**Sparen, P., L. Gustafsson, L.G. Friberg, J. Ponten, R. Bergstrom, and H.O. Adami.** 1995. Improved control of invasive cervical cancer in Sweden over six decades by earlier clinical detection and better treatment. J. Clin. Oncol. *13*:715-725.
130.**Spellman, P.T., G. Sherlock, M.Q. Zhang, V.R. Iyer, K. Anders, M.B. Eisen, P.O. Brown, D. Botstein, et al.** 1998. Comprehensive identification of cell cycle-regulated genes of the yeast *Saccharomyces cerevisiae* by microarray hybridization. Mol. Biol. Cell *9*:3273-3297.
131.**Stalsberg, H., V. Abesler, G.P. Blom, L. Bostad, E. Skarland, and G. Westgaard.** 1988. Observer variation in histologyc classification of malignant and borderline ovarian tumors. Hum. Pathol. *19*:1030-1035.
132.**Stanyon, C.A. and O. Bernard.** 1999. LIM-kinase1. Int. J. Biochem. Cell Biol. *31*:389-394.
133.**Steichen-Gersdorf, E., H.H. Gallion, D. Ford, C. Girodet, D.F. Easton, R.A. DiCioccio, G.**

Evans, M.A. Ponder, et al. 1994. Familial site-specific ovarian cancer is linked to BRCA1 on 17q12-21. Am. J. Hum. Genet. *55*:870-875.

134.**Stouffer, R.L., M.S. Grodin, J.R. Davis, and E.A. Surwit.** 1984. Investigation of binding sites for follicle stimulating hormone and chorionic gonadotrophin in human ovarian cancer. J. Clin. Endocrinol. Metab. *59*:441-446.

135.**Stromberg, K., T.J. Collins 4th, A.W. Gordon, C.L. Jackson, and G.R. Johnson.** 1992. Transforming growth factor-α acts as an autocrine growth factor in ovarian carcinoma cell lines [published erratum appears in Cancer Res. 1992. *52*:1382]. Cancer Res. *52*:341-347.

136.**Stulik, J., H. Kovarova, A. Macela, J. Bures, P. Jandik, F. Langr, A. Otto, B. Thiede, et al.** 1997. Overexpression of calcium-binding protein calgranulin B in colonic mucosal diseases. Clin. Chim. Acta *265*:41-55.

137.**Stupack, D.G., C.M. Storgard, and D.A. Cheresh.** 1999. A role for angiogenesis in rheumatoid arthritis. Braz. J. Med. Biol. Res. *32*:573-581.

138.**Sugino, N., S. Takiguchi, S. Kashida, A. Karube, Y. Nakamura, and H. Kato.** 2000. Superoxide dismutase expression in the human corpus luteum during the menstrual cycle and in early pregnancy. Mol. Hum. Reprod. *6*:19-25.

139.**Suzuki, T., N. Sugino, T. Fukaya, S. Sugiyama, T. Uda, R. Takaya, A. Yajima, and H. Sasano.** 1999. Superoxide dismutase in normal cycling human ovaries: immunohistochemical localization and characterization. Fertil. Steril. *72*:720-726.

140.**Tamada, Y., S. Iida, D. Aoki, S. Nozawa, and T. Irimura.** 1999. Carbohydrate epitopes and mucins expressed by 17 human ovarian carcinoma cell lines. Oncol. Res. *11*:233-241.

141.**Tamayo, P., D. Slonim, J. Mesirov, Q. Zhu, S. Kitareewan, E. Dmitrovsky, E.S. Lander, and T.R. Golub.** 1999. Interpreting patterns of gene expression with self-organizing maps: methods and application to hematopoietic differentiation. Proc. Natl. Acad. Sci. USA *96*:2907-2912.

142.**Tanimoto, H., Y. Yan, J. Clarke, S. Korourian, K. Shigemasa, T.H. Parmley, G.P. Parham, and T.J. O'Brien.** 1997. Hepsin, a cell surface serine protease identified in hepatoma cells, is overexpressed in ovarian cancer. Cancer Res. *57*:2884-2887.

143.**Tashiro, H., K. Miyazaki, H. Okamura, A. Iwai, and M. Fukumoto.** 1992. c-myc over-expression in human primary ovarian tumours: its relevance to tumour progression. Int. J. Cancer *50*:828-833.

144.**Tasinato, R., M. Godina, M. Biral, O. Nicoletto, and L. Griggio.** 1999. Peritoneal carcinomatosisi from ovarian cancer: cytoreductive surgery and intraperitoneal chemotherapy. J. Chemother. *11*:239-240.

145.**Tavazoie, S., J.D. Hughes, M.J. Campbell, R.J. Cho, and G.M. Church.** 1999. Systematic determination of genetic network architecture. Nat. Genet. *22*:281-285.

146.**Taylor-Papadimitriou, J., J. Burchell, D.W. Miles, and M. Dalziel.** 1999. MUC1 and cancer. Biochim. Biophys. Acta *1455*:301-313.

147.**Thall, P.F., R.E. Millikan, and H.G. Sung.** 2000. Evaluating multiple treatment courses in clinical trials. Stat. Med. *19*:1011-1028.

148.**van't Veer, L.J., R. Hermens, L.A. van den Berg-Bakker, N.C. Cheng, G.J. Fleuren, J.L. Bos, F.J. Cleton, and P.I. Schrier.** 1988. ras oncogene activation in human ovarian carcinoma. Oncogene *2*:157-165.

149.**van Hal, N.L., O. Vorst, A.M. van Houwelingen, E.J. Kok, A. Peijnenburg, A. Aharoni, A.J. van Tunen, and J. Keijer.** 2000. The application of DNA microarrays in gene expression analysis. J. Biotechnol. *78*:271-280.

150.**Van Niekerk, C.C., F.C. Ramaekers, A.G. Hanselaar, J. Aldeweireldt, and L.G. Poels.** 1993. Changes in expression of differentiation markers between normal ovarian cells and derived tumors. Am. J. Pathol. *142*:157-177.

151.**Versnel, M.A., M. Haarbrink, A.W. Langerak, P.A. de Laat, A. Hagemeijer, T.H. van der Kwast, L.A. van den Berg-Bakker, and P.I. Schrier.** 1994. Human ovarian tumors of epithelial origin express PDGF in vitro and in vivo. Cancer Genet. Cytogenet. *73*:60-64.

152.**von Lampe, B., B. Barthel, S.E. Coupland, E.O. Riecken, and S. Rosewicz.** 2000. Differential expression of matrix metalloproteinases and their tissue inhibitors in colon mucosa of patients with inflammatory bowel disease. Gut *47*:63-73.

153.**Walsh, D.A.** 1999. Angiogenesis and arthritis. Rheumatology (Oxford) *38*:103-112.

154.**Watson, J.M., J.L. Sensintaffar, J.S. Berek, and O. Martinez-Maza.** 1990. Constitutive production of interleukin 6 by ovarian cancer cell lines and by primary ovarian tumor cultures. Cancer Res.

50:6959-6965.

155. **Werness, B.A., R.A. DiCioccio, and M.S. Piver.** 1997. Identical, unique p53 mutations in a primary ovarian mucinous adenocarcinoma and a synchronous contralateral ovarian mucinous tumor of low malignant potential suggest a common clonal origin. Hum. Pathol. *28*:626-630.
156. **Whang, J.D. and J.H. Lee.** 1997. Molecular genetics of gynecologic cancer. J. Korean Med. Sci. *12*:383-389.
157. **Whitman, G.F., T.E. Nolan, and D.G. Gallup.** 1993. Cancer of the Ovary. Raven Press, New York.
158. **Whittemore, A.S.** 1992. Characteristics relating to ovarian cancer risk: collaborative analysis of 12 U.S. case-control studies. Am. J. Epidemiol. *136*:1184-1203.
159. **Wu, S., K. Rodabaugh, O. Martinez-Maza, J.M. Watson, D.S. Silberstein, C.M. Boyer, W.P. Peters, J.B. Weinberg, et al.** 1992. Stimulation of ovarian tumor cell proliferation with monocyte products including interleukin-1, interleukin-6, and tumor necrosis factor-α. Am. J. Obstet. Gynecol. *166*:997-1007.
160. **Yamashita, K., I. Azumano, M. Mai, and Y. Okada.** 1998. Expression and tissue localization of matrix metalloproteinase 7 (matrilysin) in human gastric carcinomas. Implications for vessel invasion and metastasis. Int. J. Cancer *79*:187-194.
161. **Yee, D., F.R. Morales, T.C. Hamilton, and D.D. Von Hoff.** 1991. Expression of insulin-like growth factor I, its binding proteins, and its receptor in ovarian cancer. Cancer Res. *51*:5107-5112.
162. **Yonekura, H., S. Sakurai, X. Liu, H. Migita, H. Wang, S. Yamagishi, M. Nomura, M.J. Abedin, et al.** 1999. Placenta growth factor and vascular endothelial growth factor B and C expression in microvascular endothelial cells and pericytes. Implication in autocrine and paracrine regulation of angiogenesis. J. Biol. Chem. *274*:35172-35178.
163. **Zanders, E.D., M.G. Goulden, T.C. Kennedy, and K.E. Kempsell.** 2000. Analysis of immune system gene expression in small rheumatoid arthritis biopsies using a combination of subtractive hybridization and high-density cDNA arrays. J. Immunol. Methods *233*:131-140.
164. **Zhou, D.J., N. Gonzalez-Cadavid, H. Ahuja, H. Battifora, G.E. Moore, and M.J. Cline.** 1988. A unique pattern of proto-oncogene abnormalities in ovarian adenocarcinomas. Cancer *62*:1573-1576.

9 Gene Expression Analysis of Ovarian Tumors with a cDNA Microarray Containing Ovarian Tumor-Specific Clones

Kwong-Kwok Wong[1], Rita Cheng[1], Jae-Hoon Kim[2], and Samuel C. Mok[3]

[1]Texas Children's Cancer Center, Baylor College of Medicine, Houston, TX, USA; [2]Department of Obstetrics and Gynecology, Saint Vincent Hospital and Catholic University of Korea, Kyong-Ki-Do, Korea; and [3]Brigham and Women's Hospital, Harvard Medical School, Boston, MA, USA

INTRODUCTION

Microarray technology has becoming a popular tool for the target discovery and for the molecular classification of tumors (11–13,15). Many potential diagnostic, prognostic, and therapeutic markers have been identified in a variety of cancers. Several laboratories have applied microarray technology to the study of ovarian cancer (2,5,8,16–18). Using a cDNA microarray with only 2400 genes, we have previously identified two potential molecular markers, prostasin and osteopontin (7,10,17), for the early detection of ovarian cancers. In addition, we have fabricated a cDNA microarray with 10 000 genes and have identified a few hundred differentially expressed genes in ovarian tumors. The functional role of these differentially expressed genes is analyzed with Onto-Express (http://vortex.cs.wayne.edu/Projects.html) (Figure 1) (6). Apparently, many genes involved in signal transduction are differentially expressed, followed by genes involved in cell-to-cell interaction. Such an analysis gives us insight into what biological processes or pathways are distorted in tumor cells. However, these differentially expressed genes may not be ideal tumor-specific therapeutic targets because the genes represented by most

Perspectives in Gene Expression
Edited by K. Appasani

cDNA microarrays are not specific to ovarian cancer. These markers not only express in ovarian tumors but also in other tumor types or normal cells. For instance, osteopontin, a potential early detection marker for ovarian cancer, is also up-regulated in colon cancer and breast cancer (1,4). Until a complete set of transcripts derived from the completed human genome sequence can be developed for microarray analysis, a tumor-specific library may be an alternative for the identification of tumor-specific markers.

FABRICATION OF AN OVARIAN-SPECIFIC cDNA MICROARRAY

The National Cancer Institute (NCI) Cancer Genome Anatomy Project (CGAP) has determined the gene expression profiles of normal, precancer, and cancer cells by generating more than 400 CGAP libraries and sequencing thousands of clones from each library. The CGAP Web site (http://cgap.nci.nih.gov) allows researchers to search and analyze all CGAP libraries and other cDNA libraries. Using the cDNA xProfiler tool at the CGAP Web site, we were able to identify nine ovarian tumor libraries that were derived from microdissected ovarian tumor cells. A total of 7706 sequenced clones from these nine different CGAP microdissected ovarian tumor libraries were purchased from Incyte Genomics

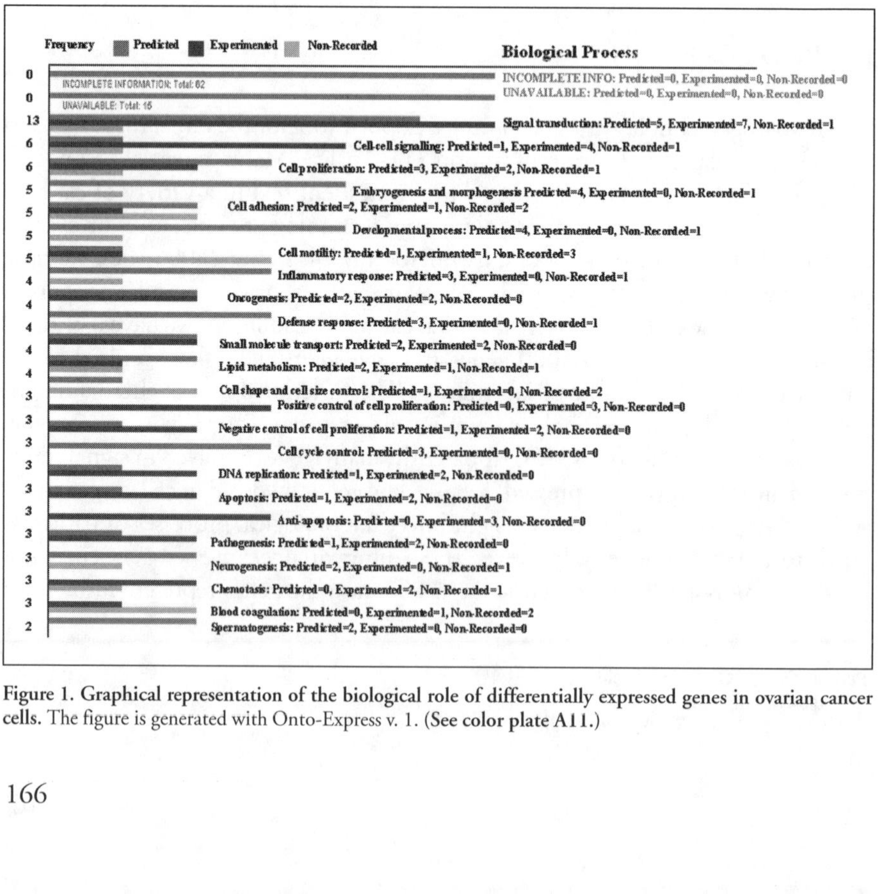

Figure 1. Graphical representation of the biological role of differentially expressed genes in ovarian cancer cells. The figure is generated with Onto-Express v. 1. (See color plate A11.)

Table 1. CGAP Microdissected Ovarian Tumor Libraries

Library Name	No. of Unique Genes[a]	Characteristics (Microdissected Tumors)
NCI_CGAP_Ov26	98	ovary papillary serous carcinoma
NCI_CGAP_Ov31	39	ovary invasive tumor
NCI_CGAP_Ov32	118	ovary invasive tumor
NCI_CGAP_Ov33	17	ovary borderline (LMP)
NCI_CGAP_Ov34	24	ovary borderline (LMP)
NCI_CGAP_Ov36	19	ovary borderline (LMP)
NCI_CGAP_Ov37	1	ovary early stage papillary serous carcinoma
NCI_CGAP_Ov39	31	ovary papillary serous ovarian metastasis to the omentum
NCI_CGAP_Ov40	74	ovary endometroid ovarian metastasis to the omentum

[a]Some of the unique genes are present in more than one of these libraries.
LMP, low malignant potential.

Table 2. Number of Genes[a] with Greater Than 2-Fold Up-Regulation in Different Histological Subtypes of Ovarian Tumors

Histological Subtype (Total Number)	Endometrioid	Clear Cell	Mucinous	Serous Low-Grade	Serous High-Grade
Endometrioid (4)	18	1	2	0	8
Clear cell (5)		2	1	1	2
Mucinous (3)			10	1	4
Serous low-grade (4)				1	1
Serous high-grade (9)					18

[a]The number of up-regulated genes are genes that are up-regulated in all the samples that have been analyzed.

(Palo Alto, CA, USA) (Table 1). By comparing the gene expression between these pools of ovarian tumor cDNA libraries with the rest of the cDNA libraries (approximately 6000), a list of 294 genes that are unique to these nine ovarian tumor libraries was identified. From the sequencing analysis, 12 of these clones have similarity to Alu repeats, and eight of these clones are no longer in the UniGene database. Using these 294 ovarian unique expressed sequence tags, 19 housekeeping genes, three *Arabidopsis* spike controls, and another 443 known genes from these libraries as well as from our own ovarian tumor library, we constructed a cDNA microarray. Clones were re-arrayed into two 384-well plates, PCR amplified, and purified. PCR products were re-sequenced and printed onto GAPS slides (Corning, Corning, NY, USA) in duplicate using a PixSys™ 5500XL arrayer (Cartesian Technologies, Irvine, CA, USA). Figure 2 shows a typical result.

GENE EXPRESSION ANALYSIS OF 25 OVARIAN TUMORS

We used this ovarian-specific cDNA microarray to analyze the expression profiles of 25 ovarian tumors, including nine serous grade III, four serous grade I, three mucinous, five clear cell, and four endometrioid histological subtypes. A pool of six independent human ovarian surface epithelial (HOSE) primary cell cultures was used as reference RNA. The tyramide signal amplification (TSA) indirect labeling method (17) was adopted for this study because of the limited amount of total RNA from the samples. Briefly, 1 μg total RNA from tumor samples was labeled with biotin-dUTP during cDNA synthesis, while the reference RNA was labeled with fluorescein-dUTP. After hybridization, the signals from the tumor and reference samples were amplified with tyramide-Cy3 and tyramide-Cy5, respectively, in the presence of HRP and hydrogen peroxide. Slides were

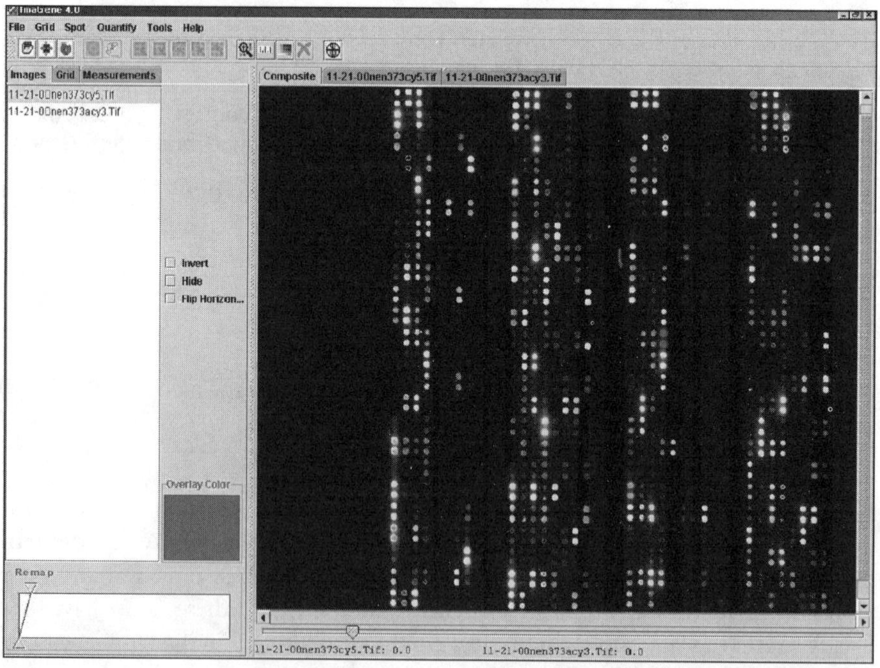

Figure 2. Quantitation of differential hybridization signals between normal ovarian surface epithelial cells (green) and ovarian cancer cells (red) from a homemade ovarian-specific cDNA microarray with 768 genes printed in duplicate. Analysis was performed with software Imagene v. 4.1. Yellow indicates almost equal signals from both samples. In this analysis, we identified over 40 differentially expressed genes. One microgram of total RNA from tumor or HOSE cells was labeled with the MICROMAX™ TSA system. **(See color plate A12.)**

Table 3. Ovarian-Specific Clones That Are Highly Expressed in Normal Ovarian Surface Epithelial Cells

UGCluster[a]	UGRepAcc	Chromosome Location[b]	Library
Hs.145671	AI265824	12q21	NCI_CGAP_Ov26
Hs.235556	AI305310	19q13.3	NCI_CGAP_Ov33
Hs.245672	AW085913	13q12	NCI_CGAP_Ov32
Hs.327062	AI250953	N.D.	NCI_CGAP_Ov36
Hs.254398	AW268344	8q24.2	NCI_CGAP_Ov26
Hs.258282	AW516738	N.D.	NCI_CGAP_Ov39
Hs.258177	AW277033	1p31	NCI_CGAP_Ov39

[a]List of ovarian tumor-specific genes was selected as consistently in the top list of highly expressed genes from the 25 replicates of hybridization signal from HOSE cells.
[b]The chromosome location is identified by homology search with BLAST against the draft human genome sequences (http://www.ncbi.nlm.nih.gov/genome/seq/HsBlast.html).
N.D., not determined.

then scanned with a dual laser scanner, ScanArray® 3000 (Perkin Elmer Life Sciences, Boston, MA, USA). Hybridization signals were quantified with the ImaGene™ v. 4.2 software (BioDiscovery, Marina Del Rey, CA, USA). Hybridization signals were normalized to the median signal of 19 housekeeping genes.

OVARIAN-SPECIFIC GENES THAT ARE HIGHLY EXPRESSED IN NORMAL OVARIAN SURFACE EPITHELIAL CELLS (HOSE)

The majority of ovarian tumors is believed to originate from ovarian surface epithelial cells. Thus, we would expect that some of the 294 ovarian-specific genes might also express strongly in HOSE cells. Total RNA extracted from a pool of six independent HOSE cells was used as the reference RNA for each of the 25 tumor samples. The number of genes that are consistently up-regulated in each of the histological subtypes was identified (Table 2). A list of genes that are highly expressed in the HOSE cells was identified (Table 3). It is not surprising to find that some of these ovarian tumor-specific genes are also highly expressed in the normal HOSE cells since ovarian tumors are believed to derive from ovarian surface epithelial cells. The expression level is comparable to some of the housekeeping genes. Thus, the result agrees with the notion that the ovarian tumors originate from the HOSE cells. By a homology search with BLAST software (9) against the human genome sequences, we were able to locate the chromosome location of five of these genes, while the remaining two did not have any hit on the current human genome sequence. These two unmapped genes are likely to be located in the sequencing gap of the human genome sequence.

OVARIAN-SPECIFIC GENES THAT ARE DIFFERENTIALLY EXPRESSED IN OVARIAN TUMORS

About 10% of the 294 genes are consistently up-regulated in ovarian tumors using HOSE cells as the reference (Table 4). The list of positive significant genes is generated by the Significance Analysis of Microarrays (SAM) software (14). SAM is similar to Student's *t* test statistics, but with permutations to calculate the false discovery rate. It computes a statistic di for each gene i, measuring the strength of the relationship between differential expression with different histological subtypes in a multi-class response analysis. The input data are the expression ratios for each of the 25 microarray experiments. More technical detail can be found in the manual that can be downloaded with the software (http://www-stat.stanford.edu/~tibs/SAM). The positive significant genes are found to be up-regulated greater than 2-fold in at least two tumor samples of the same histological subtypes. A majority of these clones is up-regulated in endometrioid and high-grade serous histological subtypes. This bias is consistent with the origin of the tumor-specific clones that are mainly derived from microdissected tumor cells of endometrioid and high-grade serous histological subtypes.

FUTURE PERSPECTIVE

Microarray analysis of ovarian cancer has facilitated the rapid identification of potential diagnostic and therapeutic targets. However, this is just the beginning of a long journey to fully understand the pathogenesis of ovarian cancer. The cDNA microarray analysis of the 294 ovarian tumor-specific cDNA clones with 25 ovarian tumors demonstrated that tumor-specific genes exist that can serve as potential targets for drug development. A problem with current chemotherapy for cancer treatment is its lack of adequate specificity. As a result, chemotherapeutic agents also cause significant damage to normal host tissues. If we can validate the tissue specificity of these tumor-specific genes and its role in the pathogenetic pathways, then we might be able to develop a more specific drug for a specific tumor type. Similar strategies can be applied to other tumors. It will be exciting to identify all the tissue-specific and tumor-specific cDNA clones using the CGAP resources (3) and use those identified clones for a tumor-specific microarray. By coupling with the recent advances in data mining and visualization tools, we might be able to gain a better understanding of the pathogenetic pathways in ovarian cancers. By defining the functional role of each ovarian tumor-

Table 4. Ovarian-Specific Clones That Are Differentially Expressed in Ovarian Tumors

UGCluster	UGRepAcc	In Unique Library	Ovarian Histological Subtype with Up-Regulation[a]
Hs.133114	AI255115	NCI_CGAP_Ov32, NCI_CGAP_Ov26	endometrioid, clear cell, and high-grade serous
Hs.133115	AI792568	NCI_CGAP_Ov33, NCI_CGAP_Ov26	endometrioid, clear cell, low-grade serous, and high-grade serous
Hs.133279	AI792515	NCI_CGAP_Ov26	low-grade serous and high-grade serous
Hs.145333	AI792894	NCI_CGAP_Ov32	endometrioid and high-grade serous
Hs.146039	AI792572	NCI_CGAP_Ov31, NCI_CGAP_Ov26	clear cell, high-grade serous
Hs.203180	AI792869	NCI_CGAP_Ov34	endometrioid and high-grade serous
Hs.208587	AI054269	NCI_CGAP_Ov26	endometrioid, clear cell, high-grade serous
Hs.209026	AI792554	NCI_CGAP_Ov26	endometriod, clear cell, low-grade serous, and high-grade serous
Hs.242088	AW086056	NCI_CGAP_Ov32	clear cell, low-grade serous, and high-grade serous
Hs.243396	AW086168	NCI_CGAP_Ov32	high-grade serous
Hs.245672	AW085913	NCI_CGAP_Ov32	endometrioid, low-grade serous, and high-grade serous
Hs.254313	AW265129	NCI_CGAP_Ov40	endometrioid, clear cell, and high-grade serous
Hs.254549	AW270514	NCI_CGAP_Ov40	clear cell and high-grade serous
Hs.254561	AW270769	NCI_CGAP_Ov40	high-grade serous
Hs.254853	AW276726	NCI_CGAP_Ov39	endometrioid, and high-grade serous
Hs.254878	AW277062	NCI_CGAP_Ov39	high-grade serous
Hs.256249	AW270399	NCI_CGAP_Ov40	endometrioid and high-grade serous
Hs.256846	AW438546	NCI_CGAP_Ov40	clear cell and high-grade serous
Hs.275021	BE049069	NCI_CGAP_Ov26	high-grade serous
Hs.278023	AW277123	NCI_CGAP_Ov39	clear cell and high-grade serous

[a]Genes are up-regulated greater than 2-fold in at least two tumor samples of the same histological subtype.

specific differentially expressed gene into the context of the altered cancer pathways, hopefully, we may one day conquer ovarian cancer.

REFERENCES

1. **Agrawal, D., T. Chen, R. Irby, J. Quackenbush, A.F. Chambers, M. Szabo, A. Cantor, D. Coppola, and T.J. Yeatman.** 2002. Osteopontin identified as lead marker of colon cancer progression, using pooled sample expression profiling. J. Natl. Cancer Inst. *94*:513-521.

2. **Bayani, J., J.D. Brenton, P.F. Macgregor, B. Beheshti, M. Albert, D. Nallainathan, J. Karaskova, B. Rosen, et al.** 2002. Parallel analysis of sporadic primary ovarian carcinomas by spectral karyotyping, comparative genomic hybridization, and expression microarrays. Cancer Res. *62*:3466-3476.
3. **Boon, K., E.C. Osorio, S.F. Greenhut, C.F. Schaefer, J. Shoemaker, K. Polyak, P.J. Morin, K.H. Buetow, et al.** 2002. An anatomy of normal and malignant gene expression. Proc. Natl. Acad. Sci. USA *99*:11287-11292.
4. **Fedarko, N.S., A. Jain, A. Karadag, M.R. Van Eman, and L.W. Fisher.** 2001. Elevated serum bone sialoprotein and osteopontin in colon, breast, prostate, and lung cancer. Clin. Cancer Res. *7*:4060-4066.
5. **Haviv, I. and I.G. Campbell.** 2002. DNA microarrays for assessing ovarian cancer gene expression. Mol. Cell. Endocrinol. *191*:121-126.
6. **Khatri, P., S. Draghici, G.C. Ostermeier, and S.A. Krawetz.** 2002. Profiling gene expression using Onto-express. Genomics *79*:266-270.
7. **Kim, J.H., S.J. Skates, T. Uede, K.K. Wong, J.O. Schorge, C.M. Feltmate, R.S. Berkowitz, D.W. Cramer, et al.** 2002. Osteopontin as a potential diagnostic biomarker for ovarian cancer. JAMA *287*:1671-1679.
8. **Konno, R.** 2001. Gene expression profiling of human ovarian epithelial tumors by oligo nucleotide microarray. Hum. Cell *14*:261-266.
9. **Madden, T.L., R.L. Tatusov, and J. Zhang.** 1996. Applications of network BLAST server. Methods Enzymol. *266*:131-141.
10. **Mok, S.C., J. Chao, S. Skates, K. Wong, G.K. Yiu, M.G. Muto, R.S. Berkowitz, and D.W. Cramer.** 2001. Prostasin, a potential serum marker for ovarian cancer: identification through microarray technology. J. Natl. Cancer Inst. *93*:1458-1464.
11. **Nielsen, T.O., R.B. West, S.C. Linn, O. Alter, M.A. Knowling, J.X. O'Connell, S. Zhu, M. Fero, et al.** 2002. Molecular characterization of soft tissue tumours: a gene expression study. J. Lancet *359*:1301-1307.
12. **Pomeroy, S.L., P. Tamayo, M. Gaasenbeek, L.M. Sturla, M. Angelo, M.E. McLaughlin, J.Y. Kim, L.C. Goumnerova, et al.** 2002. Prediction of central nervous system embryonal tumor outcome based on gene expression. Nature *415*:436-442.
13. **Selaru, F.M., T. Zou, Y. Xu, V. Shustova, J. Yin, Y. Mori, F. Sato, S. Wang, et al.** 2002. Global gene expression profiling in Barrett's esophagus and esophageal cancer: a comparative analysis using cDNA microarrays. Oncogene *21*:475-478.
14. **Tusher, V.G., R. Tibshirani, and G. Chu.** 2001. Significance analysis of microarrays applied to the ionizing radiation response. Proc. Natl. Acad. Sci. USA *98*:5116-5121.
15. **van't Veer, L.J., H. Dai, M.J. van de Vijver, Y.D. He, A.A. Hart, M. Mao, H.L. Peterse, K. van der Kooy, et al.** 2002. Gene expression profiling predicts clinical outcome of breast cancer. Nature *415*:530-536.
16. **Wang, K., L. Gan, E. Jeffery, M. Gayle, A.M. Gown, M. Skelly, P.S. Nelson, W.V. Ng, et al.** 1999. Monitoring gene expression profile changes in ovarian carcinomas using cDNA microarray. Gene *229*:101-108.
17. **Wong, K.K., R.S. Cheng, and S.C. Mok.** 2001. Identification of differentially expressed genes from ovarian cancer cells by MICROMAX cDNA microarray system. BioTechniques *30*:670-675.
18. **Wong, K.K., R.S. Cheng, R.S. Berkowitz, and S.C. Mok.** 2002. Gene expression analysis of ovarian cancer cells by cDNA microarrays, p. 127-138. *In* J.A. Warrington, R.Todd, and D. Wong (Eds.), Microarrays and Cancer Research. Eaton Publishing, Westboro, MA.

10 Microarray Applications in Developmental Toxicology

Thomas B. Knudsen, Jeffrey H. Charlap, and Kimberly Rozett Nemeth
Department of Pathology, Anatomy, and Cell Biology, Developmental Biology and Teratology Program, Jefferson Medical College, Thomas Jefferson University, Philadelphia, PA, USA

INTRODUCTION

Development of an organism involves the complex orchestration of cell divisions, migrations, interactions, death, and gene regulation and differentiation. Any agent that interferes with these cellular processes can cause birth defects. Such agents, called "teratogens," include a wide variety of drugs and environmental chemicals for which human exposures are known (28). One goal of researchers working in the area of developmental toxicology is the prevention and intervention of birth defects and other adverse reproductive outcomes in humans (early pregnancy loss, reduced birth weight, and altered functional development) caused by exposures to environmental contaminants, therapeutic and abused drugs, and other factors (25).

The emotional and fiscal impact of birth defects and developmental disabilities cannot be overstated (15). Because of growing concern that human populations are being exposed to an ever-increasing number of potentially harmful agents through the environment, the field of developmental toxicology has become a national research priority (35). How embryos react inappropriately to drugs and chemicals remains a central question in the field. The emergence of anatomical structure, which is, in itself, a phenomenon of differential gene expression and cellular diversification, requires transitory genetic signals and responses that are perturbed by teratogenic agents either directly (10,32) or through cellular damage (16,34). For this reason, the impact that drugs and chemicals have on the embryo can be studied from the perspective of gene expression. Relatively

Perspectives in Gene Expression
Edited by K. Appasani

little is known about the consequences of a teratogenic insult on the molecular homeostasis of the developing embryo (15). DNA microarray analysis, which has become a widely used tool for the generation of gene expression data on a genomic scale, represents a powerful new approach to discover these consequences.

The ultimate goal in large-scale analysis of gene expression is to group genes according to similar expression patterns and to extract functional and causal relationships between genes (45). Microarray-based analysis of RNA profiles (transcriptome) has begun to explore the expression patterns of model organisms such as *Saccharomyces cerevisiae* (12), *Caenorhabditis elegans* (27), *Drosophila melanogaster* (48), *Xenopus laevis* (2), and *Mus musculus* (37) to identify functional and causal relationships between genes as these organisms develop and adapt to changes in their environment. An overriding assumption is that genes that follow similar expression across a range of conditions or times are likely to share common molecular regulatory processes or to participate in similar or complementary functions (45). By partitioning the transcriptome into groups of genes, termed expression clusters, and solving the functional annotation of these gene ensembles within the context of genomic structure and evolution, researchers hope to deduce the molecular physiology of previously uncharacterized genes and to establish the hierarchy of physiological control mechanisms during normal and diseased states.

Such information has a clear impact on our understanding birth defects. In toxicology, the mechanism of action refers to the series of molecular events following an interaction of a drug or chemical with its cellular target(s) and the subsequent alteration(s) in target function that precedes a cascade of cellular events that ultimately leads to the observed effect (14). Because a compound's mode of action includes its primary mechanism of action plus the plethora of heterotypic cellular responses that lead to the obligatory or "critical event," changes in gene expression alone do not imply toxicity (42). Biologically meaningful interpretations of changes in gene expression profiles can only be made within the context of adverse phenotype or unique cellular outcomes; therefore, focused studies are needed in the context of developmental toxicity in general and teratogenesis in particular. A search of the Medline database using the keyword "microarray" returned 1624 published papers to date. Combining this search term with keywords "toxic" or "embryo" yielded 22 and 20 papers, respectively, none of which dealt with developmental toxicity or teratogenesis. Therefore, a major challenge is to apply high-throughput functional genomics procedures to the embryo and make sense of the computational read-outs of deviant gene expression profiles with respect to the critical control processes that underlie teratogenesis.

PRINCIPLES OF TERATOLOGY

Developmental disease may be defined as any abnormal condition of body structure or function that is considered harmful to the developing embryo, fetus, neonate, or juvenile. A hospital-based surveillance study of pregnant women admitted with detailed environmental history found no obvious cause in nearly half of the newborn patients born with congenital malformations, which is consistent with a multifactorial etiology (36). As in any disease process, teratogenesis has four basic elements: causative factors, initiating mechanisms, means of propagation, and clinical outcome. What sets teratogenesis apart from other disease processes is the dynamic nature of the embryo and the acuity by which specific precursor target cell populations move to and from states of differential susceptibility. Trajectories of cellular dysregulation and adaptation interact with developmental changes in the steady state that accompanies each advancing minute or hour. Furthermore, the mammalian embryo, unlike its cleidoic *Vertebrate* relatives, is strongly influenced by the mother's current physiological status. This not only relates to the causative factors of birth defects but also to preventive campaigns such as dietary folate supplementation (15).

In 1959, Dr. James Wilson (49) indoctrinated five major principles that he later expanded and modified that address the biological basis of differential teratogenic susceptibility. The original principles are paraphrased as follows. *Genotype:* the susceptibility of a developing organism is dependent on genetic makeup with respect to species, strain, or individual genotype. *Developmental stage:* organ systems traverse asynchronously through critical periods of susceptibility. *Pathogenetic specificity:* a wide range of initiating mechanisms converges to final limiting pathways of cellular pathogenesis. *Bioavailability:* chemical nature is important to an agent's metabolism, availability to the embryo, and effects on maternal physiology. *Dose response:* manifestations of abnormal development increase in frequency and extent, from subclinical [no observable adverse effect level (NOAEL)] toward threshold effects and complete embryotoxicity or lethality.

Today, these principles serve as a framework to understand the impact that environmental factors have on the fetus and to discover better ways of objectively assessing the health risks to children from prenatal exposure to drugs and chemicals (28). The future of risk assessment in developmental toxicology will likely come from integrating the fundamental principles with new knowledge of developmental mechanisms (25). Renewed interest in system-level understanding of biological processes motivated by progress in genomics, high-throughput data collection, and information technology (29) presents a unique opportunity to put Wilson's principles to task. High-throughput data collection (taxonomy), perturbation analysis in different teratological contexts (phenotyping), dynamic control theory (regulatory networks), and building a coherent picture of the exposure-disease continuum (hypothesis testing) are all part of the newer computational efforts to better understand teratogenic mechanisms by integrating the large-scale analysis of gene expression in the embryo under different physiological states.

This chapter focuses on the application of high-throughput taxonomy and high-throughput phenotyping through the construction of a microarray data set, provisionally referred to as developmental toxicity of the mouse embryo

(DTME), which includes gene expression data from various teratogens, doses, exposure times, developmental stages, and precursor target cell populations. The laboratory mouse is an invaluable model organism for understanding the organization and evolution of the human genome as it pertains to mechanisms of disease. Various inbred murine strains (22) and targeted gene knockout models (31) have unique differential susceptibilities to teratogen-induced malformations. The application of comprehensive gene expression profiling technologies to compare normal and diseased systems or to assess molecular alterations that arise from various experimental manipulations in the developing mouse embryo can lead to qualitative and quantitative descriptions of fundamental mechanisms in developmental toxicology. Although DTME is currently limited to experiments in our own laboratory (30), it will eventually incorporate similar kinds of data from other laboratories and model organisms, thus serving as a conceptual framework to investigate the physiological state of the embryo in silico (cyberteratology).

DTME DESCRIPTION

Merging data from different laboratories, technology platforms, and data normalization methods requires standards and practices for reporting the results of microarray-based gene expression experiments. Mimimum information about a microarray experiment (MIAME) provides guidelines that address this need (http://www.mged.org/Workgroups/MIAME/miame.html) (5). MIAME 1.1 aims to provide a conceptual structure for microarray experiment descriptions in two major sections: array design description, which includes platform type, physical characteristics, features of each element in the matrix, and source and gene expression experiment description, which includes the conditions under which RNA samples are generated, processed, and analyzed. Compliance with the MIAME standard will allow for the uniform presentation of such information in sufficient detail to permit the replication, interpretation, and querying of the data, with flexibility to changes in platform or protocol. Developmental toxicologists must remain vigilant of the need for controlled vocabularies to enable database queries and automated data analysis.

Array Design

Generally, the analysis of early embryonic samples is limited by small amounts of tissue. For this and other reasons, we started the DTME project using the two-channel MICROMAX™ system from Perkin Elmer Life Sciences (Gaithersburg, MD, USA). MICROMAX technology uses novel integrative chemistry that culminates in massive signal amplification and greatly enhanced genetic detection (1). The method incorporates unique two-channel indirect labeling of target RNA products with biotin-11-dCTP and fluorescein-12-dCTP with post-hybridization amplification through conventional immunodetection and tyramide signal amplification. Several MICROMAX probes are now available from human and murine cDNA libraries. Murine probes were unavailable when we began the project; hence, the initial studies used microchip MPS621, a human probe spotted with 4800 sequence-verified gene elements (2400 in duplicate) from over 50 different cDNA libraries and positive- and negative-control genes. The cDNAs represent more than 10 different tissue sources: 80% brain-derived and more than 40% full-length cDNA sequence and a range of human diseases, metabolic, and regulatory pathways. The probe is effective with target RNA from human, mouse, or rat biosources.

Experiment Design

For a two-channel platform in developmental toxicology, the mRNA samples may be compared in different contexts. Contrasts may be made between different tissues from the same organism to show how gene expression is correlated with differential susceptibility to teratogen-induced alterations. For example, the forebrain of an early mouse embryo has a low threshold for teratogen-induced programmed cell death (apoptosis), whereas the developing heart from the same embryo is resistant (34). This may well involve baseline differences in the expression of elements in apoptosis signaling pathways (34) or in related controls that more generally influence metabolic differentiation (20).

Alternatively, contrasts may be drawn for the same tissue within an organism after treatment with various teratogens. DTME addresses a set of interrelated questions pertaining to the impact that drugs and chemicals have on patterns of gene expression in the developing mouse embryo (30). Specific experimental conditions are anchored to a priori knowledge of the risk for malformation, guided by the basic tenet that we should try as much as possible to preserve homeostasis in the embryo by testing conditions that flank the threshold for an observable adverse effect. Thus, response to treatment is determined for experimental parameters such as the dose rate of exposure. Dose selection can be guided by the use of empirical teratological data and low-dose modeling with Benchmark Dose Software (BMDS version 1.3.1) from the U.S. Environmental Protection Agency (EPA) (http://www.epa.gov/ncea/bmds.htm). To flank the threshold for developmental toxicity, one simply needs to select the highest dose below the NOAEL, a benchmark dose modeled for an extra 5% risk of malformation (BMD_5) and a moderately teratogenic dose (BMD_{25}). The threshold for a diseased phenotype can thus be estimated as the lower 95% confidence limit of the BMD_5, also referred to as the BMDL (51). Similarly, we can model the dose effect with respect to severity. The particular phenotypes we are concerned with, for example, involve the eye and include keratolenticular dysgenesis (Peters' anomaly), microphthalmia (small eye), and missing eye (anophthalmia), which may be assigned graded scores of 1, 2, or 3, respectively, versus a score of 0 for normophthalmia (50) as a function of dosage.

Contrasts for gene expression profiling may be investigated with respect to time, either as a response through time after teratogen treatment or as a function of gestational age (stage of development). Knowledge of a system's response through time is every bit as critical as its dose response. However, time is a more difficult parameter to model unless the observations are related to a distinct biological process or cellular outcome. To investigate the realizations of a temporal process in a developing embryo, one should ideally anchor the gene expression data to a critical time, which can generally imply the post-treatment interval when dysmorphogenesis becomes obligatory (9).

Contrasts may also involve the same tissue from different organisms. For example, a comparison of embryos matched from two species that respond differently to the same teratogen can provide useful information pertaining to the basis of susceptibility and response through time. Alternatively, intraspecies comparisons may be drawn for genetically modified embryos with targeted disruption of critical genes (50).

DTME currently has approximately 100 hybridizations focused on dose response and fixed-time intervals. Most comparisons in this series used a contemporous reference sample closely matched to the test sample. Additional quality-control steps include an independent replication of each test (T) and reference (R) sample using the reverse labeling scheme (replica swap) not only relevant to incorporation of fluorescein (FL) and biotin (BN) but also the post-hybridization steps for two-channel immunodetection and Cy3 and Cy5 color development. Each hybridization in DTME has a replica twin whereby T = FL-Cy3 and R = BN-Cy5 for sample one and T = BN-Cy5 and R = FL-Cy3 for sample two.

Samples

The quality of the biological sample (biomaterial) from which the target nucleic acids are prepared ultimately controls the experiment. MIAME compliance requires the annotation of experimental biomaterials by their origin (biosource), experimental manipulation, extract preparation, and labeling. The DTME project used *M. musculus* purchased from commercial breeders (Charles River Laboratories or Jackson ImmunoResearch Laboratories). The embryos were harvested on days 8–10 post-coitus, corresponding to human development during the 3rd–4th weeks post-fertilization (24). The precise embryological age was ascertained by counting somite pairs to an accuracy of ±2 pairs per litter. Organism parts included the cranial neural folds or optic pit from Day 8 embryos (4–8 somite pairs), the prosencephalon or optic vesicle from Day 9 embryos (16–20 somite pairs), and the optic cup of Day 10 embryos (28–32 somite pairs). Tissue procurement used conventional microdissection and pooled embryos from the same litter into one sample. Cell types in these samples may include undifferentiated neuroepithelium, cranial neural crest, presumptive foregut endoderm, and occasional immature blood cells. Experiments to date used outbred CD-1 mice and inbred strains that differed in susceptibility to various teratogens. The agents themselves included a prototype antimetabolite (2-chloro-2′-deoxyadenosine), environmental pollutant (methylmercury), oxygen insufficiency (hypoxia), substance of abuse (ethanol), and a pharmacological agent that potentially corrects some or most of the alterations associated with these teratogens (PK11195) (9,38).

Hybridization extract preparation uses total cellular RNA prepared under critical conditions using the RNeasy® kit (Qiagen, Valencia, CA, USA) and DEPC-treated water (Ambion, Austin, TX, USA). This yields high-quality biomaterial when assessed by the RNA 6000 Nano LabChip® kit on the Agilent 2100 Bioanalyzer (Agilent Technologies, Palo Alto, CA, USA) (18). Another quality-control measure is to verify the biological activity of the biomaterial with standard RT-PCRs for positive-control genes (e.g., β-actin and 16S rRNA) or other target genes that are appropriate to the sample. RNA (0.5–2.0 μg) that meets rigid quality-control standards (Figure 1) is spiked with an aliquot of plant RNA specific for positive-control elements on the MPS621 microarray, and the mixture is reverse-transcribed to cDNA using avian myeloblastosis virus (AMV) reverse transcriptase and two-channel hapten incorporation, either fluorescein-12-dCTP (FL) or biotin-11-dCTP (BN). The immunoblot assay provides a quality-control measure for hapten cDNA synthesis labeling efficiency across the two channels (Figure 1).

Hybridization and Image Quantitation

Approximately 15 ng labeled cDNA from each sample is hybridized competitively to an MPS621 microarray probe in a volume of 20 μL overnight at 56°C–59°C. Stringency washes use 0.5× SSC, 0.1% SDS, followed by 0.06× SSC, 0.01% SDS, and 0.06× SSC. For immunodetection, the slide is reacted first with peroxidase-conjugated anti-fluorescein and Cy™3-tyramide, followed by peroxidase conjugated streptavidin and Cy™5-tyramide (Amersham Biosciences, Piscataway, NJ, USA). Fluorescent images of Cy3 and Cy5 channels are generated by MICROMAX Scanning and Data Processing Service (Perkin Elmer Life Sciences). Upper and lower signal limits are approximately 200 and 65 534, respectively. The images are visually inspected for artifacts such as scratches, high background, low signal, dusting, cratering, and local spot dysmorphology. Flags are added to each experiment to denote questionable entries, and the raw images are archived in JPEG format for each channel (FL-Cy3 and BN-Cy5).

SOURCES OF VARIABILITY IN GENE EXPRESSION DATA

Microarray analysis is a multistep process that has sources of error at each level. Although the sources of variation in microarray data are not fully understood, it is necessary to identify as many potential sources of error as possible.

Kerr and Churchill (26) address the main sources of variation in two-channel microarray experiments by identifying four basic factors they refer to as varieties, genes, dyes, and arrays. These factors will be discussed here within the context of teratological experiment design.

Varieties

Variety is the generic term used for test parameters under investigation (26), and its main effects may be parameter-dependent or independent. Parameter-dependent variability is the subject of interest. It derives from the abundance of target RNA molecules in the sample as a function of biosource properties (e.g., species, strain, genotype, stage, and tissue) and experimental manipulation

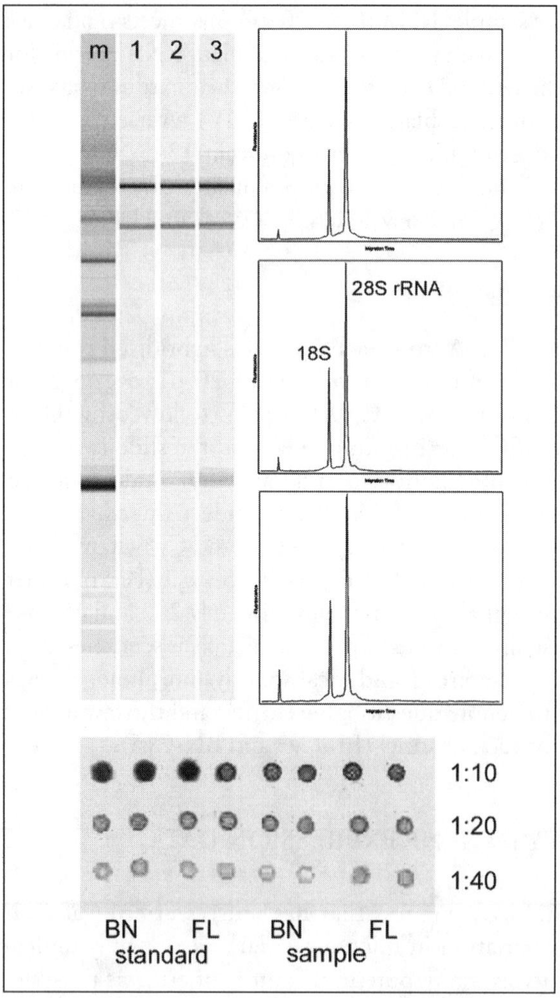

Figure 1. Quality-control assessment by the RNA 6000 nano-microchip electrophoresis on the Agilent 2100 Bioanalyzer. This is an example of three independent samples of the headfold of 8–10 pooled mouse embryos on gestation Day 8 that were collected under various experimental conditions. Qiagen (RNeasy) column purification of total cellular RNA removes small RNAs such as the 5S and transfer RNA bands (these bands are absent at 25 s) but leaves RNA molecules longer than 200 nucleotides. A high-quality mammalian total RNA sample has a 2:1 ratio of 28S:18S rRNA bands, whereas a degraded sample may be 0.5 or lower because the 28S rRNA species is less stable. The 2:1 ratio was observed in all three electropherograms, which indicates that these RNA samples are intact. High-quality embryonic RNA samples were efficiently labeled with biotin-11-dCTP (BN) or fluorescein-12-dCTP (FL) to be used in two-channel microarray hybridization. This is demonstrated by the "3rd dot" rule in a serial dilution of labeled sample versus a pre-labeled standard by quantitative immunoblot assay.

(agent, dose, time, and intervention). In contrast, parameter-independent variability is the biological variation in gene expression that remains after reasonable steps have been taken to control properly for all experimental outcome variables (21). For example, in a teratology experiment, subtle differences in maternal age, physiology, nutritional state, and health status can introduce inter-litter effects on the teratological outcome. Investigators strive to minimize such effects through animal husbandry and colony environs but will soon find that it is impractical to control or reproduce the intrinsic variation across individuals. In light of the high sensitivity of microchip technology, the potential maternal effects further complicate the interpretation of gene expression data.

To interpret gene expression data from a microarray experiment, the magnitude and complexity of genome-wide expression profiles for the system under study should be well characterized under normal (baseline) conditions. This means assessing the natural differences in gene expression for individual embryos at any one developmental stage and any changes in magnitude and complexity between stages. One problem is caused by the heterogeneity of a population of embryos collected at the same gestational age. For example, mouse embryos can vary by several hours within the same litter (24). This blocking factor can be diminished by narrowing the breeding cycle to 4–6 h (vaginal plug = Day 0 of gestation). Still, it is imperative to accurately stage embryos by specific developmental landmarks and to explicitly define the range of embryos used as a biosource for microchip analysis. Counting somite pairs, for example, can produce a narrowly defined range of embryos and, since mice add one pair of somites in 2 h, a collection of embryos within a ±2 somite pair range roughly corresponds to a temporal range of ±4 h or 1% of the 19.5-day gestation period.

Genes

In describing sources of variability in microarray data, Kerr and Churchill (26) operationally define genes as the target RNA molecules in the sample. Thus, the main gene effects occur when certain elements on the probe have different fluorescent signals overall when compared to other elements. In theory, these effects arise because of differential gene expression and the various genetic regulatory processes that govern upward or downward shifts in transcriptional activity or transcript stability. However, in practice, the main gene effects also reflect the cellular complexity of the biosource and procedural artifacts that may arise from differences in mRNA secondary structure or cross-hybridization of homologous sequences (17,23,46). In some cases, the main gene effects may reflect natural differences in gene expression between animals. For example, microarray-based analysis compared baseline levels of gene expression for several tissues procured from six male C57BL6 mice (40). Four replicate measurements were taken for each tissue, and the variance was compared across six mice for 5406 spotted elements on the microarray probe. Analysis of variance (ANOVA) revealed significant inter-animal differences in the variability of gene expression for 0.8%–3.3% of the genes in the adult tissues studied (40). To our knowledge, a study of such individual differences has not been undertaken in the developing mouse embryo.

The reproducibility of the gene expression profiles detected with the application of the Perkin Elmer integrated technology platform to early embryos can be illustrated with a simple quality-control measure. A cross-species comparison of gene expression was performed, using two-channel hybridization of target RNA populations from normal mouse or normal rat embryos hybridized to the human probe. Each biosource was embryonic headfold at the 4–8 somite stage of development, obtained independently from rat and mouse embryos. The patterns were remarkably similar in spite of the collection of RNA samples by collaborating scientists at geographically distinct institutions (Figure 2).

Dyes

The dye main effects refer to differences in the two-channel fluorescent signals (26). The problem of two-channel hapten cDNA synthesis was already discussed,

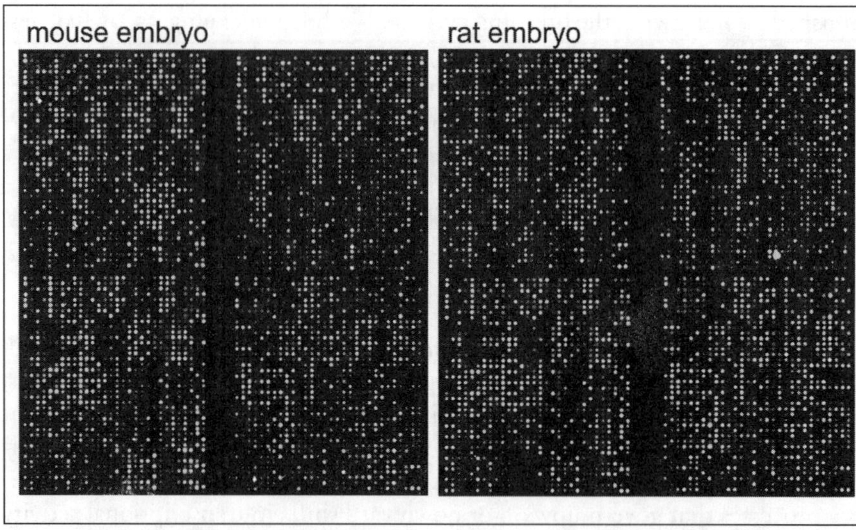

Figure 2. Quality-control assessment by hybridization of mouse and rat embryonic RNA to human probe. RNA procured from the headfold of normal mouse and rat embryos at the 4–8 somite stage of gestation was labeled for two-channel competitive hybridization to the 4800-element MICROMAX MPS621 probe, which has duplicated spots for 2400 genes, including positive (spiked RNA) and negative (unspiked) plant control genes. Each image represents a red-green channel overlay for the intraspecies comparison of normal embryos. High-fidelity hybridization is clearly reflected in: (*i*) sensitivity, shown in the high percentage of spots detected in early embryos; (*ii*) precision, reflected in the degree of similarity between overlay images of mouse-mouse to human probe (left panel) and rat-rat to human probe (right panel); (*iii*) dynamic range, reflected in the signals for duplicate elements in the array; and (*iv*) the consistency between laboratories. The RNA samples from mice were collected and processed at Thomas Jefferson University (Philadelphia, PA), whereas the RNA samples from rats were collected by Dr. C. Lau and Dr. S. Hunter at the National Health and Environmental Effects Research Laboratory, EPA (Research Triangle Park, NC), and shipped to Philadelphia for subsequent processing. **(See color plate A13.)**

and Cy3 or Cy5 fluorescence by their differences in excitation wavelengths, extinction coefficients, and quantum yield may introduce procedural variation that differs between the two channels. Tseng et al. (46) elegantly addressed the problem using aliquots of the same RNA sample and performing "same-same" co-hybridization comparison to calibrate the dye main effects. Each channel required the nonlinear normalization that ultimately depended on the dynamic response within an array slide. Failure to normalize this inter-channel variability would lead to potentially misleading results in the downstream data interpretation. For example, one might erroneously misinterpret a decrease in the Cy3:Cy5 ratio as a downward shift in relative expression, when, in fact, the difference might be an artifact of inter-channel variability. Tseng et al. (46) offer reversal of labeling assignments between test and reference samples as useful protection against the nonlinearity of label normalization, without the need to explicitly model the nonlinearity. This is sometimes referred to as self-normalization (52). Because of the added complexity of the experimental design associated with replica swaps, especially in an experimental series involving a time course or dose response, one should give adequate attention to offsetting dye main effects by appropriate experimental design (26).

An effective method for normalizing the raw data has been reported (52). Data are transformed to $\log_2$ space to make the distribution essentially normal and the variance independent of amplitude. The method compares log (test) with log (reference) by plotting the magnitude of the difference, $M = \log_2(T/R)$, against the average, $A = (1/2)\log_2(T^*R)$. Locally weighted regression (lowess) is then performed to smooth intensity-dependent ratios. The lowess smoothing offsets curvature at the low end and high end of the dynamic range. The spread of data is then standardized within each array using median absolute deviation (52), although we had success with absolute deviation of data points from their mean. Both methods assume scaling independent of outliers. After within-slide normalization, the arrays are subjected to paired-slides normalization for the replica swaps to offset the systematic variability affiliated with two-channel measurements. Unfortunately, a simple replica swap does not necessarily return the inverse absolute ratio in all cases. For example, if a particular gene is 2-fold R > T but labels 5-fold FL-Cy3 > BN-Cy5, then corresponding $\log_2$ ratios for T:R will be -5.0 and +1.25 in the swapped replicas when T = Cy5 or Cy3, respectively. These very different values should not be compared until the assignments are calibrated; hence, we equalize Cy3:Cy5 and Cy5:Cy3 ratios for each spot in the matrix to offset the specific column (k) by a constant scalar for each row (p) in the matrix. This forces a constraint on the data in which the net result of Cy3:Cy5 = Cy5:Cy3 for each gene. Figure 3 shows the impact that normalization has on the spread of microarray data.

Arrays

Array main effects measure the overall variation in fluorescent signal from microarray to microarray (26). The systematic variation between slides may be attributed to complex factors such as the print tip deformation, curvature of the support medium, different drying patterns of the gene elements, individual cDNA sequence, variations in the hybridization environment, and scratches introduced during microchip manufacture or coverslip removal. These can be substantially greater effects than the systematic variability between channels (V > D); hence, there is a need for multiple spotting as well as multiple slides to help assess systematic variation in the probe caused by spots and slides (46). For DTME, the strategy has been the repetition of similar experiments across nonidentical test conditions. Looking across many nonidentical experimental conditions, the spurious low-quality spot should have minimal effect on data interpretation. This reduces the likelihood of potentially misleading results derived from procedural artifacts associated with the microarray probe and ultimately improves the power of the technology to detect low-amplitude changes (19). There is a consensus for performing each experiment at least twice under identical conditions. Each experiment in the DTME dataset was thus replicated at minimum with independent biological samples, with the reversal of label assignments (replica swaps) applied to the independent samples.

Interactions Between Main Effects

With four major sources of potential error, varieties, genes, dyes, and arrays, there are $2^4 = 16$ possible interactions between the main effects (26). Of all possible multifactor combinations, only the varieties/gene interaction has meaning for the biological interpretation of the experiment. All other sources of variation reflect physical-chemical processes that comprise the technology. Varieties/genes interactions arise from both experimental effects (parameter-dependent) and

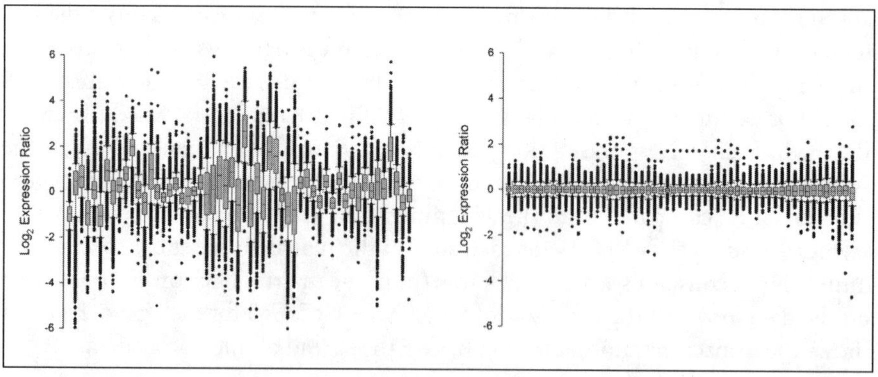

Figure 3. Effect of normalization on the spread of data from 54 DTME arrays (log_2 scale). Box and whiskers plot shows median (line), 25th–75th percentiles (box), and 5th–95th percentiles (whiskers), with outliers as individual points. Adjacent pairs are replica swaps. The left panel plots log_2 ratio before normalization. The right panel plots these same data after normalization by methods adapted from Yang et al. (52). The alignment of the median (zero) and scaling of the data brought 90% of data points within 0.25. Thus, in each case, the outliers are defined as those genes falling below or above the 5th–95th percentiles, respectively.

natural biological variation (parameter-independent); the latter defined as the variation inherent to a system under identical conditions and not caused by technical processing (21). Hence, independent replicates are needed for precision of measurement and to supply an estimate of error by which significance can be judged (n = 1 offers no degrees of freedom for error modeling), requiring adequate attention to the fundamental principles of study design (26).

DTME DATA ANALYSIS

The success of DNA microarray analysis is dependent on the quality and precision of the data, and concerns with regard to the reproducibility of microarray data have motivated ways to minimize the variance of the data. Several approaches have been designed to remove systematic errors introduced by the physics and chemistry of the technology and to provide a close approximation for the true expression ratios in replicated data sets (17,23,46).

Quality-Control Filter

The ability to restrict a gene list based on the behavior of its genes in experiments or in individual samples is an important quality-control tool. Several steps are required to filter genes with low precision, large error values, those that do not vary significantly across multiple samples or are flagged for expression levels that are close to background (23). What remains after normalization should theoretically come under statistical control; there is a practical limit to affordable repeats in microarray experiments so that traditional significance testing (e.g., α = 0.05) may not be applicable to the normalized data set. Because microarrays can monitor 10 000 genes simultaneously, a type I error rate of 5%, for example, would potentially generate 500 false positives. On the other hand, the application of arbitrary cut-off levels can miss subtle but biologically meaningful lower amplitude changes that become evident only as the number of arrays increases (19,46). Since we want to recover as much of the data as possible, the problem is to trade type II error (e.g., false negative and filters that are too conservative) for type I error (e.g., false positive and filters that are too liberal). A "global error model" can be used to help offset the small number of replicates (n = 2). In this case, GeneSpring™ 5.1 software (Silicon Genetics, Redwood City, CA, USA) accounts for two different kinds of random variation to estimate the variability in gene expression measurements: (*i*) measurement variation, corresponding to a single gene per sample (array) within each array and (*ii*) sample-to-sample variation specified by the user (replicas). Precision is based on variance component analysis that calculates the standard error and P value. From these values, we constructed an error model at progressively higher fold change thresholds (19).

An error model is built from embryonic microarray data by binning the numbers of genes passing at each designated fold change and then plotting the percentage of experiments as a function of these bins (Figure 4). For example, when such a plot was generated from normalized data in 80 microarray experiments, nearly all of the 40 replicate conditions produced two or more genes changing by 1.2-fold ($\log_2 = 0.6$). Most of these experiments produced 20 or more genes for this fold change. Contrasting the same data, only 5% of the experiments produced two or more genes at a 2-fold change ($\log_2 = 1.0$). We determined from this error model that the range from a 1.2- to 2-fold change captured the optimal balance between specificity (type I error) and sensitivity (type II error). These data were then used to construct a differential model, whereby the number of genes that passed at a given fold change was graphed by its distribution across all experimental conditions (Figure 4). For example, 941 genes spotted on the microchip cumulatively passed in 15% of the experiments at a 1.5-fold change whereas none of the genes passed in half of the experiments. Similar derivations at a 1.2- or 2-fold revealed little specificity or sensitivity, respectively, suggesting a 1.5-fold change maximizes the balance between specificity and sensitivity, which is in agreement with previous studies (19,46).

Cluster Analysis

Simple sorting and filtering of microchip data is often used to generate lists of genes that show the largest or most significant changes due to a biological pertur-

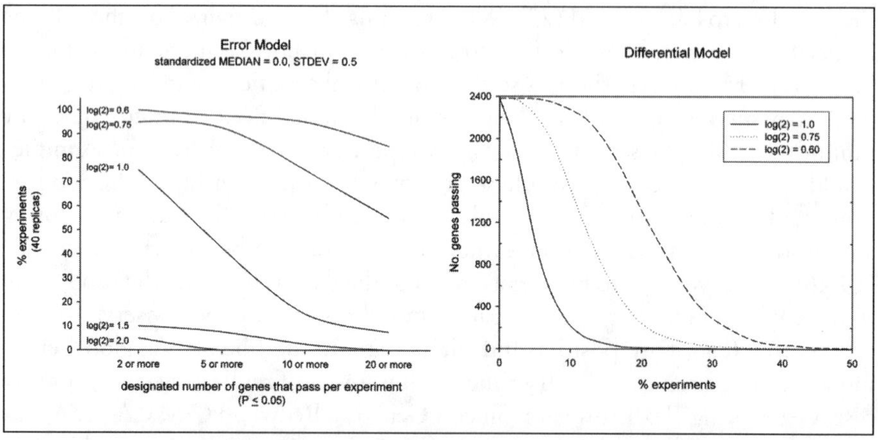

Figure 4. Gene-specific error model and differentiality model for 80 DTME arrays. The 80 arrays (40 replicas) represent diverse conditions that mostly derive from control, sub-threshold, or weakly teratogenic exposure scenarios. The left panel plots the percentage of experiments as a function of designated number of genes (of approximately 2400 in single- or duplicate-element microchips) that pass at each fold level ($P \leq 0.05$ for each replica pair with global error model). The error model is built by assigning bins, each containing a number of genes that passed Student's *t* test comparison ($P \leq 0.05$) at the designated fold change filter, and determining the percentage of experiments (19). The right panel plots the number of genes that pass at a designated fold change as a function of the experiment. The potential for type II error (false negative) and type I error (false positive) must optimize a balance between specificity and sensitivity.

bation. In many cases, the more interesting genes do not exceed a 2-fold change and require richer experimental designs (4). In cases such as developmental toxicology experiments that involve multiple parameters (genotype, stage, tissue, agent, dose, time, and intervention), no two genes are likely to display precisely the same response (44). In general, multivariate approaches are used to allow for the detection of significant correlation in gene expression profiles (45). One common approach is to group genes into a strict hierarchy of nested subsets based on the relative similarity in expression profiles (clustering). This creates a phylogenetic tree whose branch lengths represent the degree of similarity between the sets for discovering the organizing principles inherent to the data structure (13). Phenetic clustering has proven useful to classify genes during various developmental processes, metabolic transitions, and toxicological responses (12,13,19,27,33,41,47,48).

Several main features were evident in the DTME data set during the hierarchical clustering with GeneCluster version 2.1 and visualization with TreeView version 1.5 (http://rana.lbl.gov/EisenSoftware.htm). Using this approach we could easily see the effect that varying the differential filtering had on the main data structure. The high-dimensionality structure decayed when the filter was too stringent and clouded when the filter was too relaxed. The breakdown point for the critical loss of structure was assessed by examining the effect that varying the differential filter had on the clustering of rigidly co-expressing gene ensembles such as the ribosomal protein genes and glycolytic enzymes. Reasonable purity and comprehensivity were achieved when the differential filter was set to a 1.5-fold change across 15% of the experiments. The genes that passed this criterion and were subjected to hierarchical clustering partitioned into two main classes, representing mutually exclusive responses and experimental effects (Figure 5). Analysis of the hierarchical clustering several different times and in several ways produced many consistent features among the resulting dendritograms. The most robust features of the DTME dataset occurred for subsets of genes linked with threshold-level teratological effects and remedial intervention. If robustness for hierarchical inferences of classification or function applies to this analysis (39), then an interesting new concept emerges: regulation of the embryonic transcriptome during the critical period following teratogen exposure is homeostatic, indicated by the highly structured changes in genome-wide gene expression up to a defined threshold indicated by the breakdown in data structure. In the context of systems biology and dynamic control theory, this implies “graceful degradation” that leads to systems failure (29). We are currently exploring other clustering techniques that impose less rigid structure on microchip data (3,6,44).

DTME BIOINFORMATICS

In many respects, DNA microarray technology would not be possible without considerable advances in bioinformatics (45). Systematic methods such as hierarchical clustering have contributed greatly to the fields of comparative and evolutionary biology as tools for finding natural patterns in molecular and morphological data. Whereas phylogenetic systematics track the evolutionary relationships among taxa, systematic treatment of microarray data order that are intrinsic to gene expression profiles may yield insights into molecular, cellular, and tissue-level processes and functions (39). It is not intuitively obvious how similarity and character-based techniques can be applied to the molecular events that govern toxicologically related outcomes (14). Clustering of gene expression data makes sense when it is associated with known biological processes and experimental conditions; hence, a gene's key attribute is its expression level in the different samples. The phenetic trees (dendritograms) constructed by hierarchical clustering are merely visual representations of the overall degrees of similarity based on characters or attributes of the individual sample (13). For this reason, it is essential to correlate gene expression profiling data with adverse phenotype or unique cellular outcomes and to uncover the biological basis for the higher order structure in a microarray dataset. Figure 5 depicts a microarray dendrogram that can be used to frame two questions regarding the biology of the experimental system.

Question 1: which genes group together by similarity of expression across nonidentical experimental conditions? Phenetic clustering is a useful starting point for deriving new information of, or relating to, overall similarity based on gene expression profiles, without regard to evolutionary history. However, the ultimate goal is to define the functional and causal relationships between genes (13). Synexpression group (SEG) is the name given to a set of genes that share a complex "spatial" (multiple tissues) expression pattern and that function in the same process. This term has recently been extended to groups of genes that share complex "temporal" (multiple time points) or "experimental" (multiple experimental conditions) expression profiles that likely confer emergent properties during evolution (37). SEGs may account for at least part of the higher order structure in the DTME data set, as evidenced by the clustering of gene coding for glycolytic enzymes, ribosomal proteins, large macromolecular complexes such as proteasome or chromatin, mitochondrial functions, or evolutionary conserved signaling pathways and checkpoints. Animal development uses at least 17 distinct signaling pathways and several checkpoints to integrate morphogenetic processes (35). The relationship of these coherent pathways to phenetic expression clusters is an important bioinformatics issue. For example, mining DTME for such relationships following exposures that increase the risk for ocular teratogenesis revealed a global effect on the expression of components in the receptor tyrosine kinase signaling pathway. This demonstrates a functional compartmentation of the embryonic genome with respect to a signaling pathway with known importance in ocular development (see Reference 15).

The physical linkage of genes on chromosomes had an agglomerative effect under some experimental conditions. Syntenic relationships refer to blocks of genes preserved in some fashion across species such as segments of a chromosome that have the homologous genes in the same order. These "homoeologous segments" represent the plesiomorphic or shared ancestral condition and may have important bearing on our ability to interpret large-scale gene expression data for several reasons. First, they can be used as landmarks in the low-resolution transcriptome map of chromosomes. Some investigators observed a co-expression of genes that share a common chromosome area. In the developing mouse brain, for example, there is evidence for the topographical co-expression of closely

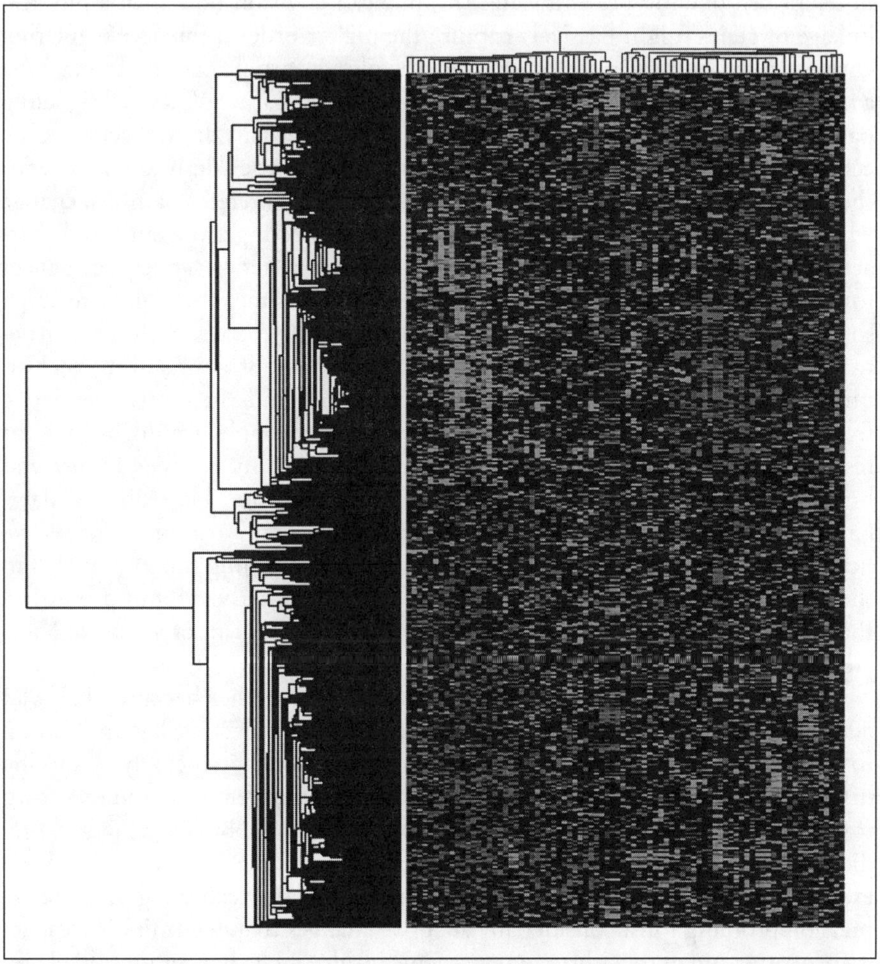

Figure 5. Hierarchical clustering of genes filtered from the DTME data set. Filtering the 2400 genes on the MPS621 microchip by the concordance error model across 40 replicated experiments in the early mouse embryo produced a list of 941 genes that passed at 1.5-fold change in 15% of the replicated samples. Ratiometric gene expression data were subjected to lowess smoothing, scaling, hierarchical clustering, and visualization. Horizontal taxa in the TreeView dendritogram are samples (k = 80), and the vertical relationships are genes (p = 2400). Color-coding: red, up-regulated; green, down-regulated; and black for no change. The intensity of the color indicates the amplitude of change. **(See color plate A14.)**

linked genes within a histological section (Dr. Gregor Eichele, Max Planck Gesellschaft, Hannover Germany, personal communication). A similar phenomenon is observed for many acute inflammation-regulated genes in the rat liver that map to a region on chromosome 6q, with conserved synteny on mouse chromosome 12 and human chromosome 14q13-q32 (43). During the characterization of the human transcriptome map, Caron et al. (8) discovered that more abundant genes tend to broadly group into what they called regions of increased density of gene expression (RIDGES). Because RIDGES vary between different cell phenotypes and states (8), one can envision how such a physical linkage of genes might passively capture the higher-order influences regulating chromatin structure. On the other hand, the plesiomorphic condition may have a more direct relevance to the evolutionary conserved nature of key cell signaling pathways. In an enlightening discussion of ancestral versus derived states of the cell, Planet et al. (39) suggest that genes that reflect the "derived characters," those attributes essentially shared by all cells that have entered a pathway, change when a cell first enters a pathway, while the genes reflecting more "ancestral characters" remain unchanged. If the physical linkage of genes actively influences functional compartmentation of the embryonic genome, then more research should be directed at understanding the relationship(s) between genomic structure-function in the context of the normal development of an organism. Such an understanding from the embryo's response to drugs and chemicals may derive from informational bridges to human disease. For example, we might look for responsive gene mapping to regions of conserved synteny between mouse and humans and search the Online Mendelian Inheritance in Man (OMIM) database for diseases that also map to the same chromosome region (http://www.ncbi.nlm.nih.gov/entrez/query.fcgi?db=OMIM). Performing this exercise led to the hypothesis of coenzyme A and biotin-dependent carboxyl group transfers that are carried out by the mitochondrion in the mechanism of action of a teratogen, which was clearly an unanticipated finding (30).

Question 2: which experimental samples group together because of similar gene expression profiles? Learning how samples group together has application toward identifying sets of biomarker genes that respond predictably to specific drugs or chemicals, characterizing the mode of action of reference and unknown toxicants, elucidating regulatory networks inherent to the biological target, predicting individual susceptibility, and so forth (6,11,47; see Reference 14 for an excellent summary of the constraints and caveats of the technology). Gene expression profiling must consider any adaptive changes that determine the capacity of a target organ to withstand teratogenic insult. Such changes most likely reflect the functionality of signal transduction pathways that integrate the morphogenetic processes with cellular homeostasis (e.g., metabolism, DNA replication and repair, transcription, and protein synthesis and turnover). The concept of "hormesis" or hormetic effects in toxicology refers to a non-monotonic dose response curve characterized by low-dose stimulation and high-dose inhibition (7). This may be applicable to adaptive responses that result in the graceful degradation of a biological system under stress, eventually leading to chaos and system failure above the threshold.

The hormetic mechanism may be a common, evolutionary-based checkpoint strategy to carefully regulate resource allocation in a definable range within the context of the reestablishment of homeostasis (7). It may have relevance as a unifying concept in teratogenesis. For example, Dr. Philip Mirkes at the University of Washington uses microarrays to find genes and associated signaling pathways that govern cellular susceptibility to teratogen-induced programmed cell death (apoptosis) in early mouse embryos. Working with the University of Washington Center for Expression Arrays (http://ra.microslu.washington.edu), the Mirkes team uses a two-color direct-labeling platform with mouse cDNA probes spotted with 15 K elements from the IMAGE consortium. Preliminary findings reveal changes in genes linked with the p53 tumor suppressor gene, a major checkpoint pathway inducing apoptosis (P. Mirkes, personal communication). Dr. Kathleen Sulik at the University of North Carolina-Chapel Hill uses microarrays to identify differences in gene expression that confer susceptibility to ethanol-induced apoptosis in early mouse embryos. Using Affymetrix oligonucleotide microchips, the Sulik team has obtained preliminary evidence for regional effects in sensitive (hindbrain) versus insensitive (midbrain) regions of the primitive brain at critical stages of sensitivity to alcohol-related birth defects (K. Sulik, personal communication). The diversity in experimental design for these and DTME-based studies shows the need for controlled vocabulary and the MIAME-compliant standards and practices that will be needed to merge data across platforms, even for research projects that are reasonably similar in biosource and objectives. Building a compendium of microarray-based experiments will require specialized software such as GeNet™ (Silicon Genetics, Redwood City, CA, USA) to merge primary data (new) with secondary (existing) information from other laboratories and technology platforms.

CONCLUSIONS

Functional genomics and computational biology provide new ways to unravel the complex responses of developing tissues to drugs and chemicals that increase or lower the risk for specific malformations. The new technologies can reveal the relationships between chemical exposure and genome-wide expression patterns. However, several questions remain. What constitutes an optimal microarray design in developmental toxicology or experimental teratogenesis? Can we effectively calibrate different technology platforms? How do we harmonize experiment annotation to make a meaningful contribution to centralized microarray repositories? What is the most meaningful way to correlate transcript signature profiles with teratological outcome? What number of genes determines the critical biology of a system? Are large-scale changes in gene expression even a primary determinant of teratogenesis? Building a comprehensive gene expression matrix that captures the biological complexity of the embryonic transcriptome and its regulation provides a computational resource for deducing the physiological state of the embryo during development and at the threshold for developmental disease (cyberteratology).

ACKNOWLEGMENTS

This publication was made possible by STAR grant no. R82744501-01 from the EPA, grant no. RO1 ES09120 from the National Institute of Environmental Health Sciences (NIEHS), and grant no. RO1 AA13205 from the National Institute on Alcoholism and Alcohol Abuse (NIAAA). K.R.N. was a fellow on training grant no. T32 ES07282 from the NIEHS. The contents of this chapter are solely the responsibility of the authors and do not necessarily represent the official views or policies of the EPA, NIEHS, or the NIAAA.

REFERENCES

1. **Alder, K., J. Broadbent, R. Garlick, R. Joseph, A. Khimani, A. Mikulskis, P. Rapiejko, and J. Killian.** 2000. MICROMAX™: a highly sensitive system for differential gene expression on microarrays, p. 221-230. *In* M. Schena (Ed.), Microarray Biochip Technology. BioTechniques Books, Eaton Publishing, Natick MA.
2. **Altmann, C.R., E. Bell, A. Sczyrba, J. Pun, S. Bekiranov, T. Gaasterland, and A.H. Brivanlou.** 2001. Microarray-based analysis of early development in *Xenopus laevis*. Dev. Biol. *236*:64-75.
3. **Aronow, B.J., B.D. Richardson, and S. Handwerger.** 2001. Microarray analysis of trophoblast differentiation: gene expression reprogramming in key gene function categories. Physiol. Genomics *6*:105-116.
4. **Audic, S. and J.M. Claverie.** 1997. The significance of digital gene expression profiles. Genome Res. *7*:986-995.
5. **Brazma, A., P. Hingamp, J. Quackenbush, G. Sherlock, P. Spellman, C. Stockert, J. Aach, W. Ansorge, et al.** 2001. Minimum information about a microarray experiment (MIAME)—toward standards for microarray data. Nat. Genet. *29*:365-371.
6. **Burczynski, M.E., M. McMillian, J. Ciervo, L. Li, J.B. Parker, R.T. Dunn II, S. Hicken, S. Farr, et al.** 2000. Toxicogenomics-based discrimination of toxic mechanism in HepG2 human hepatoma cells. Toxicol. Sci. *58*:399-415.
7. **Calabrese, E.J. and L.A. Baldwin.** 2002. Applications of hormesis in toxicology, risk assessment and chemotherapeutics. Trends Pharmacol. Sci. *23*:331-337.
8. **Caron, H., B. van Schaik, M. van der Mee, F. Baas, G. Riggins, P. van Sluis, M.-C. Hermus, R. van Asperen, et al.** 2001. The human transcriptome map: clustering of highly expressed genes in chromosomal domains. Science *291*:1289-1292.
9. **Charlap, J.H., R.J. Donahue, and T.B. Knudsen.** 2003. Exposure-disease continuum for 2-chloro-2′-deoxyadenosine, a prototype ocular teratogen. III. Intervention with PK11195. Birth Defects Res. (Part A) *67*:108-115.
10. **Cooper, M.K., J.A. Porter, K.E. Young, and P.A. Beachy.** 1998. Teratogen-mediated inhibition of target tissue response to Shh signaling. Science *280*:1603 1607.
11. **Cunningham, M.J., L.S. Fuhrman, J.J. Seilhamer, and R. Somogyi.** 2000. Gene expression microarray data analysis for toxicology profiling. Ann. NY Acad. Sci. *919*:52-67.
12. **DeRisi, J.L., V.R. Iyer, and P.O. Brown.** 1997. Exploring the metabolic and genetic control of gene

expression on a genomic scale. Science *278*:680-686.

13.**Eisen, M.B., P.T. Spellman, P.O. Brown, and D. Botstein.** 1998. Cluster analysis and display of genome-wide expression patterns. Proc. Natl. Acad. Sci. USA *95*:14863-14868.

14.**Fielden, M.R. and T.R. Zacharewski.** 2001. Challenges and limitations of gene expression profiling in mechanistic and predictive toxicology. Toxicol. Sci. *60*:6-10.

15.**Finnell, R.H., J. Gelineau-van Waes, J.D. Eudy, and T.H. Rosenquist.** 2002. Molecular basis of environmentally induced birth defects. Annu. Rev. Pharmacol. Toxicol. *42*:181-208.

16.**Fleming, A. and A.J. Copp.** 1998. Embryonic folate metabolism and mouse neural tube defects. Science *280*:2107-2109.

17.**Goryachev, A.B., P.F. MacGregor, and A.M. Edwards.** 2001. Unfolding of microarray data. J. Comput. Biol. *8*:443-461.

18.**Kuschel, M. and W. Ausserer.** 2000. Characterization of RNA quality using the Agilent 2100 Bioanalyzer. Application note, http://www.agilent.com/chem/labonachip.

19.**Hughes, T.R., M.J. Marton, A.R. Jones, C.J. Roberts, R. Stoughton, C.D. Armour, H.A. Bennett, E. Coffey, et al.** 2000. Functional discovery via a compendium of expression profiles. Cell *102*:109-126.

20.**Ibrahim, M.M., M. Razmara, D. Nguyen, R.J. Donahue, J.A. Wubah, and T.B. Knudsen.** 1998. Altered expression of mitochondrial 16S ribosomal RNA in p53-deficient mouse embryos revealed by differential display. Biochem. Biophys. Acta. *1403*:254-264.

21.**Johnson, J.R., P. Hurban, J. Woessner, and C.M. Liddell.** 2001. Statistical approaches to analyzing multi-chip data. Proceedings of SPIE *4266*:199-206.

22.**Juriloff, D.M., T.M. Gunn, M.J. Harris, D.G. Mah, M.K. Wu, and S.L. Dewell.** 2001. Multifactorial genetics of exencephaly in SELH/Bc mice. Teratology *64*:189-200.

23.**Kadota, K., R. Miki, H. Bono, K. Shimizu, Y. Okazaki, and Y. Hayashizaki.** 2001. Preprocessing implementation for microarray (PRIM): an efficient method for processing cDNA microarray data. Physiol. Genomics *4*:183-188.

24.**Kaufman, M.H.** 1989. The Atlas of Mouse Development. Academic Press, New York.

25.**Kavlock, R.J. and R.W. Setzer.** 1996. The road to embryologically based dose-response models. Environ. Health Perspect. *104 (Suppl. 1)*:107-121.

26.**Kerr, M.K. and G.A. Churchill.** 2001. Experimental design for gene expression microarrays. Biostatistics *2*:183-201.

27.**Kim, S.K., J. Lund, M. Kiraly, K. Duke, M. Jiang, J.M. Stuart, A. Eizinger, B.N. Wylie, et al.** 2001. A gene expression map for *Caenorhabditis elegans*. Science *293*:2087-2092.

28.**Kimmel, C.A., W.M. Generoso, R.D. Thomas, and K.S. Bakshi.** 1993. Contemporary issues in toxicology. A new frontier in understanding mechanisms of developmental abnormalities. Toxicol. Appl. Pharmacol. *119*:159-165.

29.**Kitano, H.** 2002. Systems biology: a brief overview. Science *295*:1662-1664.

30.**Knudsen, T.B., J.H. Charlap, B.J. Nibblo, R.C. Craig, and K. Rozett Nemeth.** 2003. Identification of an embryonic core response signature to PK11195, and an ameliorative ligand of the mitochondrial benzodiazepene receptor. Birth Defects Res. (Part A) (In press).

31.**Knudsen, T.B. and J.A. Wubah.** 1998. Transgenic animal models. Functional analysis of developmental toxicity as illustrated with the p53 suppressor model, p. 209-221. *In* W. Slikker Jr. and L.W. Chang (Eds.), Handbook of Developmental Neurotoxicology. Academic Press, San Diego.

32.**Kochhar, D.M., H. Jiang., J.D. Penner, A.T. Johnson, and R.A. Chandraratna.** 1998. The use a retinoid receptor antagonist in a new model to study vitamin A-dependent developmental events. Int. J. Dev. Biol. *42*:601-608.

33.**Miki, R., K. Kadota, H. Bono, Y. Mizuno, Y. Tomaru, P. Carninci, M. Itoh, K. Shibata, et al.** 2001. Delineating developmental and metabolic pathways in vivo by expression profiling using the RIKEN set of 18 816 full-length enriched mouse cDNA arrays. Proc. Natl. Acad. Sci. USA *98*:2199-2204.

34.**Mirkes, P.E. and S.A. Little.** 1998. Teratogen-induced cell death in postimplantation mouse embryos: differential tissue sensitivity and hallmarks of apoptosis. Cell Death Differ. *5*:592-600.

35.**National Research Council.** 2000. Scientific Frontiers in Developmental Toxicology and Risk Assessment. National Academy Press, Washington, DC.

36.**Nelson, K. and L.B. Homes.** 1989. Malformations due to presumed spontaneous mutations in newborn infants. N. Engl. J. Med. *320*:119-123.

37.**Niehrs, C. and N. Pollet.** 1999. Synexpression groups in eukaryotes. Nature *402*:483-487.

38.**O'Hara, M.F., J.H. Charlap, R.C. Craig, and T.B. Knudsen.** 2002. Mitochondrial transduction of ocular teratogenesis during methylmercury exposure. Teratology *65*:131-144.

39.**Planet, P.J., R. DeSalle, M. Siddall, T. Bael, I.N. Sarker, and S.E. Stanley.** 2001. Systematic analysis of DNA microarray data: ordering and interpreting patterns of gene expression. Genome Res. *11*:1149-1155.

40.**Pritchard, C.C., L. Hsu, J. Delrow, and P.S. Nelson.** 2001. Project normal: defining normal variance in mouse gene expression. Proc. Natl. Acad. Sci. USA *98*:13266-13271.

41.**Roberts, C.J., B. Nelson, M.J. Marton, R. Stoughton, M.R. Meyer, H.A. Bennett, Y.D. He, H. Dai, et al.** 2000. Signaling and circuitry of multiple MAPK pathways revealed by a matrix of global gene expression profiles. Science *287*:873-880.

42.**Simmons, P.T. and C.J. Portier.** 2002. Toxicogenomics: the new frontier in risk analysis. Carcinogenesis *23*:903-905.

43.**Soury, E., E. Olivier, D. Simon, P. Ruminy, K. Kitada, M. Hiron, M. Daveau, Y. Boyd, et al.** 2001. Chromosomal assignments of mammalian genes with an acute inflammation-regulated expression in liver. Immunogenetics *53*:634-642.

44.**Tamayo, P., D. Slonim, J. Mesirov, Q. Zhu, S. Kitareewan, E. Dmitrovsky, E.S. Lander, and T.R. Golub.** 1999. Interpreting patterns of gene expression with self-organizing maps: methods and application to hematopoietic differentiation. Proc. Natl. Acad. Sci. USA *96*:2907-2912.

45.**Thibault, C., L. Wang, L. Zhang, and M.F. Miles.** 2001. DNA arrays and functional genomics in neurobiology. Int. Rev. Neurobiol. *48*:219-253.

46.**Tseng, G.C., M.-K. Oh, L. Rohlin, J.C. Liao, and W.H. Wong.** 2001. Issues in cDNA microarray analysis: quality filtering, channel normalization, models of variations, and assessment of gene effects. Nucleic Acids Res. *29*:2549-2557.

47.**Waring, J.F., R.A. Jolly, R. Ciurlionis, P.Y. Lum, J.T. Praestgaard, D.C. Morfitt, B. Buratto, C. Roberts, et al.** 2001. Clustering of hepatotoxins based on mechanism of toxicity using gene expression profiles. Tox. Appl. Pharmacol. *175*:28-42.

48.**White, K.P., S.A. Rifkin, P. Hurban, and D.S. Hogness.** 1999. Microarray analysis of *Drosophila* development during metamorphosis. Science *286*:2179-2184.

49.**Wilson, J.G.** 1977. Current status of teratology, general principles and mechanisms derived from animal studies, p. 47-74. *In* J.G. Wilson and F.C. Fraser (Eds.), Handbook of Teratology, vol. 1. Plenum Press, New York.

50.**Wubah, J.A., M.M. Ibrahim, X. Gao, D. Nguyen, M.M. Pisano, and T.B. Knudsen.** 1996. Teratogen-induced eye defects mediated by p53-dependent apoptosis. Curr. Biol. *6*:60-69.

51.**Wubah, J.A., R.W. Setzer, C. Lau, J.H. Charlap, and T.B. Knudsen.** 2001. Exposure-disease continuum for 2-chloro-2′-deoxyadenosine, a prototype ocular teratogen. I. Dose-response analysis. Teratology *64*:154-169.

52.**Yang, Y.H., S. Dudoit, P. Luu, V. Peng, J. Ngai, and T.P. Speed.** 2002. Normalization for cDNA microarray data: a robust composite method addressing single and multiple slide systematic variation. Nucleic Acids Res. *30*:e15.

11 Identifying Antisense Genomic Modulators Using cDNA Microarrays

Yoon S. Cho-Chung[1], Maria Nesterova[1], Rakesh Srivastava[1], Yee Sook Cho[1], Meyoung-Kon Kim[1], Catherine Neary[1], Chris Cheadle[2], and Kevin G. Becker[2]
[1]Cellular Biochemistry Section, Basic Research Laboratory, CCR, National Cancer Institute, National Institutes of Health, Bethesda, and [2]DNA Array Unit, National Institute on Aging, National Institutes of Health, Baltimore, MD, USA

INTRODUCTION

The use of antisense oligodeoxynucleotides (ODNs) to turn off specific genes dates back to the 1970s, when ssDNA was shown to inhibit the translation of a complementary RNA in a cell-free system (59) and a short, antisense ODN could inhibit Rous sarcoma virus replication in culture (98). These investigations provided the first hints at the therapeutic utility of antisense nucleic acids, and Zamecnik was awarded a Lasker prize in 1996 in recognition of this work.

The RIα subunit of cAMP-dependent protein kinase (PKA) (38) is up-regulated in human cancer cell lines (15) and in primary tumors (7,8,33,36,47,50, 51,68,97). The expression of this subunit is also enhanced in cells transformed with the Ki-*ras* (82) oncogene or TGF-α (19) and upon stimulation of cell growth with granulocyte-macrophage colony-stimulating factor (GM-CSF) or phorbol esters (87). Conversely, a decrease in RIα expression correlates with growth inhibition induced by site-selective cAMP analogues in a broad spectrum of human cancer cell lines (14).

This chapter describes the influence of the regulatory subunits of PKA on the circuitry of cAMP signaling in cell growth and differentiation. Experimental approaches described here include antisense ODNs, gene transfer, and gene expres-

Perspectives in Gene Expression
Edited by K. Appasani

sion profiling using cDNA microarrays. An extensive amount of literature has indicated the sequence-specific antisense mechanism of action at a single gene level. This chapter describes the antisense targeted to the RIα regulatory subunit of PKA as an illustration of the antisense mechanism from both the classic view and the genomic-scale view.

TWO ISOFORMS OF PKA TYPE I AND TYPE II

cAMP activates signaling by binding to the regulatory (R) subunit of PKA and inducing the reversible phosphorylation of proteins involved in a number of cellular processes. The two types of PKA, PKA-I and PKA-II, share a common C subunit but contain distinct R subunits, RI and RII (5). Four isoforms of the R subunits (RIα, RIβ, RIIα, and RIIβ) (44,48) and three different C subunits (Cα, Cβ, and Cγ) have been identified (6,67,89,90). However, preferential co-expression of one of these C subunits with any of the R subunits has not been found (6,67). Both RI and RII contain two tandem, cAMP-binding domains at the carboxyl terminus, which is highly conserved, but they differ significantly in the amino terminus, at a proteolytically sensitive hinge region that occupies the catalytic domain of the C subunit in the holoenzyme complex (76). In this segment, RII has the sequence Arg-Arg-X-Ser (75), which can undergo autophosphorylation at Ser (64), whereas RI has the sequence Arg-Arg-X-Ala (77), which does not undergo autophosphorylation but instead participates in the high-affinity binding of ATP in the type I holoenzyme (76). RI and RII also differ in molecular weight, isoelectric point, and immunological characteristics. Other differences include the affinity for cAMP, cAMP analog selectivity, tissue specificity, subcellular localization, expression patterns during development and the cell cycle, and altered expression patterns in transformation and differentiation. These differences suggest specific roles for RI and RII, apart from inhibition of the C subunit, in the regulatory function of cAMP. PKA-I and PKA-II may therefore serve distinct physiological functions. In support of this notion, PKA-I is only transiently overexpressed in normal cells in response to proliferation signals (85,86) but is constitutively overexpressed in cancer cells (7,8,18,33,36,47,50, 51,68,97). Conversely, PKA-II is preferentially expressed in normal, differentiated tissues (17,18,81).

RIα OVEREXPRESSION IN CANCER GENESIS AND PROGRESSION

The expression of the RIα subunit of PKA increases in various primary human tumors and cell lines, including cancers of the breast (36,50,51), ovary (47,68), lung (97), and colon (7,8,33). Furthermore, overexpression of this subunit correlates with malignancy and poor prognosis in cancer patients (47,50,51,68). Increased expression of RI has been associated with both chemical and viral carcinogenesis and oncogene-induced cell transformation, as in the cases of dimethylbenz (α) anthracene-induced mammary carcinogenesis (13), the incidence of gastric adenocarcinoma N-methyl-N′-nitrosguanine, and the trophic action of gastrin on gastric carcinoma production (95).

Although only PKA-II is detected in normal 3T3 cells, PKA-I and PKA-II are both detected in spontaneously transformed cells and cells transformed with simian virus 40 (SV40) or methylcholanthrene (28,93). Despite the increase in PKA-I, the total kinase activity in these cells is equivalent to that of normal cells. RI and PKA-I expression increases in rat 3Y1 cells transformed by human adenovirus type 12, but little or no change occurs in PKA-II expression (41). Likewise, a marked increase in RI expression with a concomitant decrease in RII expression has been detected in Ha-MuSV-transformed NIH 3T3 clone 13-3B-4 cells (21,74), rat (NRK) kidney cells transformed with TGFα or v-Ki-*ras* oncogene (82), mouse mammary epithelial cells transformed by TGFα (20), and a human mammary epithelial cell line (MCF-1OA HE) transformed by a point-mutant cHa-*ras* or by the c-*erbB*-2 proto-oncogene (20). These results suggest that RI may act as a mediator of various mitogenic stimuli.

PKA-I AND PKA-II IN ONTOGENIC DEVELOPMENT AND CELL DIFFERENTIATION

Several researchers have reported a switch in predominant PKA isozymes during development. PKA-I is primarily associated with the rapid increase in protein kinase activity observed in rat testicles during the first postnatal week (43). However, PKA-II activity increases concomitantly with the onset of complete spermatogenesis. Likewise, the PKA-I:PKA-II ratio decreases during the first 20 days of postnatal uterine development and then increases again by Day 40, marking fluctuations in PKA-I activity while PKA-II remains unchanged (23). Similar isozyme changes have been observed during rat Sertoli cell maturation (24); again, the PKA-I:PKA-II ratio decreases as development progresses. Other examples have been observed in ovarian follicles (61) and steroid-dependent tissues post-castration (26). Changes in the PKA isozyme activity have also been observed in mouse heart development (35,46). Both PKA-I and PKA-II levels decrease during heart development, but the decrease in PKA-I is more pronounced, which results in a decreased ratio of PKA-I:PKA-II in the adult heart. The PKA-I:PKA-II ratio also decreases during mouse spermatogenesis from 2.0 in pachytene spermatocytes to 1.0 in round spermatids (22). Similar results have been observed in the ontogeny of the palate in dogs (45) and during human retinal development (27).

Which isozyme marks differentiation has been the point of some debate (60,91,94), although the above results clearly point to PKA-II. Overall, the available literature has not permitted any broad generalization for the roles of PKA-I and PKA-II in cell growth and differentiation. Although the evidence indicates a switch in the isozyme ratio during development, the data have not directly indicated whether differential expression of PKA-I and PKA-II is a signal for cell proliferation or differentiation. Recent studies have used gene transfer and an antisense strategy to address this question.

GENE TRANSFER THROUGH RETROVIRAL VECTORS

Antisense-directed depletion of the RIα subunit results in the compensatory upregulation of RIIβ in growth-arrested cancer cells (16,17,55). Moreover, an RIIβ antisense ODN blocks the growth inhibition and differentiation induced by cAMP in HL-60 leukemia cells (83). These studies suggest a role for RIIβ in controlling cancer cell growth. Therefore, retroviral vector-mediated overexpression of RIIβ has to be used to further explore the relationship between RIIβ and cancer cell growth.

RIIβ Overexpression Induces Reverse Transformation

Overexpression of RIIβ inhibits growth, with no sign of toxicity, in a variety of cancer cell types, including SK-N-SH neuroblastoma, MCF-7 breast carcinoma, Ki-*ras*-transformed NIH 3T3 clone DT (15,79), HL-60 leukemia cells (79), and PC12 mutant A-126 cells (80). SK-N-SH, DT, and A-126 cells also display striking changes in morphology (Figure 1) (15,79,80) similar to those observed in cells exposed to RIα antisense (71,88), including a flat phenotype that is typical of the reverted phenotype of transformed cells. As shown in SK-N-SH and HL-60 cells, overexpression of RIIβ directly induces growth arrest and reverse transformation; no further treatment with cAMP analogs is required (79). These results suggest that the RIIβ subunit acts as a tumor suppressor protein that inhibits growth and promotes differentiation and reverse transformation. A role for RIIβ as a tumor suppressor is further supported by the effects of mutant RIIβ-P overexpression. This mutant, in which the autophosphorylation site Ser 114 has been changed to alanine (42), is incapable of binding the cAMP response element, unlike wild-type RIIβ (69), and cells overexpressing RIIβ-P behave as transformed cells (9,42,58), which suggests that the mechanism of RIIβ may involve transcriptional control of cancer cell growth (69). RIIα, RIα, or Cα overexpression has no effect on cell growth in cancer cells (58).

Preferential Formation of PKA-II over PKA-I Formation on Intact Cells

Overexpression of the RII subunit induces a striking shift in the PKA isozyme distribution by reducing PKA-I levels and increasing PKA-II levels in LS-174T colon carcinoma (58) and MCF-7 breast carcinoma cells (42). PKA-I levels are almost completely eliminated in cells overexpressing RIIβ or RIIβ-P, and different species of PKA-II, which do not appear in parental cells or in cells overexpressing RIIα, are detected in these cells. In contrast, PKA-II levels are unaltered in cells overexpressing RIα and in cells overexpressing Cα. These data suggest that the R and C subunits are in equilibrium between PKA-I and PKA-II and that PKA-II formation is highly favored in intact cells. Preferential formation of PKA-II is not limited to LS-174T and MCF-7 cancer cells; such a preference has also been demonstrated in *ras*-transformed NIH 3T3 cells and in AtT20 pituitary cells (48). Most likely, RIIα and Cα associate preferentially in these cells, and PKA-I is formed only if the C subunit is present in excess over RIIα. Excess, free RIα may be degraded; therefore, an increase in RIα or PKA-I cannot occur in cells overexpressing RIα even though RIα mRNA increases (58).

Mutant RIα-P, Which Mimics RIIβ, Inhibits Growth

An autophosphorylation site has been introduced into human RIα at Ala 99 via a single nucleotide change, G→T, which replaces this alanine with serine (42).

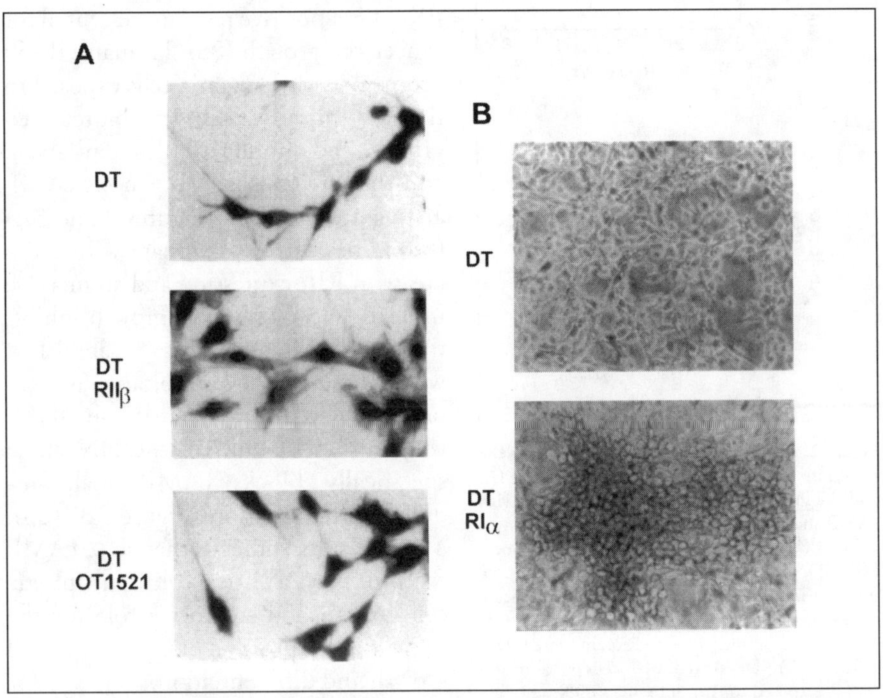

Figure 1. RIIβ and RIα overexpression. (A) RIIβ overexpression induces phenotypic change in Ki-*ras*-transformed NIH 3T3 cells (DT). Cells, uninfected control (DT-control) or infected with either MT-RIIβ retroviral vector (DT-RIIβ) or OT1521 retroviral vector (DT-OT1521) were seeded at 1×10^4 cells/mL into four-chamber tissue culture slides and grown in the presence of $CdCl_2$ (1 μM) for 40 h. Cells were lysed and stained with Wright's stain × 540. (B) RIα overexpression induces enhanced cell proliferation in Ki-*ras*-transformed NIH 3T3 cells (DT). Cells, uninfected control (DT-control) or infected with MT-RIα retroviral vector (DT-RIα) were grown in monolayer in the presence of $CdCl_2$ (1 mM) for 48 h. Unfixed cells photographed, 75×. (Reproduced with permission from Tortora et al., 1994, Cell Growth Differ. *5*:753-759.)

Similar to cells overexpressing RIIβ, cells that overexpress RIα-P require a higher concentration of cAMP to activate endogenous PKA (42). Unlike wild-type RIα, which cannot down-regulate PKA-II (48,58), RIα-P down-regulates PKA-II when it is overexpressed (42). In MCF-7 breast cancer cells, RIα-P overexpression inhibits growth and induces apoptosis (42). The dominant activity of RIα-P may arise from the ability of this mutant to trap endogenous, wild-type RIα into inactive dimers, blocking PKA-I activity and thereby inhibiting growth.

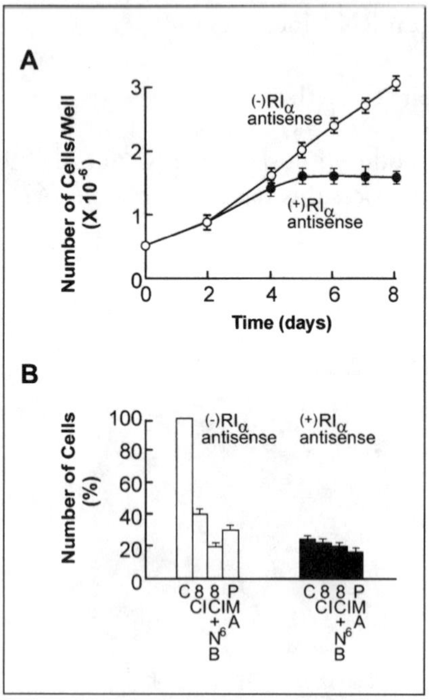

Figure 2. Effect of RIα antisense ODNs on the basal rate of growth of HL-60 leukemic cells (A) and the growth of these cells when treated with cAMP analogs or PMA (B). (A) Cells were grown without (○) or with (●) RIα antisense unmodified ODN (15 μM) in the absence of cationic lipid. At indicated times, cell counts in duplicate were performed. (B) On Day 4 of experiment A, cells exposed or unexposed to Riα antisense ODN were reseeded (Day 0) at 5×10^6 cells per dish, and cells preexposed to RIα antisense ODN were further treated with the oligomer at Day 0 and Day 2. The cAMP analogs and PMA were added once at Day 0. Cells were counted on a Coulter Counter on Day 4. 8-chloro, 8-chloro-cAMP (10 μM); 8-chloro + N^6-B, 8-chloro-cAMP (5 μM) + N^6-benzyl-cAMP (5 μM); PMA (10^{-8} M). (Reproduced from Tortora et al., 1991, Proc. Natl. Acad. Sci. USA *88*:2011-2015.)

ANTISENSE RIα—CLASSICAL VIEW

Differentiation of Leukemia Cells In Vitro

Antisense has been used to explore the possibility that the RI subunit of PKA-I is a positive regulator essential for cancer cell growth (88). Human HL-60 promyelocytic leukemia cells exposed to the RIα antisense ODN exhibit retarded growth and monocytic differentiation, and the effects of an exogenous cAMP analog are bypassed (Figures 2 and 3A) (88). These effects correlate with a decrease in RIα expression and an increase in RIIβ levels. However, growth inhibition and differentiation is abolished when these cells are simultaneously exposed to RIα and RIIβ antisense ODNs (88), and RIIβ antisense ODN alone specifically blocks cAMP analog-induced differentiation (Figure 3B) (83). These results indicate that RIα cAMP receptor plays a critical role in neoplastic cell growth and support a role for the differential expression of isozymes in cell growth and differentiation.

Sustained Inhibition of Tumor Growth In Vivo

Both unmodified and nuclease-resistant phosphorothioate (PS) ODNs of

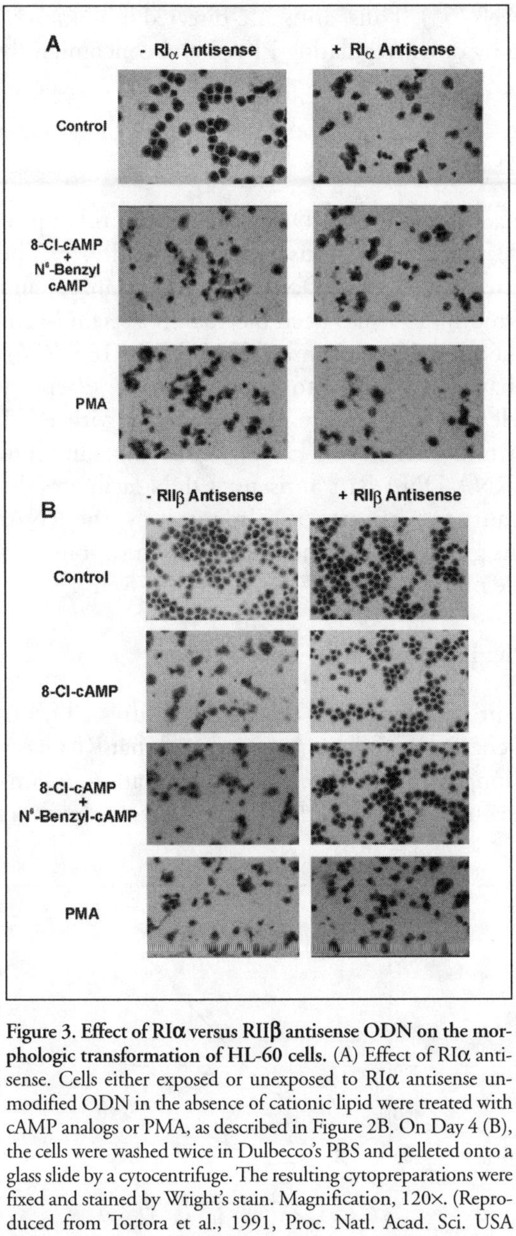

Figure 3. Effect of RIα versus RIIβ antisense ODN on the morphologic transformation of HL-60 cells. (A) Effect of RIα antisense. Cells either exposed or unexposed to RIα antisense unmodified ODN in the absence of cationic lipid were treated with cAMP analogs or PMA, as described in Figure 2B. On Day 4 (B), the cells were washed twice in Dulbecco's PBS and pelleted onto a glass slide by a cytocentrifuge. The resulting cytopreparations were fixed and stained by Wright's stain. Magnification, 120×. (Reproduced from Tortora et al., 1991, Proc. Natl. Acad. Sci. USA *88*:2011-2015.) (B) Effect of RIIβ antisense. Cells either exposed or unexposed to RIIβ antisense unmodified ODN in the absence of cationic lipid were treated with cAMP analogs or PMA, as described in Figure 2B. On Day 4 (see Figure 2B), the cells were washed twice in Dulbecco's PBS and pelleted onto a glass slide by cytocentrifuge. The resulting cytopreparations were fixed and stained by Wright's stain. Magnification, 30×. (Reproduced from Tortora et al., 1990, Proc. Natl. Acad. Sci. USA *87*:705-708.)

RIα antisense inhibit growth in breast (MCF-7), colon (LS-174T), gastric carcinoma (TMK-1), and neuroblastoma (SK-N-SH) cells in vitro with no sign of cytotoxicity (96). A single, subcutaneous injection of an antisense PS-ODN targeted to codons 8–13 of human RIα into nude mice bearing LS-174T human colon carcinoma almost completely suppressed tumor growth for several days (Figure 4) (55). Even after 14 days, tumor growth is significantly inhibited. Growth inhibition results specifically from specific targeting of RIα; the isoform, RIIα levels remained unchanged. RIα levels in tumors decrease within 24 h and remain at low levels for 2–3 days (55).

Likewise, RIα antisense treatment completely eliminates PKA-I from tumors within 24 h (55). Down-regulation of PKA-I lasted up to 5–7 days post-treatment, although RIα levels became elevated after 3 days of treatment, indicating that the RIα that appeared after its initial suppression is present mostly in subunit form (55).

Concomitant with the suppression of PKA-I, RIα antisense treatment induced changes in the PKA-II profile of tumors. On Day 3, following RIα antisense treatment, RIIβ was detected in tumors, which indicates that antisense-targeted suppression of RIα promotes a compensatory increase in RIIβ levels (55). Thus, antisense-directed inhibition of RIα expression suppresses growth by down-regulating PKA-I and concomitantly up-regulating PKA-IIβ.

RNA-DNA Mixed-Backbone, Second-Generation, Antisense RIα

Similar to other polyanionic macromolecules, ODNs interact with other proteins and growth factors (25,34), which may cause nonspecific effects. The polyanionic nature (1) of the antisense RIα PS ODN has been minimized, and the immunostimulatory (GCGT motif) (39) has been blocked in a second-generation, RNA-DNA, mixed-backbone RIα antisense ODN (HYB0165) (56). Such second-generation ODNs have been shown to improve antisense activity (49,52), is more resistant to nucleases, forms more stable duplexes with RNA (49,66), and retains the capability to induce RNase H (49). Thus, in addition to reducing nonspecific effects, the RNA-DNA RIα antisense ODN facilitates the exploration of sequence-specific antisense effects (56). Importantly, the RNA-DNA RIα antisense ODN inhibits growth and induces apoptosis in various cancer cell lines and in tumors in nude mice (3,11,16,17,56,57,70,71,78,92).

Increase in RIIβ Protein Half Life: Proof of Target Specificity

The target specificity of RIα antisense has been thoroughly addressed using RNA-DNA antisense RIα. Pulse-chase experiments have revealed that RIα has a relatively short half-life, which, along with its message down-regulation, is consistent with the rapid RIα down-regulation observed in antisense-treated tumors

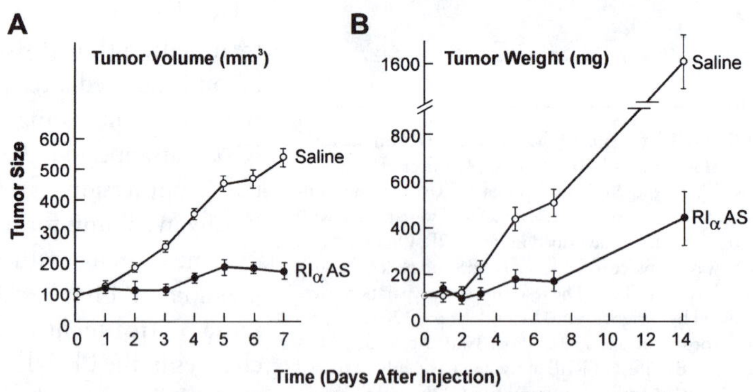

Figure 4. Inhibition of in vivo tumor growth by a single dose of RIα antisense. (A) Tumor volume obtained from daily measurement. (B) Tumor weight at the time the mice were killed. Data represent $\bar{x} \pm$ SD of 8–32 tumors in panel A and the $\bar{x} \pm$ SD of 8 tumors in panel B. These data were obtained from four separate experiments that used 24 mice each. (Reproduced from Nesterova and Cho-Chung, 1995, Nat. Med. *1*:528-533.) RIα AS, RIα antisense.

(55,57). Concomitantly, the half-life of the RIIβ protein is increased 5.5-fold (Figure 5) (57). Thus, the PKA-I:PKA-II ratio decreases in tumor cells. The half-lives of RIIα and Cα remain unchanged in antisense-treated cells. Similar results are observed in LNCaP prostate cancer cells upon treatment with antisense RIα (57); RIα expression is reduced, and growth is arrested in LNCaP cells. RIα antisense-induced stabilization of the RIIβ protein is consistent with results observed in RIβ and RIIβ knockout mice, in which the compensatory stabilization-induced elevation of the RIα protein appears in tissues that normally express the β isoforms of the R subunit (4).

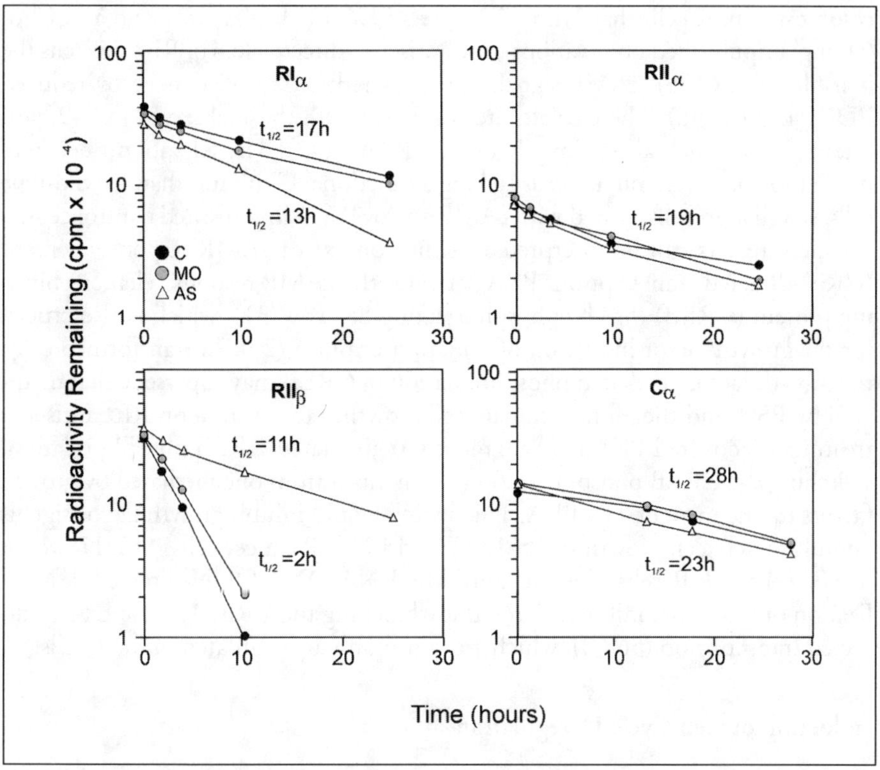

Figure 5. Stabilization of RIIβ protein in RIα antisense-treated LS-174T colon carcinoma cells. Pulse-chase analysis of RIα, RIIα, RIIβ, and Cα in cells treated with RNA-DNA second-generation RIα antisense HYB0165 (AS), mismatched control oligonucleotide HYB0295 (MO) treated (0.5 μM for four days), and untreated control (C) cells. Five sets (for five time points) of duplicate 60-mm plates (5×10^5 cells) with antisense-treated, control oligonucleotide-treated, and untreated cells were labeled with ^{35}S and chased for a total of 24 h in growth medium containing 4 mM L-methionine + L-cysteine. Lysates were made at 0, 2, 5, 10, and 24 h of chase. Lysates containing equivalent concentrations of protein were incubated with polyclonal antibodies specific for Cα, RIα, RIIα, or RIIβ proteins and immunoprecipitated with protein A-Sepharose®. The immunoprecipitates were solubilized and centrifuged. The supernatants were run on SDS-polyacrylamide gels and transferred to nitrocellulose sheets. The nitrocellulose sheets were air-dried and exposed to Kodak® XAR film (Eastman Kodak, Rochester, NY, USA) for 12–36 h at -20°C, and the radioactivities of the supernatants were counted using the liquid scintillation counter. (Reproduced from Nesterova et al., 2000, Clin. Cancer Res. *6*:3434-3441.)

Effect of RIα Antisense on cAMP-Signaling Cascade

The effects of RIα antisense RNA-DNA ODN on the cAMP-signaling cascade are dependent on the overall expression of PKA-I and PKA-II in the cell. In addition to the loss of RIα and the compensatory stabilization of RIIβ observed in both LS-174T colon cancer cells and LNCaP prostate cancer cells that express both PKA-I and PKA-II (58), RIα antisense treatment also increases the activity of the cAMP-inducible enzyme PDE4 (56). The nuclear translocation of the Cα subunit (54) in the absence of increased cellular cAMP is also triggered by RIα antisense treatment in these cells. Thus, the loss of RIα activates cAMP signaling by activating PKA-I and bypassing adenylate cyclase. However, in the case of HCT-15 MDR colon carcinoma cells that primarily express PKA-I (M. Nesterova and Y.S. Cho-Chung, unpublished observations), the antisense-directed loss of RIα shortens the half life of Cα (56). cAMP signaling is thus reduced, as evidenced by reduced PDE4 activity (56). These results are consistent with those observed in S49 lymphoma cells, which also primarily express PKA-I (73). The RI subunit becomes much more labile in mutant cells lacking a functional C subunit than in wild-type cells as well as in cells treated with cAMP analogs than in untreated control cells.

These results can be interpreted in the context of cAMP response element (CRE)-directed transcription. PKA activates the cAMP response element binding protein (CREB) (53) by phosphorylating Ser 133 (31), which is also crucial for the growth factor induction of c-*fos* transcription (29). In transformed cells, the growth factor-mediated phosphorylation of CREB may supersede that mediated by PKA and therefore stimulate cell growth. However, upon RIα antisense treatment, activated PKA may augment Cα-mediated CREB phosphorylation, switching the CREB phosphorylation mechanism from one mediated by growth factors to one mediated by PKA. This switch would inhibit growth factor signals and ultimately cell growth in LS-174T and LNCaP cancer cells (56). However, in HCT-15 MDR cells, the RIα antisense RNA-DNA ODN-directed destabilization of Cα may inhibit cell growth by blocking the CRE, Ap-1, and Sp-1-directed transcription (56,63), which are commonly up-regulated in these cells.

Induction of Cell Cycle Deregulation

The growth inhibition induced by treatment with the RIα antisense ODN accompanies changes in cell cycle phase distribution. Flow cytometric analysis has demonstrated that within 18 h, antisense treatment promotes an increase in the number of LS-174T cells in the S phase (16). Likewise, treatment of HL-60 cells with the RIα antisense ODN results in an increase in the population of cells in S phase and induction of differentiation (16). These results indicate a possible role for RIα and PKA-I in the regulation of S phase cell entry.

Because the cyclin E/CDK2 complex is essential for the initiation of DNA replication (30), the effects of RIα antisense on cyclin E levels have also been examined. In untreated control cells, cyclin E levels rise during G_1 phase, peak in the G_1/S transition, and drop sharply in the S and G_2M phases. In RIα antisense-treated cells, cyclin E remains at its peak level throughout the S and G_2M phases (12,16). In RIα antisense-treated cells, the increase of cyclin E expression appears to be delayed, occurring at the S and G_2M phases as if cells are locked in the G_1/S transition phase. Thus, antisense-directed suppression of RIα expression deregulates the cell cycle at two critical points: the entry into the S phase and the G_2M phase. This deregulation results in the accumulation of cells in the S phase, which may lead to apoptosis and differentiation. In fact, RIα antisense treatment of HL-60 cells decreases the expression of myelomonocytic antigens, such as CD11b, CD11c, and CD14 and increases the expression of the granulocyte antigen CD15, indicating a granulocytic differentiation (12,16).

Induction of Apoptosis and Changes in Cell Morphology

The RNA-DNA RIα antisense ODN has been used to explore the mechanism of action that underlies antisense-directed growth inhibition of MCF-7 hormone-dependent breast cancer cells (71). RIα antisense treatment inhibits growth in a time- and concentration-dependent manner and induces changes in cell morphology (Figure 6). Cells exhibit an elongated and flat phenotype and an increased cytoplasm-to-nucleus ratio, and treated cells grow sparingly to form much smaller cell clusters than untreated control cells. Antisense treatment also induces apoptosis, as evidenced by the fragmentation of nuclei and/or condensed chromatin (Figure 6).

Treatment with RIα antisense similarly induces growth inhibition, changes in cell morphology, and apoptosis in MDA-MB-231 hormone-independent breast cancer cells (70) and in androgen-insensitive PC3M prostate cancer cells (11). In addition, Bcl-2 hyperphosphorylation, an increase in the level of proapoptotic proteins Bax, Bak, and Bad, and Bad hypophosphorylation are observed in antisense-treated cells (11). The effects of antisense ODN treatment mirror those induced by antisense gene overexpression (11). RIα antisense, in addition to inhibiting growth, thus induces apoptosis and phenotypic changes in cancer cells of various types.

Blockade of the Tyrosine Kinase Signaling Pathway

RIα subunit expression increases in human cancers, including ovarian cancer, in which an autocrine pathway for the TGFα-EGF receptor (EGFR) is activated (19,20). RIα has been shown to interact with the ligand-activated EGFR complex by binding to the SH3 domain of the Grb2 adaptor protein (84). Treatment of OVCAR8 cells with the RNA-DNA RIα antisense ODN decreases the expression of EGFR, c-erbB-2, and c-erbB-3, but not c-erbB-4, concomitant with reduced RIα expression (3). Again, antisense treatment inhibits cell growth (3). Suppressed RIα expression and inhibited cell growth are also observed in OV-

CAR8 cells overexpressing human EGFR cDNA in the 3′→5′ orientation (3). These results indicate that RIα antisense can block both the serine-threonine and the tyrosine kinase signaling pathways in ovarian cancer cells.

ANTISENSE RIα—GENOMIC-SCALE VIEW

Modulation of Gene Expression Profile by Antisense RIα

The sequence-specific effects of RIα antisense and various unpredicted effects might reflect previously unidentified gene expression patterns in antisense-treated cells. A cDNA microarray containing 2304 nonredundant clones has been used to analyze gene expression patterns in cells treated with the RNA-DNA RIα antisense ODN and in cells overexpressing the antisense RIα gene (10). Endogenous antisense expression avoids some of the problems with exogenous antisense treatment, such as limited delivery, ODN stability, and nonspecific binding of ODNs to cellular macromolecules (72).

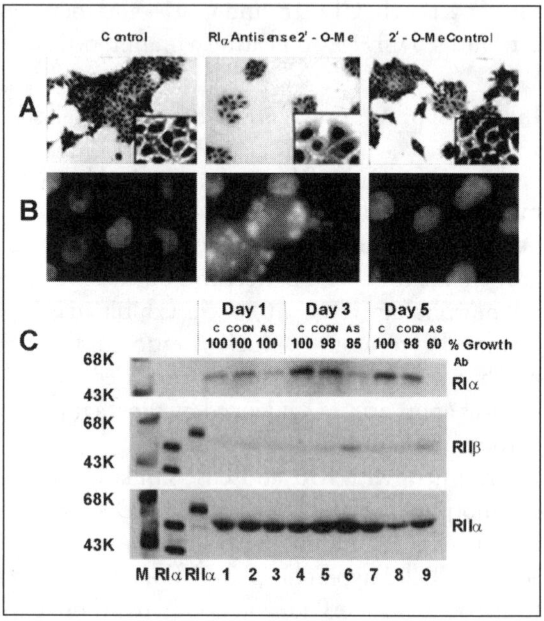

Figure 6. RIα antisense-induced changes in cell morphology and apoptosis in MCF-7 cells. Cell growth: MCF-7 cells (1 × 10^6/well) were plated in a 24-well plate containing culture medium. Then, 5 μM RIα antisense (AS) (2′-O-Me) or four mismatched control ODNs (CODNs) were added (8 h after seeding) to each of four replicate wells. To increase the delivery of the oligonucleotides into the cells, transfection reagent DOTAP (Roche Molecular Biochemicals, Indianapolis, IN, USA) was used. (A) Cell morphology: the treated or untreated cells were washed with PBS, fixed with 70% methanol for 5 min, and stained with Giemsa stain. The whole cells were visualized under an inverted microscope. × 80. Insets, 240×. (B) Apoptotic nuclei: to assay apoptotic nuclei, the cells were washed with PBS, fixed with 70% ethanol for 1 h, and stained with 1 mM Hoechst 3325B. The nuclear morphology of the cells was visualized by fluorescence microscope at 2600×. (C) Cell growth and R subunits level. Cell growth was measured by MTT assays, and R subunits content was measured by photoaffinity labeling-immunoprecipitation. (Reproduced from Cho-Chung et al., 1997, Antisense Nucleic Acid Drug Dev. *7*:217-223.)

The expression levels of approximately 240 genes, representing 10% of the total DNA elements on the array, were altered (2-fold or greater were up-regulated or down-regulated) in cells treated with antisense ODN and in cells overexpressing the antisense gene (Table 1) (10). Less than 2% of the altered expression profiles were specific to either antisense ODN-treated cells or cells overexpressing the antisense gene (Figure 7C) (10). These results indicate that antisense ODN treatment and endogenous antisense gene overexpression affected nearly identical genomic pathways. Furthermore, antisense-induced alterations in the expression profile of PC3M cells are closely comparable to those observed in LS-174T cells overexpressing the antisense gene (Figure 7F), although the two cell lines exhibit large differences in their basic patterns of expression (Figure 7D) and in the magnitude of altered expression (compare Figure 7B with Figure 7E). Thus, different cell types exhibit comparable expression profiles when treated with RIα antisense (10).

Molecular Portrait of the Reverted Phenotype

The classical view of antisense RIα demonstrates that, in addition to growth inhibition, RIα antisense treatment also induces changes in cell morphology, including a flat phenotype similar to the reverted phenotype of transformed cells (17). To identify global changes in the molecular portrait of cancer cells and tumors following antisense treatment, a hierarchical clustering methodology has been used to group genes on the basis of similarities in their expression patterns (65). The clusters of coordinately expressed genes are defined as "signatures," which are named based on the cellular process in which the component genes participate (2). The

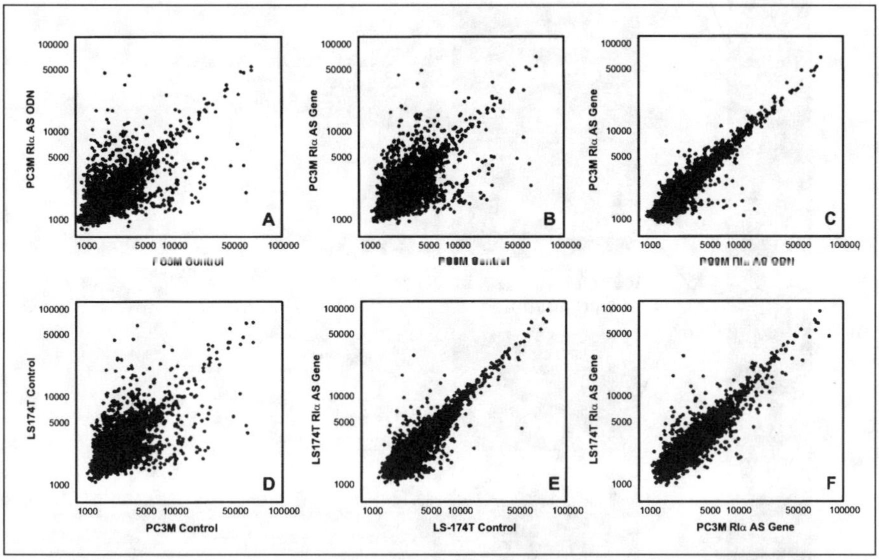

Figure 7. Scatter plots for expression profile comparison between antisense ODN treatment and antisense gene overexpression and between PC3 M cells and LS-174T cells. Expression profiles of untreated control versus antisense (AS) ODN/gene-targeted cells (A, B, and E), AS ODN versus AS gene treatment (C), PC3 M control cells versus LS-174T control cells (D), and PC3 M cells overexpressing the antisense gene versus LS-174T overexpressing the antisense gene (F) are shown as bivariate scatter plots of 2304 genes from the microarray. The values are corrected intensities that represent the levels of expression for the DNA elements of the microarrays. (Reproduced from Cho et al., 2001, Proc. Natl. Acad. Sci. USA *98*:9819-9823.)

map reveals that RIα antisense treatment affects a cluster of genes involved in proliferation and another cluster involved in differentiation (Figure 8) (10).

Genes within the proliferation signature are highly expressed in untreated cells and markedly suppressed in cells treated with antisense ODN or in cells overex-

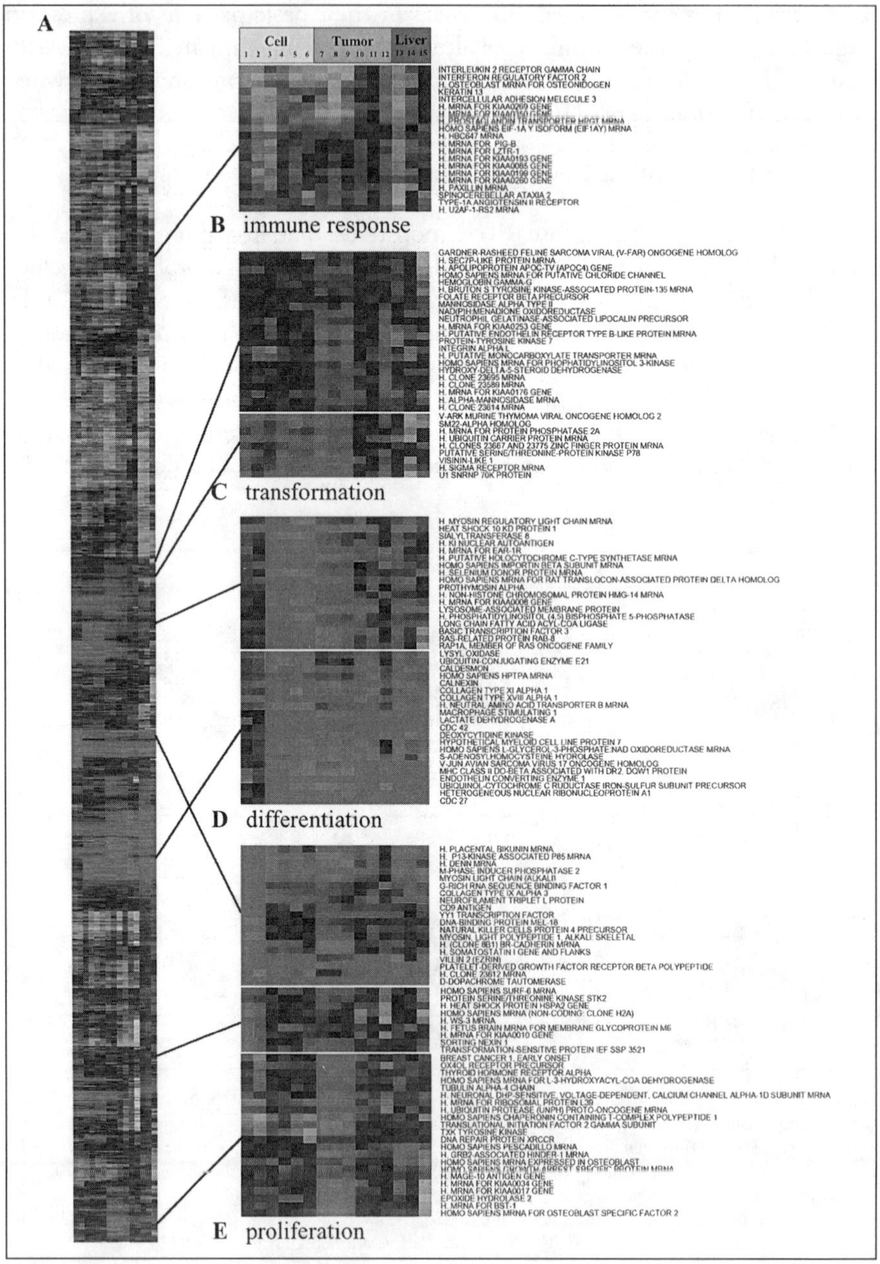

pressing the antisense gene (Figure 8E, top and middle panels). Predominant genes in this signature encode proteins evolved in cell cycle control, DNA synthesis and regulation, transcription, and translation. Conversely, genes in the differentiation and reverse transcription signature are induced by RIα antisense (Figure 8D). This cluster primarily includes genes that encode small G proteins (e.g., Cdc42 and Rapl), transcription factors, and regulatory proteins of the cytoskeleton, specifically the microtubules.

Similar proliferation and differentiation signatures are found in antisense-treated tumors (Figure 8, D and E, top and middle panels), but an expression profile distinct from that seen in antisense-treated cells is also apparent. For example, genes in the tumor-specific proliferation signature, such as those encoding TXK tyrosine kinase and Grb-2-associated protein, are markedly down-regulated in tumors but unchanged in cells (Figure 8E, bottom panel). Conversely, genes in the tumor-specific differentiation signature, which include genes encoding such developmental proteins as the wingless-type mouse mammary tumor virus integration site family and sex-determining region Y, were markedly up-regulated in tumors but not in cells (see Figure 8, legend). Similarly, genes in the transformation signature such as oncogenes and genes for tyrosine and serine/threonine kinases that are usually overexpressed in tumors are specifically down-regulated following antisense treatment (Figure 8C). These expression signatures induced by the antisense RIα—namely, the down-regulation of the proliferation signature and up-regulation of the differentiation signature—reflect the profile of the nonmalignant or reverted phenotype. This nonmalignant profile is also observed in the host livers of antisense-treated and untreated animals.

Figure 8. Molecular portrait of the reverted phenotype of PC3 M cells and tumors treated with antisense RIα. Data from control ODN-treated cells, antisense ODN/gene-treated cells, untreated control cells, antisense ODN- or control ODN-treated tumors, control tumors, and host livers were combined and clustered. Cluster analysis was performed on Z-transformed microarray data by using two separate programs available as shareware from Michael Eisen's laboratory (http://rana.lbl.gov). Each gene is represented by a single row of colored boxes; each experimental sample is represented by a single column. Columns 1 and 2, untreated control cells. Columns 3 and 4, antisense RNA-DNA hybrid ODN-treated cells. Columns 5 and 6, antisense gene overexpressing cells. Columns 7 and 8, untreated control tumors. Column 9, control ODN-treated tumor. Column 10, antisense RNA-DNA ODN-treated tumor. Column 11, mouse antisense PS-ODN-treated tumor. Column 12, antisense PS-ODN-treated tumor. Column 13, antisense PS-ODN-treated liver. Column 14, antisense RNA-DNA ODN-treated liver. Column 15, untreated control liver. Panel A shows the entire clustered image. Full gene names are shown for representative clusters containing functionally related genes involved in immune response (B), transformation (C), differentiation (D), and proliferation (E). These clusters also contain uncharacterized genes and named genes uninvolved in these processes. The up-regulation of Cdc42 in tumors is not apparent in panel D because of its high basal levels. The deviations of log ratios of expression levels from controls in antisense ODN-treated cells and tumors were 4.77 and 1.84, respectively. The expression levels of genes for the wingless type mouse mammary tumor virus integration site family and sex-determining region Y were not depicted in panel D. The deviations of log ratios of these genes from controls were 0.26 and 1.08, respectively, in antisense-treated cells and 1.90 and 2.04, respectively, in antisense-treated tumors. The size of the tumors compared with untreated (saline injected) control tumors were: 42%, 63%, 56%, and 98%, after four days of treatment with antisense RNA-DNA ODN, PS-ODN, mouse PS-ODN, and control ODN, respectively. There was no sign of systemic toxicity in treated animals, and the sizes of the liver and spleen remained unchanged. (Reproduced from Cho et al., 2001, Proc. Natl. Acad. Sci. USA *98*:9819-9823.) **(See color plate A15.)**

Table 1. Gene Expression Profile of Altered PC3M Cells with Antisense ODN Treatment or Antisense Gene Overexpression

Genes	Fold Change	
	AS ODN	AS Gene
CDC42	20.0	18.9
Myosin light chain, alkali	14.4	14.0
Kinesin heavy chain	12.7	11.1
Endothelin-converting enzyme 1	12.5	10.0
CDC27	9.7	9.2
Lysosome-associated membrane protein 1	7.9	6.4
Basic transcription factor 3	6.9	8.8
RAP1A	5.4	6.3
Ras-related protein RAB-8	5.2	5.8
IL-3 receptor, α	4.7	3.6
MHC class II	4.4	4.2
Signal recognition particle receptor	3.8	3.4
Probable G protein-coupled receptor HM74	3.7	4.0
Integrin, α-6	3.6	3.7
PDE4B, cAMP specific	2.3	3.2
Heat shock protein HSPA2	-2.3	-2.3
Transformation-sensitive protein IEF SSP 3521	-2.3	-2.9
Superoxide dismutase I (Cu/Zn)	-2.6	-2.6
Plasminogen-like protein	-2.7	-2.6
MAPKKK5	-2.9	-2.8
Phosphatidylinositol 3-kinase associated p85	-3.1	-3.6
YY1 transcription factor	-3.2	-3.2
Cytochrome p450 HE1	-3.7	-4.3
Natural killer cell protein 4	-4.0	-4.0
Collagen, type IV, α-4	-4.0	-4.5
Diacylglycerol kinase, γ (90 kDA)	-4.5	-4.5
Catalase	-6.7	-6.7
Ligase 1, DNA, ATP-dependent	-6.7	-6.7
M-phase inducer phosphatase 2	-7.7	-7.7
Crystallin Mu	-20.0	-20.0

AS ODN, antisense ODN treatment; AS gene, antisense gene overexpression. Fold change represents altered expression levels in antisense ODN/antisense gene cells, compared with control cells. (Reproduced from Cho et al., 2001, Proc. Natl. Acad. Sci. USA *98*:9819-9823.)

Clinical Relevance of the Second-Generation Antisense RIα

To verify the specificity of the antisense effects on gene expression signatures, three antisense ODNs that differed in sequence or chemical modification were used (10): a PS-ODN antisense directed against codons 8–13 of human RIα; the immunosuppressive, less cytotoxic, second-generation RNA-DNA mixed-backbone antisense; and a nonimmunostimulatory 5′-CCG containing PS-ODN, targeted to codons 8–13 of mouse RIα, which can cross-hybridize with human RIα. All three ODNs induce similar alterations in expression signatures, mirroring the above-mentioned observations, in tumors but not in host liver (10). The antisense modulation of these signatures is sequence-specific; these expression signatures are not detected in cells or tumors treated with control ODNs (10). The cDNA microarray approach thus presents the second-generation RNA-DNA mixed backbone oligonucleotides antisense RIα as a clinically relevant, nucleic acid-based medicine.

Identifying Antisense Genomic Modulators

Although most antisense ODNs efficiently inhibit their target genes (classical view), how this inhibition affects global gene expression (genomic-scale view) is poorly understood. Microarray analysis is advantageous in providing insight into the overall cellular mechanism of action of antisense ODNs. This type of analysis will narrow down the number of selected target genes and reveal new target genes for antisense therapeutics.

DNA microarray technology has demonstrated for the first time that, in a sequence-specific manner, antisense RIα modulates a wide set of genes related to cell proliferation and transformation versus other sets of genes involved in growth arrest and differentiation (10). Thus, antisense-directed depletion of RIα modulates the signal transduction signatures of multiple pathways beyond that of cAMP pathway. This approach has therefore confirmed and extended the findings of a decade's worth of classical approaches to antisense, including biochemical, molecular biology, and translational methods, and enhanced the classical view of antisense ODNs. Gene expression profiling thus presents a way of refining cancer chemotherapies in the future. Revisiting antisense-targeted gene expression on a genomic scale will facilitate the discovery of clinically appropriate antisense drugs and provide a unique perspective on the development of new chemotherapeutic combinations based on the molecular actions of these drugs.

PERSPECTIVE

The diverse effects of cAMP signaling depend highly on the different stages of development and differentiation. PKA exists in two isozymes that appear in different ratios during ontogenic development, cell differentiation, and transformation. However, the distinct functions of these isozymes have been widely debated because these isozymes exhibit identical catalytic activities in phosphorylating target proteins. Only during the past decade has experimental evidence started to reveal that intracellular, balanced expression between PKA-I and PKA-II may play a critical role in the control of cell growth and differentiation. Classical antisense experiments, which depend on the ability of the antisense to target a single gene via Watson-Crick base-paring, have provided the first direct lines of evidence that the PKA isoforms have opposite roles in cancer cell growth. PKA-I/RIα promotes cell growth, and PKA II/RIIβ inhibits growth and induces differentiation. This evidence has been further supported by gene transfer experiments showing that overexpression of RIIβ, but not of RIα, RIIα, or Cα, inhibits cancer cell growth and induces a reverted phenotype in various cancer cell lines (9,17,58).

Although most classical antisense experiments have demonstrated a strong potential for these agents as targeted, nucleic acid-based medicines, the genomic-scale view of these drugs has not been explored. Microarray technology has allowed a new vision in antisense strategies. For the first time, cDNA microarrays have revealed that in a sequence-specific manner, antisense RIα can modulate a wide set of genes related to cell proliferation and another set of genes related to differentiation. Differentiation and proliferation expression signatures are specifically up-regulated and down-regulated, respectively, in tumor cells; these signatures are quiescent and unaltered in the host livers of antisense-treated animals. This observation clearly indicates separate and distinct cAMP signaling pathways that regulate growth for normal cells versus cancer cells. For example, germline mutations in RIα are responsible for a multiple neoplasia syndrome characteristic of Carney Complex (37), which suggests that there is a role for RIα as a tumor suppressor in the normal, embryonic regulation of cell growth (37). However, an oncogene, *RET/ptc2*, is the product of a papillary thyroid carcinoma translocation event and consists of the c-*ret* proto-oncogene fused with RIα (40), which suggests that there a role for RIα as a positive regulator of tumor cell growth.

Although RIα antisense behaves as a genomic modulator, its precise molecular mechanism of action is still unknown. However, it is not surprising that antisense RIα modulates the signal transduction signatures of multiple pathways beyond that of cAMP. CRE-directed transcription (62), which is triggered by the activation of PKA via the antisense depletion of RIα (54,56), may play a central role in this process. CRE-directed transcription is central to the cell's signal transduction network, with crosstalk among signaling molecules upstream, and the transcription coactivator, CBP (CREB-binding protein) (32), working downstream, binding transcription factors involved in multiple signaling pathways. The genomic-scale view of antisense oligonucleotides has led to a new vision in antisense technology. It can define the antisense effects beyond its single target gene, on the basis of the whole cell and the whole organism, and would thus facilitate the development of nucleic acid medicines with high target specificity and minimum side effects.

ACKNOWLEDGMENTS

I thank Dr. Frances McFarland of Palladian Partners, Inc., who provided editorial support under contract no. NO2-BC-76212/C2700212 with the National Cancer Institute.

REFERENCES

1. **Agrawal, S. and Q. Zhao.** 1998. Mixed backbone oligonucleotdes: improvement in oligonucleotide-induced toxicity in vivo. Antisense Nucleic Acid Drug Dev. *8*:135-139.
2. **Alizadeh, A.A., M.B. Eisen, R.E. Davis, C. Ma, I.S. Lossos, A. Rosenwald, J.S. Boldrick, H. Sabet, et al.** 2000. Distinct types of diffuse large B-cell lymphoma identified by gene expression profiling. Nature *403*:503-511.
3. **Alper, O., N.F. Hacker, and Y.S. Cho-Chung.** 1999. Protein kinase A-Iα subunit-directed antisense inhibition of ovarian cancer cell growth: crosstalk with tyrosine kinase signaling pathway. Oncogene *18*:4999-5004.
4. **Amieux, P.S., D.E. Cummings, K. Motamed, E.P. Brandon, L.A. Wailes, K. Le, R.I. Idzerda, and G.S. McKnight.** 1997. Compensatory regulation of RIα protein levels in protein kinase A mutant mice. J. Biol. Chem. *272*: 3993-3998.
5. **Beebe, S.J. and J.D. Corbin.** 1986. Cyclic nucleotide-dependent protein kinases, p. 43-111. *In* Boyer, P.D. and E.G. Krebs (Eds.), The Enzymes, vol. 17. Academic Press, New York.
6. **Beebe, S.J., O. Oyen, M. Sandberg, A Froysa, V. Hansson, and T. Jahnsen.** 1990. Molecular cloning of a unique tissue-specific protein kinase Cg from human testis—representing a third isoform for the catalytic subunit of the cAMP-dependent protein kinase. Mol. Endocrinol. *4*:465-475.
7. **Bold, R.J., S. Alpard, J. Ishizuka, C.M. Townsend, Jr., and J.C. Thompson.** 1994. Growth-regulatory effect of gastrin on human colon cancer cell lines is determined by protein kinase a isoform content. Regul. Pept. *53*:61-70.
8. **Bradbury, A.W., D.C. Carter, W.R. Miller, Y.S. Cho-Chung, and T. Clair.** 1994. Protein kinase A (PK-A) regulatory subunit expression in colorectal cancer and related mucosa. Br. J. Cancer *69*:738-742.
9. **Budillon, A., A. Cereseto, A. Kondrashin, M. Nesterova, G. Merlo, T. Clair, and Y.S. Cho-Chung.** 1995. Point mutation of the autophosphorylation site or in the nuclear location signal causes protein kinase A RIIβ regulatory subunit to lose its ability to revert transformed fibroblasts. Proc. Natl. Acad. Sci. USA *92*:10634-10638.
10. **Cho, Y.S., M.-K. Kim, C. Cheadle, C. Neary, K.G. Becker, and Y.S. Cho-Chung.** 2001. Antisense DNAs as multisite genomic modulators identified by DNA microarray. Proc. Natl. Acad. Sci. USA *98*:9819-9823.
11. **Cho, Y. S., M.-K. Kim, L. Tan, R. Srivastava, S. Agrawal, and Y.S. Cho-Chung.** 2002. Protein kinase A RIα antisense inhibition of prostate cancer growth: Bcl-2 hyperphosphorylation, Bax upregulation, and Bad-hypophosphorylaton. Clin. Cancer Res. *8*:607-614.
12. **Cho-Chung, Y.S.** 1998. Protein kinase A-directed antisense blockade of cancer growth: single gene-based therapeutic approach, p. 263-281. *In* C.A. Stein and A.M. Krieg (Eds.), Applied Antisense Oligonucleotide Technology. John Wiley & Sons, New York.
13. **Cho-Chung, Y.S., T. Clair, and C. Shepheard.** 1983. Anticarcinogenic effect of N6,O2-dibutyryl cyclic adenosine 3′:5′-monophosphate on 7,12-dimethylbenz(α)anthracene mammary tumor induction in the rat and its relationship to cyclic adenosine 3′,5′-monophosphate metabolism and protein kinase. Cancer Res. *43*:2736-2740.
14. **Cho-Chung, Y.S., T. Clair, P. Tagliaferri, S. Ally, D. Katsaros, G. Tortora, L. Neckers, T.L. Avery, et al.** 1989. Site-selective cyclic AMP analogs as new biological tools in growth control, differentiation and proto-oncogene regulation. Cancer Inv. *7*:161-177.
15. **Cho-Chung, Y.S., T. Clair, G. Tortora, and H. Yokozaki.** 1991. Role of site-selective cAMP analogs in the control and reversal of malignancy. Pharmacol. Ther. *50*:1-33.
16. **Cho-Chung, Y.S., M. Nesterova, A. Kondrashin, K. Noguchi, R.K. Srivastava, and S. Pepe.** 1997.

Antisense-protein kinase A: a single-gene-based therapeutic approach. Antisense Nucleic Acid Drug Dev. *7*:217-223.

17. **Cho-Chung, Y.S., M. Nesterova, S. Pepe, G.R. Lee, K. Noguchi, R.K. Srivastava, A.R. Srivastava, O. Alper, et al.** 1999. Antisense DNA-targeting protein kinase A-RIA subunit: a novel approach to cancer treatment. Front Biosci. *4*:D898-907.

18. **Cho-Chung, Y.S., S. Pepe, T. Clair, A. Budillon, and M. Nesterova.** 1995. cAMP-dependent protein kinase: role in normal and malignant growth. Crit. Rev. Oncol. Hematol. *21*:33-61.

19. **Ciardiello, F., S. Pepe, C. Bianco, G. Baldassarre, A. Ruggiero, M.P. Selvam, A.R. Bianco, and G. Tortora.** 1993. Down-regulation of RIα subunit of cAMP-dependent protein kinase induces growth inhibition of human mammary epithelial cells transformed by c-Ha-ras and c-erbB-2 proto-oncogenes. Int. J. Cancer *53*:438-443.

20. **Ciardiello, F., G. Tortora, N. Kim, T. Clair, S. Ally, D.S. Salomon, and Y.S. Cho-Chung.** 1990. 8-Chloro-cAMP inhibits transforming growth factor a transformation of mammary epithelial cells by restoration of the normal mRNA patterns for cAMP-dependent protein kinase regulatory subunit isoforms which show disruption upon transformation. J. Biol. Chem. *265*:1016-1020.

21. **Clair, T., S. Ally, P. Tagliaferri, R.K. Robins, and Y.S. Cho-Chung.** 1987. Site-selective cAMP analogs induce nuclear translocation of the RII cAMP receptor protein in Ha-MuSV-transformed NIH/3T3 cells. FEBS Lett. *224*:377-384.

22. **Conti, M., R. Geremia, and V. Monesi.** 1979. Adenosine 3′,5′-cyclic monophosphate dependent protein kinase activity in differentiating germ cells of the mouse testis. Mol. Cell Endocrinol. *13*:137-148.

23. **Eppenberger, U., W. Roos, D. Fabbro, A. Sury, J. Weber, E. Bechtel, P. Huber, and R.A. Jungmann.** 1979. Ontogeny of the adenosine-3′:5′-phosphate-dependent protein-kinase system during early uterine development. Inverse relationship with the estrogen receptors. Eur. J. Biochem. *98*:253-259.

24. **Fakunding, J.L. and A.R. Means.** 1977. Characterization and follicle stimulating hormone activation of Sertoli cell cyclic AMP-dependent protein kinases. Endocrinology *101*:1358-1368.

25. **Fennewald, S.M. and R.F. Rando.** 1995. Inhibition of high affinity basic fibroblast growth factor binding by oligonucleotides. J. Biol. Chem. *270*:21718-21721.

26. **Fuller, D.J., C.V. Byus, and D.H. Russell.** 1978. Specific regulation by steroid hormones of the amount of type I cyclic AMP-dependent protein kinase holoenzyme. Proc. Natl. Acad. Sci. USA *75*:223-227.

27. **Gentleman, S., B.A. Hemmings, P. Russell, and G.J. Chader.** 1989. Developmental expression of the RI subunit of cyclic AMP-dependent protein kinase in retina. Exp. Eye Res. *48*:717-731.

28. **Gharrett, A.J., A.M. Malkinson, and J.R. Sheppard.** 1976. Cyclic AMP-dependent protein kinases from normal and SV40-transformed 3T3 cells. Nature *264*:673-675.

29. **Ginty, D.D., A. Bonni, and M.E. Greenberg.** 1994. Nerve growth factor activates a ras-dependent protein kinase that stimulates *c-fos* transcription via phosphorylation of CREB. Cell *77*:713-725.

30. **Gong, J., F. Traganos, and Z. Darzynkiewicz.** 1993. Expression of cyclins B and E in individual MOLT-4 cells and in stimulated human lymphocytes during their progression through the cell cycle. Int. J. Oncol. *3*:1037-1042.

31. **Gonzalez, G.A., W. Biggs, III, W.W. Vale, and M.R. Montminy.** 1989. A cluster of phosphorylation sites on the cyclic AMP-regulated nuclear factor CREB predicated by its sequence. Nature *337*:749-752.

32. **Goodman, R.H. and S. Smolik.** 2000. CBP/p300 in cell growth, transformation, and development. Genes Dev. *14*:1553-1577.

33. **Gordge, P.C., M.J. Hulme, R.A. Clegg, and W.R. Miller.** 1996. Elevation of protein kinase A and protein kinase C activities in malignant as compared with normal human breast tissue. Eur. J. Cancer *32A*:2120-2126.

34. **Guvakova, M.A., L.A. Yakubov, I. Vlodavsky, J.L. Tonkinson, and C.A. Stein.** 1995. Phosphorothioate oligodeoxynucleotides bind to basic fibroblast growth factor, inhibit its binding to cell surface receptors, and remove it from low affinity binding sites on extracellular matrix. J. Biol. Chem. *270*:2620-2627.

35. **Haddox, M.K., W.R. Roeske, and D.H. Russell.** 1979. Independent expression of cardiac type I and II cyclic AMP-dependent protein kinase during murine embryogenesis and postnatal development. Biochim. Biophys. Acta *585*:527-534.

36. **Handschin, J.S. and U. Eppenberger.** 1979. Altered cellular ratio of type I and type II cyclic AMP-dependent protein kinase in human mammary tumors. FEBS Lett. *106*:301-304.

37.**Kirschner, L.S., J.A. Carney, S.D. Pack, S.E. Taymans, C. Giatzakis, Y.S. Cho, Y.S. Cho-Chung, and L.A. Stratakis.** 2000. Mutations of the gene encoding the protein kinase A type I-α regulatory subunit in patients with the Carney complex. Nat. Genet. *26*:89-92.

38.**Krebs, E.G.** 1972. Protein kinases. Curr. Top. Cell. Regul. *5*:99-133.

39.**Krieg, A.M., A.K. Yi, S. Matson, T.J. Waldschmidt, G.A. Bishop, R. Teasdale, G.A. Koretzky, and D. Klinman.** 1995. CpG motifs in bacterial DNA trigger direct B-cell activation. Nature *374*:546-549.

40.**Lanzi, C., M.G. Borrello, I. Bongarzone, A Migliazza, A. Fusco, M. Grieco, M. Santoro, R.A. Gambetta, et al.** 1992. Identification of the product of two oncogenic rearranged forms of the RET protooncogene in papillary thyroid carcinomas. Oncogene *7*:2189-2194.

41.**Ledinko, N. and I.-J. Chan.** 1984. Increase in type I cyclic adenosine 3′,5′-monophosphate-dependent protein kinase activity and specific accumulation of type I regulatory subunits in adenovirus type 12-transformed cells. Cancer Res. *44*:2622-2627.

42.**Lee, G.R., S.N. Kim, K. Noguchi, S.D. Park, S.H. Hong, and Y.S. Cho-Chung.** 1999. Ala99ser mutation in RIα regulatory subunit of protein kinase A causes reduced kinase activation by cAMP and arrest of hormone-dependent breast cancer cell growth. Mol. Cell Biochem. *195*:77-86.

43.**Lee, P.C., D. Radloff, J.S. Schweppe, and R.A. Jungmann.** 1976. Testicular protein kinase: characterization of multiple forms and ontogeny. J. Biol. Chem. *251*:914-921.

44.**Levy, F.O., O. Oyen, M. Sandberg, K. Tasken, W. Eskild, V. Hansson, and T. Jahnsen.** 1988. Molecular cloning, complementary deoxyribonucleic acid structure and predicted full-length amino acid sequence of the hormone-inducible regulatory subunit of 3′,5′-cyclic adenosine monophosphate-dependent protein kinase from human testis. Mol. Endocrinol. *2*:1364-1373.

45.**Linask, K.K. and R.M. Greene.** 1989. Ontogenetic analysis of embryonic palatal type I and type II cAMP-dependent protein kinase isozymes. Cell Differ. Dev. *28*:189-202.

46.**Malkinson, A.M., L. Hogy, A.J. Gharrett, and T.J. Gunderson.** 1978. Ontogenetic studies of cyclic AMP-dependent protein kinase enzymes from mouse heart and other tissues. J. Exp. Zool. *205*:423-431.

47.**McDaid, H.M., M.T. Cairns, R.I. Atkinson, S. McAleer, D.P. Harkin, P. Gilmore, and P.G. Johnston.** 1999. Increased expression of the RIα subunit of the cAMP-dependent protein kinase A is associated with advanced stage of ovarian cancer. Br. J. Cancer *79*:933-939.

48.**McKnight, G.S., C.H. Clegg, M.D. Uhler, J.C. Chrivia, G.G. Cadd, L.A. Correll, and A.D. Otten.** 1988. Analysis of the cAMP-dependent protein kinase system using molecular genetic approaches. Recent Prog. Horm. Res. *44*:307-335.

49.**Metelev, V., J. Liszlewicz, and S. Agrawal.** 1994. Study of antisense oligonucleotide phosphorothioates containing segments of oligodeoxynucleotides and 2′-O-methyloligoribonucleotides. Bioorg. Medicinal Chem. Lett. *4*:2929-2934.

50.**Miller, W.R., M.J. Hulme, Y.S. Cho-Chung, and R.A. Elton.** 1993. Types of cyclic AMP binding proteins in human breast cancers. Eur. J. Cancer *29A*:989-991.

51.**Miller, W.R., D.M.A. Watson, W. Jack, U. Chetty, and R.A. Elton.** 1993. Tumor cyclic AMP binding proteins: an independent prognostic factor for disease recurrence and survival in breast cancer. Breast Cancer Res. Treat. *26*:89-94.

52.**Monia, B.P., E.A. Lesnik, C. Gonzalez, W.F. Lima, and D. McGee.** 1993. Evaluation of 2′-modified oligonucleotides containing 2′-deoxygaps as antisense inhibitors of gene expression. J. Biol. Chem. *268*:14514-14522.

53.**Montminy, M.R. and L.M. Bilezikjian.** 1987. Binding of a nuclear protein to the cyclic-AMP response element of the somatostatin gene. Nature *328*:175-178.

54.**Neary, C.L. and Y.S. Cho-Chung.** 2001. Nuclear translocation of the catalytic subunit of protein kinase A induced by an antisense oligonucleotide directed against the RIα regulatory subunit. Oncogene *20*:8019-8024.

55.**Nesterova, M. and Y.S. Cho-Chung.** 1995. A single-injection protein kinase A-directed antisense treatment to inhibit tumor growth. Nature Med. *1*:528-533.

56.**Nesterova, M. and Y.S. Cho-Chung.** 2000. Oligonucleotide sequence-specific inhibition of gene expression, tumor growth inhibition, and modulation of cAMP signaling by an RNA-DNA hybrid antisense targeted to protein kinase A RIα subunit. Antisense Nucleic Acid Drug Dev. *10*:423-433.

57.**Nesterova, M., K. Noguchi, Y.G. Park, Y.N. Lee, and Y.S. Cho-Chung.** 2000. Compensatory stabilization of RIIβ protein, cell cycle deregulation, and growth arrest in colon and prostate carcinoma cells by antisense-directed down-regulation of protein kinase A RIα protein. Clin. Cancer Res. *6*:3434-3441.

58. **Nesterova, M.V., H. Yokozaki, L. McDuffie, and Y.S. Cho-Chung.** 1996. Overexpression of RIIβ regulatory subunit of protein kinase A in human colon carcinoma cell induces growth arrest and phenotypic changes that are abolished by site-directed mutation of RIIβ. Eur. J. Cancer *253*:486-494.
59. **Paterson, B.M., B.E. Roberts, and E.L. Kuff.** 1977. Structural gene identification and mapping by DNA-mRNA hybrid-arrested cell-free translation. Proc. Natl. Acad. Sci. USA *74*:4370-4374.
60. **Prasad, K.N.** 1975. Differentiation of neuroblastoma cells in culture. Biol. Rev. Camb. Philos. Soc. *50*:129-165.
61. **Richards, J.S. and A.I. Rolfes.** 1980. Hormonal regulation of cyclic AMP binding to specific receptor proteins in rat ovarian follicles. Characterization by photoaffinity labeling. J. Biol. Chem. *255*:5481-5489.
62. **Roesler, W.J., G.R. Vandenbark, and R.W. Hanson.** 1988. Cyclic AMP and the induction of eukaryotic gene transcription. J. Biol. Chem. *263*:9063-9066.
63. **Rohlff, C. and R.I. Glazer.** 1995. Regulation of multidrug resistance through the cAMP and EGF signaling pathways. Cell. Signal. *7*:431-443.
64. **Rosen, O.M. and J. Erlichman.** 1975. Reversible autophosphorylation of a cyclic 3′,5′-AMP-dependent protein kinase from bovine cardiac muscle. J. Biol. Chem. *250*:7788-7794.
65. **Schena, M., D. Shalon, R.W. Davis, and P.O. Brown.** 1995. Quantitative monitoring of gene expression patterns with a complementary DNA microarray. Science *270*:467-470.
66. **Shibahara, S., S. Mukai, H. Morisawa, H. Nakashima, S. Kobayashi, and N. Yamamoto.** 1989. Inhibition of human immunodeficiency virus (HIV-1) replication by synthetic oligo-RNA derivatives. Nucleic Acids Res. *17*:239-252.
67. **Showers, M.O. and R.A. Maurer.** 1986. A cloned bovine cDNA encodes an alternate form of the catalytic subunit of cAMP-dependent protein kinase. J. Biol. Chem. *261*:16288-16291.
68. **Simpson, B.J.B., A.D. Ramage, M.J. Hulme, D.J. Burns, D. Katsaros, S.P. Langdon, and W.R. Miller.** 1996. Cyclic adenosine 3′,5′-monophosphate-binding proteins in human ovarian cancer: correlations with clinicapathological features. Clin. Cancer Res. *2*:201-206.
69. **Srivastava, R.K., Y.N. Lee, K. Noguchi, Y.G. Park, M.J. Ellis, J.S. Jeong, S.N. Kim, and Y.S. Cho-Chung.** 1998. The RIIβ regulatory subunit of protein kinase A binds to cAMP response element: an alternative cAMP signaling pathway. Proc. Natl. Acad. Sci. USA *95*:6687-6692.
70. **Srivastava, R.K., A.R. Srivastava, Y.G. Park, S. Agrawal, and Y.S. Cho-Chung.** 1998. Antisense depletion of RIα subunit of protein kinase A induces apoptosis and growth arrest in human breast cancer cells. Breast Cancer Res. Treat. *49*:97-107.
71. **Srivastava, R.K., A.R. Srivastava, P. Seth, S. Agrawal, Y.S. Cho-Chung.** 1999. Growth arrest and induction of apoptosis in breast cancer cells by antisense depletion of protein kinase A-RI α subunit: p53-independent mechanism of action. Mol. Cell. Biochem. *195*:25-36.
72. **Stein, C.A.** 1995. Does antisense exist? Nat. Med. *1*:1119-1121.
73. **Steinberg, R.A. and D.A. Agard.** 1981. Turnover of regulatory subunit of cyclic AMP-dependent protein kinase in S49 mouse lymphoma cells. Regulation by catalytic subunit and analogs of cyclic AMP. J. Biol. Chem. *256*:10731-10734.
74. **Tagliaferri, P., T. Clair, M.E. Debortoli, and Y.S. Cho-Chung.** 1985. Two classes of cAMP analogs synergistically inhibit p21 ras protein synthesis and phenotypic transformation of NIH/3T3 cells transfected with Ha-MuSV DNA. Biochem. Biophys. Res. Commun. *130*:1193-1200.
75. **Takio, K., S.B. Smith, E.G. Krebs, K.A. Walsh, and K. Titani.** 1984. Amino acid sequence of the regulatory subunit of bovine type II adenosine cyclic 3′,5′-phosphate dependent protein kinase. Biochemistry *23*:4200-4206.
76. **Taylor, S.S., J. Bubis, J. Toner-Webb, L.D. Saraswat, E.A. First, J.A. Buechler, D.R. Knighton, and J. Sowadski.** 1988. cAMP-dependent protein kinase: prototype for a family of enzymes. FASEB J. *2*:2677-2685.
77. **Titani, K., T. Sasagawa, L.H. Ericsson, S. Kumar, S.B. Smith, E.G. Krebs, and K.A. Walsh.** 1984. Amino acid sequence of the regulatory subunit of bovine type I adenosine cyclic 3′,5′-phosphate dependent protein kinase. Biochemistry *23*:4193-4199.
78. **Tortora, G., R. Bianco, V. Damiano, G. Fontanini, S. De Placido, A.R. Bianco, and F. Ciardiello.** 2000. Oral antisense that targets protein kinase A cooperates with taxol and inhibits tumor growth, angiogenesis, and growth factor production. Clin. Cancer Res. *6*:2506-2512.

79. **Tortora, G., A. Budillon, H. Yokozaki, T. Clair, S. Pepe, G. Merlo, C. Rohlff, and Y.S. Cho-Chung.** 1994. Retroviral vector-mediated overexpression of the RIIβ subunit of the cAMP-dependent protein kinase induces differentiation in human leukemia cells and reverts the transformed phenotype of mouse fibroblasts. Cell Growth Differ. *5*:753-759.
80. **Tortora, G. and Y.S. Cho-Chung.** 1990. Type II regulatory subunit of protein kinase restores cAMP-dependent transcription in a cAMP-unresponsive cell line. J. Biol. Chem. *265*:18067-18070.
81. **Tortora, G. and F. Ciardiello.** 2000. Targeting of epidermal growth factor receptor and protein kinase A: molecular basis and therapeutic applications. Ann. Oncol. *11*:777-783.
82. **Tortora, G., F. Ciardiello, S. Ally, T. Clair, D.S. Salomon, and Y.S. Cho-Chung.** 1989. Site-selective 8-chloroadenosine 3′,5′-cyclic monophosphate inhibits transformation and transforming growth factor alpha production in Ki-*ras*-transformed rat fibroblasts. FEBS Lett. *242*:363-367.
83. **Tortora, G., T. Clair, and Y.S. Cho-Chung.** 1990. An antisense oligodeoxynucleotide targeted against the type RIIβ regulatory subunit mRNA of protein kinase inhibits cAMP-induced differentiation in HL-60 leukemia cells without affecting phorbol ester effects. Proc. Natl. Acad. Sci. USA *87*:705-708.
84. **Tortora, G., V. Damiano, C. Bianco, G. Baldassarre, A.R. Bianco, L. Lanfrancone, P.G. Pelicci, and F. Ciardello.** 1997. The RIα subunit of protein kinase A (PKA) binds to Grb2 and allows PKA interaction with the activated EGF-receptor. Oncogene *14*:923-928.
85. **Tortora, G., S. Pepe, C. Bianco, G. Baldassarre, A Budillon, T. Clair, Y.S. Cho-Chung, A.R. Bianco, et al.** 1994. The RIα subunit of protein kinase A controls serum dependency and entry into cell cycle of human mammary epithelial cells. Oncogene *9*:3233-3240.
86. **Tortora, G., S. Pepe, A.M Grafici, F. Ciardello, A. Porcellini, T. Clair, G. Colletta, Y.S. Cho-Chung, et al.** 1993. Thyroid-stimulating hormone-regulated growth and cell cycle distribution of thyroid cells involve type I isozyme of cyclic AMP-dependent protein kinase. Cell Growth Differ. *4*:359-365.
87. **Tortora, G., S. Pepe, H. Yokozaki, S. Meissner, and Y.S. Cho-Chung.** 1991. Cooperative effect of 8-Cl-cAMP and rhGM-CSF on the differentiation of HL-60 human leukemia cells. Biochem. Biophys. Res. Commun. *177*:1133-1140.
88. **Tortora, G., H. Yokozaki, S. Pepe, T. Clair, and Y.S. Cho-Chung.** 1991. Differentiation of HL-60 leukemia cells by type I regulatory subunit antisense oligodeoxynucleotide of cAMP-dependent protein kinase. Proc. Natl. Acad. Sci. USA *88*:2011-2015.
89. **Uhler, M.D., D.F. Carmichael, D.C. Lee, J.C. Chrivia, E.G. Krebs, and G.S. McKnight.** 1986. Isolation of cDNA clones coding for the catalytic subunit of mouse cAMP-dependent protein kinase. Proc. Natl. Acad. Sci. USA *83*:1300-1304.
90. **Uhler, M.D., J.C. Chrivia, and G.S. McKnight.** 1986. Evidence for a second isoform of the catalytic subunit of cAMP-dependent protein kinase. J. Biol. Chem. *261*:15360-15363.
91. **Walter, U., M.R. Costa, X.O. Breakefield, and P. Greengard.** 1979. Presence of free cyclic AMP receptor protein and regulation of its level by cyclic AMP in neuroblastoma-glioma hybrid cells. Proc. Natl. Acad. Sci. USA *76*:3251-3255.
92. **Wang, H., Q. Cai, X. Zeng, D. Yu, S. Agrawal, and M.Q. Zhang.** 1999. Antitumor activity and pharmacokenetics of a mixed-backbone antisense oligonucleotide targeted to the RIα subunit of protein kinase A after oral administration. Proc. Natl. Acad. Sci. USA *96*:13989-13994.
93. **Wehner, J.M., A.M. Malkinson, M.F. Wiser, and J.R. Sheppard.** 1981. Cyclic AMP-dependent protein kinases from Balb 3T3 cells and other 3T3 derived lines. J. Cell Physiol. *108*:175-184.
94. **Wittmaack, F.M., W. Weber, and H. Hilz.** 1983. Isoenzymes of cAMP-dependent protein kinase in developing rat liver and in malignant hepatic tissues. Eur. J. Biochem. *129*:669-674.
95. **Yasui, W. and E. Tahara.** 1985. Effect of gastrin on gastric mucosal cyclic adenosine 3′,5′-monophosphate-dependent protein kinase activity in rat stomach carcinogenesis induced by *N*-methyl-*N*-nitro-*N*-nitrosoguanidine. Cancer Res. *45*:4763-4767.
96. **Yokozaki, H., A. Budillon, G. Tortora, S. Meissner, S.L. Beaucage, K. Miki, and Y.S. Cho-Chung.** 1993. An antisense oligodeoxynucleotide that depletes RIα subunit of cyclic AMP-dependent protein kinase induces growth inhibition in human cancer cells. Cancer Res. *53*:868-872.
97. **Young, M.R.I., M. Montpettit, Y. Lozano, A. Djordjevic, S. Devata, I.P. Matthews, S. Yedavilli, and G. Chejfec.** 1995. Regulation of Lewis lung carcinoma invasion and metastasis by protein kinase A. Int. J. Cancer *61*:104-109.
98. **Zamecnik, P. and M. Stephenson.** 1978. Inhibition of Rous sarcoma virus replication and cell transformation by a specific oligodeoxynucleotide. Proc. Natl. Acad. Sci. USA *75*:280-284.

Section IV

Oligochips

12 Applicability and Practice of Genome-Wide Expression Profiling: Insights from Studies of Skeletal Muscle Diversity and Disease Mechanisms

John D. Porter[1–4], Patrick Leahy[5], Sangeeta Khanna[1,4], Georgiana Cheng[1,4], and Martina L. Veigl[5]

Departments of [1]Ophthalmology, [2]Neurology, and [3]Neurosciences, the [4]Visual Sciences Research Center, and [5]The Gene Expression Array Core Facility and The Comprehensive Cancer Center, Case Western Reserve University and The Research Institute of University Hospitals of Cleveland, Cleveland, OH, USA

INTRODUCTION

The evolution of knowledge in the field of skeletal muscle biology has been substantial (19), largely because of the innovative application of a wide range of techniques from electron microscopy to molecular biology to isolated muscle contractile studies. We now have considerable understanding of how individual muscle properties interact to determine the key tissue-specific traits of contraction and fatigability, how these properties are regulated during development, and how they are perturbed in disease states. However, data from the integrated application of these approaches suggest that some key muscle biology concepts represent oversimplifications. For example, while the red (I), intermediate (IIA and IIX/D), and white (IIB) fiber type classification schemes (8,10,39) have contributed immeasurable value to the understanding of muscle biology in health and disease, it has become recognized that specialized skeletal muscles often do not fit the traditional

Perspectives in Gene Expression
Edited by K. Appasani

"color" scheme (42). Moreover, in axial and appendicular musculature, there may be adaptations among muscle fibers of the same apparent type that are required for muscle group functional role specificity. Therefore, the skeletal muscle field needs the appropriate tools for the study of the global coordination of traits that contribute to contraction speed and fatigue resistance and to facilitate an understanding of the developmental mechanisms and structure-function adaptations, behind which may be a rainbow of skeletal muscle fiber types.

In the same way, studies of skeletal muscle disease have used the tools available to each generation of scientists and focused on disease genes in the inherited myopathies or dystrophies or on tightly defined cellular/molecular cascades responsible for muscle pathology. Again, this approach has allowed for the rapid accumulation of knowledge of disease mechanisms. In the muscular dystrophy field alone, the initial discovery of dystrophin as the Duchenne muscular dystrophy gene in 1987 (25) led to the current generalized dystrophy model that is based on an elaborate membrane-associated dystrophin-glycoprotein complex, for which mutations in many components either cause a distinct type of dystrophy or are embryo lethal (4,12). However, no one will claim that the understanding of the muscular dystrophies is complete, as there are many facets of the disease that remain unexplained by existing models, and most dystrophies have no effective treatment. Thus, as is the case with our knowledge of the basic muscle phenotype, a cohesive understanding of the muscular dystrophies and other muscle diseases requires broader delineation of the mechanisms that operate within and on the skeletal muscle fiber. These include the interaction of primary and secondary mechanisms in the etiology and progression of disease and the operation of common and disease-specific mechanisms in diverse disease states.

An important technological breakthrough that may resolve many global issues in skeletal muscle biology is genome-wide expression profiling. The near-complete determination of sequence information for both the human genome and those of a number of important animal research models is ending a highly revolutionary period in genomics. This revolution of information has made possible a new, second wave of intense research effort, now largely directed toward genome-wide expression profiling using DNA microarray (53) and serial analysis of gene expression (SAGE) (63). Expression profiling techniques have had a dramatic impact on biomedical research in general; as of August 2002, the National Center for Biotechnology Information (NCBI) PubMed listed 3191 publications containing the term "DNA microarray" and 570 publications containing the term "SAGE." Placed in this context, the application of profiling technology to skeletal muscle biology is only in its infancy.

Gene expression profiling requires collaborations among skilled personnel with a variety of backgrounds, calls for implementation of careful quality-control measures, and generates overwhelming amounts of new data, as illustrated in studies of oncology (22) and the cell cycle (16). Data are accumulating globally, nationally, and within individual institutions at a rate that has caught almost

Table 1. Publications in Skeletal Muscle Biology Using Genome Profiling

Year	DNA Microarray	SAGE
1999	2	1
2000	1	1
2001	8	3
2002	23	2
Total to Date	34	7

Gene expression profiling publications in the primary muscle biology literature, as tabulated from an NCBI PubMed database search in August 2002.

everyone ill-prepared for the myriad problems of sample acquisition, processing, data analysis, and data warehousing. Moreover, many traditional statistical tools are inappropriate for large databases, necessitating interactions among statisticians, computer scientists, and molecular biologists to arrive at meaningful data endpoints (50). Although acquired through great expense, gene expression profiling data form a resource of valuable biological information that is destined to remain untapped unless it is archived carefully and mined diligently.

This chapter presents an overview of the brief history of expression profiling in the field of skeletal muscle biology. We also consider methodology and general aspects of the application of genome-wide profiling technology, as applied in our laboratories and shared core facilities at Case Western Reserve University and The Research Institute of the University Hospitals of Cleveland. Finally, we describe the specific application of this technology to the muscle biology field, including the evolution of our strategies for experimental design and data analysis.

PRELUDE: A BRIEF HISTORY OF GENOME PROFILING AND SKELETAL MUSCLE

While genome-profiling technologies largely arose from the oncology field, their impact is being felt in all disciplines. In skeletal muscle biology, the use of this technology has been very limited (Table 1). Results from the initial DNA microarray and SAGE studies of skeletal muscle appeared in 1999 (31,40,65). The appearance of these techniques in muscle publications did not accelerate until 2001–2002 and still only represent approximately 1% of all DNA microarray and SAGE publications. Many of the early gene profiling studies in skeletal muscle were directed toward the characterization of muscle or muscle group-specific transcripts (13,40,45,65) or the study of sarcopenia (muscle aging) (64,66,67). The broad nature of expression profiling data has had a particular impact in delineating differences in red versus white skeletal muscles (13) and in appendicular versus the highly specialized extraocular muscles (15,21,45). In a recent trend, expression profiling has focused on muscle disease models (2,5,14,23,24,46,60, 61,70). Although single gene mutations represent the immediate cause of a variety of myopathies and dystrophies, the ensuing pathogenesis typically involves a complex array of interactive mechanisms that are readily dissectible using profiling techniques. Because expression profiling is ideally suited to address multidimensional processes, DNA microarray data are providing a better understanding of the roles of secondary disease mechanisms (46) and regeneration (46,61,70) in muscular dystrophy.

Online databases provide ready access to expression profiling data, but, at present, these resources are very limited for skeletal muscle biology. Normative skeletal muscle DNA microarray data from Eric Hoffman's laboratory is available at the NCBI Gene Expression Omnibus (GEO) repository (http://www.ncbi.nlm.nih.gov/geo/) (GenBank® accession nos. GSM1339–1350). Skeletal muscle SAGE libraries also are available in the NCBI GEO database (young human skeletal muscle, GSM824; old human skeletal muscle, GSM819; and rat extraocular muscle, GSM581) or from a muscle laboratory Web page (http://www.urmc.rochester.edu/smd/CRC/Swindex.html). In an alternative profiling strategy, a large muscle transcript database generated from cDNA libraries includes relative abundance data based on expressed sequence tag (EST) frequencies in non-normalized libraries (database available at http://telethon.bio.unipd.it/GETProfiles/) (6). Finally, the Weizmann Institute of Science GeneCards database (http://bioinfo. weizmann.ac.il/cards/), a useful resource for gene function and disease annotation data, has compiled cross-tissue expression profiling data, including heart and skeletal muscle, from both DNA microarray and cDNA libraries.

PROFILING STRATEGIES IN EXTRAOCULAR MUSCLE BIOLOGY

The extraocular muscles encompass six skeletal muscles that originate from the posterior and medial walls of the orbit, insert on sclera of the eye, and are responsible for a complex array of voluntary and reflexive eye movements. Accumulating evidence suggests that the extraocular muscle phenotype is distinctive among skeletal muscles and that this novel muscle group is differentially involved in a variety of metabolic and neuromuscular diseases (42). Since skeletal muscle is an elegant structure-function model that is responsive to changes in activation patterns, the extraocular muscle phenotype likely reflects adaptations to the diverse functional demands of eye movement control.

The central goal of the Porter laboratory is to develop a global understanding of extraocular muscle biology in health and disease. Our major objectives are to: (*i*) fully characterize the extraocular muscle phenotype, identifying important structure-function divergences from the axial/appendicular muscle stereotype; (*ii*) determine the developmental regulatory mechanisms behind the novel phenotype; and (*iii*) relate the unique pattern of sparing or involvement of this muscle group to properties of the distinctive eye muscle phenotype. In the area of muscle disease, we are particularly interested in the mechanisms responsible for eye muscle sparing in the muscular dystrophies because these data will both contribute to the understanding of the extraocular muscle phenotype and may yield insights into new treatment directions for these devastating neuromuscular disorders. Given the broad scope of the questions in our laboratory and the relatively few investigators with research programs dedicated to extraocular muscle biology, genome-wide profiling techniques represent an ideal discovery strategy.

We utilize two key tools, DNA microarray and SAGE, in the determination of broad gene expression patterns in skeletal muscle. While each has its own particular advantages, the complementary use of these gene profiling technologies has allowed important advances; one clear example is in the retinal cell biology field (3,33,54). Our principal technique for expression profiling is DNA microarray using the Affymetrix GeneChip® system (Santa Clara, CA, USA). DNA microarray studies rely on the differential binding of labeled target transcripts to either cDNA or oligonucleotide probes that are arrayed on a chip. Strengths of the microarray approach include its considerable resolving capacity, its ability to operate in a high-throughput mode, and its relatively low cost when data yield is taken into account. This technology is limited by the quality and number of gene-specific probes that have been designed into the array. In this regard, high-density commercial arrays still evaluate only a portion of the entire mammalian genome, and probe quality and identity may bias low-density or specialized arrays. Available cDNA libraries used in custom arrays require normalization and/or careful validation before being printed to arrays. Finally, while DNA microarrays are renowned for their ability to quantify same-gene changes in expression level, they do not provide a readily portable measure of the level of gene expression and do not allow reliable cross-gene expression-level comparisons.

SAGE, on the other hand, potentially allows the detection and quantification of any transcript, without pre-existing bias by the investigator or chip designer, and has equally precise quantification of both same-gene and cross-gene expression levels. It is the more useful of the two techniques for identifying the complete content of a tissue/cell type transcriptome. SAGE is limited in that the 14-base tags that it utilizes do not always provide unambiguous identification of single transcripts, and not all genes have a recognized *Nla*III SAGE tag. Data obtained with SAGE do have the important trait of being fully portable; that is, provided appropriate controls are undertaken (e.g., GC content), a SAGE database can be used for cross-laboratory comparisons. By contrast, microarray data sets are inherently dependent on factors such as microarray type, probe design, transcript isolation, purification, labeling, hybridization conditions, and scanner photomultiplier tube sensitivity, thereby limiting the direct comparison of data obtained from different laboratories. Gene expression data can be obtained more rapidly with DNA microarrays than with SAGE (i.e., SAGE is a lower throughput technology), and many investigators are restrained by the per library sequencing costs (the sequencing costs alone for one 50 000-tag SAGE library are equivalent to an Affymetrix microarray purchase and processing costs for approximately 10 samples).

There have been few direct comparative studies of SAGE and DNA microarray. Evans and colleagues (20) recently showed a high degree of correspondence (>80% detection by Affymetrix) between Affymetrix GeneChip and SAGE data obtained for transcripts detected at an abundance level ≥0.008% of the total SAGE tags analyzed. However, based on data that Affymetrix detection rates fell to ≤65% for lower abundance SAGE tags (2–5 tags/76 790 total), these authors further argued that SAGE is superior for the quantification of low abundance transcripts. This argument was made despite the authors' concession that transcripts detected at levels of less than five SAGE tags are unreliable. Thus, one can only conclude from the data of Evans et al. (20) that there is a high degree of correspondence between the two techniques.

For the reasons discussed earlier, DNA microarray has become our principal tool for muscle expression profiling, with SAGE used only when we desire broader transcriptome data from specific studies. The next sections of this chapter will overview the approach taken by our laboratory and core facilities to assess skeletal muscle expression profiling using the Affymetrix GeneChip system.

METHODOLOGY AND RESOURCES

The Affymetrix GeneChip System

Affymetrix documentation claims that GeneChip analysis can detect as few as 4–5 copies of mRNA per cell and can reproducibly identify from 2- to 3-fold changes in mRNA levels, while interrogating between 8000 and 12 000 gene products per microarray (with some redundancy). Each transcript is interrogated by a set of 16–20 different immobilized, paired "perfect match" (PM) and "mismatch" (MM) 25-mer oligonucleotide probes. Lockhart et al. (34) chose 25-mer probes as the best balance between signal intensity and the ability to discriminate between the correct target and other closely related sequences. PM probes are designed from unique regions within the 3′-end of the gene to allow for the optimal detection of individual transcripts within gene families. The probe pairs are restricted to the 3′-end of each gene to match the target characteristics that result from the sample amplification and labeling step. However, probes are widely spaced to interrogate various regions of each transcript, which helps to ensure the robust detection of each expressed gene. MM probes contain a homomeric mutation at the central base of the 25-mers and are used to gauge the level of nonspecific binding that occurs on the PM probes. The MM probes are arrayed as pairs alongside their respective PM partners on the chip. However, the 16–20 different probe pairs for a specific gene are not contiguous but are scattered throughout the microarray. This random tiling practice eliminates the possibility that a gene's signal will fail to be detected due to technical difficulties (i.e., array imperfection, dust, or bubbles that impede homogeneous washing or staining).

Affymetrix's use of the PM versus MM probe strategy is somewhat controversial. One view argues that the MM probe hybridization signal may not represent a good measure of PM specificity. Moreover, MM probes may inappropriately hybridize with unintended/undefined transcripts in the sample. Finally, some biostatisticians working on issues pertaining specifically to Affymetrix microarrays question the use of MM probes. Affymetrix counters with empirical data (Affymetrix Technical Note No. 701009) that indicate that the PM-MM system adds considerable sensitivity, particularly at the low end of the transcript concentration range. This point continues to be debated. However, at this time, MM data play a significant role in the use of the Affymetrix system and their standard algorithm.

Sample Acquisition and Initial Processing

Inbred mouse and rat strains are used in most of our studies of muscle biology. These animal models offer several very clear advantages when used in skeletal muscle studies focused on answering broad molecular questions. Advantages include the tightly controlled genetic background, the wide variety of available disease models, and the ability to rapidly obtain and process entire muscles or muscle groups at any developmental or pathogenic stage. The conservation of genetic background and the ready availability of rodents through managed breeding programs also allow for pooling of tissue across animals (we always pool tissues from 3–15 animals/replicate; exact number depending on experiment, tissue, and age). This strategy minimizes the variation among individuals, which is generally larger than other sources of variation in array studies (55), and thereby improves the consistency of transcript measures.

In humans, DNA microarray studies must rely on small biopsies of single, often variable, muscles from patients with diverse genetic backgrounds. Moreover, human tissue samples can be obtained only at highly restricted times (i.e., at the time of diagnostic biopsy in patients or at intervals approved by the Institutional Review Boards in volunteers). A complication of this strategy in heterogeneous diseases such as muscular dystrophy is that small muscle biopsies may not reflect the full status of the pathology in a given skeletal muscle. This potential sample bias requires the investigator to consider the tradeoffs between single versus pooled samples in experimental design (1,14,24). Microarray analyses of human muscle disease demonstrate considerable merit (for review, see Reference 24) provided that experiments are properly designed with sufficient numbers of independent samples. In contrast to its benefits in studies of inbred animal strains, the pooling of heterogeneous human samples in microarray studies may mask biologic variability that is important to an understanding of the disease process.

The initial treatment of skeletal muscle material for DNA microarray studies is handled within the Porter laboratory. Rodents are killed, tissues are rapidly removed by dissection, and muscles are snap-frozen in liquid nitrogen and stored at -80°C. Individual muscle samples are pooled (from a minimum of three animals matched by strain, age, and sex) before RNA extraction. Starting with the raw tissues and ending with data analyses, a minimum of three independent replicates per experimental condition are used to increase the statistical power of our studies. Total RNA is extracted using TRIZOL (Invitrogen, Carlsbad, CA, USA), according to the manufacturer's instructions. RNA pellets are resuspended at 1 μg RNA/μL DEPC-treated water. At this stage of processing, a yield of 100 μg RNA is preferred, but we have been successful with as low as 20 μg total RNA before the Qiagen cleanup (before cleanup, the 260/280 ratio typically is 1.6 to 1.7). Our experience is that isolated RNA is reproducibly stable at -80°C for at least six months.

Our questions in the areas of muscle phenotype, development, and pathology are in the realm of gene expression as it relates to systems biology and, therefore, require an in vivo model. The caveat of this approach is that transcripts from several other cell classes are represented in the sample (e.g., fibroblasts, endothelium, vascular smooth muscle, and Schwann cells). While cell culture allows for the evaluation of homogenous populations and is desirable for many questions (since it allows use of homogeneous cell populations as opposed to the mixed cell types of muscle tissue), this approach is inappropriate here. As protocols for laser microdissection microscopy and linear transcript amplification intersect with advances in microarray techniques that require less starting material, in vitro array studies will, as necessary, become more focused.

Evaluation of Samples on Affymetrix Microarrays

Following the initial sample processing steps, we utilize an on-campus gene expression array core facility that is operated by two of us (P. Leahy and M.L. Veigl). The core handles all steps from RNA cleanup to GeneChip scanning. This approach offers numerous advantages, not the least of which is the experience the core has obtained from processing approximately 2000 microarrays over its two- and a half-year period of operation and nearly 1000 microarrays processed in the past year alone. Based on data obtained from different species, tissues, and treatments for several different Affymetrix gene expression microarray types, our DNA microarray core has established very powerful baseline descriptive statistics that, in turn, serve to ensure the highest quality-control standards. The economic expedients in obtaining volume discounts are also commensurate with good quality-control practices; for example, crucial reagents are purchased in bulk, thereby avoiding lot-to-lot variability. Individual samples are processed with pooled reagents, a practice that minimizes the effects of sample-to-sample variability and small-volume pipetting errors.

Before converting total RNA to cRNA, the Core Facility routinely cleans up total RNA using Qiagen (Valencia, CA, USA) spin columns. The first step in the standard Affymetrix protocol converts the message contained in 5–8 μg high-quality total RNA into double-stranded cDNA using SuperScript® II reverse transcriptase (Invitrogen) and an oligo-dT primer with a T7 RNA polymerase promoter linked to its 5′-end. The resulting cDNA is cleaned and used in an in vitro transcription reaction to generate adequate amplified, biotin-labeled cRNA

(50–80 μg). The cRNA is fragmented, and 15 μg cRNA are added to 300 μL hybridization cocktail. The hybridization cocktail is a multicomponent, complex, buffered environment, with BSA and herring-sperm DNA present as carrier molecules. Four in vitro transcription products of bacterial genes are added at final concentrations ranging from 1.5 to 100 pM. These spiked transcripts serve as hybridization controls and are always monitored as part of our quality-control regimen. The fragmented biotin-labeled cRNA is then added, and the final volume is adjusted to 300 μL with molecular biology-grade water.

Immediately before hybridization, sample cocktails are preconditioned by heating to 99°C for 5 min, and then 45°C for an additional 5 min, followed by centrifugation (21 000× *g*) at maximum speed for 5 min in a bench-top centrifuge. The centrifugation step pellets any denatured proteins and debris that may have accumulated during the preheating session. Meanwhile, the microarrays are warmed to room temperature. The 300-μL capacity hybridization chambers are pre-wetted with 200 μL 1× hybridization buffer [100 mM MES, 1 M NaCl, 20 mM EDTA, 0.01% Tween 20, 0.1 mg/mL herring sperm DNA, 0.5 mg/mL BSA (acetylated), and Affymetrix IVT controls as per manufacturer] and incubated at 45°C, with agitation for 10–15 min. After pre-wetting, the 1× buffer is removed from the hybridization chamber of the microarray and replaced by 200 μL of the sample cocktail. Samples are hybridized to microarrays in a rotating hybridization oven (Affymetrix Model 640). The total amount of biotin-labeled cRNA incubated with each microarray is 10 μg. The hybridization process consumes only negligible amounts of target; thus, following the overnight hybridization, samples may be retrieved and used again.

After sample retrieval, hybridized microarrays are subjected to a series of increasingly stringent washes to remove nonspecific binding of biotin-labeled cRNA. We use a GeneChip Fluidics Station 400 (Affymetrix) for standard post-hybridization wash and double-stain protocols. The hybridization chamber in the microarray is flooded with a solution of streptavidin-conjugated phycoerythrin dye to tag the biotin associated with the bound cRNA for visualization. The signal is then amplified in a second staining protocol by incubation with a biotin-tagged streptavidin antibody, followed by another round of staining with the streptavidin-conjugated phycoerythrin dye. We use a GeneArray® Scanner (Affymetrix) for the laser scanning of the microarrays, and the resulting images are captured by a Microarray Suite 5.0 (Affymetrix) and saved as data image (.dat) files. We routinely only use scans from the antibody stain microarrays, in which the enhanced signal allows for the detection of low copy number transcripts.

Microarray Quality Control

We have found that a good quality-control indicator is the percentage of transcripts that are called present (P) by the Affymetrix algorithm. In our hands, this parameter reliably varies for each type of microarray (Figure 1A), and the tight standard deviations allow us to use the %P calls as one among a number of diagnostic criteria. The usefulness of %P as a diagnostic tool can be seen in Figure 1B, which shows the P call percentages for a series of 18 mouse skeletal muscle samples. Here there is considerable consistency in %P across the samples, but sample 10 is clearly an outlier that is caused by either sample or chip problems. As another quality-control measure, independent evaluations of a sample hybridized to two separate microarrays has shown a high degree of correlation in results, which is consistent with the effectiveness of our standardized protocols. Other quality-control parameters routinely evaluated include the background/noise and expression levels of the bacterial transcript spikes and the housekeeping genes specific to each organism. However, we have found, from evaluations of hundreds of samples, the %P call metric provides the most reproducible parameter for each type of microarray and, consequently, can be used to reliably assess data quality.

As discussed earlier, the probe pairs interrogating expression are localized to the 3′ region of the gene. However, the Affymetrix microarrays also contain probe pair sets to the 5′ and middle regions for the housekeeping genes, β-actin, and GAPDH. These probe pairs allow the user to evaluate the integrity of the biotin-labeled cRNA target. Briefly, a ratio of the signal intensities measured for the 3′ versus 5′ region of a gene greater than one could suggest that the starting RNA was fragmented or that the cRNA and/or the in vitro transcription reactions did not progress to completion. However, note that tissue-specific expression patterns and isoforms do impact this ratio. Therefore, before a sample is included in a data set, these ratios are compared for the housekeeping genes, and samples that deviate from the group are eliminated.

In muscle biology studies, the routine assessment of sample-to-sample variability is an additional quality-control measure. Independent replicate samples representing pooled material from genetically homogeneous mouse strains should show a high degree of correlation in the absence of processing or hy-

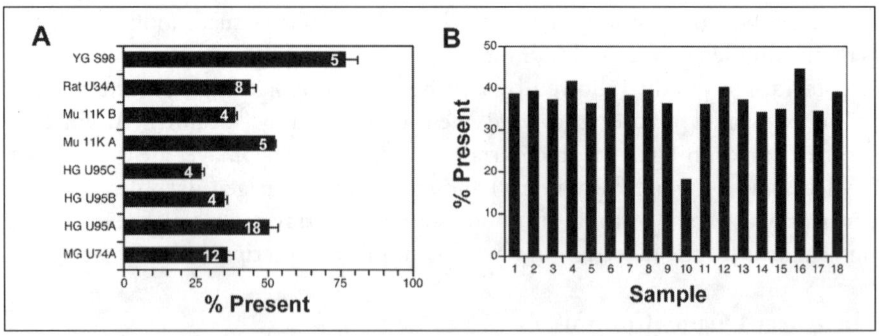

Figure 1. Quality-control measures used in the analysis of Affymetrix GeneChips. (A) Percentage of transcripts called present, as obtained by our DNA microarray core facility, with a variety of Affymetrix GeneChip types (*n* indicated on each bar). Note the tight standard deviations obtained for each chip. (B) Representation of percent present (P) calls obtained by the analysis of 18 independent skeletal muscle samples using the Affymetrix U74Av1 arrays and Microarray Suite. Note that sample 10 is an outlier among an otherwise tightly clustered group.

bridization aberrations. Figure 2 shows a plot of transcript-by-transcript signal values for two fully independent extraocular muscle samples (i.e., muscles were isolated from 10 C57BL/10 SnJ mice/sample and were independently processed and hybridized to separate U74Av1 arrays). Genes designated as present in both samples ($n = 4180$) were plotted and yielded exceptional consistency ($r = 0.993$).

Data Reduction and Basic Analytical Approaches

The management of the large databases arising from high-density oligonucleotide arrays requires the intersection of skills in statistics, computer science, and molecular biology. Many of the tools, even those commercially available, are neither easy to learn nor user-friendly and may not be worth the investment for a single laboratory unless DNA microarray is used extensively. Our DNA microarray core provides data analysis or training in many of the analytic tools described below, but it is impractical for a single facility to both complete the bench work and provide analytic support for all of the more than 1000 GeneChip analyses per year. Thus, in the areas of vision science and eye muscle biology, one of us (J.D. Porter) has established a bioinformatics core facility that provides shared data analysis services.

There is no single generally accepted means to analyze Affymetrix microarray data. The most straightforward analyses of GeneChip data begin with the proprietary Affymetrix Microarray Suite software to obtain the numeric expression-level values ("signal") and the text qualifiers ["detection;" P, M (marginal), and A (absent)] for each transcript. The numeric signal values are weighted means of the multiple PM-MM values in each probe set (weighting represents an attempt to reduce the vulnerability of signal values to outlier probes). Despite this feature of the algorithm, one often encounters transcript signals with values in the tens of thousands that are deemed absent, while other, more modest signals, in the sub-100 range are called present. These apparent inconsistencies are attributable to the second part of the Affymetrix algorithm that considers the consistency of behavior (vis-à-vis PM-MM differ-

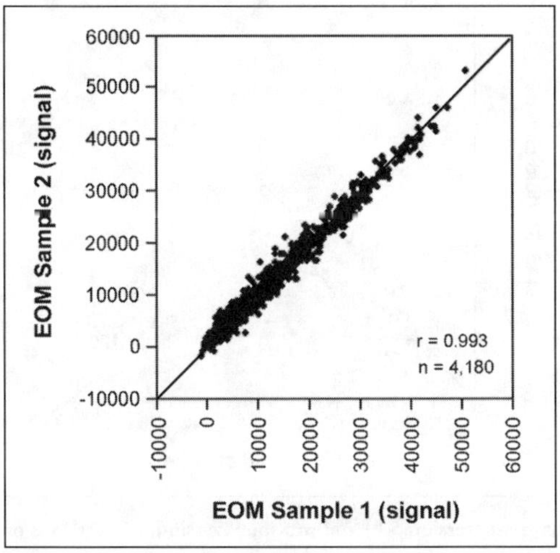

Figure 2. Quality-control plot showing the comparison of Affymetrix Microarray Suite-generated signal values for two independent replicates of wild-type mouse extraocular muscle. All transcripts called present on both chips were plotted. Solid line with unity slope indicates where values would fall if the signals were equal in both GeneChips for all probes. EOM, extraocular muscle.

entials) across a complete probe set when making the absent-present calls.

Once signal values are generated from the software, there are broad choices for chip-to-chip comparative analyses. Many published studies have then undertaken binary comparisons using the Affymetrix Microarray Suite software. This approach provides definitive increase (I), decrease (D), and no change (NC) calls and corresponding fold change ("signal log ratio") values. In a typical muscle biology experiment, we insist that all of three to five independent replicate comparisons show identical I or D calls and fold change values that are ≥2 for a transcript to be considered differentially regulated (for an example, see Reference 45). The Microarray Suite algorithm is structured such that an increase or decrease in transcript expression must be registered by a sufficiently large proportion of all the probe pairs in the probe set. In this way, increases in signal values attributable to the performance of a small number of probe pairs are not construed as genuine changes. When encountered, such incongruities should bolster confidence since the algorithm is successfully screening against high-probability false changes. Affymetrix has recently upgraded their Microarray Suite software (from versions 4.0 to 5.0) to correct several problems in the previous version, including the occasional association of I calls with negative fold change values. We have evaluated the impact of the algorithm on the results from a binary muscle

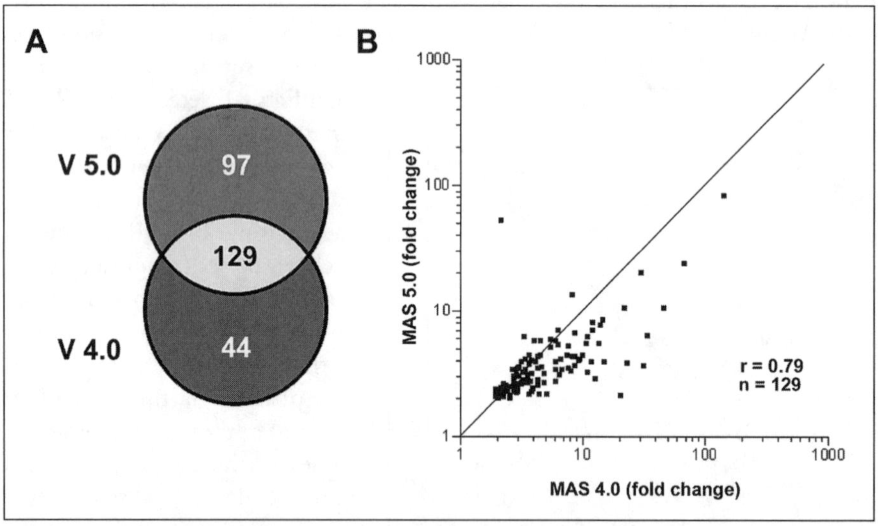

Figure 3. Comparison of results obtained with current (version 5.0) and previous (version 4.0) versions of the Affymetrix Microarray Suite algorithm. Analyses followed our requirement that software return a consistent call (I or D) across all replicates for a transcript to be considered differentially regulated. (A) Each component of the Venn diagram represents differentially regulated genes from a rat extraocular muscle versus leg muscle comparison (U34A chip; *n* = 3 independent replicates/muscle group). Genes in the intersection represent those detected by both algorithms; note the higher total number detected by version 5.0 (226 transcripts) compared to version 4.0 (173 transcripts). (B) Transcript-by-transcript comparison of the respective fold differences for those 129 genes detected by both algorithms. Note the tight correlation in fold change values. MAS, Microarray Suite.

type comparison, and the outcomes are shown in Figure 3. In general, Microarray Suite version 5.0 appears to detect a higher number of differentially regulated transcripts than version 4.0, although an analysis of fold change values of those genes detected by both program versions indicates a tight correlation.

Advanced Analytical Approaches

There has been a proliferation of commercial and individual lab-generated tools for the analysis of DNA microarray data. Affymetrix has issued caveats concerning the management of their data, but it is common to encounter individuals who have ignored these caveats in specialized analyses. In particular, approaches based on the statistical treatment of the Affymetrix signal values often ignore the text qualifiers (i.e., A versus P and I versus NC versus D) that are generated within the Microarray Suite platform. Moreover, cautions (55) have been issued in regard to the use of other statistical tools, such as cluster analysis, indicating that these should be used only with an understanding of their underlying mathematic basis.

Two commonly applied commercial tools for microarray data analysis are GeneSpring® (Silicon Genetics, Redwood City, CA, USA) and DecisionSite (Spotfire, Framingham, MA, USA). These programs provide integrated packages containing a range of analytical tools, including commonly used hierarchical and *k*-means cluster analyses, self-organizing maps (SOMs), and filters that allow for the use of the Affymetrix text qualifier information. More sophisticated statistical tools have been developed to circumvent the problems encountered when one applies traditional statistical measures to large databases. One of these, the significance analysis of microarrays (SAM) (62), relies on the relationship between the change in signal level and the standard deviation among replicate controls and experimental signals to assign a score to each gene. Changes in the expression of genes with scores above a certain threshold are accompanied by estimates of false detection rate. Our studies have applied a combination of Microarray Suite and SAM analyses, using two different strategies (45,46). When we intersected genes found to be differentially regulated using a stringent application of the Affymetrix algorithm (using our 5/5 consistent call plus ≥2-fold criteria) with those determined by SAM, only a small number in the Affymetrix Microarray Suite were dropped as false positives (46).

Statistical treatments aside, the most compelling value of DNA microarray and, at the same time, a daunting task for the investigator, is the delineation of patterns among the often hundreds of differentially expressed genes. While it is difficult to draw definitive conclusions from single differentially regulated transcripts without an adequate follow-up (e.g., real-time quantitative PCR, Northern or slot blot analysis, in situ hybridization, or protein measures), the elucidation of patterns of differentially regulated genes that comprise specific metabolic or signal transduction pathways allows the investigator to exploit the full potential of the gene profiling strategy. The provision of simple lists of differentially regulated genes is no longer the standard for the publication of expression profiling data because now significant data mining is required.

Commercial (e.g., GeneSpring) and public domain (e.g., GenMAPP; http://www.genmapp.org/) (17) tools can greatly assist in the process of mapping data onto pathways. However, the precise name/function of a specific gene is often a moving target that requires the investigator to use microarray probe accession numbers to obtain the most current functional/disease annotations before publication. The NCBI LocusLink (http://www.ncbi.nlm.nih.gov/LocusLink/) and UniGene (http://www.ncbi.nlm.nih.gov/UniGene/) databases represent the best starting point for this information. Both databased allow for the update of the gene name and symbol associated with probe accession numbers, and LocusLink also provides brief functional information and transcript-specific links to the NCBI Online Mendelian Inheritance in Man for additional function and disease data. A public domain tool, UnChip (http://www.unchip.org), provides batch analysis access to some of this information using Affymetrix probe identifiers as the starting point. Further annotation of gene ontology/function is available from GeneCards (http://bioinformatics.weizmann.ac.il./cards) and Jackson ImmunoResearch Laboratories (West Grove, PA, USA) Mouse Genome Informatics (http://www.informatics.jax.org/).

In the following sections, we describe projects that are representative of those in the Porter laboratory that use genome-wide profiling to address questions in muscle biology. Our goal here is 2-fold: (*i*) to review progress in muscle biology in health and disease that has been derived from profiling technology and (*ii*) to illustrate representative experimental and analytical strategies that we have used to derive the most value from the DNA microarray methodology.

FUNDAMENTAL DIFFERENCES BETWEEN "TRADITIONAL" SKELETAL MUSCULATURE AND THE HIGHLY SPECIALIZED EXTRAOCULAR MUSCLES

The historical argument that extraocular muscle represents a fundamentally distinct class of skeletal muscle relies on the seemingly anecdotal and inefficient accumulation of identified gene/protein expression differences between the appendicular and extraocular muscle groups that has occurred over the past 20 years (for review, see References 42, 43, and 59). For example, adult extraocular muscles exhibit a tissue-specific myosin heavy chain isoform, express α-cardiac myosin, and retain myosin isoforms that are normally repressed during development (9,27,51,52,68,69). While it is clear that extraocular muscle presents other significant divergences from skeletal muscle (7,29,35-37), it has been difficult to develop a unified structure/function model of this novel muscle group on the basis of these fragmented data.

Understanding extraocular muscle biology is an ideal application for expression profiling. The first systematic attempt to explore the potentially broad range of transcript differences between extraocular and hindlimb musculature used differential display PCR (38); this study focused on identifying novel ESTs that were highly expressed in eye muscle rather than on obtaining a broad muscle profile. Genome-wide profiling using DNA microarray offered a highly efficient means of going beyond these data to develop a panoramic view of the depth and breadth of basic differences between extraocular and the more traditional limb musculature. In our studies, we first compared three presumptive classes of skeletal muscle, extraocular muscle, masticatory, and hindlimb muscle in adult mice (45). Masticatory muscle was included because it has been proposed that, like extraocular muscle, it is a distinct skeletal muscle class (26).

We executed pairwise comparisons with the Affymetrix algorithm and employed a highly conservative data acceptance criteria (consistency with Affymetrix calls across all replicates and a ≥2-fold cut-off) to compare extraocular muscle with both of the other skeletal muscle groups. Using this approach, we identified a signature consisting of 287 genes that distinguished eye muscle from the other skeletal muscle groups. In such analyses, we believe that there is no alternative other than to study sufficient numbers of each experimental class; repeated analyses of the same sample and iterative comparisons among a small group of experimental and control samples are not adequate substitutes. Surprisingly, we detected no differences between the jaw and hindlimb muscles, but both are principally comprised of the same skeletal muscle fiber type (IIB).

The nature and extent of the differences between extraocular and other skeletal muscles can be assessed and visualized in several ways. Figure 4A illustrates the extent of the muscle group differences, shown by a schematic plot of numbers of differentially regulated genes that are arranged by functional class. This approach shows that expression differences lie across a wide range of functional categories. Notably, the eye muscle pattern is considerably broader than that observed in earlier red versus white skeletal muscle comparisons (13), which largely reflected the contractile protein and metabolic differences that define the skeletal muscle fiber types. Illustrated in other ways, both a hierarchical cluster analysis (Figure 4B) and a dendogram (Figure 4C) of the three muscle groups show a clear similarity of jaw and hindlimb muscles and the segregation of extraocular muscles from both of these muscle groups.

Our findings have subsequently been replicated by similar microarray studies in rat by Fischer et al. (21) and ourselves (unpublished data). Moreover, in independent studies (15), we have used SAGE to address the identical divergence question for eye muscle versus leg muscle. The rat SAGE data reinforce and extend the distinctive gene expression pattern of extraocular muscle, including extraocular muscle-specific patterns in contractile, energy metabolism, ribosomal, and cytoskeletal protein gene expression.

Extraocular muscle is known to share properties that are generally considered to be specific to cardiac muscle; however, until now, we have not been aware of the extent of the conservation of traits across these muscle groups. Figure 5 illustrates the use of a program intended for phylogenetic analyses (PHYLIP version 3.573c; available from J. Felsenstein at http://www.phylip.com/) to compare microarray data from the major muscle tissues (e.g., cardiac, skeletal, which is represented by hindlimb muscle, and smooth muscle) with extraocular muscle (three independent replicates/muscle type). Such distance analyses may be more effective in the elucidation of group differences than hierarchical clustering, which may lose key distinctions in the noise of the larger number of unchanged genes (55). The distance tree, constructed by the tabulation of the pairwise distances between samples (number of differentially regulated genes weighted by their respective binary comparison fold change values; for another application, see Reference 18), illustrates the relationship of the two skeletal muscles, extraocular and leg, the relationship between the three striated muscles, eye, leg, and heart, and the separation between striated (eye, leg, and heart) and smooth muscle. Moreover, these findings support the sharing of traits between cardiac

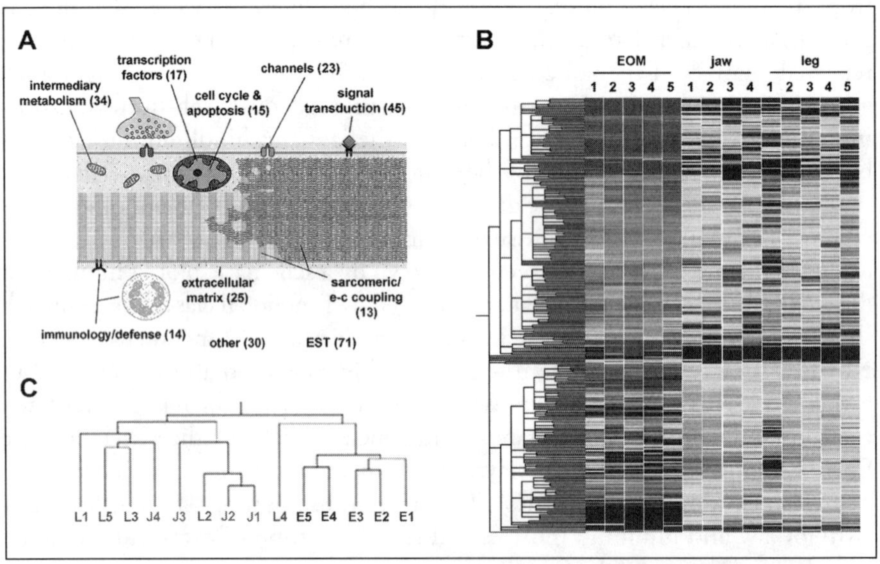

Figure 4. Gene expression profiling of extraocular muscle in comparison with jaw (masseter) and hindlimb (combined gastrocnemius/soleus) muscles in mouse. [Data replotted from Porter et al. (45)]. (A) Muscle fiber schematic showing the number of transcripts by gene class that are differentially regulated in extraocular muscles versus other skeletal muscles. Note the breadth of the functional class differences. (B) Hierarchical cluster analysis showing normalized replicate signal values for the 287 genes that are differentially regulated in extraocular muscle (EOM) versus jaw or leg muscles (n = 4–5 replicates/muscle group; red denotes high expression and blue denotes low expression). Note the similarities between the jaw and leg muscles and the differences in both from the extraocular muscle. Also note that the major differences between extraocular muscles and other muscle groups lie predominately in the level of gene expression, as opposed to genes that are uniquely expressed in eye muscle. (C) Dendogram showing the distribution of skeletal muscle samples when compared across all differentially regulated genes shown in panel B. Extraocular muscle replicates cluster together and separate from jaw and leg replicates, while the jaw and leg replicates are intermixed. E, extraocular muscle; J, jaw; and L, leg. (**See color plate A16.**)

and extraocular muscle, which suggests that this relationship should be explored further to understand eye muscle biology.

In an additional analysis of rat extraocular versus hindlimb muscle data, we obtained results that were very similar to those we reported previously in the mouse. However, we show here that microarray data are heavily dependent on data analysis tools and strategies. To illustrate this point, the same data set was evaluated using both Microarray Suite version 5.0 and GeneSpring version 4.2.1. We found 53% more differentially regulated transcripts with GeneSpring (selection criteria: dropping consistently absent probes, ≥2-fold change, a 0.05 statistical cut-off, and no multiple correction testing) than with Microarray Suite (selection criteria: consistent I or D call across all replicates and ≥2-fold change) (Figure 6). The wide difference in endpoints emphasizes the clear need for further investigation of the impact of the Affymetrix text qualifiers, particularly the call criterion (decision for I versus NC versus D), which is not taken into account in GeneSpring.

Among the expression differences detected in our previous mouse study (45), our unpublished rat data, and the subsequent analysis by another group (21), extraocular muscle was found to be low in transcripts for enzymes and regulators related to glycogen metabolism. This finding is atypical for a fast-twitch skeletal muscle but is consistent with previous findings that glycogen content is exceptionally low in extraocular muscle (59). Other data from our studies further suggest that there are additional differences in the energy metabolism mechanisms in eye muscle. These patterned differences are best understood by plotting microarray data onto metabolic pathways, as in Porter et al. (45). Based on poorly developed mechanisms for the glycogen metabolism in eye muscle, we suspected that there may be metabolic similarities with cardiac muscle in which β-oxidation of fatty acids is a key energy source. However, the microarray-facilitated evaluation of the comparative patterns in mitochondrial β-oxidation enzyme transcripts suggests that this is not the case. Instead, we suggest that eye muscle must utilize glucose directly without expending energy to first store it as glycogen, as in other fast-twitch muscles. Taken together, microarray data that identify patterned changes in a particular metabolic or signal transduction

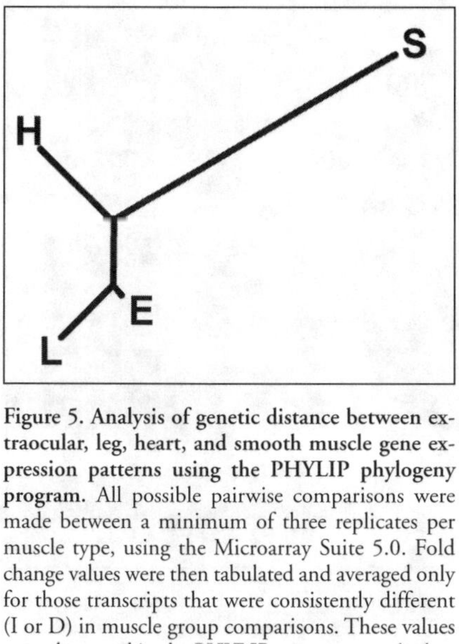

Figure 5. Analysis of genetic distance between extraocular, leg, heart, and smooth muscle gene expression patterns using the PHYLIP phylogeny program. All possible pairwise comparisons were made between a minimum of three replicates per muscle type, using the Microarray Suite 5.0. Fold change values were then tabulated and averaged only for those transcripts that were consistently different (I or D) in muscle group comparisons. These values were then used in the PHYLIP program to calculate and plot the genetic distance between muscle types, as represented here. E, extraocular; L, leg; H, heart; and S, smooth muscle from stomach.

pathway are immensely more valuable that simple lists of differentially regulated genes. The ability to identify the contribution of specific pathways to a particular tissue or process represents the clearest strategic advantage of the microarray technique and contributes toward the overall confidence in the findings.

In a departure from our usual focus on patterns in microarray data, we used a dramatic eye muscle versus leg muscle difference in a single transcript, myomesin 2 (*Myom2*), as a trigger, which has identified a new direction in our laboratory. *Myom2* is an integral component of the sarcomeric M-line, which bisects the sarcomere, interconnects myosin filaments, and serves as an anchorage for muscle-specific creatine kinase (MCK). By microarray, *Myom2* expression levels were 60× higher in hindlimb muscle than in extraocular muscle (45), suggesting that the M-line might be absent in the extraocular muscle group. Conceptually, this would represent a significant divergence from known skeletal muscle biology because fast-twitch muscles rely on phosphocreatine metabolism by MCK, localized immediately adjacent to the contractile apparatus. In a follow-up to the *Myom2* data, we confirmed by electron microscopy, that an M-line is absent from the eye muscles and, by real-time quantitative PCR, that M-line and MCK transcripts are expressed at low levels (unpublished data). The notion that extraocular muscle might utilize alternatives for high-energy phosphate metabolism is currently under study. Note that this significant divergence from other

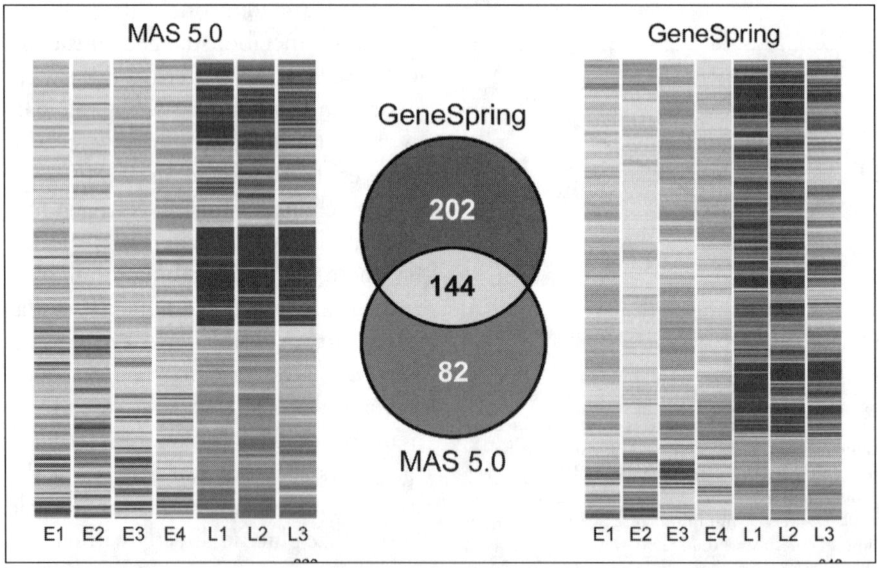

Figure 6. Comparison of differential results obtained from the same sample set (rat extraocular and leg muscles run on U34A chip) with Affymetrix Microarray Suite 5.0 and GeneSpring 4.2.1. Selection criteria used with both programs are given in text. Note the higher number of differentially regulated genes determined by GeneSpring (right panel) versus Microarray Suite (left panel). The Venn diagram in the center indicates that, while there is substantial overlap between techniques, a large number of genes are detected as differentially regulated by only one of the two programs. MAS, Microarray Suite; E, extraocular; and L, leg. (**See color plate A17.**)

skeletal muscles would not have been suspected without the insight provided by expression profiling data.

In summary, microarray data established a distinctive molecular signature for extraocular muscle, demonstrating that these muscles are more different from any other skeletal muscle than the extremes of typical skeletal muscle, red and white, are from each other. We suggest that extraocular muscle must exploit the full range of plasticity of muscle tissue to serve its distinctive role in eye movement control. This point can only be appreciated from the global perspective of DNA microarray analyses. Collectively, the patterned differences in gene expression define extraocular as a distinctive skeletal muscle class and may ultimately explain the unique response of these muscles in metabolic and neuromuscular diseases.

DIFFERENTIAL MUSCLE GROUP SENSITIVITY TO MUSCULAR DYSTROPHY

In contrast to other skeletal muscles, the extraocular muscles are completely spared in Duchenne muscular dystrophy and in several other dystrophies involving the sarcolemmal-associated dystrophin-glycoprotein complex (28,30,32,41, 44,47-49). Understanding the mechanisms that underlie extraocular muscle protection in the muscular dystrophies is important because they may yield insights into the basic biology of this novel muscle group and suggest new treatment strategies for patients with these devastating neuromuscular disorders.

To understand eye muscle protection mechanisms in dystrophin-deficient mice (*mdx*), we have systematically tested predicted muscle-sparing mechanisms that are based on existing disease models (4,12) using more traditional cell/molecular biology approaches. Other than an increase in the expression of a dystrophin homolog, utrophin (48), which does not appear to participate in eye muscle sparing, the extraocular muscles do not adapt in predictable ways to manage dystrophin deficiency. In DNA microarray studies, we sought to take advantage of the broad view of the technique and determine whether eye muscle might use adaptive strategies that were not predicted by any current model of muscular dystrophy. Following our routine approach in Affymetrix GeneChip studies, we compared *mdx* and wild-type mice at an active stage of the disease (eight weeks postnatal) using five independent replicate analyses of each strain, 10 mice per replicate, and strict data acceptance criteria (5/5 consistent calls with the Microarray Suite algorithm and false-positive rejection with SAM analysis).

For the *mdx* hindlimb muscle, we observed the expected pattern of the up-regulation of transcripts associated with fibrosis, inflammation, and muscle regeneration. Overall, 242 genes were differentially expressed in *mdx* hindlimb, 75% of which had not been previously linked to either human Duchenne muscular dystrophy or any of its animal models (Figure 7) (46). The implications of such a high number of previously unrecognized transcripts in a well-studied disease process are profound. As these data demonstrate, disease gene products can no longer be seen only in terms of simple relationships with their local networks of interacting proteins, a notion that considerably enhances the value of global expression profiling. For example, while inflammation has been considered a consequence rather than a contributing factor in the pathogenesis of muscular dystrophy, the magnitude and apparent coordination of inflammatory response transcripts (30% of all up-regulated genes) suggested that inflammation is a contributing factor and, therefore, a potential therapeutic target in a disease that is currently without effective treatment. We are further evaluating the coordinated pattern of cytokine and chemokine signaling, leukocyte adhesion and diapedesis, invasive cell type-specific markers, and complement system activation seen by microarray, using traditional gene/protein biology approaches and temporal expression profiling. If these data support the contentions that inflammation is an active participant in, rather than a bystander effect of, the pathogenesis of muscular dystrophy (11, 56–58), then a new treatment avenue may be opened.

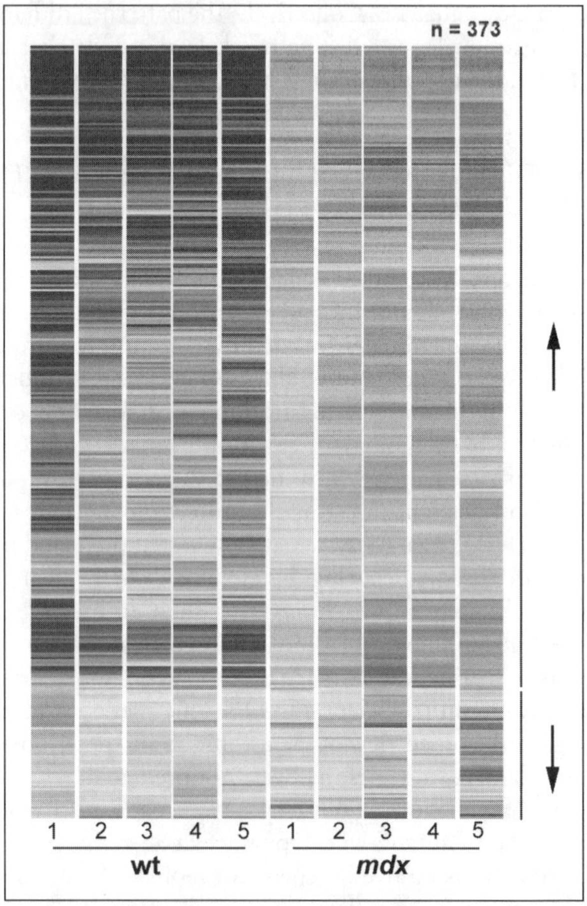

Figure 7. Hierarchical cluster analysis showing differentially regulated genes in comparisons between wild-type and dystrophic mouse hindlimb musculature at eight weeks of age. [Data replotted from Porter et al. (46)]. Data were initially processed with Affymetrix Microarray Suite 5.0, filtering for those genes with both consistent I or D calls across all five replicate comparisons and a ≥2-fold average change. Cluster analysis was then performed on this subset using GeneSpring. The majority of differential regulated genes showed increased expression (indicated by the up arrow; down-regulated transcripts are denoted by the down arrow) in dystrophic leg muscle. wt, wild-type; *mdx*, dystrophic mouse. (**See color plate A18.**)

By contrast, very different results were obtained for *mdx* extraocular muscle. Only seven genes were consistently differentially expressed in dystrophic versus wild-type mouse eye muscle (three were up-regulated and four were down-regulated). Collectively, transcript changes did not constitute a recognizable pattern, which leads us to the conclusion that extraocular muscle does not adapt, but rather is constitutively spared in Duchenne muscular dystrophy. Because we know that there are broad differences between extraocular and traditional skeletal muscle (discussed earlier), we are using additional molecular biologic and bioinformatic data reduction approaches to determine what constitutive eye muscle properties are instrumental in their protection in the muscular dystrophies.

CONCLUSION

The panoramic view available with genome-wide profiling technology is ideally suited for the study of: (*i*) phenotypic diversity in areas that have received little attention but are of substantial importance in the function of the organism and (*ii*) disease processes in which prior research emphasis has been on tightly focused mechanisms revolving around a specific mutation or a narrow pathogenic cascade. In both of these situations, expression profiling serves to broaden the investigators' scope and to identify previously unrecognized pathways and mechanisms. Since their development in 1995, DNA microarray and SAGE have rapidly progressed beyond the stage in which simple, unprocessed gene lists constitute a solid publication. Instead, we are now entering a considerably more valuable phase in which a convergence of multidisciplinary expertise, tools, and insights in expression profiling may allow a truly integrated understanding of the cell in health and disease.

ACKNOWLEDGMENTS

The technical assistance of Anita Merriam and data analysis and illustration preparation by Jonathan Law, Jason Feuerman, and Elizabeth Shurell are much appreciated. We also thank Alan Beggs, Despina Sanoudou, Judith Haslett, and Louis Kunkel (Children's Hospital and Harvard Medical School, Boston, MA) for many helpful discussions of analysis strategies. This work was supported by grants from The Research Institute of University Hospitals of Cleveland (J.D.P.), Research to Prevent Blindness (departmental grant; Senior Scientific Investigator Award to J.D.P.), the Knights-Templar Eye Research Foundation (S.K.), the Muscular Dystrophy Association USA (J.D.P.), and National Institutes of Health grant nos. EY09834 and EY12779 (J.D.P.). The Gene Expression Array Core Facility is supported by the Case Western Reserve University/University Hospitals of Cleveland Ireland Comprehensive Cancer Center and National Cancer Institute grant no. P30 CA43703, and the Bioinformatics Module is supported by the National Eye Institute Core Facility grant no. P30 EY11373. J.D.P. also receives support from the Carl F. Asseff, M.D. Professorship in Ophthalmology.

REFERENCES

1. **Bakay, M., Y.W. Chen, R. Borup, P. Zhao, K. Nagaraju, and E.P. Hoffman.** 2002. Sources of variability and effect of experimental approach on expression profiling data interpretation. BMC Bioinformatics *3*:4.
2. **Bakay, M., P. Zhao, J. Chen, and E.P. Hoffman.** 2002. A web-accessible complete transcriptome of normal human and DMD muscle. Neuromuscul. Disord. *12*:S125-141.
3. **Blackshaw, S., R.E. Fraioli, T. Furukawa, and C.L. Cepko.** 2001. Comprehensive analysis of photoreceptor gene expression and the identification of candidate retinal disease genes. Cell *107*:579-589.
4. **Blake, D.J., A. Weir, S.E. Newey, and K.E. Davies.** 2002. Function and genetics of dystrophin and dystrophin-related proteins in muscle. Physiol. Rev. *82*:291-329.
5. **Boer, J.M., E.J. de Meijer, E.M. Mank, G.J.B. van Ommen, and J.T. den Dunnen.** 2002. Expression profiling in stably regenerating skeletal muscle of dystrophin-deficient *mdx* mice. Neuromuscul. Disord. *12*:S118-124.
6. **Bortoluzzi, S., F. d'Alessi, C. Romualdi, and G.A. Danieli.** 2000. The human adult skeletal muscle transcriptional profile reconstructed by a novel computational approach. Genome Res. *10*:344-349.
7. **Briggs, M.M., J. Jacoby, J. Davidowitz, and F.H. Schachat.** 1988. Expression of a novel combination of fast and slow troponin T isoforms in rabbit extraocular muscles. J. Muscle Res. Cell Motil. *9*:241-247.
8. **Brooke, M.H. and K.K. Kaiser.** 1970. Muscle fiber types: how many and what kind? Arch. Neurol. *23*:369-379.
9. **Brueckner, J.K., O. Itkis, and J.D. Porter.** 1996. Spatial and temporal patterns of myosin heavy chain expression in developing rat extraocular muscle. J. Muscle Res. Cell Motil. *17*:297-312.
10. **Burke, R.E., D.N. Levine, and F.E. Zajac III.** 1971. Mammalian motor units: physiological-histochemical correlation in three types in cat gastrocnemius. Science *174*:709-712.
11. **Cai, B., M.J. Spencer, G. Nakamura, L. Tseng-Ong, and J.G. Tidball.** 2000. Eosinophilia of dystrophin-deficient muscle is promoted by perforin-mediated cytotoxicity by T cell effectors. Am. J. Pathol. *156*:1789-1796.
12. **Campbell, K.P.** 1995. Three muscular dystrophies: loss of cytoskeleton-extracellular matrix linkage. Cell *80*:675-679.
13. **Campbell, W.G., S.E. Gordon, C.J. Carlson, J.S. Pattison, M.T. Hamilton, and F.W. Booth.** 2001. Differential global gene expression in red and white skeletal muscle. Am. J. Physiol. Cell Physiol. *280*:C763-768.
14. **Chen, Y.W., P. Zhao, R. Borup, and E.P. Hoffman.** 2000. Expression profiling in the muscular dystrophies: identification of novel aspects of molecular pathophysiology. J. Cell Biol. *151*:1321-1336.
15. **Cheng, G., and J.D. Porter.** 2002. Transcriptional profile of rat extraocular muscle by serial analysis of gene expression. Invest. Ophthalmol. Vis. Sci. *43*:1048-1058.
16. **Cho, R.J., M.J. Campbell, E.A. Winzeler, L. Steinmetz, A. Conway, L. Wodicka, T.G. Wolfsberg, A.E. Gabrielian, et al.** 1998. A genome-wide transcriptional analysis of the mitotic cell cycle. Mol. Cell *2*:65-73.
17. **Dahlquist, K.D., N. Salomonis, K. Vranizan, S.C. Lawlor, and B.R. Conklin.** 2002. GenMAPP, a new tool for viewing and analyzing microarray data on biological pathways. Nat. Genet. *31*:19-20.
18. **Enard, W., P. Khaitovich, J. Klose, S. Zollner, F. Heissig, P. Giavalisco, K. Nieselt-Struwe, E. Muchmore, et al.** 2002. Intra- and interspecific variation in primate gene expression patterns. Science *296*:340-343.
19. **Engel, A.G. and C. Franzini-Armstrong.** 1994. Myology. McGraw-Hill, New York. 1129 pp.
20. **Evans, S.J., N.A. Datson, M. Kabbaj, R.C. Thompson, E. Vreugdenhil, E.R. De Kloet, S.J. Watson, and H. Akil.** 2002. Evaluation of Affymetrix Gene Chip sensitivity in rat hippocampal tissue using SAGE analysis. Eur. J. Neurosci. *16*:409-413.
21. **Fischer, M.D., J.R. Gorospe, E. Felder, S. Bogdanovich, F. Pedrosa-Domellof, R.S. Ahima, N.A. Rubinstein, E.P. Hoffman, et al.** 2002. Expression profiling reveals metabolic and structural components of extraocular muscles. Physiol. Genomics *9*:71-84.
22. **Golub, T.R., D.K. Slonim, P. Tamayo, C. Huard, M. Gaasenbeek, J.P. Mesirov, H. Coller, M.L. Loh, et al.** 1999. Molecular classification of cancer: class discovery and class prediction by gene expression monitoring. Science *286*:531-537.
23. **Greenberg, S.A., D. Sanoudou, J.N. Haslett, I.S. Kohane, L.M. Kunkel, A.H. Beggs, and A.A. Amato.** Molecular profiles of inflammatory myopathies. Neurology *59*:1170-1182.
24. **Haslett, J. and L. Kunkel.** 2002. Microarray analysis of normal and dystrophic skeletal muscle. Int. J. Dev. Neurosci. *20*:359.
25. **Hoffman, E.P., R.H. Brown, Jr., and L.M. Kunkel.** 1987. Dystrophin: the protein product of the

Duchenne muscular dystrophy locus. Cell *51*:919-928.

26. **Hoh, J.F.Y., S. Hughes, G. Hugh, and I. Pozgaj.** 1989. Three hierachies in skeletal muscle fibre classifiction: allotype, isotype, and phenotype, p. 15-26. *In* L.H. Kedes and F.E. Stockdale (Eds.), Cellular and Molecular Biology of Muscle Development. Alan R. Liss, New York.
27. **Jacoby, J., K. Ko, C. Weiss, and J.I. Rushbrook.** 1990. Systematic variation in myosin expression along extraocular muscle fibres of the adult rat. J. Muscle Res. Cell Motil. *11*:25-40.
28. **Kaminski, H.J., M. al-Hakim, R.J. Leigh, M.B. Katirji, and R.L. Ruff.** 1992. Extraocular muscles are spared in advanced Duchenne dystrophy. Ann. Neurol. *32*:586-588.
29. **Kaminski, H.J., L.L. Kusner, and C.H. Block.** 1996. Expression of acetylcholine receptor isoforms at extraocular muscle endplates. Invest. Ophthalmol. Vis. Sci. *37*:345-351.
30. **Karpati, G., S. Carpenter, and S. Prescott.** 1988. Small-caliber skeletal muscle fibers do not suffer necrosis in mdx mouse dystrophy. Muscle Nerve *11*:795-803.
31. **Khan, J., M.L. Bittner, L.H. Saal, U. Teichmann, D.O. Azorsa, G.C. Gooden, W.J. Pavan, J.M. Trent, et al.** 1999. cDNA microarrays detect activation of a myogenic transcription program by the PAX3-FKHR fusion oncogene. Proc. Natl. Acad. Sci. USA *96*:13264-13269.
32. **Khurana, T.S., R.A. Prendergast, H.S. Alameddine, F.M. Tome, M. Fardeau, K. Arahata, H. Sugita, and L.M. Kunkel.** 1995. Absence of extraocular muscle pathology in Duchenne's muscular dystrophy: role for calcium homeostasis in extraocular muscle sparing. J. Exp. Med. *182*:467-475.
33. **Livesey, F.J., T. Furukawa, M.A. Steffen, G.M. Church, and C.L. Cepko.** 2000. Microarray analysis of the transcriptional network controlled by the photoreceptor homeobox gene Crx. Curr. Biol. *10*:301-310.
34. **Lockhart, D.J., H. Dong, M.C. Byrne, M.T. Follettie, M.V. Gallo, M.S. Chee, M. Mittmann, C. Wang, et al.** 1996. Expression monitoring by hybridization to high-density oligonucleotide arrays. Nat. Biotechnol. *14*:1675-1680.
35. **McLoon, L.K. and J.D. Wirtschafter.** 1996. N-CAM is expressed in mature extraocular muscles in a pattern conserved among three species. Invest. Ophthalmol. Vis. Sci. *37*:318-327.
36. **McLoon, L.K. and J.D. Wirtschafter.** 2002. Continuous myonuclear addition to single extraocular myofibers in uninjured adult rabbits. Muscle Nerve *25*:348-358.
37. **Missias, A.C., G.C. Chu, B.J. Klocke, J.R. Sanes, and J.P. Merlie.** 1996. Maturation of the acetylcholine receptor in skeletal muscle: regulation of the AChR gamma-to-epsilon switch. Dev. Biol. *179*:223-238.
38. **Niemann, C.U., T.O. Krag, and T.S. Khurana.** 2000. Identification of genes that are differentially expressed in extraocular and limb muscle. J. Neurol. Sci. *179*:76-84.
39. **Peter, J.B., R.J. Barnard, V.R. Edgerton, C.A. Gillespie, and K.E. Stempel.** 1972. Metabolic profiles of three fiber types of skeletal muscle in guinea pigs and rabbits. Biochemistry *11*:2627-2633.
40. **Pietu, G., E. Eveno, B. Soury-Segurens, N.A. Fayein, R. Mariage-Samson, C. Matingou, E. Leroy, C. Dechesne, et al.** 1999. The genexpress IMAGE knowledge base of the human muscle transcriptome: a resource of structural, functional, and positional candidate genes for muscle physiology and pathologies. Genome Res. *9*:1313-1320.
41. **Porter, J.D.** 1998. Commentary: extraocular muscle sparing in muscular dystrophy: a critical evaluation of potential protective mechanisms. Neuromuscul. Disord. *8*:198-203.
42. **Porter, J.D. and R.S. Baker.** 1996. Muscles of a different "color": the unusual properties of the extraocular muscles may predispose or protect them in neurogenic and myogenic disease. Neurology *46*:30-37.
43. **Porter, J.D., R.S. Baker, R.J. Ragusa, and J.K. Brueckner.** 1995. Extraocular muscles: basic and clinical aspects of structure and function. Surv. Ophthalmol. 39:451-484.
44. **Porter, J.D. and P. Karathanasis.** 1998. Extraocular muscle in merosin-deficient muscular dystrophy: cation homeostasis is maintained but is not mechanistic in muscle sparing. Cell Tissue Res. *292*:495-501.
45. **Porter, J.D., S. Khanna, H.J. Kaminski, J.S. Rao, A.P. Merriam, C.R. Richmonds, P. Leahy, J. Li, et al.** 2001. Extraocular muscle is defined by a fundamentally distinct gene expression profile. Proc. Natl. Acad. Sci. USA *98*:12062-12067.
46. **Porter, J.D., S. Khanna, H.J. Kaminski, J.S. Rao, A.P. Merriam, C.R. Richmonds, P. Leahy, J. Li, et al.** 2002. A chronic inflammatory response dominates the skeletal muscle molecular signature in dystrophin-deficient *mdx* mice. Hum. Mol. Genet. *11*:263-272.
47. **Porter, J.D., A.P. Merriam, A.A. Hack, F.H. Andrade, and E.M. McNally.** 2001. Extraocular muscle is spared despite the absence of an intact sarcoglycan complex in gamma- or delta-sarcoglycan-deficient mice. Neuromuscul. Disord. *11*:197-207.

48.**Porter, J.D., J.A. Rafael, R.J. Ragusa, J.K. Brueckner, J.I. Trickett, and K.E. Davies.** 1998. The sparing of extraocular muscle in dystrophinopathy is lost in mice lacking utrophin and dystrophin. J. Cell Sci. *111*:1801-1811.

49.**Ragusa, R.J., C.K. Chow, D.K. St. Clair, and J.D. Porter.** 1996. Extraocular, limb, and diaphragm muscle group-specific antioxidant enzyme activity patterns in control and *mdx* mice. J. Neurol. Sci. *139*:180-186.

50.**Rao, J.S. and M. Bond.** 2001. Microarrays : managing the data deluge. Circ. Res. *88*:1226-1227.

51.**Rubinstein, N.A. and J.F. Hoh.** 2000. The distribution of myosin heavy chain isoforms among rat extraocular muscle fiber types. Invest. Ophthalmol. Vis. Sci. *41*:3391-3398.

52.**Rushbrook, J.I., C. Weiss, K. Ko, M.H. Feuerman, S. Carleton, A. Ing, and J. Jacoby.** 1994. Identification of α-cardiac myosin heavy chain mRNA and protein in extraocular muscle of the adult rabbit. J. Muscle Res. Cell Motil. *15*:505-515.

53.**Schena, M., D. Shalon, R.W. Davis, and P.O. Brown.** 1995. Quantitative monitoring of gene expression patterns with a complementary DNA microarray. Science *270*:467-470.

54.**Sharon, D., S. Blackshaw, C.L. Cepko, and T.P. Dryja.** 2002. Profile of the genes expressed in the human peripheral retina, macula, and retinal pigment epithelium determined through serial analysis of gene expression (SAGE). Proc. Natl. Acad. Sci. USA *99*:315-320.

55.**Simon, R., M.D. Radmacher, and K. Dobbin.** 2002. Design of studies using DNA microarrays. Genet. Epidemiol. *23*:21-36.

56.**Spencer, M.J., E. Montecino-Rodriguez, K. Dorshkind, and J.G. Tidball.** 2001. Helper (CD4(+)) and cytotoxic (CD8(+)) T cells promote the pathology of dystrophin-deficient muscle. Clin. Immunol. *98*:235-243.

57.**Spencer, M.J. and J.G. Tidball.** 2001. Do immune cells promote the pathology of dystrophin-deficient myopathies? Neuromuscul. Disord. *11*:556-564.

58.**Spencer, M.J., C.M. Walsh, K.A. Dorshkind, E.M. Rodriguez, and J.G. Tidball.** 1997. Myonuclear apoptosis in dystrophic *mdx* muscle occurs by perforin-mediated cytotoxicity. J. Clin. Invest. *99*:2745-2751.

59.**Spencer, R.F. and J.D. Porter.** 1988. Structural organization of the extraocular muscles. Rev. Oculomot. Res. *2*:33-79.

60.**Tezak, Z., E.P. Hoffman, J.L. Lutz, T.O. Fedczyna, D. Stephan, E.G. Bremer, I. Krasnoselska-Riz, A. Kumar, et al.** 2002. Gene expression profiling in DQA1*0501+ children with untreated dermatomyositis: a novel model of pathogenesis. J. Immunol. *168*:4154-4163.

61.**Tseng, B.S., P. Zhao, J.S. Pattison, S.E. Gordon, J.A. Granchelli, R.W. Madsen, L.C. Folk, E.P. Hoffman, et al.** 2002. Regenerated *mdx* mouse skeletal muscle shows differential mRNA expression. J. Appl. Physiol. *93*:537-545.

62.**Tusher, V.G., R. Tibshirani, and G. Chu.** 2001. Significance analysis of microarrays applied to the ionizing radiation response. Proc. Natl. Acad. Sci. USA *98*:5116-5121.

63.**Velculescu, V.E., L. Zhang, B. Vogelstein, and K.W. Kinzler.** 1995. Serial analysis of gene expression. Science *270*:484-487.

64.**Weindruch, R., T. Kayo, C.K. Lee, and T.A. Prolla.** 2001. Microarray profiling of gene expression in aging and its alteration by caloric restriction in mice. J. Nutr. *131*:918S-923S.

65.**Welle, S., K. Bhatt, and C.A. Thornton.** 1999. Inventory of high-abundance mRNAs in skeletal muscle of normal men. Genome Res. *9*:506-513.

66.**Welle, S., K. Bhatt, and C.A. Thornton.** 2000. High-abundance mRNAs in human muscle: comparison between young and old. J. Appl. Physiol. *89*:297-304.

67.**Welle, S., A. Brooks, and C.A. Thornton.** 2001. Senescence-related changes in gene expression in muscle: similarities and differences between mice and men. Physiol. Genomics *5*:67-73.

68.**Wieczorek, D.F., M. Periasamy, G.S. Butler-Browne, R.G. Whalen, and B. Nadal-Ginard.** 1985. Co-expression of multiple myosin heavy chain genes, in addition to a tissue-specific one, in extraocular musculature. J. Cell Biol. *101*:618-629.

69.**Winters, L.M., M.M. Briggs, and F. Schachat.** 1998. The human extraocular muscle myosin heavy chain gene (MYH13) maps to the cluster of fast and developmental myosin genes on chromosome 17. Genomics *54*:188-189.

70.**Zhao, P., S. Iezzi, E. Carver, D. Dressman, T. Gridley, V. Sartorelli, and E.P. Hoffman.** 2002. Slug is a novel downstream target of MyoD: temporal profiling in muscle regeneration. J. Biol. Chem. *277*:30091-30101.

13 Applications of Gene Expression Microarrays in the Discovery of p53 Target Genes

Suxing Liu[1] and Luquan Wang[2]
[1]Departments of Tumor Biology and [2]Discovery Technology, Schering-Plough Research Institute, Kenilworth, NJ, USA

INTRODUCTION

The p53 protein is a tumor suppressor that is involved in the regulation of the cell cycle and apoptosis (33). p53 is a sequence-specific transcription factor that binds to DNA as a tetramer and thereby transcriptionally regulates a large number of genes. El-Deiry et al. (6) used p53 to select DNA fragments from genomic DNA and found that selected fragments have two copies of the decamer motif 5′-RRRCWWGYYY-3′ separated by 0–13 bp of random sequence. In the motif, R = G or A, W = T or A, Y = C or T, and N = any base. The consensus decamer contained an internal symmetry, with two oppositely oriented half-sites of the form 5′-RRRCW-3′. One copy of the motif was not sufficient for binding. Missense mutations in the DNA binding domain of p53 occur with high frequency in most major forms of human cancer (13). The central role of p53 as a tumor suppressor protein has attracted substantial interest in defining its mechanism of action. The ultimate goal is to define the complete network of p53-regulated genes. In facing this challenge, the availability of high-density cDNA microarray technology is an invaluable resource. The complete sequence of the human genome, when combined with expression profiling analysis, makes it possible to define complete gene regulation networks. Here we summarize recent work in our laboratory on the discovery of p53 target genes using microarray technology integrated with a global p53 DNA binding site analysis (23,34,35).

Perspectives in Gene Expression
Edited by K. Appasani

A Model System for Microarray Experiments

Although the approach of differential hybridization analysis is powerful, a good model system is an essential prerequisite. To use the adenovirus vector-mediated transduction of wild-type p53, the human ovarian cancer cell line 2774qw1 that carries a mutant p53 was chosen based on its infectivity (86%–89%) with adenovirus (35). To choose an optimal concentration of adenovirus-expressing p53 (rAd-p53), temporal apoptotic effects were monitored, following the infection of cells with 1×10^9, 5×10^9, or 1×10^{10} particle/mL rAd-p53 or control vector. The results indicated that 1×10^{10} particle/mL rAd-p53 infection exhibits the greatest pro-apoptotic effect, while the same concentration of rAd-empty vector showed minimal viral effect (Figure 1). Therefore, 1×10^{10} particle/mL rAd-p53 was chosen to infect 2774qw1 cells for microarray experiments. In 2774qw1 cells infected with 10^{10} particle/mL rAd-p53, exogenous p53 became detectable at the mRNA level 4 h post-infection and at the protein level 6 h post-infection (22).

To correlate expression profiles with p53 biological activities, an apoptosis assay and flow cytometry analysis were carried out on infected cells at the same time points that RNA was isolated for microarray hybridization. The cells infected with rAd-p53 showed no obvious morphological change until 16 h post-infection. DNA fragmentation, a biochemical marker for the early events of apoptosis, became detectable 12 h post-infection and reached a peak 20 and 24 h post-infection (23). Furthermore, we double-stained the cells infected with rAd-p53 at various time points with the cell-impermeable dye 7-AAD and annexin V-PE (Guava Technologies, Hayward, CA, USA) (Figure 2A). At 12 h post-infection with rAd-p53, the percent of annexin V-positive cells (5% to 6%) was similar to that for control cells, while 20 and 24 h post-infection, most annexin

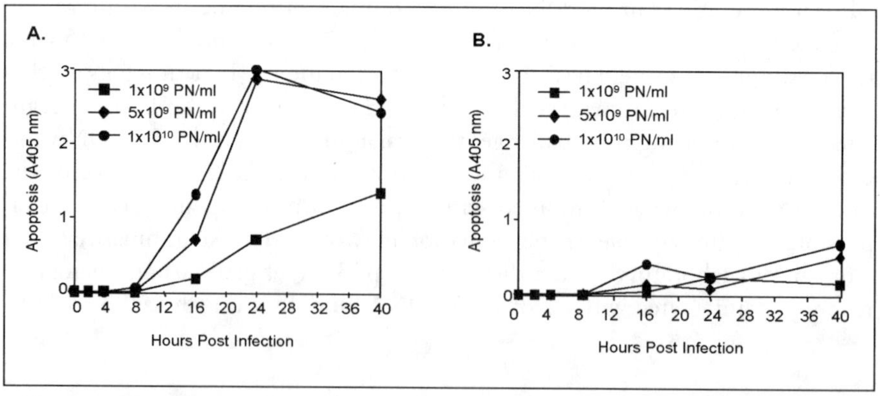

Figure 1. Evaluation of apoptosis induced by rAd-p53 and rAd empty vector. Apoptotic response of 2774qw1 cells infected with 1×10^9, 5×10^9, and 1×10^{10} particle/mL rAd-p53 (A) and rAd-empty vector (B), respectively. Apoptosis response was determined 0, 2, 4, 8, 16, 24, or 40 h post-infection by a DNA fragmentation assay as described in the text. PN, particle number.

V-positive cells (80% to 85%) were also 7-AAD-positive, suggesting that the cells were undergoing late apoptosis. An intermediate annexin V-positive cell population (approximately 12% of the total cells) was observed 16 h post-infection. At that time, half of annexin V-positive cells were also 7-AAD-positive. Taken together, the following apoptosis stages were defined. No apoptosis occurred 4 and 8 h post-infection, apoptosis was initiated at 12 and 16 h post-infection, and cells experienced late apoptosis 20 and 24 h post-infection. The effect of rAd-p53 on cell cycle progression was also assessed using flow cytometry (35). Cells demonstrated a significant G1 arrest ($P < 0.05$) 20 and 24 h post-infection with rAd-p53 in comparison with control cells (Figure 2B). During the same period of time, infection with rAd-p53 also significantly decreased the S-phase population (Figure 2B).

Furthermore, a pilot experiment was carried out by quantitative RT-PCR analysis with five apoptosis- or cell cycle-related genes, $p21^{WAF1/Cip1}$, caspase 3, bax, cyclin B1, and bcl-2, before the hybridization of RNA on the microarrays (35). Interestingly, we observed that caspase 3, not a p53 target gene, was activated at an early time point (4 h post-infection with rAd-p53) compared to cells infected with the control virus, which suggests that the secondary response to p53 expression occurred before the initiation of apoptosis. After 16 h post-infection, all five genes showed differential expression. We wanted to identify p53-responsive genes whose RNA levels varied over time during the process of apoptosis and cell cycle progression. Hence, we collected mRNA at six time points (4, 8, 12, 16, 20, and 24 h post-infection of rAd-p53 with rAd-empty vector) for microarray hybridization.

Microarray Analysis of Human Genes in Response to p53

To obtain an overall profile of gene expression, we screened a total of six GeneAlbum microarrays (Incyte Genomics, Palo Alto, CA, USA) (approximately 60 000 cDNAs) that represented 33 615 individual genes. At each time point, the cDNA prepared from purified mRNA of rAd-p53-infected cells was labeled with Cy5-dUTP and mixed with the corresponding reference probe, which consisted of cDNA made from purifying the mRNA of cells infected with the rAd-empty vector labeled with Cy3-dUTP. The two cDNA probes were mixed and simultaneously hybridized to the microarray. The hybridization results from all six microarrays were compiled and analyzed based on fold change compared to the empty vector control. Changes that were 2.5-fold or greater were scored as significant changes. The genes (1501; 4.4%) that met the criteria were selected for further analysis. Among them, 217 genes were activated and 1205 genes were repressed by p53. Interestingly, the number of the genes differentially expressed in response to p53 reached a peak after the onset of apoptosis (12 and 16 h post-infection) and declined at a later stage of apoptosis (20 and 24 h post-infection) (Table 1). Several known p53 target genes were expressed as expected, such as

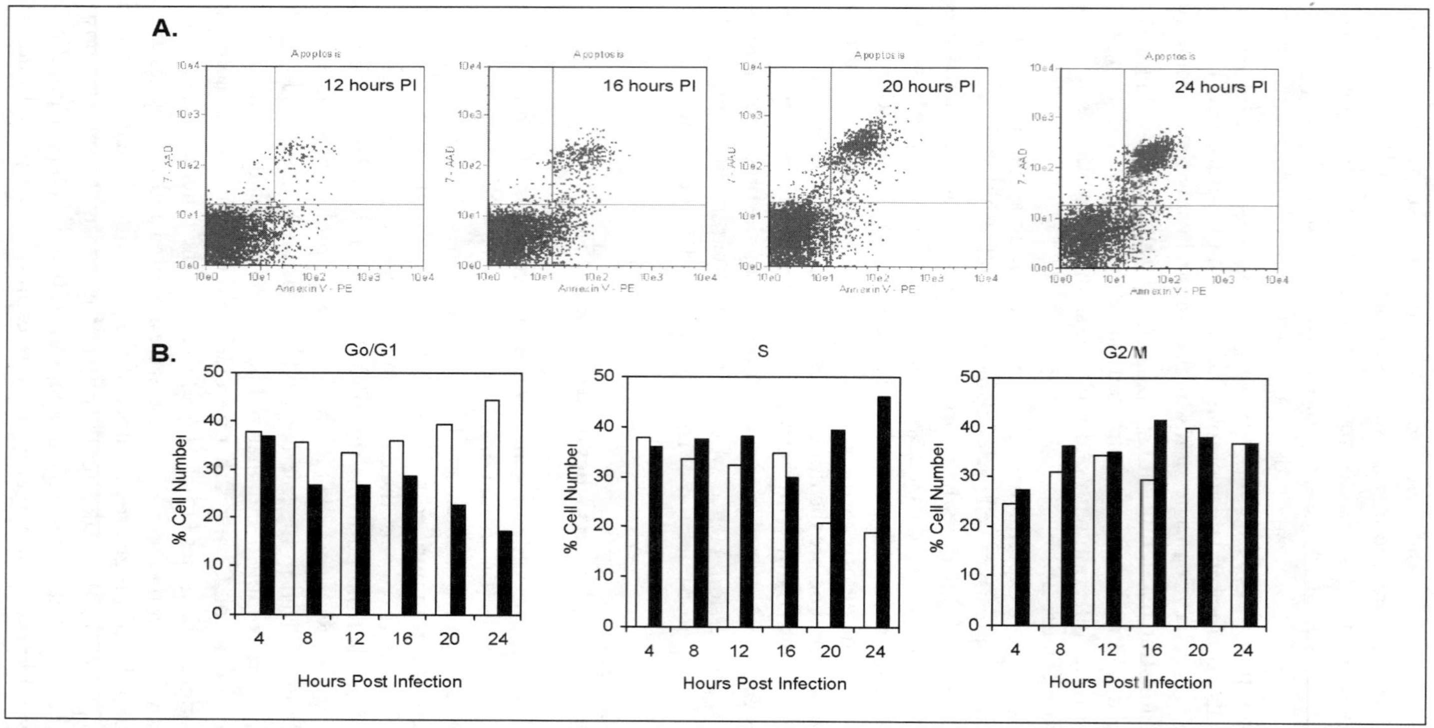

Figure 2. Temporal apoptotic and cell cycling profiles of 2774qw1 cells in response to p53. (A) p53-induced apoptosis. 2774qw1 cells were infected with rAd-p53 or control virus for 1 h and stained with Guava Annexin-PE and 7-AAD at the post-infection times indicated. Late-stage apoptotic and dead cells are defined in the upper-right quadrant gated by the quadrant markers (in pink) [Annexin V-PE (+) and 7-AAD (+)], viable cells are defined in the lower-left quadrant [Annexin V-PE (-) and 7-AAD (-)], and cells in the early stages of apoptosis in the lower-right quadrant [Annexin V-PE (+) and 7-AAD (-)]. Representative results from one of triplicate experiments are displayed. PI, post-infection. (B) Effects of rAd-p53 on cell cycle progression. Population distribution throughout the cell cycle in 2774qw1 cells infected with rAd-p53 (open bars) or control virus (closed bars) was determined using flow cytometry. Results presented are the mean of two independent experiments.

Table 1. Temporal Distribution of the Genes in Response to rAd-p53 or rAd-p21

Hours Post-Infection	rAd-p53 Infection: No. of Elements Responsive to p53	rAd-p53 Infection: No. of Elements Repressed by p53	No. of Elements Responsive to both p53 and p21	rAd-p21 Infection: No. of Elements Responsive to p21	rAd-p21 Infection: No. of Elements Repressed by p21
4	30	16	2	27	8
8	100	64	5	47	30
12	716	648	13	38	34
16	767	660	20	79	75
20	591	457	23	82	73
24	383	305	13	76	71

pig 3, $p21^{WAF1/Cip1}$, *gadd 45*, and cyclin B1. In addition, the evaluation of the differential expression of genes that had duplicate or multiple microarray elements indicated good consistency among the signals observed for the different cDNA clones on the microarrays.

Interestingly, the number of p53-repressed genes (1205 genes) was much larger than the number of p53-activated genes (217 genes). The high frequency of transcriptional repression by p53 in 2774qw1 cells was unexpected. Reverse fluorescence hybridization experiments were carried out on one microarray to evaluate whether this effect was due to a bias of dye labeling. In one experiment, the cDNA from rAd-p53 infected sample was labeled with Cy5-dUTP, and the cDNA from the control sample with Cy3-dUTP. In another experiment, the same probe pair was reversely labeled with Cy3-dUTP and Cy5-dUTP, respectively. The hybridization results indicated that 118 elements exhibited differential expression in both experiments, and all elements showed the same trend of differential expression in response to p53 when the probes were reverse fluorescent labeled. We observed that more p53-repressed genes (103 genes) were detected more frequently than p53-activated genes (15 genes). This suggests that the high frequency of the repression of gene expression in response to p53 is not due to an artifact of color bias.

Integration of the Microarray Data with the p53 Target Database

Although expression profiling identified a large number of genes whose expression changed in response to p53, this read-out includes both direct and indirect p53 target genes. To identify direct p53 target genes, we were interested in evaluating the genomic distribution of p53 DNA binding sites (34). To do so, the "FindPatterns" program (Wisconsin Sequence Analysis Package; Genetics Computer Group, Madison, WI, USA) (10) was used to search for human promoter sequences that matched the p53 consensus sequence, 5′-RRRCWWGYYY(N = 0–13)RRRCWWGYYY-3′ (6). In this motif, R = G or A, W = T or A, Y = C or T, and N = any base. Some p53 target genes (human mdm2, *igfbp3*, *gadd 45*, p53R2, and killer/*dr5*) have functional p53 response elements in their intron region (3,12,15,28,29). The database search therefore includes intronic sequences. The sequences of the promoters (2000 base pairs upstream of the transcription start site) and the first three introns of all available genes were searched (34).

On the basis of the statistics of known p53 target genes (3), no more than three mismatches to the canonical consensus sequence were permitted in the selection. We collected 21 nucleotide sequences known to be functional p53-responsive elements in 16 genes; 39 of 42 decamers bear a cytosine at position 4, and 41 of 42 decamers have a guanine at position 7 (34). Hence, the sequence matches were further sorted with no mismatch allowed at the C at position 4 or the G at position 7. In total, we have identified 4852 genes in the human genome that have at least one potential p53 DNA binding site in the regulatory region. This collection was designated as the p53 Target Database (34).

To validate the in silico prediction, 13 genes were randomly selected from the p53 Target Database, and 11 of them showed p53 responsiveness in 2774qw1 cells infected with rAd-p53 using quantitative RT-PCR analysis (34). Several features of these novel p53-responsive genes are notable. For example, it was found that p53 repressed the expression of α-fetoprotein. A recent report demonstrated that the mouse homologue of α-fetoprotein is also transcriptionally down-regulated by p53 by a p53 DNA binding site in the promoter (19). The in silico prediction identified that α-1-acid glycoprotein 2 and complement component C1-inhibitor have p53 DNA consensus sequences in their introns. The quantitative RT-PCR analysis indicated that these genes were activated by p53, suggesting that sequences of introns should be included in searches for the p53 DNA binding consensus sequences.

Of 4852 individual genes in the p53 Target Database, 3387 (70%) are represented on six GeneAlbum microarrays. The microarray data integrated with the p53 Target Database gave candidates for direct p53 target genes. The p53-responsive genes (361 total; 67 activated and 294 repressed genes) were found in the p53 Target Database. Two known p53 target genes, *pig 3* and *gadd 45*, were included. $p21^{WAF1/Cip1}$ was not included because its p53 response element is located more than 2000 bp from its transcriptional start site, the cut-off size used in the computational search (34). Most reported microarray or serial analysis of gene expression (SAGE) analyses have shown that more genes were activated than repressed by p53 (20,21,26,36,37). These studies were focused on early p53 responsiveness before the onset of apoptosis. Indeed, our data showed more activated genes (19 genes) than repressed genes (eight genes) before the onset of apoptosis (4 and 8 h post-infection). Interestingly, a much larger number of direct p53 target genes (325 genes) were identified after the onset of apoptosis.

Among them, 234 genes were repressed and 31 genes were activated after the onset of apoptosis (12 and 16 h post-infection), while 34 were repressed and 17 were activated at late apoptosis (20 and 24 h post-infection). This suggests that p53 directly targets genes even after the onset of apoptosis.

To validate the in silico prediction, 16 p53-responsive genes (six p53-activated and 10 p53-repressed genes) were selected for quantitative chromatin immunoprecipitation (ChIP) assay, which detects transcription factors recruited in vivo to the DNA region that contains the predicted p53 DNA binding site. The data indicated that p53 occupies the predicted p53 binding sites in six of six p53-activated genes (34) and eight of ten p53-repressed genes in vivo (23). If we assume that approximately 20% (59 of 294) of the p53-repressed genes and none of the 67 activated genes are false positives, then the number of p53 target genes is approximately 300. Tokino et al. (31) estimated the total number of human p53-tagged sites in the whole human genome, ranging from 200 to 300, using a yeast-based screening method. This is in the range of the number of p53 target genes that we identified.

Hierarchy Structure of p53 Target Genes

p21$^{WAF1/Cip1}$ is a primary transcriptional target of p53 that mediates cell cycle arrest (11). Therefore, genes downstream of p21$^{WAF1/Cip1}$ may be in the p53 pathway as well. Identification of these genes will help to build a hierarchy of genes downstream of p53. Parallel microarray experiments were performed using mRNA isolated from 2774qw1 cells 4, 8, 12, 16, 20, and 24 h post-infection of 10^{10} particle/mL adenovirus-expressing p21$^{WAF1/Cip1}$ (rAd-p21) or rAd-empty vector (35). A similar pilot experiment was performed with five apoptosis- and cell cycle-related genes using RNA samples from the infected cells (35). In contrast to rAd-p53-infected samples, rAd-p21-infected cells showed substantial induction of p21$^{WAF1/Cip1}$, which resulted from transgene expression and repression of cyclin B1. No alternation of caspase 3, bax, and bcl-2 mRNAs was detected in the rAd-p21-infected samples compared to the control. The performance of the hybridization was confirmed by analyses of the genes that have multiple cDNA clones represented on the microarrays. As expected, the expression of p21$^{WAF1/Cip1}$ was the most induced in the microarray experiments. In total, the expression of 159 genes was altered in a time-dependent response to rAd-p21 infection using 2.5-fold change as a cut-off. Because p53 induces the the expression of p21, we anticipated that the genes whose expression was altered in response to both p53 and p21$^{WAF1/Cip1}$ would not be primary transcriptional targets of p53. The integration of the two sets of microarray data gave 76 cDNAs (representing 66 genes) that were downstream of both p53 and p21$^{WAF1/Cip1}$. Surprisingly, 22 of them can be found in the p53 Target Database.

Although p21$^{WAF1/Cip1}$ was induced during p53-mediated apoptosis, its expression did not appear to be required for p53-mediated apoptosis (5). Functions of most p21$^{WAF1/Cip1}$-regulated genes (i.e., anillin, hMCM2, ribonucleoside-diphosphate reductase M1, and Cyr61) are associated with cell proliferation, which is consistent with the primary role of p21$^{WAF1/Cip1}$ (4,7,25,32). Note that a member of the inhibitors of apoptosis (IAP) family, HIAP2, is the gene downstream of both p53 and p21$^{WAF1/Cip1}$. The repression of HIAP2 in response to both suggests that it may play a dual role in apoptosis and cell cycle control, like its family member, survivin (18). Interestingly, some of these genes such as transcobalamin II, mki67a, and ribonucleoside-diphosphate reductase M1, were activated in five to eight of eight human breast cancer cell lines but not in the normal counterparts (S. Liu, unpublished data). It is likely that these genes controlling cell proliferation are silent in the normal counterparts. In cancer cells that have undergone malignant transformation and loss of p53/p21$^{WAF1/Cip1}$, these genes are constitutively expressed. p53 represses these genes, likely by p21$^{WAF1/Cip1}$, to mediate cell cycle arrest in cancer cells. In addition, it was reported that high mobility group protein (HMG-1) could directly interact with p53 and enhance p53 DNA binding (14). The repression of HMG-1 by p53 and p21$^{WAF1/Cip1}$ suggests that it may play a role in a negative feedback loop to modulate the function of p53 and/or p21$^{WAF1/Cip1}$.

Global Gene Regulatory Network of p53

The complexity of the transcriptional response to p53 has important implications for understanding how the pathways of apoptosis and cell cycling are coordinated to execute the p53 function. An early feature of cell death is the disruption of electron transport (10). A consequence of the loss of electron transport should be an increased production of superoxides. Superoxides and lipid peroxidation are increased during apoptosis. It was reported that p53 transcriptionally activates several reduction/oxidation-related genes during the process of p53-induced apoptosis (26). As expected, several reduction/oxidation-related genes, including a known p53 target gene, *pig 3*, showed differential expression in response to p53 (23). Interestingly, before the onset of apoptosis, creatine transporter and ABC transporter were activated by p53, and after the onset of apoptosis, many other transporters were repressed by p53. These are the calcium-binding transporter, transferrin receptor 2 α, calcium-binding mitochondrial carrier protein aralar2, an organic anion transporter (isoform 2), cation-dependent mannose-6-phosphate receptor precursor, and sialin. This suggests that p53 may trigger apoptosis by modulating the expression of transporters.

Several proteases and protease inhibitors were transcriptionally regulated by p53 after the onset of apoptosis and late apoptosis. It was reported that the transfection of various proteases into cells can induce apoptosis, although it remains unclear whether these proteases ultimately activate caspases (30). Other transcripts encoding annexins, annexin I, annexin V, and annexin VII isoform 2 were repressed by p53 in the microarray experiments after the onset of apoptosis. Dysregulation in annexin expression and activity have been correlated with human diseases including cancers (8), and p53 may induce apoptosis by repressing the expression of these annexins. In addition, we observed that p53 transcriptionally regulates a large number of intracellular regulatory proteins, such as kinases and phosphotases.

Although the proteins required for apoptosis and cell growth arrest may be constitutively expressed in cancer cells, the execution of these processes appears to be regulated at the level of transcription. Many genes associated with transcription were repressed by p53. Among them were 19 transcription factors (mostly enhancers), eight proteins associated with general transcription, and six proteins required for RNA processing after 12 h post-infection. This may explain why many p53 indirect target genes, secondary responses to p53, were repressed by p53 after the onset of apoptosis. The direct interaction of p53 with TATA binding protein (TBP) has been implicated in p53-mediated transcriptional repression (27). TBP-associated factor 172 was repressed by p53 at late time points, which may reflect a negative feedback loop in the regulation of the p53 function. p53 also regulated genes involved in the post-transcriptional regulation of gene expression, including the genes involved in translation, such as eukaryotic translation initiation factor 3 subunit 7, UNR protein, translation initiation factor eIF-4A I, eukaryotic peptide chain release factor subunit 1 (ERF1), and ribosomal protein L11. Therefore, the transcriptional repression of mRNA, blocking the processing of mRNA, and inhibition of protein involved in translation by p53 may be important events during apoptosis and/or cell growth arrest in 2774qw1 cells. Large numbers of genes showing alterations in gene expression in response to p53 suggests that a complex network is required to ensure the precise execution of the p53 function, such as decision of apoptosis and/or continuation of cell proliferation or cell cycle arrest. In other words, many altered regulatory and functional circuits are integrated to determine the phenotypic behavior of the cells in response to p53.

Exploring Mechanism of p53-Mediated Gene Repression

It is well-established that p53 transcriptionally activates gene expression via sequence-specific DNA binding. In contrast, the molecular mechanism of transcriptional repression by p53 remains unclear. This is because most genes reported to be repressed by p53 have no classical p53 binding sites in their promoter (9,16,17). A general mechanism of p53-dependent repression was ascribed to protein-protein interactions with transcription factors, such Sp1, TBP, and CCAAT-binding factor (1,2,27). In addition, recent studies indicated that transcriptional repression by p53 might result from the interaction of histone deacetylases (22,24). However, our data indicated that 295 genes were repressed by p53 and have at least one p53 consensus DNA binding sequence in their regulatory region, which suggests that some of the p53-repressed genes might be regulated via p53-specific DNA binding.

ChIP is a sensitive technique that analyzes the occupancy of transcription factors on their cognate binding elements in vivo. To further address the possibility of the p53-mediated repression of gene expression, 11 genes repressed by p53 were selected for ChIP assay (23). The differential expression of 10 of 11 genes in response to p53 detected in the microarray experiments was confirmed by quan-

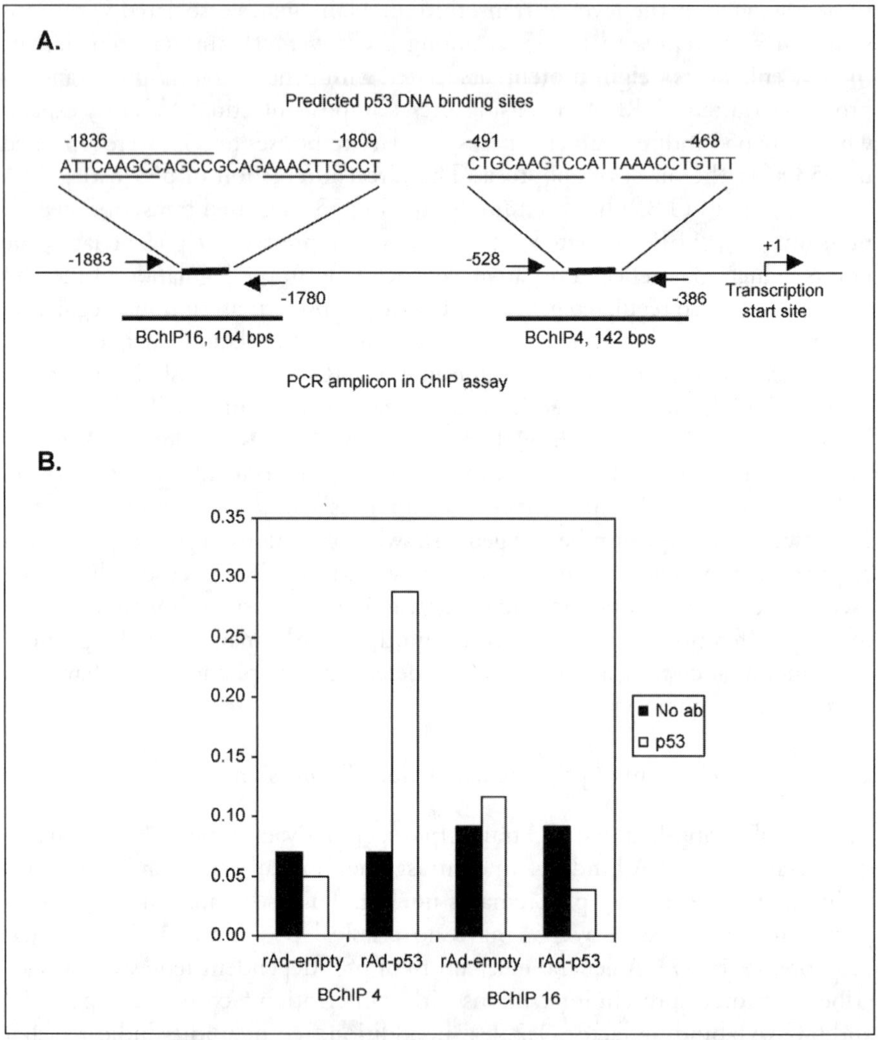

Figure 3. p53 binds to zeta-crystalline promoter in vivo during apoptosis. (A) Schematic diagram indicates the predicted p53 DNA binding sites and PCR products in quantitative ChIP assays. (B) Formaldehyde cross-linked chromatin was prepared from 2774qw1 cells 16 h post-infection with rAd-p53 and rAd-empty vector. Cross-linked chromatin from each sample was incubated with antibody to p53 (open bar) or in the absence of antibody (closed bar). Immunoprecipitates from each antibody were analyzed by TaqMan® analysis with p53 DNA binding site-specific primer sets BChIP 4 and BChIP16. The ChIP assays were repeated in five independent experiments with similar results. Representative data are shown. rAd-empty, rAd-empty vector. No ab, with no antibody.

titative RT-PCR analysis. The only exception was α-tubulin. Next, we tested the binding of p53 to the predicted p53 DNA binding sites of these 10 genes in cells 16 h post-infection with rAd-p53 and rAd-empty vector using the ChIP technique. Taking ζ-crystalline gene as an example, three putative p53 binding sites are located in the promoter of the gene (Figure 3A). Two of them overlapped each other at positions -1832 and -1836. Two DNA regions including the p53 DNA binding sites were examined by ChIP. In separate samples, the cells infected with rAd-p53 or control virus were cross-linked with formaldelhyde. Chromatin was immunoprecipitated with p53-specific antibody, followed by quantitative PCR analysis of the promoters. We included two genes as positive and negative controls in the ChIP assays. One was the proliferating cell nuclear antigen (PCNA) gene containing a known p53 binding site in the promoter. The other is α-tubulin, which has a putative p53 DNA binding site in the promoter and did not respond to p53 in real-time quantitative RT-PCR analysis. Both gave the expected results. The amount of DNA containing the DNA binding site at position -491 in the ζ-crystalline promoter following p53 expression was detected 4.3-fold higher than in the control virus-infected sample (Figure 3B). Nevertheless, not all predicted p53 binding sequences are bound by p53 in vivo. We did not detect a significant amount of the promoter fragments that contained the DNA binding site at positions -1832 and -1836 (Figure 3B). As a control, chromatin was immunoprecipated in the absence of p53-specific antibody. In these samples, no significant amounts of the promoter fragments were detected (Figure 3B). The ChIP assays were repeated in five independent experiments. The in vivo p53 DNA interaction was considered positive if the enrichment of DNA fragments containing the predicted p53 binding sequence(s) was 3-fold greater than the control with no antibodies in at least four of five independent experiments. Our data indicated that eight of 10 p53-repressed genes that were tested exhibited in vivo DNA-specific binding of p53, suggesting that a certain portion of p53-repressed genes in the human genome are transcriptionally regulated by p53 via direct DNA binding.

CONCLUSIONS

The availability of the high-density cDNA microarray technology to monitor genome-wide mRNA expression has coincided with the completion of the human genome sequence, offering a new way to study the global cellular regulatory network. No single technology is yet powerful enough to crack the complexity of p53 biology, and an integrated approach is essential to extract biological meaningful information. The key for reducing the complexity of the observed expression patterns is identifying DNA sequence elements that may act as binding sites for p53, which coordinate the altered expression of genes by p53 in whose regulatory region they appear. Here we show the identification of direct p53 target genes through the integration of microarray data and the p53 Target Database, which suggests that 300–400 direct p53 target genes are present in the human genome. In addition, recognizing a direct role of p53 as a DNA binding-dependent transcriptional repressor provides a new mechanistic view of p53-mediated transcriptional repression. The integration of parallel microarray experiments with RNA isolated from 2774qw1 cells infected with rAd-p53 or rAd-p21 built a hierarchy of p53 target genes. The work described here establishes a foundation to define a complete gene regulatory network of p53 and to provide potential therapeutic targets that bypass the requirement of p53 for the treatment of cancers. Continued expansion of these approaches to other cellular systems or other oncogenic transcription factors will provide new approaches for the strategies and evaluation of cancer therapy.

ACKNOWLEDGMENTS

Thanks to W. Robert Bishop for the critical reading and comments on this manuscript. We also thank colleagues for their contribution to the studies described here.

REFERENCES

1. **Agoff, S.N., J. Hou, D.I. Linzer, and B. Wu.** 1993. Regulation of the human hsp70 promoter by p53. Science *259*:84-87.
2. **Borellini, F. and R. Glazer.** 1993. Induction of Sp1-p53 DNA-biding heterocomplexes during graulocyte/macrophage colony-stimulating factor-dependent proliferation in human erythroleukemia cell line TF-1. J. Biol. Chem. *268*:7923-7928.
3. **Bourdon, J.-C., V. Deguin-Chambon, J.C. Lelong, P. Dessen, P. May, B. Debuire, and E. May.** 1997. Further characterization of the p53 responsive element—identification of new candidate genes for transactivation by p53. Oncogene *14*:85-94.
4. **Chen, S., B. Zhou, F. He, and Y. Yen.** 2000. Inhibition of human cancer cell growth by inducible expression of human ribonucleotide reductase antisense cDNA. Antisense Nucleic Acid Drug Dev. *10*:111-116.
5. **Deng, C., P. Zhang, J.W. Harper, S.J. Elledge, and P. Leder.** 1995. Mice lacking p21/CIP1/WAF1 undergo normal development, but are defective in G1 checkpoint control. Cell *82*:675-684.
6. **El-Deiry, W.S., S.E. Kern, J.A. Pietenpol, K.W. Kinzler, and B. Vogelstein.** 1992. Definition of a consensus binding site for p53. Nat. Genet. *1*:45-49.
7. **Fujita, M., C. Yamada, T. Tsurum, F. Hanaoka, K. Matsuzawa, and M. Inagaki.** 1998. Cell cycle- and chromatin binding state-dependent phosphorylation of human MCM heterohexameric complexes: a role for cdc2 kinase. J. Biol. Chem. *273*:17095-17101.
8. **Gerke, V. and S.E. Moss.** 2002. Annexins: from structure to function. Physiol. Rev. *82*:331-371.
9. **Gopalkrishnan, R.V., E.W.F. Lam, and C. Kedinger.** 1998. The p53 tumor suppressor inhibits transcription of the TATA-less mouse DP1 promoter, J. Biol. Chem. *273*:10972-10978.
10. **Green, D.R. and J.C. Reed.** 1998. Mitochondria and apoptosis. Science *281*:1309-1312.
11. **Harada, K. and G.R. Ogden.** 2000. An overview of the cell cycle arrest protein, p21^{WAF1}, Oral Oncol. *36*:3-7.
12. **Hollander, M.C., I. Alamo, J. Jackman, M.G. Wang, O.W. McBride, and A.J. Fornace, Jr.** 1993. Analysis of the mammalian gadd 45 gene and its response to DNA damage. J. Biol. Chem. *268*:24385-24393.
13. **Hussain, S.P. and C.C. Harris.** 1998. Molecular epidemiology of human cancer: contribution of mutation spectra studies of tumor suppressor genes. Cancer Res. *58*:4023-4037.

14. **Jayaraman, M., N.C. Moorthy, K.G. Murthy, J.L. Manley, M. Bustin, and C. Prives.** 1998. High mobility group protein-1 (HMG-1) is a unique activator of p53. Genes Dev. *12*:462-472.

15. **Juven, T., Y. Barak, A. Zauberman, D.L. George, and M. Oren.** 1993. Wild type p53 can mediate sequence-specific transactivation of an internal promoter within the mdm2 gene. Oncogene *8*:3411-3416.

16. **Kastanos, E.K., M. Zajac-Kaye, P.A. Dennis, and C.J. Aleegra.** 2001. Downregulation of p21/WAF1 expression by thymidylate synthase. Biochem. Biophys. Res. Commun. *285*:195-200.

17. **Krause, K., M. Wasner, W. Reinhard, U. Haugwitz, C.L. Dohna, J. Mossner, and K. Engeland.** 2000. The tumor suppressor protein p53 can repress transcription of cyclin B. Nucleic Acids Res. *28*:4410-4418.

18. **LaCasse E.C., R.G. Korneluk, and A.E. MacKenzie.** 1998. The inhibitors of apoptosis (IAPs) and their emerging role in cancer. Oncogene *17*:3247-3259.

19. **Lee, K.C., A.J. Crowe, and M.C. Barton.** 1999. p53-mediated repression of α-fetoprotein gene expression by specific DNA binding. Mol. Cell. Biol. *19*:1279-1288.

20. **Madden, S.L., E.A. Galella, J. Zhu, A.H. Bertelsen, and G.A. Beaudry.** 1997. SAGE transcript profiles for p53-dependent growth regulation. Oncogene *15*:1079-1085.

21. **Maxwell, S. and G.E. Davis.** 2000. Differential gene expression in p53-mediated apoptosis-resistant vs. apoptosis-sensitive tumor cell lines. Proc. Natl. Acad. Sci. USA *97*:13009-13014.

22. **Mirza, A., M. McGuirk, T.N. Hockenberry, Q. Wu, H. Ashar, S. Black, S. Wen, L. Wang, et al.** 2002. Human survivin is negatively regulated by wild-type p53 and participates in p53-dependent apoptotic pathway. Oncogene *21*:2613-2622.

23. **Mirza, A., Q. Wu, L. Wang, T. McClanahan, W.R. Bishop, F. Gheyas, W. Ding, B. Hutchins, et al.** 2003. Global transcriptional program of p53 target genes during the process of apoptosis and cell cycle progression. Oncogene (In press).

24. **Murphy, M., J. Ahn, K.K. Walker, W.H. Hoffman, R.M. Evans, A.J. Levine, and D.L. George.** 1999. Transcriptional repression by wild-type p53 utilize histone deacetylases, mediated by interaction with mSin3a. Genes Dev. *13*:2490-2501.

25. **Oegema, K., M.S. Savoian, T.J. Mitchison, and C.M. Field.** 2000. Functional analysis of a human homologue of the drosophila actin binding protein anillin suggests a role in cytokinesis. J. Cell Biol. *150*:539-552.

26. **Polyak, K., Y. Xia, J.L. Zweier, K.W. Kinzler, and B. Vogelstein.** 1997. A model for p53-induced apoptosis. Nature *389*:300-305.

27. **Seto E, A. Usheva, G.P. Zambetti, J. Momand, N. Horikoshi, R. Weinmann, A.J. Levine, T. Shenk.** 1992. Wild-type p53 binds to the TATA-binding protein and repress transcription. Proc. Natl. Acad. Sci. USA *89*:12028-12032.

28. **Tanaka, H., H. Arakawa, T. Yamaguchi, K. Shiraishi, S. Fukuda, K. Matsui, Y. Takei, and Y. Nakamura.** 2000. A ribonucleotide reductase gene involved in a p53-dependent cell-cycle checkpoint for DNA damage. Nature *404*:42-49.

29. **Teramoto, T., A. Kiss, and S.S. Thorgeirsson.** 1998. Induction of p53 and Bax during TGF-β1 initiated apoptosis in rat liver epithelial cells. Biochem. Biophys. Res. Commun. *251*:56-60.

30. **Thompson, E.B.** 1998. Special topic: apoptosis. Annu. Rev. Physiol. *60*:525-532.

31. **Tokino, T., S. Thiagalingam, W.S. El-Deiry, T. Waldman, K.W. Kinzler, and B. Vogelstein.** 1994. p53 tagged sites from human genomic DNA. Hum. Mol. Genet. *3*:1537-1542.

32. **Tong, X., D. Xie, J. O'Kelly, C.W. Miller, C. Muller-Tidow, and H.P. Koeffler.** 2001. Cyr61, a member of CCN family, is a tumor suppressor in non-small cell lung cancer. J. Biol. Chem. *276*:47709-47714.

33. **Vogelstein, B.L. and A.J. Levine.** 2000. Surfing the p53 network. Nature *408*:307-310.

34. **Wang, L., Q. Wu, P. Qiu, A. Mirza, M. McGuirk, P. Kirschmeier, J.R. Greene, Y. Wang, et al.** 2001. Analyses of p53 target genes in the human genome by bioinformatic and microarray approaches. J. Biol. Chem. *276*:43604-43610.

35. **Wu, Q., P. Kirschmeier, T. Hockenberry, T.-Y. Yang, D.L. Brassard, L. Wang, T. McClanahan, S. Black, et al.** 2002. Transcriptional regulation during p21/WAF1/Cip1-induced apoptosis in human ovarian cancer cells. J. Biol. Chem. *277*:36329-36337.

36. **Yu, J., L. Zhang, P.M. Hwang, C. Rago, K.W. Kinzer, and B. Vogelstein.** 1999. Identification of classification of p53-regulated genes. Proc. Natl. Acad. Sci. USA *96*:14517-14522.

37. **Zhao, R., K. Gish, M. Murphy, Y. Yin, D. Notterman, W.H. Hoffman, E. Tom, D.H. Mack, et al.** 2000. Analysis of p53-regulated gene expression patterns using oligonucleotide arrays. Genes Dev. *14*:981-993.

14 The CodeLink™ Bioarray System Applied to Human Liver Cancers

José L. Walewski[1], Karolyn A. Wanat[2], Myron E. Schwartz[3], Swan N. Thung[1], E.J. Fernandez[1], Randall Lockner[2], David R. Dorris[4], Bryan A. Moore[5], Teresa J. Raich[2], and Andrea D. Branch[1,3]
[1]Mount Sinai School of Medicine, New York, NY; [2]Amersham Biosciences, Piscataway, NJ; [3]Recanati/Miller Transplantation Institute, New York, NY; [4]Ambion, Austin, TX; and [5]Serologicals, Gaithersburg, MD, USA

INTRODUCTION

CodeLink Microarrays

A new microarray platform, CodeLink™ (Amersham Biosciences, Piscataway, NJ, USA), has recently been introduced, whose performance characteristics were assessed in a study of transcriptional profiles from human liver cancers. This platform uses 30 base-long oligonucleotide probes that are covalently attached to a 3-D acrylamide gel matrix on the surface of the slide. The matrix allows for hybridization to take place several microns from the surface of the glass slide, permitting the full length of the oligomer probe to bind labeled target molecules, which results in enhanced specificity and selectivity. Because the oligomers are synthesized by conventional chemistry, their composition and purity can be verified by quality-control tests, including capillary electrophoresis and/or mass spectrophotometric analysis, before their attachment to the matrix. This ensures that the sequence and length of the probe are correct.

The development of the CodeLink platform and the various analytical tools used to characterize its performance characteristics have been described elsewhere (7). In this report, our goal was to determine the reproducibility of the datasets obtained with replicate hybridizations of a labeled sample and the

Perspectives in Gene Expression
Edited by K. Appasani

concordance of the CodeLink expression data with results obtained by real-time quantitative PCR studies. The version of the human UniSet bioarray used in these studies had probes for 9200 human genes and contained an additional 384 control probes that were designed to detect bacterial genes. Positive control mRNAs were included in the labeling step to confirm the sensitivity and dynamic range of the assay. These studies were performed on paired sets of total RNA samples (tumor and surrounding nontumor liver tissues) from four men with chronic hepatitis C virus (HCV) infection undergoing liver surgery for hepatocellular carcinoma (HCC).

Microarrays and Molecular Diagnostics of HCC

HCC was selected for this study because it is one of the most common malignant tumors in the world, accounting for an estimated 1 million deaths worldwide each year. Although clinical studies show that HCCs differ in their natural history, current diagnostic methods provide only limited information about important prognostic features of the tumor, such as the likelihood of recurrence, spread to distant sites, or responsiveness to various treatments. Recent studies of leukemia and breast cancer indicate that gene expression profiling can be used to divide tumors into clinically relevant categories (1,6). A similar molecular diagnostics approach for HCC promises to augment existing histologic tests and is expected to suggest an improved classification system that combines traditional techniques and gene expression profiling. The potential impact of microarray technology on the diagnosis of liver diseases has been previously reported (2).

METHODOLOGY

Sample Collection and RNA Extraction

Liver tumors and surrounding nontumor tissues were obtained from four male patients with HCV infection, using a protocol approved by the institutional review board of the Mount Sinai School of Medicine (New York, NY). To obtain high-quality RNA samples, 1–2 g resected tissue were immediately diced and placed into 5 mL/g RNA*later*™ (Ambion, Austin, TX, USA) and incubated for several hours at room temperature to allow for the penetration of the solution into the tissue. Subsequently, specimens were stored at 4°C for 24–48 h and then frozen at -70°C until extraction. Total RNA was extracted by the homogenization of 1 g each tissue in 15 mL TRIZOL® LS and 4 mL water for 5 min with a Polytron® homogenizer. After phase separation with chloroform ($CHCl_3$) and centrifugation (3220× *g* at 4°C for 10 min), the RNA in the aqueous phase was precipitated in 25% isopropanol with one-quarter volume high-salt buffer (0.8 M sodium citrate and 1.2 M NaCl). The RNA pellet was re-extracted in 15 mL TRIZOL LS and 4 mL water and re-precipitated in 25% isopropanol with 25%

high-salt buffer. This RNA pellet was resuspended in the RLT lysis buffer (Qiagen, Valencia, CA, USA) and loaded onto an RNeasy® Maxi column (Qiagen) for RNA clean-up and on-column DNase treatment. The RNA was eluted in RNase-free water, and quantification was performed at A_{260}/A_{280} by UV spectrometry. The integrity of the total RNA was verified by denaturing agarose gel electrophoresis (Figure 1).

Biotin-Labeled Target Preparation

Redevelopment of the target preparation protocol has been described elsewhere (3). Figure 2 shows a flow diagram for the preparation of the targets. For cDNA synthesis, 10 µg of total RNA were mixed with a T7 primer and SuperScript® II reverse transcriptase (manufactured by Invitrogen [Carlsbad, CA, USA] and supplied by Amersham Biosciences) and incubated at 42°C for 1 h. The second strand of cDNA was generated by mixing the first strand with *E. coli* DNA polymerase I and RNase H and incubating at 16°C for 2 h. The double-stranded cDNA was purified using the QIAquick™ kit (Qiagen), according to the manufacturer's protocol. The purified cDNA was eluted in 10 mM Tris-HCl, pH 8.5, and dried in a SpeedVac®. For cRNA synthesis, the purified cDNA was incubated with biotin-labeled 11-UTP and 11-CTP in an overnight in vitro transcription reaction at 37°C (using the T7 MEGAscript™ kit manufactured by Ambion and supplied by Amersham Biosciences). The RNA was purified and recovered by RNeasy column purification, according to the manufacturer's protocol. RNA was eluted in RNase-free water and quantified by UV spectrophotometry at 260 nm. The size distribution of the biotin-labeled cRNA was verified on a Bioanalyzer (Agilent Technologies) and ranged from 500 to 2000 nucleotides in length. The

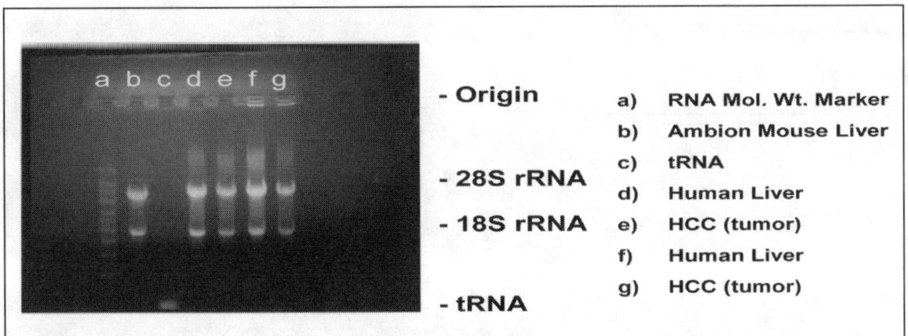

Figure 1. Agarose gel electrophoresis of glyoxal-denatured RNA from the Artzt serum and tissue bank. Total RNA samples were extracted from the corresponding HCC or surrounding nontumorous liver samples as described in the text. Five micrograms of total RNA were denatured with glyoxal and separated on a 1% agarose gel. The 28S and 18S bands are clearly visible in each RNA sample from human liver, with the proportion of each comparable to that seen in the mouse liver total RNA from Ambion. The high molecular weight material above the 28S band (lanes d–g) is RNA because it is resistant to DNase digestion and is very sensitive to RNase digestion. Mol. wt. marker, molecular weight marker (9.0→0.5 kb).

average yield of cRNA was approximately 30 μg per reaction. Biotin-labeled cRNA was fragmented by heating at 94°C for 20 min in the presence of magnesium, which yielded pieces approximately 50–200 nucleotides in length.

Hybridization and Processing

Ten micrograms of fragmented biotin-labeled cRNA were mixed with the CodeLink hybridization buffer (260 μL total volume) and injected into the CodeLink UniSet Human I bioarrays. The arrays were loaded onto a tray and hybridized overnight at 37°C, with shaking at 300 rpm in an Innova™ incubator (New Brunswick Scientific, Edison, NJ, USA). The arrays were rinsed individually and then placed in the stringency wash for 1 h. After the stringency wash, a streptavidin-alexa detection step was utilized. Each slide was incubated in 3.4 mL streptavidin-alexa 647 (Molecular Probes, Eugene, OR, USA) working solution for 30 min. After this incubation, the slides were washed four times in 240 mL TNT buffer (0.1 M Tris-HCl, pH 7.6, 0.1 M NaCl, 0.05% Tween®20). For the final washes, the slides were rinsed twice in 240 mL reagent-grade water. The slides were then dried with an N_2 gas stream.

Data Collection and Analysis

Spot detection was performed with the GenePix™ Series A scanner (Axon In-

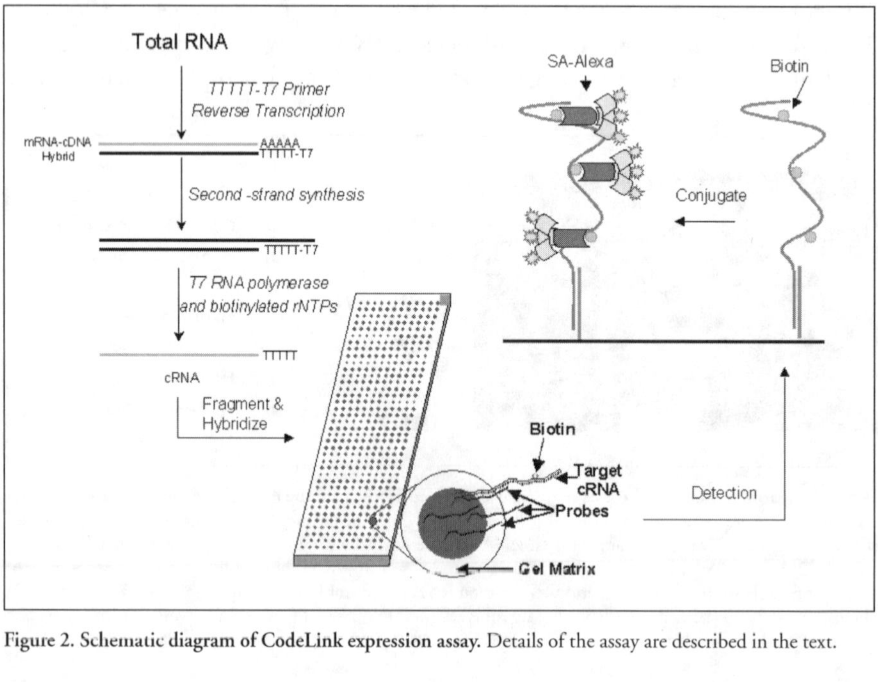

Figure 2. Schematic diagram of CodeLink expression assay. Details of the assay are described in the text.

struments, Union City, CA, USA), and spot quantitation was performed using CodeLink software. Key quantitation parameters are briefly described here. Signal intensities for each spot were calculated by the summation of the pixel intensities for each spot, and then the local background (based on the median pixel intensity of the area surrounding each spot) was subtracted. Whole array data normalization was performed independently for each slide by dividing each spot's intensity (after background subtraction) by the median signal intensity of all test probes (7).

Reproducibility Analysis

This is a calculation of the reproducibility and noise in the system. Only spots with intensities above the negative control threshold of the slide are used for minimum fold comparisons. The percent within 2-fold lists the percentage of spots with ratios less than two for each slide-to-slide comparison. The correlation coefficient is calculated as the Pearson's correlation coefficient for each pairwise comparison, and differential expression is reported as the $\log_2$ ratio of two spot values.

Cluster Analysis

The raw signals from each array were scaled individually based on the median intensity of every test probe. The statistically significant 200 most up- or down-regulated genes were selected for further analysis by comparing the median signal from the tumor and cell line samples with the normal tissue samples. The signals from the tumor or cell line samples were compared to normal liver signals by a two-tailed Student's t-test, and $P < 0.05$ was considered to be statistically significant. The normal liver RNAs were from a commercial source (Stratagene, La Jolla, CA, USA). Nontumor samples were from the same individuals who donated the tumor samples. HepG2 cells are a well-differentiated, immortal human HCC cell line. The signals from these 200 genes were hierarchically clustered by sample, using the Pearson's correlation coefficient, and by gene, using Euclidian distance (GeneSpring™; Silicon Genetics, Redwood City, CA, USA).

Real-Time Quantitative PCR by TaqMan®

The TaqMan One-Step RT-PCR Master Mix Reagent Kit (Applied Biosystems, Foster City, CA, USA) was used with each custom designed, gene-specific primer/probe set to amplify each transcript. Optimal primer/probe sets were selected using Primer Express software version 1.0 B6 (Applied Biosystems). Reactions (25 μL) contained 100 ng total RNA, 300 nM forward and reverse primers, 200 nM TaqMan probe, 12.5 μL 2× Master Mix without the enzyme uracil DNA glycosylase (UNG), 0.625 mL MultiScribe™ and RNase Inhibitor Mix, and 6.875 μL RNase-free water. RT-PCR amplification and real-time detection were performed using an ABI PRISM® 7700 Sequence Detection System (Applied Biosystems) for 30 min at 48°C (reverse transcription), 10 min at 95°C (AmpliTaq® Gold activation), 40 cycles of denaturation (15 s at 95°C), and annealing/extension (60 s at 60°C). Data were analyzed using ABI PRISM Sequence Detection Software version 1.6.3 and then further processed using Microsoft® Excel®. Cyclophilin served as the endogenous control for the normalization of input target RNA.

RESULTS

RNA Integrity

To test the performance of the platform on human specimens, RNA was prepared from a series of liver tissues. To provide a relatively narrow spectrum of samples to reduce biological variation, specimens were obtained from four men chronically infected with HCV who presented HCC and underwent hepatic resection for their tumor. Total RNA was prepared from both the surrounding

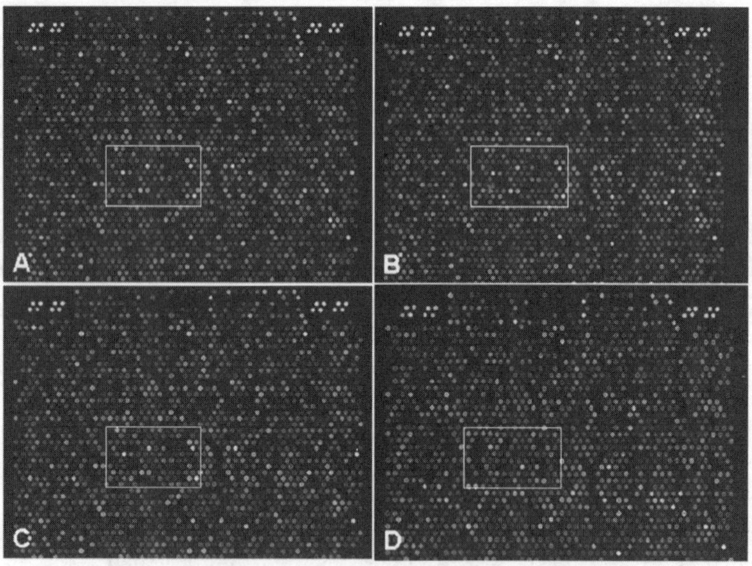

Figure 3. CodeLink UniSet human I bioarray images. Each bioarray image results from an individual hybridization. Each sample was run on triplicate arrays, and all the images were scanned on a Series A scanner. The images shown are the top section of each array (approximately 20% of the total array). The images (A–C) represent each of the triplicate arrays from a nontumor sample, and panel D represents one of the triplicate arrays from the corresponding patient's tumor sample. The white boxes in each panel indicate identical regions of the four arrays, where remarkable similarities in gene expression are seen in panels A–C, and distinct differences are noted between the tumor (D) and the three nontumor samples. (**See color plate A19.**)

Table 1. CodeLink Array Reproducibility as Assessed by %CV, % of Points within a 2-Fold Difference, and Correlation Coefficients of Triplicate Hybridizations

Sample No.[a]	Overall %cv[b]	% of Points within 2-Fold Difference	Correlation Coefficient
1-NT	9.56	99.0	0.995
1-T	10.52	99.1	0.995
2-NT	12.70	98.1	0.990
2-Ta	8.19	99.6	0.996
2-Tb	7.07	99.4	0.996
3-NT	9.95	99.6	0.996
3-T	12.27	98.2	0.979
4-NT	8.76	98.8	0.995
4-T	8.41	99.3	0.997
Nontumor[c]	NA	NA	0.918
Tumor[c]	NA	NA	0.681

[a]Samples 1-NT and 1-T; 2-NT, 2-Ta, and 2-Tb; 3-NT and 3-T; and 4-NT and 4-T are nontumorous surrounding liver and tumor specimens, each numbered set collected from the same patient.

[b]The overall %cvs for all the spots on the bioarray above negative threshold is low, and the correlation coefficients for the results obtained by triplicate hybridizations are excellent, which indicates that the results are highly reproducible.

[c]Comparisons between samples from different patients, after the replicate data from each patient were averaged. These results indicate that the nontumor samples from the four patients were similar, while the tumors showed more diversity in their gene expression profiles.

nontumor tissue and from one or two tumors from the same patient. RNA integrity was verified by denaturing agarose gel electrophoresis, as illustrated in Figure 1. RNA samples were transcribed into biotin-labeled cRNA, and several performance tests were carried out on the collection of samples. The target preparation, hybridization, and detection protocols are diagrammed in Figure 2.

Bioarray-to-Bioarray Reproducibility of the CodeLink Platform

To assess the scatter in the data, target cRNA that represented each sample was hybridized to three identical human UniSet 1 arrays (approximately 9.2 K human gene probes). Figure 3, panels A–C, shows CodeLink array images of a particular nontumor tissue on triplicate arrays produced with a GenePix Series A scanner. The top section of each array (approximately 20%) is shown. The image in Figure 3D is from an array hybridized with the patient's tumor sample. In each image, the distinct and even background facilitates individual spot detection and quantitation. In addition to probes for human genes, each slide contains a standard set of negative and positive control spots. Examples of the positive control spots can be seen in the upper right-hand and left-hand corners of each image (four spot pairs/group). The reproducibility of the data obtained in these experiments was examined by various methods. The CV was calculated among the triplicate hybridizations for each sample (Figure 4).

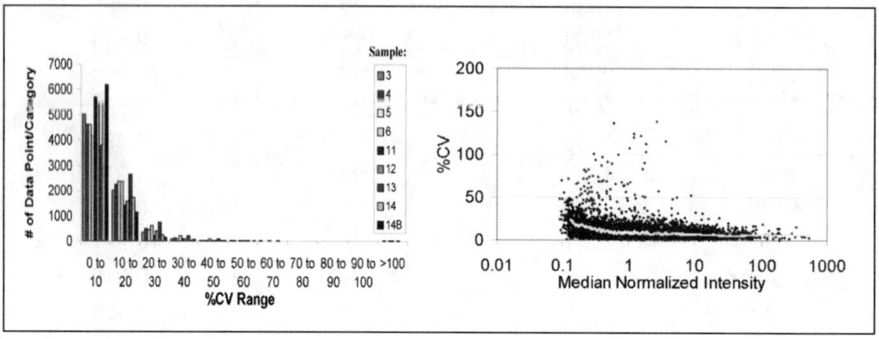

Figure 4. Array-to-array reproducibility. (A) The CV distribution chart displays the frequency of the CV values for all samples. The distribution is skewed toward the lower range of numbers where the majority of the CVs are located. (B) The CV was plotted against the average median normalized intensity for sample 13. The yellow trendline represents a 100-point moving average of the CVs. The 100-point moving average demonstrates the consistency of the platform throughout all ranges of intensities. **(See color plate A20.)**

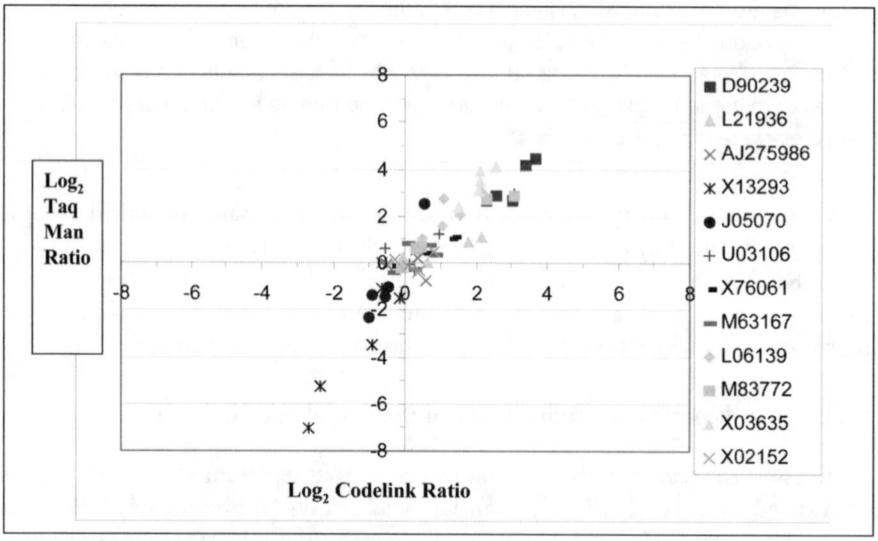

Figure 5. Log_2 ratio scatter plot comparing expression results from the CodeLink platform to results obtained by real-time quantitative PCR (TaqMan). Twelve genes were compared on both platforms, with a strong correlation between the two, with an r^2 value of 0.8957. The GenBank accession numers are shown. Reverse transcription and PCR were conducted in one step using the TaqMan PCR Reagent Kit and gene-specific primers. Data were collected and analyzed using the TaqMan 7700 Sequence Detector (Applied Biosystems). Cyclophilin served as the endogenous control in these sets of experiments for the normalization of target mRNA results. **(See color plate A20.)**

Table 2. Genes Differentially Expressed in HCC versus Nontumorous Liver in this Study and Confirmed by other Investigators Using a cDNA Microarray Platform

Gene Name	GenBank® Accession No.	Platform
Up-Regulated Genes		
Cyclin G1	X77794	cDNA
Survivin	U75285	cDNA
CD34	S53911	cDNA
α1 chain of type IV collagen	M26576	cDNA
lamin B2	X79683	cDNA
3 β-hydroxysteroid dehydrogenase	M38180	cDNA
Down-Regulated Genes		
Complement component C1r	M14058	cDNA
Complement component C9 mRNA	K02766	cDNA
Serum albumin (ALB) gene	M12523	cDNA
cytochrome P450 4A (CYP4A)	L04751	cDNA
nicotinamide N-methyltransferase (NNMT)	U08021	cDNA

Of the published genes in Okabe et al. (5), genes present on the CodeLink arrays and reported to be either up-regulated or down-regulated in HCC versus nontumor liver from the same patient showed high concordance between the two platforms.

Reproducibility was also evaluated by the calculation of the percentage of points that were within 2-fold of each other for the same samples. Ideally, all points should be within 2-fold. The lower the variability and noise of the system, the closer the ratios will get to the value of one. We also used Pearson's correlation coefficients for each pairwise comparison to examine reproducibility (Table 1). The overall CV, the average of the percentages within 2-fold, and the average of the CVs are displayed for each sample. Between the samples, the mean correlation coefficient for the nontumor samples was 0.918, and for tumor samples, the mean correlation coefficient was 0.681. These results suggest greater heterogeneity between the tumor samples.

Validation Studies of Changes in Gene Expression in HCC versus Nontumor Tissues Using Real-Time Quantitative PCR

The results obtained with the CodeLink platform were compared to those obtained with real-time quantitative PCR and are shown in Figure 5. Twelve genes were compared on both platforms: D90239 (glycine decarboxylase), L21936 [succinate dehydrogenase flavoprotein subunit (SDHA)], AJ275986 (transcription factor/SMIF gene), X13293 (B-myb gene), J05070 (type IV collagenase), U03106 [wild-type p53-activated fragment-1 (WAF1)], X76061 (p130), M63167 (rac protein kinase α), L06139 [receptor protein-tyrosine kinase (TEK)], M83772 [flavin-containing monooxygenase form II (FMO2)], X03635 (mRNA for estrogen receptor), and X02152 [mRNA for lactate dehydrogenase-A (LDH-A)]. The correlation of the results of the two assays (TaqMan vs. CodeLink) was very high (r^2 = 0.8548; r = 0.9246) for all the genes measured on the two platforms, indicating that the results obtained with the microarray platform were accurate.

Verification of CodeLink Platform-Based Gene Expression Results in Liver Cancer with cDNA Microarrays

Because HCC is one of the most common malignancies in the world, it is not surprising that other investigators have used microarray platforms to examine changes in gene expression in HCC (4,5). In microarray studies that have included a comparison of gene expression between tumor and nontumor liver samples from the same patients (5,8,9), the concordance of the results (direction of changes in gene expression) between CodeLink and other microarray platforms is high and supports the reliability of the data obtained with this platform (Table 2).

Cluster Analysis

Using the CodeLink platform, we have isolated approximately 200 of the most differentially regulated genes in human liver cancer. These expression patterns are shown in the dendrogram in Figure 6. The initial cluster analysis of the data from each individual slide in the study indicated that each replicate set clustered together, demonstrating the reproducibility of the data (data not shown). For further analysis, the data from each replicate set (three hybridizations/sample) were averaged, and these means were compared to the historical data from normal liver samples and data from HepG2 cells

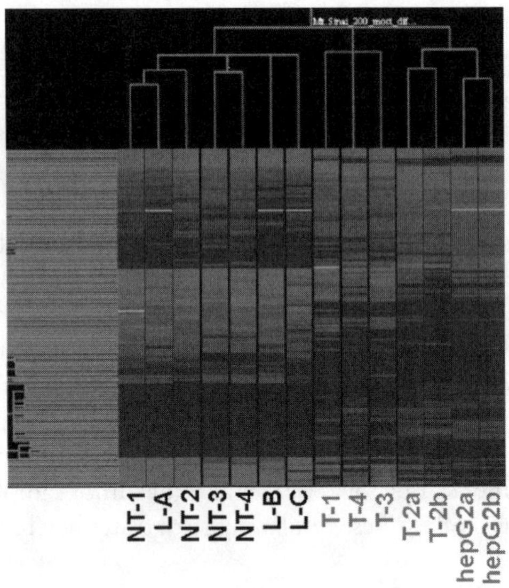

Figure 6. Clustering of the 200 most differentially expressed genes. Each sample is represented by the average of the replicates. Arrays were clustered with Smooth Pearson correlation coefficients, and the genes were clustered with distance algorithms by GeneSpring software. Up-regulated genes are shown in red, and down-regulated genes are shown in blue. In addition, normal liver and HepG2 samples from a previously performed experiment were included in this analysis. The normal liver samples clustered with the nontumor samples (black text), and the HepG2 samples clustered with the tumor samples (red text). **(See color plate A21.)**

harvested on different dates (Figure 7). This analysis shows that the nontumor liver samples from those individuals harboring tumors were similar ($r = 0.92$ between samples) and, along with the normal liver samples, form a large branch of the hierarchical tree that is distinct and separate from the tumor and cell line samples. While the HCV-infected nontumor liver samples are similar to normal liver, one expects clear differences in the expression of individual genes to be apparent in the two sets as the numbers in each group grow. The other major branch in the tree is formed by the tumor samples (which show considerable diversity in gene expression; $r = 0.68$ between four tumor samples) and the HepG2 cells, which indicated that the transformed tissues share changes in gene expression that distinguish them from nontransformed liver samples. Note that the HepG2 cells harvested on different days clustered together, and these cells clearly grouped together with two tumors from the same patient (T-2a and T-2b).

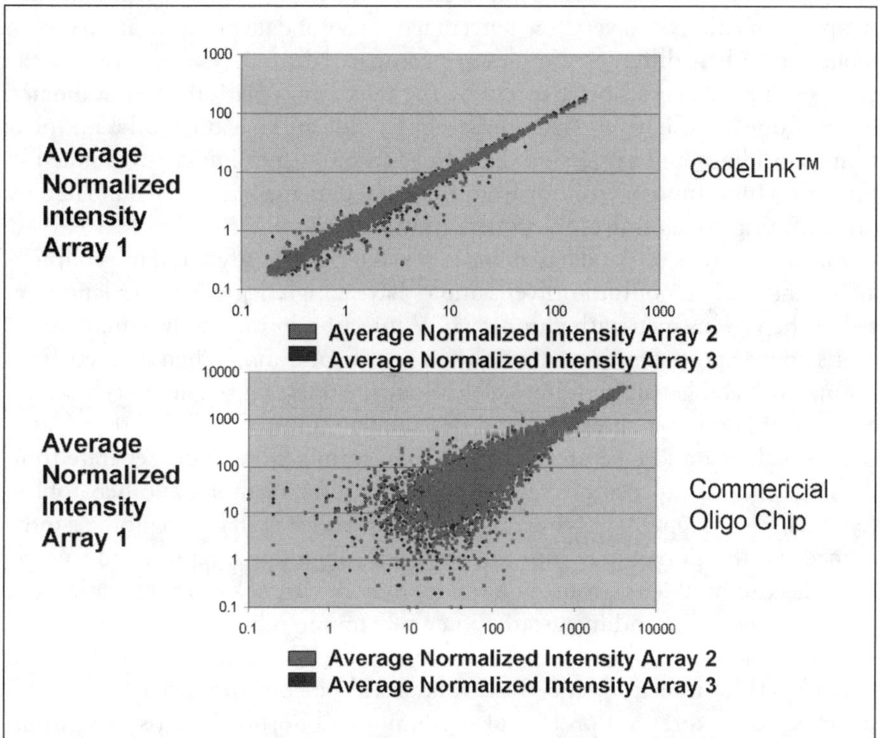

Figure 7. Reproducibility of replicate arrays, CodeLink versus commercial oligonucleotide bioarray. cRNA synthesized from Universal Human Reference RNA was labeled with biotin, according to each chip manufacturer's instructions, in three separate reactions for each platform. Each target preparation was hybridized to a single chip, the data were collected with OEM software, and the array-to-array reproducibility was assessed by comparing the scatter in the data. The closer the expression data points are to the line of identity, the more reproducible the platform becomes. **(See color plate A22.)**

SUMMARY

We have selected the CodeLink platform technology for the study of gene expression in HCC and nontumorous liver from men with chronic HCV infection. This project has had a number of distinct stages, each was critical to the success of the final outcome. These intermediate challenges included the establishment of robust protocols to extract high-quality total RNA samples from dramatically diverse tissue types and the selection of a sensitive and reproducible microarray platform. The data from these pilot studies indicated the sensitivity and reproducibility of the CodeLink platform. In addition, data from another experiment, in which three separate target labeling and hybridization reactions using the Universal Human Reference RNA (Stratagene) were compared between CodeLink and another commercial oligonucleotide-based platform, also demonstrated the platform's reproducibility (Figure 7).

The results obtained in this study to date indicate that the CodeLink platform generates highly reproducible data, as demonstrated by the remarkable within-sample correlations and very low percentage of CVs of data point means between replicates. When different samples are compared (tumor vs. nontumor), the changes in gene expression detected by the micorarray platform were validated by real-time quantitative PCR. These results indicate that the CodeLink platform is yielding valid expression data. In addition, expression changes noted in this study (nontumor vs. tumor from the same patient) have been validated by other investigators, using cDNA microarray analysis.

Further analysis of the datasets indicates that biologically relevant groupings are emerging. The nontumor liver samples have a high degree of correlation between the various men with chronic HCV infection in this study, while the tumor samples show significant diversity in gene expression. When datasets from normal liver and hepatoblastoma cell lines are included, cluster analysis based on a subset of 200 genes shown to be differentially expressed in HCC reveals that the normal liver and nontumor liver samples form a large clade, separate from the large clade composed of the corresponding tumor samples and hepatoblastoma cell lines (HepG2). As the number of datasets for each group grows, a more refined classification of the tumors into various subgroups is expected to emerge. We expect to be able to group tumors not only by degree of differentiation but also by invasiveness and metastatic potential, among other clinical parameters. Future studies will evaluate expression profiles in HCC resulting from hepatitis B virus (HBV) infection, and these results will be compared to age- and sex-matched gene expression profiles of nontumor and normal liver tissues chronically infected with HCV. The promise of molecular diagnostics is that these studies will hopefully lead to novel treatment insights and/or regimens based on gene expression profiles of the particular HCC under treatment.

ACKNOWLEDGMENTS

This work was supported in part by the Artzt Family Foundation and by RO1 DK52071 to A.D.B. The authors would like to thank the editor for his kind invitation to contribute this chapter.

REFERENCES

1. **Alizadeh, A.A., D.T. Ross, C.M. Perou, and M. van de Rijn.** 2001. Towards a novel classification of human malignancies based on gene expression patterns. J. Pathol. *195*:41-52.
2. **Branch, A.D. and J.L. Walewski.** 2001. The coming impact of gene expression profiling on the diagnosis and treatment of HCV-associated liver disease. Antiviral. Res. *52*:173-179.
3. **Dorris, D.R., R. Ramakrishnan, D. Trakas, F. Dudzik, R. Belval, C. Zhao, A. Nguyen, et al.** 2002. A highly reproducible, linear, and automated sample preparation method for DNA microarrays. Genome Res. *12*:976-984.
4. **Iizuka, N., M. Oka, H. Yamada-Okabe, N. Mori, T. Tamesa, T. Okada, N. Takemoto, A. Tangoku, et al.** 2002. Comparison of gene expression profiles between hepatitis B virus- and hepatitis C virus-infected hepatocellular carcinoma by oligonucleotide microarray data on the basis of a supervised learning method. Cancer Res. *62*:3939-3944.
5. **Okabe, H., S. Satoh, T. Kato, O. Kitahara, R. Yanagawa, Y. Yamaoka, T. Tsunoda, Y. Furukawa, et al.** 2001. Genome-wide analysis of gene expression in human hepatocellular carcinomas using cDNA microarray: identification of genes involved in viral carcinogenesis and tumor progression. Cancer Res. *6*:2129-2137.
6. **Perou, C.M., T. Sorlie, M.B. Eisen, M. van de Rijn, S.S. Jeffrey, C.A. Rees, J.R. Pollack, D.T. Ross, et al.** 2000. Molecular Portraits of Human Breast Tumors. Nature *406*:747-752.
7. **Ramakrishnan, R., D. Dorris, A. Lublinsky, A. Nguyen, M. Domanus, A. Prokhorova, L. Gieser, E. Touma, et al.** 2002. An assessment of Motorola CodeLink microarray performance for gene expression profiling applications. Nucleic Acids Res. *30*:e30.
8. **Shirota Y., S. Kaneko, M. Honda, H.F. Kawai, and K. Kobayashi.** 2001. Identification of differentially expressed genes in hepatocellular carcinoma with cDNA microarrays. Hepatology *33*:832-840.
9. **Xu, X.R., J. Huang, Z.G. Xu, B.Z. Qian, Z.D. Zhu, Q. Yan, T. Cai, X. Zhang, et al.** 2001. Insight into hepatocellular carcinogenesis at transcriptome level by comparing gene expression profiles of hepatocellular carcinoma with those of corresponding noncancerous liver. Proc. Natl. Acad. Sci. USA *98*:15089-15094.

15 Expression Profiling in *Helicobacter pylori* Infection

D. Scott Merrell and Stanley Falkow
Department of Microbiology and Immunology, Stanford University School of Medicine, Stanford, CA, USA

INTRODUCTION

Bacteria have the amazing ability to colonize virtually every environmental niche the world has to offer. From thermal vents, to human-made chemical wastelands, to our very own bodies, microbes by the trillions make a home in areas we might logically consider sterile. With its isolation from gastric biopsy specimens in 1982 by Marshall and Warren (75), the small, spiral Gram-negative bacterium *Helicobacter pylori* proved itself to be capable of living and thriving in such an environment. Although the stomach lumen reaches a pH of 2.0 during digestion, this barrier does not prevent colonization and persistent infection of *H. pylori* in over half of the world's population (reviewed in 120). While its remarkable ability to live in such a tumultuous environ is in itself fascinating, the strong association of *H. pylori* infection with gastric disease has made it the subject of intense research. The number of research articles and studies of *H. pylori* has grown exponentially, although a true understanding of its ability to live within the gastric niche and subsequently cause disease remains in its infancy

Since its discovery, persistent colonization of *H. pylori* in the gastric mucosa has been associated with numerous forms of gastric disease. The first link to human pathogenesis actually came with the original discovery of *H. pylori* in patients suffering from antral gastritis. At this time, Marshall and Warren proposed a causal relationship between the two events (75). While this suggestion was considered controversial for several years, the subsequent discovery that the majority of children suffering from gastritis also tested positive for *H. pylori* carriage and that this could be resolved by the eradication of *H. pylori* from the gastric mucosa (40,131,139) led to the acceptance of *H. pylori* as more than just an opportunistic pathogen with the ability to colonize inflamed tissue.

Perspectives in Gene Expression
Edited by K. Appasani

Chronic infection by *H. pylori* is marked by enhanced interleukin 8 (IL-8) secretion by the gastric epithelia, which leads to sustained inflammation and a marked increase in the presence of polymorphonuclear leukocytes and lymphocytes (reviewed in 41,46). In addition, atrophy and increased gastric-epithelial cell turnover often occur. Stages of increased cellular proliferation are associated with increased permeability across the epithelial membrane barrier, and it is perhaps this factor that leads to increased access of luminal acid and pepsin to the underlying tissue and the subsequent development of ulcer disease (19,32,46, 101,132). *H. pylori* infection is associated with both duodenal and gastric ulcers, and 90% and 75% of these disease cases, respectively, are attributable to gastric colonization by the organism (28). The lifetime risk of developing peptic ulcer disease in an *H. pylori*-infected individual is approximately 10%. However, this incidence has been greatly reduced by antimicrobial therapy.

Gastric cancer is the world's second most common cause of cancer-related morbidity and mortality (91). In addition to its role in the development of peptic ulcers, it is now well established that *H. pylori* infection is causally associated with two forms of gastric malignancy, MALT lymphoma and adenocarcinoma (16,41,97). The relationship was proposed as a result of epidemiological, cross-sectional, and prospective studies of *H. pylori* infections that showed that infection is associated with a 2.7- to 12-fold increased risk of developing gastric cancer (17,18). Because of this, *H. pylori* was classified as a class I carcinogen in 1994 by the World Health Organization and to date remains the only bacterium to receive such status (8).

DIRTY LITTLE SECRETS: *H. PYLORI* VIRULENCE DETERMINANTS

We lack a thorough understanding of the process of infection and pathogenesis of *H. pylori,* but, within the last three decades, progress in elucidating key components of the infection process has been made. The bacterium enters its human host from an unknown reservoir, survives exposure to acidic pH within the stomach lumen, and reaches the mucus layer that overlays the gastric epithelium. Because the bacterium requires such extreme levels of pH to survive, acid resistance (AR) has been the focus of considerable study. AR has been shown to be intricately linked with the ability of *H. pylori* to produce copious amounts of urease (86). Urease is encoded by a seven-gene cluster, of which *ureAB* encode the structural components of the enzyme and *ureIEFGH* encode accessory genes (33,67). The hydrolysis of urea to carbon dioxide and ammonia helps to maintain a proton motive force across the inner membrane of the bacterium (82,107). This activity is crucial for virulence because urease is essential for colonization and persistence in the stomach (6,42,47,127). While it was previously believed that cell-surface-localized urease created a neutral microenvironment that was conducive to bacterial survival (99), recent work has indicated that the intracellular enzyme appears to play a more important role in AR (107).

The fact that internally localized urease is required for AR would seem to present an interesting conundrum for the bacterium because its substrate, urea, is externally localized. However, recent work using an artificial *Xenopus* oocyte model has shed light on this dilemma because it was shown that UreI actually acts as an inner membrane proton-gated, urea-specific channel (133). This work demonstrated that the UreI pore opens as the pH of the medium drops below 6.5 and that the cytoplasmically localized urease is able to gain access to its urea substrate. As urease activity neutralizes the local environment, the pore closes, and urea transport stops, thus providing a regulated level of enzyme activity. UreI has at least six key, periplasmic amino acid residues that are important for transport activity. Protonation of a subset of these amino acid residues, as a result of increased acidity, presumably leads to a conformational change that opens the UreI pore (102,134). More recent work has shown that UreI functions similarly within the actual context of the *H. pylori* membrane, with the exception that one of the *Xenopus*-required residues is not needed for transport activity in *H. pylori* (20).

Other urease-independent factors have been shown to be important for AR. These factors include *wbcJ*, which is involved in the conversion of GDP-mannose to GDP-fucose and the subsequent assembly of the lipopolysaccharide layer (LPS). *H. pylori wbcJ* mutants are unable to assemble O-antigen and Lewis X or Lewis Y determinants and show increased sensitivity to acid stress (80). The importance of LPS is also highlighted by the recent finding that the *pldA* gene undergoes phase variation that results in the production of active phospholipase A. This leads to changes in the lipid composition of the outer membrane that increase survival at low pH and lead to the release of increased concentrations of urease and the vacuolating cytotoxin, VacA (119). Other *H. pylori* genes that have been implicated in AR include *recA*, which is involved in DNA repair (123); *rocF*, which is required for arginase activity (78); *atpD* and *atpF*, which encode subunits of the F_1F_0-ATPase (14,79); and *fur*, which encodes the ferric uptake regulator. Fur was recently identified as the first known transcriptional regulator required for growth in acidic conditions (15).

After entering the stomach lumen, *H. pylori* penetrates and colonizes the mucus layer that overlays the gastric epithelium. The immediate steps that follow and lead to long-term colonization and disease are poorly understood. With respect to the latter, presumably the bacteria produce "virulence factors" that are crucial to the onset of disease. While the identity and function of many of these virulence factors remain a mystery, highly pathogenic strains are known to produce a powerful vacuolating cytotoxin (VacA) (reviewed in 105). VacA is a secreted protein that has been purified as an oligomeric toxin that dissociates into a 87-kDa monomer upon exposure to low pH (29,30,71,137). The secreted toxin is internalized by eukaryotic cells and exerts its action in the cytoplasm (76). The toxic activity of VacA has been shown to produce vacuolar degeneration of gastric epithelial cells in vitro (30,69). The degeneration results in the accumulation of large, nonfunctional endosomal-lysosomal hybrids that contain Rab7 and lgp110 (87,96). This disruption of vesicular membranes has been shown to interfere with antigen processing in immune cells and to prevent lysosomal degradation of surface receptors in epithelial cells (88). The toxin was shown to act by the formation of chloride conducting channels in lipid bilayers (56,125), and more recent work has suggested that VacA actually produces transmembrane pores that allow for the selective diffusion of urea across the membrane (126).

The true biological implication of VacA intoxication on the disease process remains elusive, but the vacuolization of cells in human biopsy samples has been observed (24,45). In addition, intragastric administration of purified toxin results in epithelial damage in a murine model of infection, suggesting that toxin activity in vitro is relevant in an animal model. *VacA* mutants have an infectious dose (ID_{50}) that is 100-fold higher than an isogenic wild-type strain and show a significant colonization defect when co-infected with a wild-type strain (106). These data show that VacA plays a role in establishing colonization, although the nature of this role is not immediately evident.

Recent work by Amieva et al. (5) suggests that VacA-independent vacuolation of eukaryotic cells may provide a complex benefit to infecting *H. pylori* strains. Utilizing a tissue culture model, it was observed that *H. pylori* induced the formation of numerous large vacuoles within the cell monolayer (Figure 1). While this was no surprise, the fact that many of the vacuoles contain *H. pylori* is both surprising and intriguing (5). Time-lapse video microscopy and gentamycin protection assays were used to access the viability of the intracellular organisms and showed that the bacteria were viable and had the ability to exit the vacuoles and reinfect the tissue culture monolayer. This suggests the possibility that *H. pylori*, which had previously been considered to be an extracellular pathogen, has the ability to enter a bacterial-induced compartment within the host cell, to reside intracellularly for a period of time, and then exit the host cell. Depending on the duration of this period, this intracellular component of the bacterial life cycle has important implications for the disease process. Intracellular persistence is especially significant when one considers that antibiotic therapies that are designed for clearance of bacterial infection might not be thorough enough to eliminate intracellular organisms, thus leaving a microbial reservoir that would be capable of reseeding the bacterial infection to the gastric epithelium.

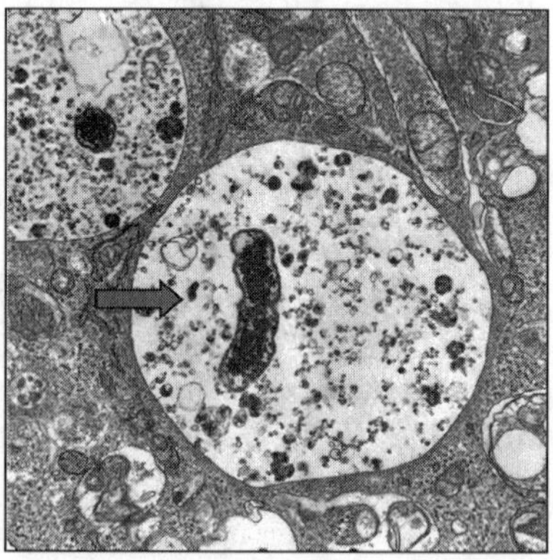

Figure 1. Transmission electron microscopy of intracellular *H. pylori*. The arrow indicates an *H. pylori* cell that has been internalized by an MDCK cell. The bacterium resides within a large multivesicular vacuole.

The most significant advance in understanding *H. pylori* pathogenesis resulted from the discovery of a pathogenicity island (PAI) that encodes a novel type IV secretion apparatus in some strains of *H. pylori* (2,25). This PAI is composed of a 40-kb locus that contains at least 27 genes. *H. pylori* strains carrying the PAI are far more likely to be associated with a serious manifestation of *H. pylori* infection (136). However, it should be noted that strains lacking the PAI are still associated with chronic gastritis and occasionally with more severe disease (28). Several genes contained on the PAI show homology to genes that encode secretion machinery in other pathogenic bacteria (28). Mutations in *cagE, F, G, H, L,* and *I* were shown early on to abrogate the strong IL-8 induction caused by *H. pylori* infection of cultured gastric cancer cells (AGS) (108,110). A more recent systematic study of each of the genes encoded within the PAI showed that mutations in 14 of the 27 genes resulted in a loss of IL-8 induction (gene designations: HP0522, HP0523, HP0525, HP0527–532, HP0537, HP0539, HP0541, HP0544, and HP0546) (48).

While the origin of the PAI is unknown, numerous studies have shown that components of the island function in the delivery of CagA, which is encoded by HP0547 and is also part of the island, to the host cell. It has been known for several years that most people infected with PAI strains develop antibodies to CagA. It became evident only recently that CagA was actually delivered to the cytoplasm or plasma membrane of the eukaryotic cell. This was elucidated after the observation that PAI and CAG+ strains induced the tyrosine phosphorylation of a 145-kDa protein within the infected host cell; strains lacking CAG did not (9,92,109,117). Mass spectrometric analysis of this 145-kDa protein revealed that it was CagA, and confocal microscopy showed that the bacterial CagA protein was inserted into the plasma membrane of host cells and, in some cases, is phosphorylated (109). It has been demonstrated recently that Src-like protein tyrosine kinases are apparently responsible for this phosphorylation event (111,116). Other studies have shown that of the 27 genes encoded on PAI, all of them are required for the efficient translocation and/or phosphorylation of CagA into the host cell, with the exception of HP0520, HP0521, HP0534–536, and HP0543 (48). The bacterial attachment and subsequent injection of CagA into the host cell has been shown to induce a striking morphological change in the infected cell. This "hummingbird phenotype" is marked by the elongation and spreading of the cells (109). CagA was shown to be necessary and sufficient for these morphological changes because AGS cells transfected with *cagA* expression constructs displayed the marked elongation and spreading (53). Tyrosine phosphorylation of CagA is required for this event, as expression of phosphorylation-deficient CagA results in no apparent change to cell morphology, although it localizes properly to the plasma membrane (53).

In addition to showing the necessary and sufficient nature of CagA in mediating the morphological changes associated with the *H. pylori* attachment to host cells, this study also identified one of the first targets of CagA within the eukaryotic cell. Segal et al. (109) and Higashi et al. (53) both noted that CagA-mediated morphological changes showed striking similarity to changes induced by exposure to hepatocyte growth factor (HGF). It had previously been shown that a cytoplasmic tyrosine phosphatase, SHP-2, played a major role in the HGF-induced changes (53,66). Higashi et al. (53) showed conclusively that CagA binds to SHP-2 within the context of gastric epithelial cells in a tyrosine-phosphorylation-dependent manner, and the immunodepletion of SHP-2 from AGS cell lysates effectively cleared all phosphorylated CagA from the lysates, thus suggesting stoichiometric binding of the two components. SHP-2 activity was also shown to be required for induction of cell elongation as phosphatase-defective SHP-2 and the inhibition of phosphatase activity with the inhibitor calpeptin both negated the ability of CagA to induce morphological cell changes. Finally, Higashi and co-workers demonstrated that even in the absence of CagA, membrane-targeted SHP-2 resulted in the indistinguishable induction of cell elongation, thus suggesting that the true mode of action of CagA was the sequestration of SHP-2 at the plasma membrane and deregulation of phosphatase activity (53). It remains to be seen which downstream targets of SHP-2 phosphatase activity will be affected in the *H. pylori*-targeted disruption of cellular activities, and what role these targets will play in the clinical outcome of infection.

EXPRESSION STUDIES: THAT WAS THEN, THIS IS NOW

The real challenge in revealing the complexity of host/pathogen interactions comes in trying to develop a thorough understanding of how each responds to the other. However, defining those respondent components is a complicated issue. The use of multiple reporter systems (36,62,63) and the development of techniques, such as in vivo expression technology (IVET), differential fluorescent induction (DFI), recombination-based in vivo expression technology (RIVET), and differential display have provided powerful tools for the elucidation of transcriptional changes in numerous bacterial species in response to various stimuli (21,73,130,135). Unfortunately, these techniques and reporters provide only a narrow picture of the many complicated responses of the organism. As the ultimate goal of mRNA expression analysis is to develop a snapshot of every mRNA molecule in an organism at a given time, these techniques fall considerably short of achieving that goal. With the onset of large-scale sequencing projects and the elucidation of the complete genome of many bacterial species, it was only a matter of time until the development of more powerful techniques to evaluate gene transcription on a genome-wide scale. The reality of microarray technology was brought to fruition only a few years ago, but since that time, a number of investigators have used this technique to analyze both genetic content and gene expression patterns of their organism of interest. The remainder of this chapter will focus on the initial utilization of microarray technology to gain a more thorough understanding of the intimate host/pathogen interactions employed by *H. pylori* during colonization of its host.

LIFE IN A TEST TUBE

Much of our understanding of virulence gene function and regulation has been learned through studies of gene expression in vitro, in conditions designed to mimic the host environment. Host environs are complex, and the true multifaceted nature of the location and active processes that colonizing bacteria undergo largely remain a mystery. Because of this, researchers have relied on the ability to simulate certain aspects of the host environment in vitro to provide a partial picture of bacterial responses within the host milieu. While this is a somewhat primitive look at the many stimuli encountered, these techniques have been very successful due to the fact that many genes with similar functions are co-regulated and regulated in response to multiple signals (98).

COME ON IN BOYS, THE WATER'S FINE: ACID-REGULATED GENE EXPRESSION

The most obvious environmental parameter encountered by *H. pylori* during the process of colonization and long-term carriage within the stomach is acid. As mentioned earlier, the ability to resist acid stress is thus intimately linked to the process of pathogenesis. It is not surprising then that two studies have already used microarrays to examine acid-regulated gene expression (3,7). The study conducted by Allan et al. (3) used nylon-based spotted arrays of 3354 pBluescript® cloned fragments of *H. pylori* DNA and radioactive cDNA probes to examine genes differentially expressed upon exposure to citrate buffer, pH 4.0, for 30 min, as compared to exposure to PBS, pH 7.0, for the same period of time. Eleven differentially regulated genes were found, of which eight were up-regulated from 1.3- to 2.2-fold and three were down-regulated from 1.9- to 6.2-fold. Ang et al. also used nylon based arrays containing 1534 predicted open reading frames and biotinylated cDNA probes to examine differentially regulated genes of *H. pylori* grown for 48 h on Columbia agar plates pH 5.5 or 7.2. They found 84 genes that were differentially regulated by 5-fold or greater, with 80 of these genes up-regulated and four down-regulated. There was no overlap between the genes identified in either study. Perhaps this is due to the differences in the pHs used, length of exposure to the stimulus, and the media matrix (broth versus agar).

We have recently performed a more comprehensive search for *H. pylori* genes that are affected by acidic pH using DNA spotted arrays (D.S. Merrell and S. Falkow, submitted). The *H. pylori* microarray contains spotted PCR products of the average size of approximately 800 base pairs and represents 98.9% of the unique open reading frames found in both of the sequenced *H. pylori* strains (4,124). Each DNA fragment is additionally spotted in duplicate on the array, thus allowing for technical replicates within the context of a single hybridization (104). This study (Figure 2) examined the response to pH as a function of time,

Table 1. Representative List of Acid-Regulated Genes of *H. pylori*

Gene Designation[a]	Predicted Function/Gene Name
Repressed	
HP0009	putative outer membrane protein
HP0010	chaperone and heat shock protein, *groEL*
HP0109	chaperone and heat shock protein 70, *dnaK*
HP0110	co-chaperone and heat shock protein, *grpE*
HP0144	cytochrome c oxidase, (CBB3-TYPE), *fixN*
HP0229	outer membrane protein, *hopA*
HP0265	cytochrome c biogenesis protein, *ccdA*
HP0317	membrane protein, *omp9*
HP0547	cag PAI protein, *cagA*
HP0590	ferredoxin oxidoreductase, b subunit, *oorB*
HP0632	quinone-reactive Ni/Fe hydrogenase, large subunit, *hyaB*
HP0633	quinone-reactive Ni/Fe hydrogenase, cytochrome b subunit, *hyaC*
HP0638	outer membrane protein, *omp13*
HP0653	nonheme iron-containing ferritin, *pfr*
HP0722	putative outer membrane protein
HP0725	putative outer membrane protein
HP0887	vacuolating cytotoxin, *vacA*
HP0912	outer membrane protein, *hopC*
HP0913	outer membrane protein, *hopB*
HP0915	iron-regulated outer membrane protein, *frpB*
HP0916	iron-regulated outer membrane protein, *frpB*
HP1027	ferric uptake regulation protein, *fur*
HP1167	putative outer membrane protein
HP1177	outer membrane protein, *omp27*
HP1243	outer membrane protein, *babA*
HP1469	outer membrane protein, *omp31*
HP1501	outer membrane protein, *omp32*
HP1512	iron-regulated outer membrane protein, *frpB*
HP1561	iron (III) ABC transporter, periplasmic iron-binding protein, *ceuE*
HP1562	iron (III) ABC transporter periplasmic iron-binding protein, *ceuE*
HP1588	conserved hypothetical protein
Induced	
HP0067	urease accessory protein, *ureH*

Table 1. Continued

Gene Designation[a]	Predicted Function/Gene Name
HP0068	urease accessory protein, *ureG*
HP0069	urease accessory protein, *ureF*
HP0071	urease accessory protein, *ureI*
HP0072	urease beta subunit, *ureB*
HP0073	urease α subunit, *ureA*
HP0115	flagellin B, *flaB*
HP0166	response regulator
HP0228	conserved hypothetical integral membrane protein
HP0294	aliphatic amidase, *aimE*
HP0295	flagellin B homolog, *fla*
HP0318	conserved hypothetical protein
HP0325	flagellar basal-body L-ring protein, *flgH*
HP0380	glutamate dehydrogenase, *gdhA*
HP0471	glutathione-regulated potassium-efflux system protein, *kefB*
HP0751	polar flagellin, *flaG*
HP0752	flagellar hook-associated protein 2, *fliD*
HP0870	flagellar hook, *flgE*
HP1021	response regulator
HP1031	flagellar motor switch protein, *fliM*
HP1119	flagellar hook-asociated protein 1 (HAP1), *flgK*
HP1225	conserved hypothetical integral membrane protein
HP1238	aliphatic amidase energy metabolism, *aimE*
HP1332	co-chaperone and heat shock protein, *dnaJ*
HP1444	small protein, *smpB*
HP1475	lipopolysaccharide core biosynthesis protein, *kdtB*
HP1507	conserved hypothetical ATP-binding protein
HP1558	flagellar basal-body rod protein, *flgC*
HP1559	flagellar basal-body rod protein, *flgB*
HP1582	pyridoxal phosphate biosynthetic protein J, *pdxJ*
HP1583	pyridoxal phosphate biosynthetic protein A, *pdxA*

[a]Gene designations using the gene number assigned by The Institute for Genomic Research.

as RNA samples were harvested after 30, 60, 90, and 120 min of exposure to pH 5.0. More than 180 genes were found to be differentially induced or repressed across the time course (Figure 2 and Table 1). This is considerably more than found in either of the previous studies and is probably a reflection of the increased sensitivity of the spotted arrays, the larger number of open reading frames represented, and the analysis of expression profiles at multiple time points after the shift to the acidic environment.

Of the 183 factors shown to be differentially regulated by the shift to acidic pH, almost half showed reduced levels of expression. Down-regulated genes (Table 1) include the virulence factors *vacA* and *cagA* and the ferric uptake regulator (fur), which was recently shown to be involved in AR and repressed at low pH (15). The reduced expression of these factors is perhaps expected because it would not be advantageous for *H. pylori* to produce high levels of virulence factors within the lumen of the stomach where they would normally encounter extremely acidic pHs. Instead, these products would be produced subsequent to entry into the mucous layer overlaying the gastric epithelia where a more neutral environment exists (100). It should be noted that previous studies using a different *H. pylori* strain and a *xylE* reporter system indicated that *cagA* expression was induced at a lower pH in a growth-phase-dependent manner (63). Direct comparison of the results from this and the previous study are complicated by the different medias used, lengths of exposure to acidified media, sensitivity of the assay, and strain differences. The result suggests that there may be differences in *cagA* expression that are intricately linked to the strain being examined.

Analysis of the 73 repressed genes reveals the presence of multiple membrane proteins and iron utilization/acquisition/storage factors. Utilization of Lexical Analysis (C.C. Kim and S. Falkow, submitted), a program that searches a partic-

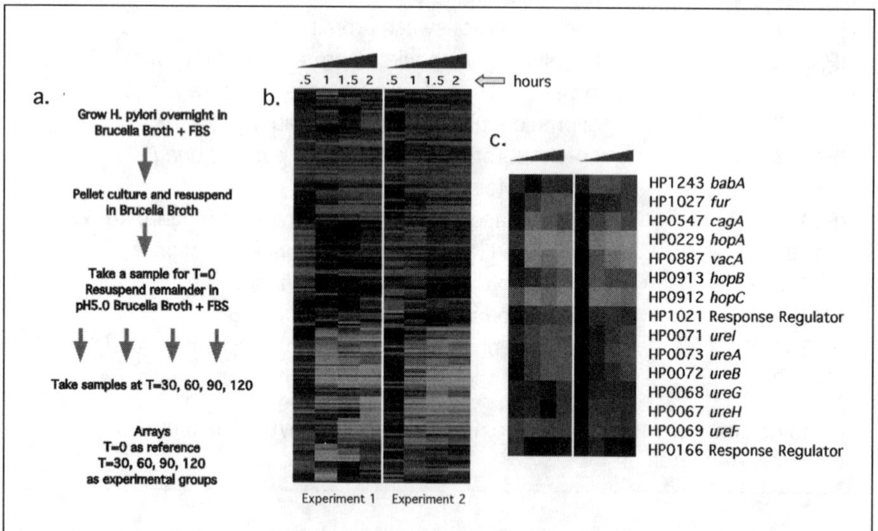

Figure 2. Low pH regulated genes of *H. pylori*. (a) Schematic depiction of experimental design by which genes affected by a shift to low pH were determined. (b) Hierarchical cluster of all 183 genes from two independent experiments that passed the filtering criteria as explained in the text. Red indicates increased expression, green represents decreased expression, and gray indicates spots for which data were not reliable. The progression of time is indicated by the arrow located above each experimental set. (c) Enlargement of a subset of the genes contained within panel b. Gene designations and names are indicated to the right of the corresponding cluster diagram. (See color plate A23.)

ular data set using a list of user-defined terms to search for trends or bias within the data, revealed a strong bias within the acid-repressed components for membrane and iron-associated factors. Lexical Analysis of the down-regulated data set revealed that 21 of 73 genes matched the membrane search terms used (terms were OMP and membrane), while an average of 4–6 matches fit the same criteria using randomly selected factors from the "genome annotation." A similar bias is exhibited in the iron analysis (search terms iron, fe, ferric, ton, and heme). While it is unclear exactly why *H. pylori* would evolve to greatly alter the protein composition of the bacterial membrane and limit the uptake of iron into the bacterial cell, it is easy to imagine that this is intricately linked to the maintenance of pH homeostasis. Alteration of the membrane composition would affect permeability and, thus, the accessibility of damaging agents to the bacterial cytoplasm. In addition, as the presence of iron is intricately linked to proton motive force via the cytochromes involved in electron transport, the alteration of the availability of iron within the cell is perhaps used by *H. pylori* to maintain the proton gradient across the bacterial membrane.

Factors shown to be induced by low pH include components of the urease gene cluster, which are known to be important for acid stress survival, flagellar genes, and two of the five two-component signal transduction (TCST) regulators found in *H. pylori* (Table 1). Sensing of the local pH and subsequent induction of the flagellar genes of *H. pylori* potentially provides a mechanism whereby the bacterium can determine its relative location within the context of the stomach and, by using increased motility, move from the extremely low pH environ of the lumen to the more neutral mucous layer overlying the gastric mucosa. This localization to a suitable niche would be an important step in the colonization process. In fact, amotile mutants of *H. pylori* have been demonstrated to show decreased colonization within the murine gastric tract (43,94). The induction of HP0166 and HP1021 as TCST regulators is of keen interest because the regulatory networks controlling AR, colonization, and pathogenesis remain ill defined. Interestingly, HP0166 was shown to be essential by two separate groups, suggesting that the factors controlled by this system are required for in vitro growth (11,77).

ONE TO GROW ON: GROWTH-PHASE-DEPENDENT GENE EXPRESSION

A core set of genes is consistently expressed throughout the life cycle of a bacterial cell. Yet, many bacteria alter gene expression as a function of their growth phase, expressing highly divergent and specific sets of genes in logarithmic and stationary phases. This is a well-studied paradigm in *E. coli*, in which the transition from rapid growth to the quiescent state of stationary phase results in the up-regulation of genes that are considered to be part of the canonical RpoS or stationary phase sigma factor, regulon (reviewed in 52). These genes then function in diverse pathways that are involved in nutrient acquisition, stress survival, and alteration of the cell envelope. Little is known about growth-phase-dependent transcription in *H. pylori*, and, interestingly, there is no obvious *rpoS* homologue contained within either of the two sequenced strains.

A thorough analysis of growth-phase-dependent gene expression of *H. pylori* Sydney strain 1 (SS1) was recently completed and showed that, similar to other bacteria, divergent sets of genes are expressed depending on the overall growth state of the cell (L. Thompson et al., submitted). In this study, Brucella Broth cultures of SS1 were inoculated from overnight cultures, and samples were taken periodically for assessment of viable colony-forming units (cfu) and isolation of total RNA. SS1 showed a characteristic bacterial growth curve, exhibiting a lag phase and period of exponential growth, followed by entry into stationary phase where viable cfu declined. Analysis of global temporal gene expression was conducted using an *H. pylori* microarray by hybridization of cDNA generated from total RNA harvested throughout the growth cycle with reference cDNA generated from the starting culture (T = 0). Raw data were subsequently normalized by mathematical transformation such that the relative abundance of each transcript was assessed by comparison to the 6-h time point. This time point was chosen as it represents the end of the lag phase, relative synchronization of the culture, and entry into the logarithmic growth phase. Analysis of the resulting data showed that approximately 85% of the *H. pylori* genes showed no significant fluctuation in expression patterns throughout the growth curve. However, the remaining 15% that did show major differences throughout the growth cycle could be assigned to four different gene groups: (*i*) induced from mid-log to stationary phase, (*ii*) induced in stationary phase, (*iii*) repressed from mid-log to stationary phase, and (*iv*) repressed in stationary phase (Figure 3).

A closer look at the functional groups of genes that show significant repression during the transition from logarithmic to stationary phase growth reveals that the largest single class is composed of ribosomal proteins. This comes as no surprise, as bacterial growth and the requirement for high levels of protein expression greatly decrease as cells enter into stationary phase. Concomitant with this, a large number of additional factors involved in translation, transcription, and DNA synthesis are down-regulated. In addition to these intuitive changes, a large number of genes that are involved in iron uptake are significantly repressed in stationary phase (Table 2). These include *fecA* and *frpB*; the latter of the two known to be controlled by Fur (37).

As noted earlier, iron is a critical requirement for bacterial growth, and as such, most bacteria have developed multiple systems for iron uptake. The iron uptake systems of SS1 are repressed in a growth-phase-dependent manner. A potential clue as to why *H. pylori* might choose to do this comes with the identification of the *pfr* gene as being significantly up-regulated at the same time that iron transport is being repressed. Pfr functions within the bacterial cytoplasm to remove free ionic iron (13,38,49). This is important because it acts as an intracellular com-

Table 2. Selection of Genes Whose Expression Is Significantly Increased or Decreased at the Logarithmic-Stationary Phase Switch as Assessed by Unpaired SAM

Gene Name	Gene Identification	Gene Description	Ratio (r_i)[b]
pfr	HP0653	nonheme iron-containing ferritin	3.51
napA	HP0243	neutrophil-activating protein (bacterioferritin)	1.93
hydA	HP0631	quinone-reactive Ni/Fe hydrogenase, small subunit	1.85
hydB	HP0632	quinone-reactive Ni/Fe hydrogenase, large subunit	1.56
C-553	HP1227	cytochrome c553	0.94
flaA	HP0601	flagellin A	0.95
cag1	HP0520	cag pathogenicity island protein	0.84
dnaK	HP0109	chaperone and heat shock protein 70 (Hsp70)	1.61
traG	HP1006	conjugal transfer protein	1.16
unknown	HP1322		1.05
unknown	HP0023		1.24
unknown	JHP0955		0.82
aimE	HP0294	aliphatic amidase energy metabolism	-3.69
frpB	HP0876	iron-regulated outer membrane protein	-3.19
fecA	HP0686	iron (III) dicitrate transport protein	-2.24
ureA	HP0073	urease α subunit (urea amidohydrolase)	-0.99
ureB	HP0072	urease β subunit (urea amidohydrolase)	-0.76
aspB	HP0672	solute-binding signature and mitochondrial signature protein	-1.13
pdxA	HP1583	pyridoxal phosphate biosynthetic protein A	-1.22
cag23	HP0544	cag PAI, DNA transfer protein (agrobacterium VirB4 homolog)	-0.76
dnaJ	HP1332	co-chaperone and heat shock protein	-0.63
mraY	HP0493	phospho-N-acytylmuramoyl-pentapeptide-transferase	-0.88
kpsF	HP1429	polysialic acid capsule expression protein	-1.02
map	HP1299	methionine amino peptidase	-1.20
gppA	HP0278	guanosine pentaphosphate phosphohydrolase	-1.16
rpoA	HP1293	DNA-directed RNA polymerase, α subunit	-1.66
rpoB	HP1198	DNA-directed RNA polymerase, β subunit	-1.30

[a]Gene identification using the gene number assigned by The Institute for Genomic Research.
[b]r_i is the relative level of gene expression in log space.

partment for the storage of iron in a nonreactive state, which implies that as cells enter into stationary phase, the need for the prevention of metal toxicity and oxidative stress that occur due to the presence of excess free iron becomes crucial. In agreement with this, *sodB*, *dna*K, *grpE*, and *hspR*, all genes known to be involved in stress response, are also induced by entry into stationary phase.

Growth-phase-dependent expression of pathogenicity factors occurs in many different bacteria, and virulence is intricately linked to growth state in a number of bacterial species (68,81). The analysis by Thompson and colleagues (manuscript submitted) showed that many of the known virulence factors of *H. pylori* are also regulated in a growth-phase-dependent manner. *cagA*, which is a component of the PAI that is delivered to the host cell; *napA*, which is the neutrophil-activating protein that induces adhesion of infiltrating neutrophils to the gastric mucosa; *flaA*, which is the primary structural component of the flagella and required for motility; and *hpaA*, which codes for the receptor-binding subunit of the N-acetylneuraminyllactose-binding fibrillar hemagglutinin (NLBH), all show relatively low levels of expression at the beginning of the growth cycle but then show maximal expression in the late log/stationary phase. Accordingly, *H. pylori* in this growth phase appear to be most capable of efficient delivery of *cagA* to cultured AGS cells and subsequent induction of cellular elongation (M. Amieva, personal communication), which suggests

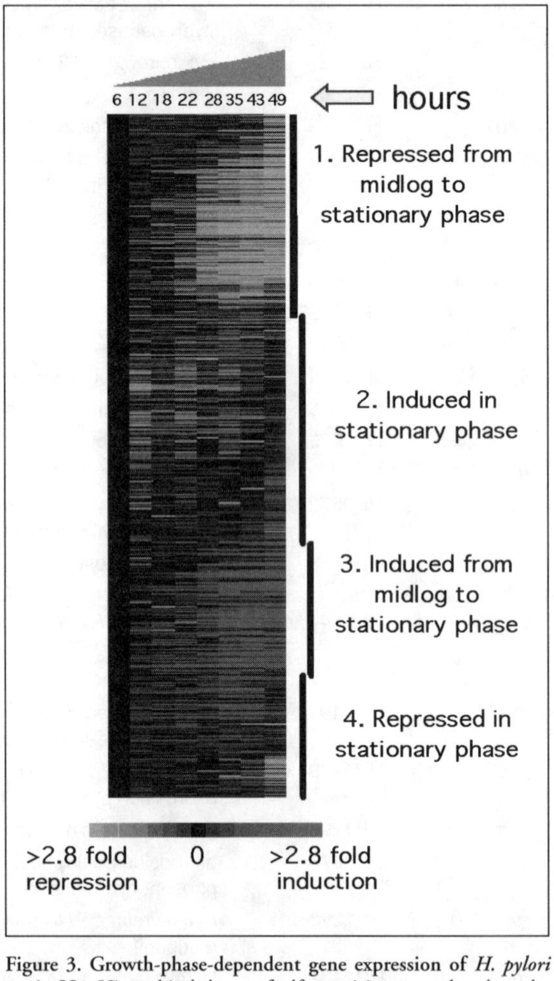

Figure 3. Growth-phase-dependent gene expression of *H. pylori* strain SS1. Hierarchical cluster of self organizing maps that show the temporal dependence of gene expression changes during the time course for all genes that passed the filtering criteria described within the text. The major classes of expression patterns are indicated, and the progression of time is indicated by the triangle above the figure. **(See color plate A24.)**

that, as with other bacteria, virulence is linked to growth phase. Of note, components of the PAI displayed differences in expression patterns throughout the growth cycle, implying differential regulation. Subsequent studies to identify the basis and components involved in this regulation may shed a greater understanding on *H. pylori* colonization and subsequent disease progression.

THEN THEY MEET: AH, THE THINGS YOU DO TO ME

Knowledge of bacterial products that are responsible for the sequelae of infection, while often informative and useful, provides a one-sided, largely incomplete view of pathogenesis. The true challenge to understanding microbial-induced disease and disease progression is the development of a thorough understanding of the dynamic nature of the host/pathogen interface. The advent of the age of genomics has made it possible for researchers to utilize microarray technology to begin to dissect the complex interactions and signaling cascades that occur as a result of the interaction of *H. pylori* within the context of both human tissue culture cells and animal models.

STUDIES IN TISSUE CULTURE

A number of different groups have used the human gastric epithelial adenocarcinoma cell line AGS to assess changes in gene expression that occur as a result of infection with *H. pylori* (10,27,72,112). In the first such study, Chiou et al. (27) utilized an Atlas human cDNA expression array (Clontech) containing DNA fragments that represent 588 genes to examine changes in AGS transcription that occurred after 24 h of infection with *H. pylori* strain K8. The K8 strain, a clinical isolate from an ulcerative patient, was demonstrated to be both CagA positive and VacA positive. Analysis of the host cell expression patterns revealed that 38 genes (6.5%) were differentially expressed in the infected monolayers; 21 were induced and 17 were repressed. Not surprisingly, a large number of the affected genes act as transcription factors (*c-jun*, *fra-1*, BTEB 2, and ETR101) or are involved in host cell signal transduction pathways (MAP kinase kinase 3 and 4, MAPK/ERK kinase I, and IL5 receptor). *H. pylori*-induced differential expression of multiple host transcription factors probably occurs at early stages and is a means of long-term alteration of host cell expression patterns and cellular behavior. For instance, both C-JUN and Fra-1 are known components of the transcription factor AP-1 that has been proposed to effect IL-8 expression (1) and act as an early factor that leads to apoptosis (58). The increased expression of IL-8 is a hallmark of *H. pylori* infection (50,113), and increasing evidence suggests that *H. pylori* induces an apoptotic response in gastric epithelial cells (74,89) as well as in macrophages (51). In addition, expression of Id-2, ERCC3, and NM23-

H2, all of which were shown to be down-regulated in the infected AGS cells, was shown to also be significantly repressed in human gastric mucosal biopsies of *H. pylori*-infected individuals as compared to uninfected patients (27). This suggests that many tissue culture/microarray-identified factors may indeed have some role within the context of an actual human infection.

In addition to the basic question of which host genes are expressed in response to *H. pylori* infection, a more pertinent question may be the elucidation of which host genes are specifically affected by the presence of *H. pylori* virulence determinants. At least three studies have attempted to address this by direct comparison of responses produced by strains containing or lacking the PAI (10,31,51a). Work conducted by Cox et al. (31) used I.M.A.G.E. (Washington University-Merck Pharm), splenic and inflammatory cDNA arrays to examine differences in gene expression induced as a result of infection with CAG+ (NCTC 11637) and CAG- (G50) *H. pylori* strains. Kato 3 gastric epithelial cell monolayers were infected with either 11637 or G50, and total RNA was harvested after 45 min, 3 h, and 24 h. A comparison of the expression patterns of uninfected Kato 3 cells to the *H. pylori*-infected cells revealed that 256 genes were differentially expressed. Similar to the previously mentioned study, these included a large number of factors involved in transcription, signaling pathways, apoptosis, growth factors, and inflammatory responses. Marked differences in the expression of 107 genes was found in comparison of host profiles induced by CAG+ and CAG- strains, implying that a large number of the host genes affected by *H. pylori* infection are due to the presence of the PAI. Unfortunately, the direct interpretation of the data is confounded by the fact that the two *H. pylori* strains used for the study were not isogenic. As *H. pylori* strains show a remarkable degree of genetic variation that reaches well beyond the presence or absence of the PAI (104), it is impossible to know how many of the 107 host genes are actually expressed differently due to strain differences that are completely unrelated to the presence or absence of the Cag PAI.

Bach et al. (10) utilized a wild-type CAG+ strain and its isogenic *cagA* mutant to more thoroughly address *cag*-dependent host-induced change. Using Atlas human cDNA arrays, the authors compared expression profiles of AGS cells infected for 4.5 h with *H. pylori* strain P12 to profiles obtained from cells infected with an isogenic *cagA* mutant (P17). Marked differences in gene expression were seen between the experiments as LIM kinase (LIMK), bone morphogenetic protein 1 (BMP-1), and macrophage inflammatory protein 2α (MIP-2α) were substantially up-regulated in the wild-type infected cells but not in *cagA* mutant infected cells. In addition, the expression of three genes, focal adhesion kinase (FAK), Y-box binding protein 1 (YB-1), and *c-myc*, were significantly up-regulated in AGS cells infected with the *cagA* mutant but not with wild-type *H. pylori*, suggesting *cagA*-dependent repression of expression of these factors within the context of a wild-type infection. Northern blot analysis verified that LIMK expression was induced greater than 3-fold in AGS cells infected with wild-type

H. pylori in comparison to uninfected cells, whereas only a slight increase was seen upon infection with the *cagA* mutant. In contrast, FAK expression was decreased approximately 5-fold in AGS cells infected with wild-type *H. pylori* but only slightly reduced upon infection with the *cagA* mutant. Similar results for both kinases were also obtained upon infection with a *cagE* mutant, which functions as an essential part of the secretion apparatus. This strongly suggests that translocation of CagA to the host cell via the Type IV secretion apparatus is required for the observed effects on AGS expression. While the long-term effect of the differential expression of these kinases is not completely clear, some clues may develop from the understanding that attachment and delivery of CagA results in the spreading and drastic elongation of AGS cells to produce the hummingbird phenotype (109), which overshadows the stress fiber-associated (SFA) morphology induced by both CAG+ and CAG- strains. These cellular rearrangements have been linked to the activation of Rac (95), which also has been clearly shown to induce the activation of LIMK (93,138). In addition, FAK is a peripheral membrane protein that is localized to focal adhesions and has been shown to be involved in the integrin-mediated activation of Rac and actin cytoskeleton organization (23). Perhaps, all of this information combined suggests that FAK and LIMK are involved in the formation of the SFA morphology and the hummingbird phenotype, respectively, although a more thorough investigation is required.

The most extensive microarray analysis to examine *cag*-dependent effects on AGS gene expression to date was recently conducted using human 24 K cDNA arrays, *H. pylori* wild-type G27, and isogenic mutants in *cagA*, *cagN*, *virB4*, and *cag*PAI (51a). Mutations within *virB4* and *cag*PAI, the latter of which actually deletes the entire PAI, result in the loss of a functional Type IV secretion apparatus. However, the *cagN* mutant is fully competent for the delivery of *cagA* and was chosen because of increasing evidence that it may indeed encode an additional secreted effector (K. Guillemin and S. Falkow, unpublished data). Total RNA was harvested from uninfected monolayers of AGS cells and monolayers infected with all five of the previously mentioned strains, after 0, 1, 3, 6, 12, and 24 h of infection. Analysis of the resulting 720 000 data points by hierarchical clustering (44) and significance analysis for microarrays (SAM) (129) revealed that on a genome-wide scale, striking similarities were observed between responses induced by the wild-type and the *cagN* mutant strains. Additionally, the *virB4* and *cag*PAI mutants elicited similar responses to each other, and the *cagA* mutant displayed an intermediate phenotype between the two classes, which showed similar patterns of expression with distinct subsets of the genes affected by the two groups.

A closer look at the *cag*-dependent changes showed a particularly interesting cluster of genes that was greatly induced early in the wild-type and *cagN* mutants but remained uninduced in any of the strains unable to deliver *cagA* (Figure 4). Included within this cluster were the plasminogen activator, claudin 4, and the Cdc42 effector protein 2, which is also known as Borg1. Borg1 was previously identified as a Cdc42 effector molecule through yeast two hybrid screens and has subsequently been shown to mediate cell elongation in multiple cell types (54,59,60).

In addition to claudin 4, the increased expression of claudin 1 was demonstrated. Both claudin 1 and claudin 4 are members of the larger claudin family, which is a family of proteins that is intricately linked to barrier functions of epithelia; they are the primary transmembrane constituents of tight junctions (128). Examination of *H. pylori* cells associating with MDCK cells reveals that the majority of bacteria appears to localize to the junctions of the MDCK cells. Further analysis has shown that the infecting *H. pylori* co-localize with ZO1, a tight junction-associated protein (M.R. Amieva et al., submitted). However, this co-localization is CagA-dependent, as both *cagA* mutants and strains deficient for CagA delivery show a random distribution of adherent bacteria on the MDCK cell surface. The biological implications of the junctional association remain elusive, but, due to its intricate requirement for the presence of CagA, suggest that downstream events will perhaps be linked to disease development and progression.

WELCOME TO THE REAL WORLD: LIFE IN THE STOMACH

While valuable information can be obtained from the utilization of tissue culture cells and in vitro growth conditions designed to mimic in vivo environments, these conditions cannot fully reflect the complex interactions that occur within the setting of an intact host. Therefore, a major push in scientific research is to develop techniques that will allow an understanding of changes that occur within the more complex environment of an actual infection. Two such studies have attempted to assess expression changes that occur within the context of the murine stomach upon *Helicobacter* infection (83,89a).

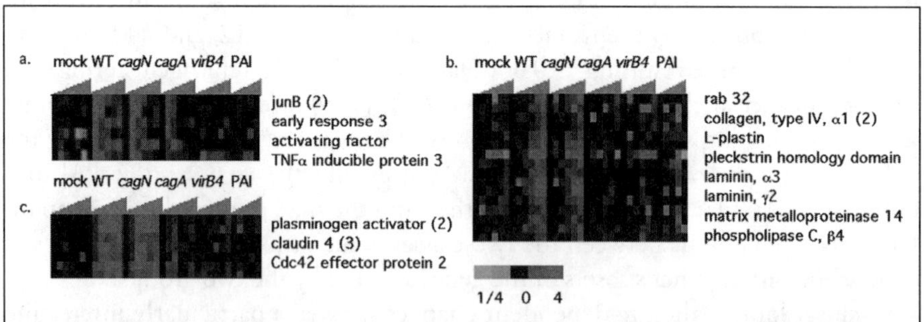

Figure 4. AGS transcriptome analysis. Gene expression during infection time courses with wild-type *H. pylori* and isogenic mutants lacking individual PAI genes or the entire PAI as indicated. Each row represents an individual gene, and each column represents an individual condition, with increasing time during the infection (0, 1, 3, 6, 12, and 24 h) represented by the width of the gray triangle. The fold induction relative to the zero hour time point is indicated colorimetrically according to the scale bar. Three groups of co-regulated genes are shown (panels a-c), with representative gene names indicated. (**See color plate A25.**)

A recent report by Mills et al. (83) utilized Mu11K GeneChips® (Affymetrix, Santa Clara, CA, USA) containing approximately 11 000 mouse genes and expressed sequence tags to investigate gene expression of gastric cells from germ-free mice. This work also identified changes in expression within different gastric cell populations subsequent to infection with *H. pylori.* Stomachs from germ-free animals and *H. pylori*-infected animals were excised, and the glandular mucosa was recovered. Different gastric cell populations were separated based on the fact that parietal cells (PC), which are specialized acid-secreting cells that are abundant throughout the gastric epithelium, express unique surface glycans that can be used for purification by lectin panning. While it is well understood that PC are an important constituent of acid peptic disease in humans, the effect of *H. pylori* upon PC within the context of the human stomach remains poorly defined. Therefore, PC+ and PC- populations were purified using the lectin panning approach, and the expression patterns were investigated.

The direct comparison of the transcriptome of the PC+ fraction to the PC- fraction revealed approximately 240 genes that were preferentially expressed within the PC. Not surprisingly, this data set was enriched for factors involved in acid secretion/vesicular trafficking, cellular energy metabolism, and free radical scavenging (83). Next, comparisons were made to the PC+ and PC- fractions harvested from ex-germ-free mice that had been colonized with the well-characterized *H. pylori* clinical isolate Hp1 (104). Interestingly, direct comparison of the PC+ fraction from the infected and uninfected mice revealed no genes that were both increased in the PC+ population of *H. pylori* infected animals (as compared to the germ-free animals) and preferentially expressed in the PC+ compared to PC- population of infected animals.

In contrast, the expression of 275 probe sets was increased in Hp1-infected PC- fractions compared to germ-free PC- fractions. (These 275 also fit the criterion of being preferentially expressed in Hp1-infected PC- fractions as compared to Hp1-infected PC+ fractions.) While the vast majority of the 275 probe sets encode transcripts for immunoglobulins, induction of 99 unique genes was also shown. Of these 99, only 16 had previously been identified as *H. pylori*-induced genes. The remaining 83 were grouped into eight functional categories: (*i*) interferon signaling related; (*ii*) LPS-regulated responses; (*iii*) actin/cytoskeleton associated; (*iv*) extracellular matrix/cell motility related; (*v*) factors involved in protein turnover; (*vi*) signal transduction components; (*vii*) miscellaneous; and (*viii*) genes with unknown function (83). Among the factors involved in signal transduction were the CD79a antigen, CD22 antigen, *rac2*, GEF1, JAK3, and Pim-1 proto-oncogene.

Pim-1 is a serine/threonine kinase that was recently independently identified as being induced within the context of *H. pylori*-infected AGS monolayers (112). The Pim serine/threonine kinase family has three members and was first identified as a common proviral insertion site in murine T and B cell lymphomas (26,34). Transcription of the Pim kinases is induced by T cell antigen receptor cross-linking and by cytokines such as IL-4, IL-6, and IFN-γ, and forced expression of Pim-1 has been shown to reconstitute thymic cellularity in mice lacking IL-7 (35,39,57, 70,140). This strongly suggests that Pim-1 plays a downstream role in cytokine-mediated signaling. Recent work has additionally shown that CD40 is able to activate increased expression of *pim-1* in an NFKB-dependent manner. Increased levels of Pim-1 then lead to increased cell survival, presumably through anti-apoptotic pathways. Based on this information, it is not a big leap to propose that increased cellular survival and proliferation of certain cell types could play an important role in disease progression stimulated by *H. pylori* infection.

Recent work conducted (89a) has investigated the disease progression in BALB/C mice infected with various *Helicobacter* species. In this study, mice were infected with the *Helicobacter* species of interest and maintained for a period of 24 months. Animals from each infection group were sacrificed at 6, 12, 18, 20, 22, and 24 months after infection, and the stomachs were removed, bisected, and half was used for histopathological analysis and the remainder was snap-frozen for RNA preparation and subsequent microarray analysis. Different species of *Helicobacter* showed marked differences in the time of onset and progression of disease in the murine model. Notably, *H. pylori*, while inducing moderate levels of lymphocytic infiltration and lymphoepithelial lesions, required a significantly longer infection time for disease symptoms to develop and failed to induce lymphoma within the time frame of the experiment. However, *H. felis* and *H. heilmannii* showed significant levels of infiltration and lesion formation at time points much earlier in the infection cycle.

Due to the large numbers showing histopathological and lymphoepithelial lesions in the *H. heilmannii*-infected animals, these were chosen for subsequent microarray analysis. Total RNA was harvested from the frozen portion of the intact stomachs from the infected animals and from age-matched, uninfected animals, and each was hybridized to a murine microarray containing representations of 38 000 transcriptional units. Unsupervised hierarchical clustering of the resulting 41 arrays and analysis of the dendogram showed that, with few exceptions, arrays from uninfected animals tended to cluster tightly together and in a separate node from arrays from infected animals (Figure 5a). In addition, many differences in gene expression between the two groups were visually evident. SAM analysis of the resulting data set was conducted to identify: (*i*) statistically significant genes showing differential regulation in infected animals that exhibited a mild lymphocytic response versus uninfected animals; (*ii*) statistically significant genes showing differential regulation in animals exhibiting moderate lymphocytic response; and (*iii*) statistically significant genes showing differential regulation in animals exhibiting severe lymphocytic response or lymphoma. Distinct genes showing increased and decreased expression fell into each of these functional categories, and a subset will be discussed (Figure 5b).

Ceacam10 (Cca10) and the small proline-rich protein 2A were both identified as belonging to the first class of significantly induced genes. Cea10 exists as a member of the carcinoembryonic antigen (CEA) family, which belongs to the larger immunoglobulin superfamily (114,121,122). Members of the CEA family have diverse roles, but several have been implicated as important for cellular adhesion in vitro. In addition, it has been noted that overexpression of some CEA members seems to correlate with cancer progression (90). Cea10, which is highly expressed in the colon and to a lesser extent in the small intestine, cecum, and stomach, has previously been characterized as having no role in early tumor development, as steady-state levels of Cea10 mRNA were equivalent in normal colonic mucosa and adenomatous polyps of mice (64). The statistically significant increased expression of *cea10* within the context of the murine stomach in *Helicobacter*-infected animals displaying mild lymphocytic responses suggests that Cea10 does play a role in early disease progression within this model and perhaps suggests a degree of tissue or cancer tropism. The small proline-rich protein 2A (Sprr2a), in addition to being significantly increased within the *Heli-*

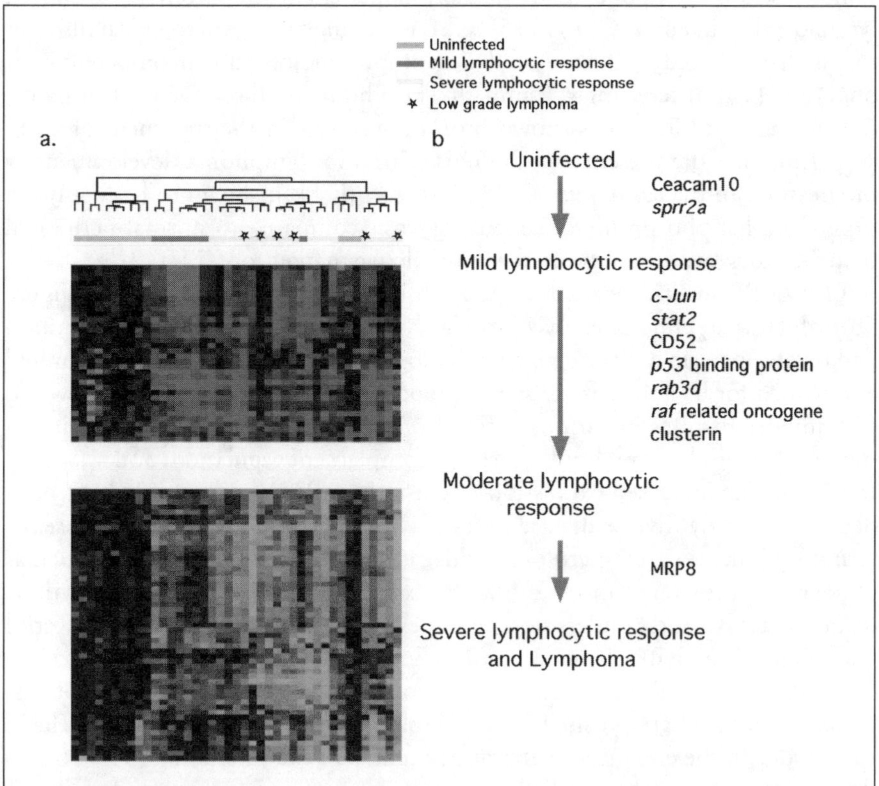

Figure 5. Unsupervised hierarchical clustering of 41 *H. heilmannii* infected and uninfected murine arrays. Infection and histopathology grading are indicated below the dendogram. Distinct nodes of infected and uninfected are seen as, with few exceptions, arrays from uninfected animals cluster tightly together and in a separate node from arrays from infected animals. (b) Schematic depiction of basis for SAM of the arrays shown in panel a. A selection of representative induced genes is shown to the right. (**See color plate A25.**)

cobacter study described here, was recently shown to be the most highly induced component when germ-free mice were infected with *Bacteroides thetaiotaomicron* (55). Microarray analysis showed a 280-fold increase in villus epithelial expression levels of Sprr2a. Members of the sprr family contribute to the barrier functions of squamous epithelia, as components of the cornified cell envelope and as cross-bridging proteins linked to desmosomal desmoplakin (118). Thus, increased expression suggests that upon microbial infection the host attempts to increase fortification of the epithelial barrier in direct response to colonization. However, this increase may have nothing to do with subsequent disease progression, per se, as *B. thetaiotaomicron* exists primarily as a commensal organism.

The composition of the second class of regulated genes is highly enriched for a significant number of signal transduction and oncogene-related proteins (Figure 5b). These include *c-jun*, *stat2*, CD52, *p53*-binding protein, *rab3d*, the *raf*-related oncogene, and clusterin. CD52 is a cell surface antigen that has been shown to be present on lymphocytes of virtually all patients with chronic B and T cell lymphocytic leukemia (CLL) (115). CLL is a malignant hematological disorder that is characterized by the accumulation of immunologically incompetent lymphocytes (115). Interestingly, Campath-1H, a humanized monoclonal antibody, directed against CD52 has shown promising results in the treatment of CLL, suggesting that the presence of CD52 is critical for lymphoma development or progression. In the same vein, CD24 has recently been shown to be highly expressed on lympho-proliferative leukocytes as compared to normal peripheral blood leukocytes and can be used as a predictive marker for CLL (12).

One of the most interesting of the second class of genes is clusterin (Figure 5b), which is an enigmatic glycoprotein showing ubiquitous tissue distribution and involvement in a diverse range of biological processes. Originally discovered and named for its ability to cluster red blood cells, clusterin has been shown to play important roles in various pathophysiological processes such as tissue remodeling, reproduction, lipid transport, complement regulation, and apoptosis (84,103). The increased expression of clusterin mRNA and protein have been detected in vastly diverse disease states (61). These include Alzheimer's disease, retinitis pigmentosa, glomerulonephritis, atherosclerosis, and in cancers such as gliomas and prostrate cancer. While the exact role of clusterin in each of these disease states is poorly understood, recent evidence has highlighted the powerful anti-apoptotic activity of clusterin in carcinoma models and has shown that increased levels of clusterin accelerate carcinoma cell growth by inhibition of Fas-mediated-induced apoptosis (85,141). Prolonged survival of the cancer cells in turn results in the enhanced metastatic potential of the carcinoma (84). The induced expression of this, as well as other anti-apoptotic factors such as Pim-1 (discussed earlier), may help to explain the deregulation of the host immune response and the subsequent *Helicobacter* induced development of gastric cancers.

The major constituent of the third class of genes was the macrophage migration inhibiting factor-related protein 8 (MRP8). MRP8 is a calcium-binding protein of the S100 family that is expressed during myeloid differentiation, is abundant in both granulocytes and monocytes, and forms a heterodimeric complex with MRP14, another S100 family member (65). MRP8-positive phagocytes generally belong to the early infiltrate and dominate acute inflammatory lesions. Elevated serum levels of MRP8 and MRP14 have been found in cystic fibrosis, rheumatoid arthritis, and chronic bronchitis, perhaps not surprisingly, as all of these are characteristic inflammatory disorders. The exact biological function of MRP8 is unclear, but increasing evidence indicates a regulatory role in the transendothelial migration of human monocytes (65). Therefore, perhaps increased expression at the later stages of *Helicobacter*-induced disease is linked to the strong and continued inflammatory response that is a hallmark of *Helicobacter* infection.

In addition to the comprehensive, unsupervised clustering analysis and SAM analysis conducted on these data sets, Mueller et al. (89a) grouped data sets based on histopathology and then completed two additional analyses where p values were calculated for every gene using a paired Student's *t* test. In the first analysis, uninfected samples were compared to all infected animals, and all genes with p values less than 0.001 were chosen for subsequent cluster analysis. Approximately 300 genes fit these criteria, and striking differences are evident between infected and uninfected arrays as each group clusters on a separate node on the dendogram. In the second analysis, comparisons were made only between infected animals, and p values were determined between animals showing mild lymphocytic responses and animals showing severe lymphocytic responses. Once again, only those genes with p values less than 0.001 were carried forward for cluster analysis (approximately 280). The resulting dendogram tightly clusters uninfected animals into their own node and further segregates the low- and high-grade pathology animals into distinct nodes (Figure 6). Visual analysis also revealed two clusters where animals exhibiting mild and severe lymphocytic response differ significantly from one another, which presents the exciting possibility that genes within these two clusters may be predictive for the development and progression of *Helicobacter*-induced lymphoma.

THE WRAP UP

With the inception and development of microarray technology, the scientific community has been given a valuable tool for the analysis of diverse systems and the elucidation of complex answers. Where once examination of gene expression was conducted at the level of a single or handful of genes at a time, it is now possible to simultaneously assess expression patterns of virtually every mRNA molecule in organisms for which the genomic sequence has been determined. This ability has proven useful within the context of both eukaryotic and prokaryotic systems, but nowhere is the possibility for understanding more remarkable than within the context of the host/pathogen interface. Where once the determination of co-regulated genes and elucidation of virulence factors was a laborious and time-consuming process (reviewed in 22), the age of genomics has ushered in a wealth of new possibilities for the discovery of the repertoire of genes employed by a pathogen as it interacts with and establishes disease in its particular host. The same is true of the host perspective because one can now more readily assess global changes in gene expression in response to an infection.

As we continue to settle into this post-genomic era, researchers will be faced with two major challenges. The first of these will be to advance to the next level of discovery in host/pathogen interactions; that is, the elucidation of pathogen and host gene expression profiles within the context of an actual human infection. While the first transcriptome of a bacterium shed from an infected human was recently reported (81), the majority of pathogens are not amenable to this type of analysis because relatively few organisms can be obtained from a bona fide infection. In addition, the amount of material that can be obtained from biopsies of infected human tissues remains relatively small, and host transcriptome analysis within the context of the actual infection will thus be faced with the same limitations. Therefore, to achieve this goal, we will be required to develop more sophisticated and sensitive techniques for expression profiling.

The second and perhaps more challenging

Figure 6. Supervised hierarchical clustering of 41 *H. heilmannii* infected and uninfected murine arrays. A paired Student's *t* test was used to calculate p values for every gene contained on the murine array. Comparisons were made only between infected animals, and p values were determined for genes between animals showing mild lymphocytic responses and animals showing severe lymphocytic responses. The 280 genes with p values less than 0.001 were carried forward for cluster analysis. The resulting dendogram tightly clusters uninfected animals into their own node and further segregates the low-grade and high-grade pathology animals into distinct nodes. Potentially predictive clusters where animals exhibiting low grade and high-grade pathology differ significantly from one another are marked by a vertical black bar on the left side of the figure. **(See color plate A26.)**

barrier to overcome will be brought about as the result of the "jackpot era" of gene discovery that we have now entered. While it is an amazing fact that a single, well-devised experiment can give you a list of hundreds of genes that are affected or differentially regulated by the condition of interest, the real challenge becomes understanding the role of each of those genes within the context of the bacterial/host life cycle. Microarray technology doesn't readily provide this information and, more and more, as scientists, we will be called on to take this glut of information and assemble it into interdigitating networks that incorporate the function, regulation, and role of each gene involved. Perhaps the most difficult job we have lies before us; however, this will undoubtedly be the most exciting and rewarding of our tasks.

ACKNOWLEDGMENTS

We thank M. Amieva, A. Mueller, L. Thompson, and K. Guillemin for sharing unpublished data and figures and for crucial discussion and review. Additionally, we thank A. Camilli, S. Fisher, and E.A. Joyce for critical reading of the manuscript. D.S.M. is supported by funds from The Damon Runyon Walter Winchell Foundation and Training grant no. 5T32AI07502. Research in the laboratory of S. Falkow is supported by National Institutes of Health grant nos. CA92229 and AI38459.

REFERENCES

1. **Aihara, M., D. Tsuchimoto, H. Takizawa, A. Azuma, H. Wakebe, Y. Ohmoto, K. Imagawa, M. Kikuchi, et al.** 1997. Mechanisms involved in *Helicobacter pylori*-induced interleukin-8 production by a gastric cancer cell line, MKN45. Infect. Immun. *65*:3218-3224.
2. **Akopyants, N.S., S.W. Clifton, D. Kersulyte, J.E. Crabtree, B.E. Youree, C.A. Reece, N.O. Bukanov, E.S. Drazek, et al.** 1998. Analyses of the cag pathogenicity island of *Helicobacter pylori*. Mol. Microbiol. *28*:37-53.
3. **Allan, E., C.L. Clayton, A. McLaren, D.M. Wallace, and B.W. Wren.** 2001. Characterization of the low-pH responses of *Helicobacter pylori* using genomic DNA arrays. Microbiology *147*:2285-2292.
4. **Alm, R.A., L.S. Ling, D.T. Moir, B.L. King, E.D. Brown, P.C. Doig, D.R. Smith, B. Noonan, et al.** 1999. Genomic-sequence comparison of two unrelated isolates of the human gastric pathogen *Helicobacter pylori*. Nature *397*:176-180.
5. **Amieva, M.R., N. Salama, L. Tompkins, and S. Falkow.** 2002. *Helicobacter pylori* enter and survive within multivesicular vacuoles of epithelial cells. Cell. Microbiol. *4*:677-690.
6. **Andrutis, K.A., J.G. Fox, D.B. Schauer, R.P. Marini, J.C. Murphy, L. Yan, and J.V. Solnick.** 1995. Inability of an isogenic urease-negative mutant stain of *Helicobacter mustelae* to colonize the ferret stomach. Infect. Immun. *63*:3722-3755.
7. **Ang, S., C.Z. Lee, K. Peck, M. Sindici, U. Matrubutham, M.A. Gleeson, and J.T. Wang.** 2001. Acid-induced gene expression in *Helicobacter pylori*: study in genomic scale by microarray. Infect. Immun. *69*:1679-1686.
8. **Anonymous.** 1994. Schistosomes, liver flukes and *Helicobacter pylori*. IARC Monogr. Eval. Carcinog. Risks Hum. *61*:1-241.
9. **Asahi, M., T. Azuma, S. Ito, Y. Ito, H. Suto, Y. Nagai, M. Tsubokawa, Y. Tohyama, et al.** 2000. *Helicobacter pylori* CagA protein can be tyrosine phosphorylated in gastric epithelial cells. J. Exp. Med. *191*:593-602.
10. **Bach, S., A. Makristathis, M. Rotter, and A.M. Hirschl.** 2002. Gene expression profiling in AGS cells

stimulated with *Helicobacter pylori* isogenic strains (*cagA* positive or *cagA* negative). Infect. Immun. *70*:988-992.

11.**Beier, D. and R. Frank.** 2000. Molecular characterization of two-component systems of *Helicobacter pylori.* J. Bacteriol. *182*:2068-2076.

12.**Belov, L., O. de la Vega, C.G. dos Remedios, S.P. Mulligan, and R.I. Christopherson.** 2001. Immunophenotyping of leukemias using a cluster of differentiation antibody microarray. Cancer Res. *61*:4483-4489.

13.**Bereswill, S., U. Waidner, S. Odenbreit, F. Lichte, F. Fassbinder, G. Bode, and M. Kist.** 1998. Structural, functional and mutational analysis of the pfr gene encoding a ferritin from *Helicobacter pylori.* Microbiology *144*:2505-2516.

14.**Bijlsma, J.J., A.L.M. Lie, I.C. Nootenboom, C.M. Vandenbroucke-Grauls, and J.G. Kusters.** 2000. Identification of loci essential for the growth of *Helicobacter pylori* under acidic conditions. J. Infect. Dis. *182*:1566-1569.

15.**Bijlsma, J.J., B. Waidner, A.H. Vliet, N.J. Hughes, S. Hag, S. Bereswill, D.J. Kelly, C.M. Vandenbroucke-Grauls, et al.** 2002. The *Helicobacter pylori* homologue of the ferric uptake regulator is involved in acid resistance. Infect. Immun. *70*:606-611.

16.**Blaser, M.J.** 1998. *Helicobacter pylori* and gastric diseases. BMJ *316*:1507-1510.

17.**Blaser, M.J., P.H. Chyou, and A. Nomura.** 1995. Age at establishment of *Helicobacter pylori* infection and gastric carcinoma, gastric ulcer, and duodenal ulcer risk. Cancer Res. *55*:562-565.

18.**Blaser, M.J. and J. Parsonnet.** 1994. Parasitism by the "slow" bacterium *Helicobacter pylori* leads to altered gastric homeostasis and neoplasia. J. Clin. Invest. *94*:4-8.

19.**Borch, K., C. Sjostedt, U. Hannestad, J.D. Soderholm, L. Franzen, and S. Mardh.** 1998. Asymptomatic *Helicobacter pylori* gastritis is associated with increased sucrose permeability. Dig. Dis. Sci. *43*:749-753.

20.**Bury-Mone, S., S. Skouloubris, A. Labigne, and H. De Reuse.** 2001. The *Helicobacter pylori* UreI protein: role in adaptation to acidity and identification of residues essential for its activity and for acid activation. Mol. Microbiol. *42*:1021-1034.

21.**Camilli, A., D. Beattie, and J. Mekalanos.** 1994. Use of genetic recombination as a reporter of gene expression. Proc. Natl. Acad. Sci. USA *91*:2634-2638.

22.**Camilli, A., D.S. Merrell, and J.J. Mekalanos.** 2001. Strategies to identify bacterial pathogenicity factors, p. 134-177. *In* E.A. Groisman (Ed.), Principles of Bacterial Pathogenesis. Academic Press, San Diego.

23.**Cary, L.A., D.C. Han, and J.L. Guan.** 1999. Integrin-mediated signal transduction pathways. Histol. Histopathol. *14*:1001-1009.

24.**Caselli, M., N. Figura, L. Trevisani, P. Pazzi, P. Guglielmetti, M.R. Bovolenta, and G. Stabellini.** 1989. Patterns of physical modes of contact between *Campylobacter pylori* and gastric epithelium: implications about the bacterial pathogenicity. Am. J. Gastroenterol. *84*:511-513.

25.**Censini, S., C. Lange, Z. Xiang, J.E. Crabtree, P. Ghiara, M. Borodovsky, R. Rappuoli, and A. Covacci.** 1996. cag, a pathogenicity island of *Helicobacter pylori,* encodes type I-specific and disease-associated virulence factors. Proc. Natl. Acad. Sci. USA *93*:14648-14653.

26.**Chen, X.P., J.A. Losman, S. Cowan, E. Donahue, S. Fay, B.Q. Vuong, M.C. Nawijn, et al.** 2002. Pim serine/threonine kinases regulate the stability of Socs-1 protein. Proc. Natl. Acad. Sci. USA *99*:2175-2180.

27.**Chiou, C.C., C.C. Chan, D.L. Sheu, K.T. Chen, Y.S. Li, and E.C. Chan.** 2001. *Helicobacter pylori* infection induced alteration of gene expression in human gastric cells. Gut *48*:598-604.

28.**Covacci, A., J.L. Telford, G. Del Giudice, J. Parsonnet, and R. Rappuoli.** 1999. *Helicobacter pylori* virulence and genetic geography. Science *284*:1328-1333.

29.**Cover, T.L.** 1996. The vacuolating cytotoxin of *Helicobacter pylori.* Mol. Microbiol. *20*:241-246.

30.**Cover, T.L. and M.J. Blaser.** 1992. Purification and characterization of the vacuolating toxin from *Helicobacter pylori.* J. Biol. Chem. *267*:10570-10575.

31.**Cox, J.M., C.L. Clayton, T. Tomita, D.M. Wallace, P.A. Robinson, and J.E. Crabtree.** 2001. cDNA array analysis of cag pathogenicity island-associated *Helicobacter pylori* epithelial cell response genes. Infect. Immun. *69*:6970-6980.

32.**Curtis, G.H. and D.G. Gall.** 1992. Macromolecular transport by rat gastric mucosa. Am. J. Physiol. *262*:G1033-G1040.

33.**Cussac, V., R.L. Ferrero, and A. Labigne.** 1992. Expression of *Helicobacter pylori* urease genes in *Es-*

cherichia coli grown under nitrogen-limiting conditions. J. Bacteriol. *174*:2466-2673.

34.**Cuypers, H.T., G. Selten, W. Quint, M. Zijlstra, E.R. Maandag, W. Boelens, P. van Wezenbeek, C. Melief, et al.** 1984. Murine leukemia virus-induced T-cell lymphomagenesis: integration of proviruses in a distinct chromosomal region. Cell *37*:141-150.

35.**Dautry, F., D. Weil, J. Yu, and A. Dautry-Varsat.** 1988. Regulation of *pim* and *myb* mRNA accumulation by interleukin 2 and interleukin 3 in murine hematopoietic cell lines. J. Biol. Chem. *263*:17615-17620.

36.**de Vries, N., E.J. Kuipers, N.E. Kramer, A.H. van Vliet, J.J. Bijlsma, M. Kist, S. Bereswill, C.M. Vandenbroucke-Grauls et al.** 2001. Identification of environmental stress-regulated genes in *Helicobacter pylori* by a *lacZ* reporter gene fusion system. Helicobacter *6*:300-309.

37.**Delany, I., A.B. Pacheco, G. Spohn, R. Rappuoli, and V. Scarlato.** 2001. Iron-dependent transcription of the *frpB* gene of *Helicobacter pylori* is controlled by the Fur repressor protein. J. Bacteriol. *183*:4932-4937.

38.**Doig, P., J.W. Austin, and T.J. Trust.** 1993. The *Helicobacter pylori* 19.6-kilodalton protein is an iron-containing protein resembling ferritin. J. Bacteriol. *175*:557-560.

39.**Domen, J., N.M. van der Lugt, P.W. Laird, C.J. Saris, A.R. Clarke, M.L. Hooper, and A. Berns.** 1993. Impaired interleukin-3 response in Pim-1-deficient bone marrow-derived mast cells. Blood *82*:1445-1452.

40.**Drumm, B., P. Sherman, E. Cutz, and M. Karmali.** 1987. Association of *Campylobacter pylori* on the gastric mucosa with antral gastritis in children. N. Engl. J. Med. *316*:1557-1561.

41.**Dunn, B.E., H. Cohen, and M.J. Blaser.** 1997. *Helicobacter pylori.* Clin. Microbiol. Rev. *10*:720-741.

42.**Eaton, K.A. and S. Krakowka.** 1994. Effect of gastric pH on urease-dependent colonization of gnotobiotic piglets by *Helicobacter pylori.* Infect. Immun. *62*:3604-3607.

43.**Eaton, K.A., S. Suerbaum, C. Josenhans, and S. Krakowka.** 1996. Colonization of gnotobiotic piglets by *Helicobacter pylori* deficient in two flagellin genes. Infect. Immun. *64*:2445-2448.

44.**Eisen, M.B., P.T. Spellman, P.O. Brown, and D. Botstein.** 1998. Cluster analysis and display of genome-wide expression patterns. Proc. Natl. Acad. Sci. USA *95*:14863-14868.

45.**el-Shoura, S. M.** 1995. *Helicobacter pylori*: I. ultrastructural sequences of adherence, attachment, and penetration into the gastric mucosa. Ultrastruct. Pathol. *19*:323-333.

46.**Ernst, P.B. and B.D. Gold.** 2000. The disease spectrum of *Helicobacter pylori*: the immunopathogenesis of gastroduodenal ulcer and gastric cancer. Annu. Rev. Microbiol. *54*:615-640.

47.**Ferrero, R.L., V. Cussac, P. Courcoux, and A. Labigne.** 1992. Construction of isogenic urease-negative mutants of *Helicobacter pylori* by allelic exchange. J. Bacteriol. *174*:4212-4217.

48.**Fischer, W., J. Puls, R. Buhrdorf, B. Gebert, S. Odenbreit, and R. Haas.** 2001. Systematic mutagenesis of the *Helicobacter pylori* cag pathogenicity island: essential genes for CagA translocation in host cells and induction of interleukin-8. Mol. Microbiol. *42*:1337-1348.

49.**Frazier, B.A., J.D. Pfeifer, D.G. Russell, P. Falk, A.N. Olsen, M. Hammar, T.U. Westblom, et al.** 1993. Paracrystalline inclusions of a novel ferritin containing nonheme iron, produced by the human gastric pathogen *Helicobacter pylori*: evidence for a third class of ferritins. J. Bacteriol. *175*:966-972.

50.**Gionchetti, P., D. Vaira, M. Campieri, J. Holton, M. Menegatti, A. Belluzzi, E. Bertinelli, M. Ferretti, et al.** 1994. Enhanced mucosal interleukin-6 and -8 in *Helicobacter pylori*-positive dyspeptic patients. Am. J. Gastroenterol. *89*:883-887.

51.**Gobert, A. P., Y. Cheng, J. Y. Wang, J. L. Boucher, R. K. Iyer, S. D. Cederbaum, R. A. Casero, Jr., J. C. Newton, and K. T. Wilson.** 2002. *Helicobacter pylori* induces macrophage apoptosis by activation of arginase II. J Immunol *168*:4692-700.

51a.**Guillemin, K., N.R. Salama, L.S. Tompkins, and S. Falkow.** 2002. Cag pathogenicity island specific responses of gastric epithelial cells to *Helicobacter pylori* infection. Proc. Natl. Acad. Sci. USA (In press).

52.**Hengge-Aronis, R.** 2000. The general stress response in *Escherichia coli*, p. 161-178. *In* G. Storz, and R. Hengge-Aronis (Ed.), Bacterial Stress Response, 1st ed. ASM Press, Washington D.C.

53.**Higashi, H., R. Tsutsumi, S. Muto, T. Sugiyama, T. Azuma, M. Asaka, and M. Hatakeyama.** 2002. SHP-2 tyrosine phosphatase as an intracellular target of *Helicobacter pylori* CagA protein. Science *295*:683-686.

54.**Hirsch, D.S., D.M. Pirone, and P.D. Burbelo.** 2001. A new family of Cdc42 effector proteins, CEPs, function in fibroblast and epithelial cell shape changes. J. Biol. Chem. *276*:875-883.

55.**Hooper, L.V., M.H. Wong, A. Thelin, L. Hansson, P.G. Falk, and J.I. Gordon.** 2001. Molecular analysis of commensal host-microbial relationships in the intestine. Science *291*:881-884.

56.**Iwamoto, H., D.M. Czajkowsky, T.L. Cover, G. Szabo, and Z. Shao.** 1999. VacA from *Helicobacter pylori*: a hexameric chloride channel. FEBS Lett. *450*:101-104.
57.**Jacobs, H., P. Krimpenfort, M. Haks, J. Allen, B. Blom, C. Demolliere, A. Kruisbeek, H. Spits, et al.** 1999. PIM1 reconstitutes thymus cellularity in interleukin 7- and common gamma chain-mutant mice and permits thymocyte maturation in Rag- but not CD3gamma-deficient mice. J. Exp. Med. *190*:1059-1068.
58.**Jehn, B.M. and B.A. Osborne.** 1997. Gene regulation associated with apoptosis. Crit. Rev. Eukaryot. Gene Expr. *7*:179-193.
59.**Joberty, G., R.R. Perlungher, and I.G. Macara.** 1999. The Borgs, a new family of Cdc42 and TC10 GTPase-interacting proteins. Mol. Cell. Biol. *19*:6585-6597.
60.**Joberty, G., R.R. Perlungher, P.J. Sheffield, M. Kinoshita, M. Noda, T. Haystead, and I.G. Macara.** 2001. Borg proteins control septin organization and are negatively regulated by Cdc42. Nat. Cell Biol. *3*:861-866.
61.**Jones, S.E. and C. Jomary.** 2002. Clusterin. Int. J. Biochem. Cell Biol. *34*:427-431.
62.**Joyce, E.A., J.V. Gilbert, K.A. Eaton, A. Plaut, and A. Wright.** 2001. Differential gene expression from two transcriptional units in the cag pathogenicity island of *Helicobacter pylori*. Infect. Immun. *69*:4202-4209.
63.**Karita, M., M.K. Tummuru, H.P. Wirth, and M.J. Blaser.** 1996. Effect of growth phase and acid shock on *Helicobacter pylori cagA* expression. Infect. Immun. *64*:4501-4507.
64.**Keck, U., P. Nedellec, N. Beauchemin, J. Thompson, and W. Zimmermann.** 1995. The *cea10* gene encodes a secreted member of the murine carcinoembryonic antigen family and is expressed in the placenta, gastrointestinal tract and bone marrow. Eur. J. Biochem. *229*:455-464.
65.**Kerkhoff, C., M. Klempt, and C. Sorg.** 1998. Novel insights into structure and function of MRP8 (S100A8) and MRP14 (S100A9). Biochim. Biophys. Acta *1448*:200-211.
66.**Kodama, A., T. Matozaki, A. Fukuhara, M. Kikyo, M. Ichihashi, and Y. Takai.** 2000. Involvement of an SHP-2-Rho small G protein pathway in hepatocyte growth factor/scatter factor-induced cell scattering. Mol. Biol. Cell *11*:2565-2575.
67.**Labigne, A., V. Cussac, and P. Courcoux.** 1991. Shuttle cloning and nucleotide sequences of *Helicobacter pylori* genes responsible for urease activity. J. Bacteriol. *173*:1920-1931.
68.**Lee, C. A. and S. Falkow.** 1990. The ability of *Salmonella* to enter mammalian cells is affected by bacterial growth state. Proc. Natl. Acad. Sci. USA *87*:4304-4308.
69.**Leunk, R.D., P.T. Johnson, B.C. David, W.G. Kraft, and D.R. Morgan.** 1988. Cytotoxic activity in broth-culture filtrates of *Campylobacter pylori*. J. Med. Microbiol. *26*:93-99.
70.**Lilly, M., T. Le, P. Holland, and S.L. Hendrickson.** 1992. Sustained expression of the pim-1 kinase is specifically induced in myeloid cells by cytokines whose receptors are structurally related. Oncogene *7*:727-732.
71.**Lupetti, P., J.E. Heuser, R. Manetti, P. Massari, S. Lanzavecchia, P.L. Bellon, R. Dallai, R. Rappuoli, et al.** 1996. Oligomeric and subunit structure of the *Helicobacter pylori* vacuolating cytotoxin. J. Cell Biol. *133*:801-807.
72.**Maeda, S., M. Otsuka, Y. Hirata, Y. Mitsuno, H. Yoshida, Y. Shiratori, Y. Masuho, M. Muramatsu, et al.** 2001. cDNA microarray analysis of *Helicobacter pylori*-mediated alteration of gene expression in gastric cancer cells. Biochem. Biophys. Res. Commun. *284*:443-449.
73.**Mahan, M.J., J.M. Slauch, and J.J. Mekalanos.** 1993. Selection of bacterial virulence genes that are specifically induced in host tissues. Science *259*:686-688.
74.**Mannick, E.E., L.E. Bravo, G. Zarama, J.L. Realpe, X.J. Zhang, B. Ruiz, E.T. Fontham, R. Mera, et al.** 1996. Inducible nitric oxide synthase, nitrotyrosine, and apoptosis in *Helicobacter pylori* gastritis: effect of antibiotics and antioxidants. Cancer Res. *56*:3238-3243.
75.**Marshall, B.J. and J.R. Warren.** 1984. Unidentified curved bacilli in the stomach of patients with gastritis and peptic ulceration. J. Lancet *1*:1311-1315.
76.**McClain, M.S., W. Schraw, V. Ricci, P. Boquet, and T.L. Cover.** 2000. Acid activation of *Helicobacter pylori* vacuolating cytotoxin (VacA) results in toxin internalization by eukaryotic cells. Mol. Microbiol. *37*:433-442.
77.**McDaniel, T.K., K.C. Dewalt, N.R. Salama, and S. Falkow.** 2001. New approaches for validation of lethal phenotypes and genetic reversion in *Helicobacter pylori*. Helicobacter *6*:15-23.
78.**McGee, D.J., F.J. Radcliff, G.L. Mendz, R.L. Ferrero, and H.L. Mobley.** 1999. *Helicobacter pylori rocF* is required for arginase activity and acid protection in vitro but is not essential for colonization of

mice or for urease activity. J. Bacteriol. *181*:7314-7322.

79.**McGowan, C.C., T.L. Cover, and M.J. Blaser.** 1997. Analysis of F1F0-ATPase from *Helicobacter pylori.* Infect. Immun. *65*:2640-2647.

80.**McGowan, C.C., A. Necheva, S.A. Thompson, T.L. Cover, and M.J. Blaser.** 1998. Acid-induced expression of an LPS-associated gene in *Helicobacter pylori.* Mol. Microbiol. *30*:19-31.

81.**Merrell, D.S., S.M. Butler, F. Qadri, N.A. Dolganov, A. Alam, M.B. Cohen, S.B. Calderwood, G.K. Schoolnik, et al.** 2002. Host-induced epidemic spread of the cholera bacterium. Nature *417*:642-645.

82.**Meyer-Rosberg, K., D.R. Scott, D. Rex, K. Melchers, and G. Sachs.** 1996. The effect of environmental pH on the proton motive force of *Helicobacter pylori.* Gastroenterology *111*:886-900.

83.**Mills, J.C., A.J. Syder, C.V. Hong, J.L. Guruge, F. Raaii, and J.I. Gordon.** 2001. A molecular profile of the mouse gastric parietal cell with and without exposure to *Helicobacter pylori.* Proc. Natl. Acad. Sci. USA *98*:13687-13692.

84.**Miyake, H., M.E. Gleave, S. Arakawa, S. Kamidono, and I. Hara.** 2002. Introducing the clusterin gene into human renal cell carcinoma cells enhances their metastatic potential. J. Urol. *167*:2203-2208.

85.**Miyake, H., S. Hara, T. Zellweger, S. Kamidono, M.E. Gleave, and I. Hara.** 2001. Acquisition of resistance to Fas-mediated apoptosis by overexpression of clusterin in human renal-cell carcinoma cells. Mol. Urol. *5*:105-111.

86.**Mobley, H.L., M.D. Island, and R.P. Hausinger.** 1995. Molecular biology of microbial ureases. Microbiol. Rev. *59*:451-480.

87.**Molinari, M., C. Galli, N. Norais, J.L. Telford, R. Rappuoli, J.P. Luzio, and C. Montecucco.** 1997. Vacuoles induced by *Helicobacter pylori* toxin contain both late endosomal and lysosomal markers. J. Biol. Chem. *272*:25339-25344.

88.**Molinari, M., M. Salio, C. Galli, N. Norais, R. Rappuoli, A. Lanzavecchia, and C. Montecucco.** 1998. Selective inhibition of Ii-dependent antigen presentation by *Helicobacter pylori* toxin VacA. J. Exp. Med. *187*:135-140.

89.**Moss, S.F., J. Calam, B. Agarwal, S. Wang, and P.R. Holt.** 1996. Induction of gastric epithelial apoptosis by *Helicobacter pylori.* Gut *38*:498-501.

89a.**Mueller, A., J. O'Rourke, J. Grimm, K. Guillemin, M.F. Dixon, A. Lee, and S. Falkow.** 2002. Distinct gene expression profiles characterize histopathological stages of disease progression in *Helicobacter* induced MALT Lymphoma. Proc. Natl. Acad. Sci. USA (In press).

90.**Nedellec, P., C. Turbide, and N. Beauchemin.** 1995. Characterization and transcriptional activity of the mouse biliary glycoprotein 1 gene, a carcinoembryonic antigen-related gene. Eur. J. Biochem. *231*:104-114.

91.**Neugut, A.I., M. Hayek, and G. Howe.** 1996. Epidemiology of gastric cancer. Semin. Oncol. *23*:281-291.

92.**Odenbreit, S., J. Puls, B. Sedlmaier, E. Gerland, W. Fischer, and R. Haas.** 2000. Translocation of *Helicobacter pylori* CagA into gastric epithelial cells by type IV secretion. Science *287*:1497-1500.

93.**Ohashi, K., K. Nagata, M. Maekawa, T. Ishizaki, S. Narumiya, and K. Mizuno.** 2000. Rho-associated kinase ROCK activates LIM-kinase 1 by phosphorylation at threonine 508 within the activation loop. J. Biol. Chem. *275*:3577-3582.

94.**Ottemann, K.M. and A.C. Lowenthal.** 2002. *Helicobacter pylori* uses motility for initial colonization and to attain robust infection. Infect. Immun. *70*:1984-1990.

95.**Palovuori, R., A. Perttu, Y. Yan, R. Karttunen, S. Eskelinen, and T.J. Karttunen.** 2000. *Helicobacter pylori* induces formation of stress fibers and membrane ruffles in AGS cells by rac activation. Biochem. Biophys. Res. Commun. *269*:247-253.

96.**Papini, E., M. de Bernard, E. Milia, M. Bugnoli, M. Zerial, R. Rappuoli, and C. Montecucco.** 1994. Cellular vacuoles induced by *Helicobacter pylori* originate from late endosomal compartments. Proc. Natl. Acad. Sci. USA *91*:9720-9724.

97.**Parsonnet, J., G.D. Friedman, D.P. Vandersteen, Y. Chang, J.H. Vogelman, N. Orentreich, and R.K. Sibley.** 1991. *Helicobacter pylori* infection and the risk of gastric carcinoma. N. Engl. J. Med. *325*:1127-1131.

98.**Peterson, K.M. and J.J. Mekalanos.** 1988. Characterization of the *Vibrio cholerae* ToxR regulon: identification of novel genes involved in intestinal colonization [published erratum appears in Infect. Immun., 1989, *57*:660]. Infect. Immun. *56*:2822-2829.

99.**Phadnis, S.H., M.H. Parlow, M. Levy, D. Ilver, C.M. Caulkins, J.B. Connors, and B.E. Dunn.** 1996. Surface localization of *Helicobacter pylori* urease and a heat shock protein homolog requires bacterial autolysis. Infect. Immun. *64*:905-912.

100.**Quigley, E.M. and L.A. Turnberg.** 1987. pH of the microclimate lining human gastric and duodenal mucosa *in vivo.* Studies in control subjects and in duodenal ulcer patients. Gastroenterology *92*:1876-1884.

101.**Rabassa, A.A., R. Goodgame, F.M. Sutton, C.N. Ou, C. Rognerud, and D.Y. Graham.** 1996. Effects of aspirin and *Helicobacter pylori* on the gastroduodenal mucosal permeability to sucrose. Gut *39*:159-163.

102.**Rektorschek, M., A. Buhmann, D. Weeks, D. Schwan, K.W. Bensch, S. Eskandari, D. Scott, G. Sachs, et al.** 2000. Acid resistance of *Helicobacter pylori* depends on the UreI membrane protein and an inner membrane proton barrier. Mol. Microbiol. *36*:141-152.

103.**Rosenberg, M.E. and J. Silkensen.** 1995. Clusterin: physiologic and pathophysiologic considerations. Int. J. Biochem. Cell Biol. *27*:633-645.

104.**Salama, N., K. Guillemin, T.K. McDaniel, G. Sherlock, L. Tompkins, and S. Falkow.** 2000. A whole-genome microarray reveals genetic diversity among *Helicobacter pylori* strains. Proc. Natl. Acad. Sci. USA *97*:14668-14673.

105.**Salama, N.R., K.M. Ottemann, and S. Falkow.** 2002. Toxins, tropisms and travels: *H. pylori* and host cells, p. 173-201. *In* Y. Yamamoto, H. Friedman, and P. Hoffman (Eds.), *Helicobacter* Infections and Immunity. Kluwer Academic/Plenum Publishers, New York.

106.**Salama, N.R., G. Otto, L. Tompkins, and S. Falkow.** 2001. Vacuolating cytotoxin of *Helicobacter pylori* plays a role during colonization in a mouse model of infection. Infect. Immun. *69*:730-736.

107.**Scott, D.R., D. Weeks, C. Hong, S. Postius, K. Melchers, and G. Sachs.** 1998. The role of internal urease in acid resistance of *Helicobacter pylori.* Gastroenterology *114*:58-70.

108.**Segal, E.D.** 1997. Consequences of attachment of *Helicobacter pylori* to gastric cells. Biomed. Pharmacother. *51*:5-12.

109.**Segal, E.D., J. Cha, J. Lo, S. Falkow, and L.S. Tompkins.** 1999. Altered states: involvement of phosphorylated CagA in the induction of host cellular growth changes by *Helicobacter pylori.* Proc. Natl. Acad. Sci. USA *96*:14559-14564.

110.**Segal, E.D., C. Lange, A. Covacci, L.S. Tompkins, and S. Falkow.** 1997. Induction of host signal transduction pathways by *Helicobacter pylori.* Proc. Natl. Acad. Sci. USA *94*:7595-7599.

111.**Selbach, M., S. Moese, C.R. Hauck, T.F. Meyer, and S. Backert.** 2002. Src is the kinase of the *Helicobacter pylori* CagA protein *in vitro* and *in vivo.* J. Biol. Chem. *277*:6775-6778.

112.**Sepulveda, A.R., H. Tao, E. Carloni, J. Sepulveda, D.Y. Graham, and L.E. Peterson.** 2002. Screening of gene expression profiles in gastric epithelial cells induced by *Helicobacter pylori* using microarray analysis. Aliment Pharmacol. Ther. *16 (Suppl 2)*:145-157.

113.**Shimada, T. and A. Terano.** 1998. Chemokine expression in *Helicobacter pylori*-infected gastric mucosa. J. Gastroenterol. *33*:613-617.

114.**Shively, J.E. and J.D. Beatty.** 1985. CEA-related antigens: molecular biology and clinical significance. Crit. Rev. Oncol. Hematol. *2*:355-399.

115.**Sorokin, P.** 2001. Campath-1H. Clin. J. Oncol. Nurs. *5*:65-66.

116.**Stein, M., F. Bagnoli, R. Halenbeck, R. Rappuoli, W.J. Fantl, and A. Covacci.** 2002. c-Src/Lyn kinases activate *Helicobacter pylori* CagA through tyrosine phosphorylation of the EPIYA motifs. Mol. Microbiol. *43*:971-980.

117.**Stein, M., R. Rappuoli, and A. Covacci.** 2000. Tyrosine phosphorylation of the *Helicobacter pylori* CagA antigen after cag-driven host cell translocation. Proc. Natl. Acad. Sci. USA *97*:1263-1268.

118.**Steinert, P.M. and L.N. Marekov.** 1999. Initiation of assembly of the cell envelope barrier structure of stratified squamous epithelia. Mol. Biol. Cell *10*:4247-4261.

119.**Tannaes, T., N. Dekker, G. Bukholm, J.J. Bijlsma, and B.J. Appelmelk.** 2001. Phase variation in the *Helicobacter pylori* phospholipase A gene and its role in acid adaptation. Infect. Immun. *69*:7334-7340.

120.**Taylor, D.N. and J. Parsonnet.** 1995. Epidemiology and natural history of *H. pylori* infections, p. 551-564. *In* M.J. Blaser, P.F. Smith, J. Ravdin, H. Greenberg, and R.L. Guerrant (Eds.), Infections of the Gastrointestinal Tract. Raven Press, New York.

121.**Teglund, S., A. Olsen, W.N. Khan, L. Frangsmyr, and S. Hammarstrom.** 1994. The pregnancy-specific glycoprotein (PSG) gene cluster on human chromosome 19: fine structure of the 11 PSG genes and identification of 6 new genes forming a third subgroup within the carcinoembryonic antigen

(CEA) family. Genomics *23*:669-684.

122. **Thompson, J.A., F. Grunert, and W. Zimmermann.** 1991. Carcinoembryonic antigen gene family: molecular biology and clinical perspectives. J. Clin. Lab. Anal. *5*:344-366.

123. **Thompson, S. A. and M.J. Blaser.** 1995. Isolation of the *Helicobacter pylori recA* gene and involvement of the *recA* region in resistance to low pH. Infect. Immun. *63*:2185-2193.

124. **Tomb, J.F., O. White, A.R. Kerlavage, R.A. Clayton, G.G. Sutton, R.D. Fleischmann, K.A. Ketchum, H.P. Klenk, et al.** 1997. The complete genome sequence of the gastric pathogen *Helicobacter pylori.* Nature *388*:539-547.

125. **Tombola, F., C. Carlesso, I. Szabo, M. de Bernard, J.M. Reyrat, J.L. Telford, R. Rappuoli, C. Montecucco, et al.** 1999. *Helicobacter pylori* vacuolating toxin forms anion-selective channels in planar lipid bilayers: possible implications for the mechanism of cellular vacuolation. Biophys. J. *76*:1401-1409.

126. **Tombola, F., L. Morbiato, G. Del Giudice, R. Rappuoli, M. Zoratti, and E. Papini.** 2001. The *Helicobacter pylori* VacA toxin is a urea permease that promotes urea diffusion across epithelia. J. Clin. Invest. *108*:929-937.

127. **Tsuda, M., M. Karita, M.G. Morshed, K. Okita, and T. Nakazawa.** 1994. A urease-negative mutant of *Helicobacter pylori* constructed by allelic exchange mutagenesis lacks the ability to colonize the nude mouse stomach. Infect. Immun. *62*:3586-3589.

128. **Tsukita, S. and M. Furuse.** 2000. The structure and function of claudins, cell adhesion molecules at tight junctions. Ann. NY Acad. Sci. *915*:129-135.

129. **Tusher, V.G., R. Tibshirani, and G. Chu.** 2001. Significance analysis of microarrays applied to the ionizing radiation response. Proc. Natl. Acad. Sci. USA *98*:5116-5121.

130. **Valdivia, R.H. and S. Falkow.** 1996. Bacterial genetics by flow cytometry: rapid isolation of *Salmonella typhimurium* acid-inducible promoters by differential fluorescence induction. Mol. Microbiol. *22*:367-378.

131. **Valle, J., K. Seppala, P. Sipponen, and T. Kosunen.** 1991. Disappearance of gastritis after eradication of *Helicobacter pylori.* A morphometric study. Scand. J. Gastroenterol. *26*:1057-1065.

132. **Vera, J.F., M. Gotteland, E. Chavez, M.T. Vial, E. Kakarieka, and O. Brunser.** 1997. Sucrose permeability in children with gastric damage and *Helicobacter pylori* infection. J. Pediatr. Gastroenterol. Nutr. *24*:506-511.

133. **Weeks, D.L., S. Eskandari, D.R. Scott, and G. Sachs.** 2000. A H+-gated urea channel: the link between *Helicobacter pylori* urease and gastric colonization. Science *287*:482-485.

134. **Weeks, D.L. and G. Sachs.** 2001. Sites of pH regulation of the urea channel of *Helicobacter pylori.* Mol. Microbiol. *40*:1249-1259.

135. **Wong, K.K. and M. McClelland.** 1994. Stress-inducible gene of *Salmonella typhimurium* identified by arbitrarily primed PCR of RNA. Proc. Natl. Acad. Sci. USA *91*:639-643.

136. **Xiang, Z., S. Censini, P.F. Bayeli, J.L. Telford, N. Figura, R. Rappuoli, and A. Covacci.** 1995. Analysis of expression of CagA and VacA virulence factors in 43 strains of *Helicobacter pylori* reveals that clinical isolates can be divided into two major types and that CagA is not necessary for expression of the vacuolating cytotoxin. Infect. Immun. *63*:94-98.

137. **Yahiro, K., T. Niidome, T. Hatakeyama, H. Aoyagi, H. Kurazono, P.I. Padilla, A. Wada, and T. Hirayama.** 1997. *Helicobacter pylori* vacuolating cytotoxin binds to the 140-kDa protein in human gastric cancer cell lines, AZ-521 and AGS. Biochem. Biophys. Res. Commun. *238*:629-632.

138. **Yang, N., O. Higuchi, K. Ohashi, K. Nagata, A. Wada, K. Kangawa, E. Nishida, and K. Mizuno.** 1998. Cofilin phosphorylation by LIM-kinase 1 and its role in Rac-mediated actin reorganization. Nature *393*:809-812.

139. **Yeung, C.K., K.H. Fu, K.Y. Yuen, W.F. Ng, T.M. Tsang, F.J. Branicki, and H. Saing.** 1990. *Helicobacter pylori* and associated duodenal ulcer. Arch. Dis. Child. *65*:1212-1216.

140. **Yip-Schneider, M.T., M. Horie, and H.E. Broxmeyer.** 1995. Transcriptional induction of *pim-1* protein kinase gene expression by interferon gamma and posttranscriptional effects on costimulation with steel factor. Blood *85*:3494-3502.

141. **Zellweger, T., H. Miyake, L.V. July, M. Akbari, S. Kiyama, and M.E. Gleave.** 2001. Chemosensitization of human renal cell cancer using antisense oligonucleotides targeting the antiapoptotic gene clusterin. Neoplasia *3*:360-367.

Section V

Bioinformatics

16 Correspondence Analysis with Microarray Data

Kurt Fellenberg[1,2], Martin Vingron[1,3], Nicole C. Hauser[2], and Jörg D. Hoheisel[2]
Departments of [1]Theoretical Bioinformatics and [2]Functional Genome Analysis, German Cancer Research Center, Heidelberg, and [3]Department of Computational Molecular Biology, Max-Planck Institute for Molecular Genetics, Berlin, Germany

INTRODUCTION

Microarray technology provides insight into the transcriptional state of the cell, simultaneously measuring RNA levels for thousands of genes. As such, it has become one of the workhorses of functional genomics. However, the technology results in large amounts of data, the results of which are difficult to interpret.

Microarray Technology

Cells accomplish metabolic processes, comply with their own growth program, adapt to changing environments, or communicate with other cells by accurately controlled expression of the appropriate proteins. The protein portfolio is tailored to the particular requirements of the cell. To this end, gene-coding sequence in the DNA is transcribed and leaves the cell nucleus as mRNA. In the surrounding cytoplasm, each mRNA molecule conducts the synthesis of the particular protein that is encoded. The presence and amount of a particular mRNA regulate the presence and amount of the encoded protein. For a particular cell status, the level of mRNA can be measured in parallel for thousands of genes by DNA chip technology. RNA is prepared from the cells and is reverse-transcribed into more stable DNA, simultaneously incorporating radioactive or fluorescent labels. The labeled DNA is then applied to a DNA chip, also referred to as a DNA array or microarray.

Perspectives in Gene Expression
Edited by K. Appasani

The microarray consists of support material, onto which DNA fragments of different sequence, representing genes, have been spotted. These spots serve to measure the level of the applied DNA obtained from the RNA sample. For DNA of the same kind, complementary single strands will hybridize, resulting in ds-DNA. Therefore, the level of labeled DNA bound to a particular spot will correspond to the level of the particular kind of mRNA in the cells. The amount of label is measured for each spot. Thus, microarray technology provides insight into the transcriptional status of the cell or transcriptome, simultaneously measuring RNA levels for thousands of genes (11,17,38,43). In this way, microarray technology has become an important tool in functional genomics (3,11,20,43).

In addition to the parallel investigation of many genes, the benefit of the technique lies in its broad range of applications. Its applications range from the study of organisms with a particular gene inactivated (e.g., knockout mutants) to the investigation of the adaptation of the cells to different environmental conditions (e.g., time series). Microarrays can be used in pharmaceutical studies to obtain information for the discovery and design of effective substances for medical therapy (7,16). Moreover, microarrays can be employed for diagnostic purposes, such as for the detection of dispositions for hereditary diseases (18). In combination with mass spectrometry, microarrays enable the identification of SNPs (57). For cancer research, genome-wide transcription is measured to classify tumor samples; that is, to predict the class membership of new samples (29,60).

Ongoing sequencing projects promise to yield complete gene sets for most model organisms in the near future, which can then be mounted on DNA chips. This will increase the number and diversity of the transcriptional assays that are performed.

Experimental setting. A microarray consists of DNA fragments immobilized on a solid support. Common support materials are nylon, polypropylene, or glass. The surface is often chemically treated before DNA immobilization to improve DNA binding properties. Immobilization on the support is accomplished either by spotting the cDNA fragments or by synthesizing the oligonucleotides on the surface (41,42,49–51). Transcriptional profiling with microarrays involves the following several steps. mRNA is prepared from cells growing under certain experimental conditions. For each condition, the prepared mRNA is separately subjected to reverse transcription with radioactively or fluorescent tag-labeled nucleotides. Radioactive labeling is often carried out in combination with nylon or polypropylene supports, while fluorescence-labeled targets are primarily hybridized to glass chips. Whereas radioactive labeling with more than one kind of label per hybridization has not yet been reported, fluorescent dyes of different color offer the opportunity to yield more than one channel. With two-channel fluorescent tag labeling, each hybridization involves an additional application of a differently labeled cDNA, stemming from a control condition. Subsequently, the labeled cDNA mixture is hybridized to the microarray. After the detection of the signals, image analysis programs are used to determine spot intensities (64).

The peculiarities of microarray data. We will refer to a set of conditions as a multi-conditional experiment when all hybridizations are done with reference to one and the same control condition. Data thus produced by image analysis may be regarded as a table, with each row representing a gene and each column representing an experimental condition. However, multiple measurements for each condition, involving repeated sampling, labeling, and hybridization offer the opportunity to extract more robust signals. For the simple case of one channel per hybridization and with repeatedly performed hybridizations for each experimental condition, we will call the individual data set a hybridization and represent it by a separate column in the table. One condition of a multi-conditional experiment can thus comprise several columns. However, the intensity measurements in this table must not be taken at face value. Different levels of background may result in additive offsets, or different amounts of mRNA or different label incorporation rates may lead to multiplicative distortions among the hybridizations. Therefore, the columns of the table have to undergo a normalization procedure.

Another problem is that the ratios are unreliable at low levels of intensity. For biological interpretation, it would be useful to cluster genes that show the same expression behavior but are expressed at different absolute levels. However, the majority of genes are frequently not expressed to a measurable amount, and their signals flicker with measurement noise. The set of untranscribed genes is often large enough to hold several genes for any profile under study, therefore it is advisable to filter out these genes. In spite of these measures, single gene profiles obtained by microarray hybridization are often verified by Northern blot analysis before any further investigation of the genes. Let the expression profiles be prices in a supermarket. Thus, a certain profile (e.g., a high expression level in the first condition and no abundance in the two remaining conditions) is represented by 2.49 Euro. This price will reliably identify the shampoo if only four articles have been purchased. However, a microarray experiment is more like buying the entire store. Resulting data tables are normally only a few hybridizations wide, but perhaps tens of thousands of genes long, with the shape of very long supermarket bill. Given a certain amount of noise, any gene profile will have a high probability of occurring elsewhere in the list by chance.

Therefore, a microarray experiment is of little value if it does not comprise several repetitions. Repeated measurements allow the signals to be filtered according to reproducibility measures before analysis. Investigating the reproducibility of single signals by means of repeated measurements will lead to the identification and subsequent elimination of outliers.

Given a thoroughly preprocessed data set, one expects to be able to tackle the biological questions of data interpretation. However, filtering the genes by applying the above constraints still results in large amounts of data. Furthermore, microarray data do not consist of the transcription intensities alone. Data sets should also include information about the immobilized DNA fragments and a detailed description of the experimental steps performed, which imposes large demands on both data analysis and storage. The storage should be interconnected with the analysis; that is, it should hold the data in a format suitable for computer-based analysis because visual inspection is impractical due to the large volume of these data.

Current Methods of Data Storage and Analysis

Data storage. The full potential of correspondence analysis (CA) cannot be exploited without controlling the data format. To enable the interpretation of large data sets, the data produced need to be stored in a suitable way to allow for global comparison (3). For rapid and simple access, data should be stored in a common format, such as in a database, rather than in unequally structured flat files. Database repositories provide the convenience of a consistent view, defined interfaces, and increased access performance. Built-in methods for multi-user operation and a centralized administration also enable high standards for data security. The advantages of standardized storage apply not only to the signal intensities for each item in an array but also to all available descriptions of the sample from which the RNA has been derived and all details of its treatment. Several database projects are currently addressing these questions. While ExpressDB [Harvard University, Cambridge, MA, USA (1)] attempts to integrate data from nearly all available platforms, such as cDNA, oligonucleotide chips, and serial analysis of gene expression (SAGE), a different focus has been to develop systems for a consistent description of the samples used and the genes mounted on the array [e.g., in GeneX (National Center for Genome Resources; NCGR), http://www.ncgr.org/research/genex/; GEO (National Center for Biotechnology Information; NCBI), http://www.ncbi.nlm.nih.gov/geo/; ArrayDB (National Human Genome Reseach Institute; NHGRI) (23); ArrayExpress (European Bioinformatics Institute; EBI) (9); and RAD (University of Pennsylvania; UPenn) (54)], the last one combining both objectives.

Data analysis. Most methods recently applied to microarray data fall into one of three groups, namely classification, clustering, or projection methods. Classification methods take a grouping of objects as input and delineate characteristic features common and discriminative to the objects in the groups. The characteristic features are referred to as classifiers. For new objects, the classifier can be used to determine the appropriate group. For cancer research, these objects may consist of different tumor cell lines or of tumor samples of different tumor type, stage, or grade, often supplemented by the normal tissue of the particular organ (29). Examples of classification methods range from linear discriminant analysis (26) to support vector machines (12) or classification and regression trees (CART) (10,19). Clustering allows for the investigation of genes or hybridizations that appear to be different and transcription profiles that appear to be similar. Examples of clustering techniques are *k*-means clustering (58), hierarchical clustering (22), and self-organizing maps (56). Clustering tends to be more exploratory than classification. No group affiliations have to be known in advance.

However, parameters such as the topology for the self-organizing maps or the expected number of clusters for the *k*-means clustering have to be selected. Varying parameters may result in altered output and inappropriate parameterization in uninformative results.

Projection methods produce a low-dimensional projection of an originally high-dimensional data set. One can, for example, represent genes as numerical vectors with the number of elements for each vector representing the number of hybridizations involved. Therefore, those vectors could be plotted as points in hybridization dimensional space, if only the number of dimensions was small enough for visualization. Methods such as multidimensional scaling (MDS) (5) or principal component analysis (PCA) (36,40), as well as the technique mentioned in this study, project these points into a 2- or 3-D subspace so that they can be plotted. Embedding attempts to represent objects so that distances among points in the projection resemble their original distances in the high-dimensional space as closely as possible. Examples of the above-mentioned objects are hybridizations as vectors in gene space (Figure 1). Conversely, the rows of the data table, or *n* genes, can be represented in *m*-dimensional hybridization space. MDS or PCA can be used to visualize either the former or the latter method. CA (14,25,39) and related techniques such as Gabriel's biplot (13) are able to visualize both.

A projection plot is an exploratory way to visualize the underlying structure of a data set. It shows which clusters are well separated and whether cluster borders are well defined. It also allows for the visual judgment of the number of clusters.

Interaction of Storage and Analysis

Statistical thinking, while being largely dispensable for small-scale experimental settings, is necessary for both the design and interpretation of microarray experiments (63). Large amounts of data and high levels of detail in data annotation render the visual inspection of these data impractical. They have to be investigated by automated computer-aided analysis. This emphasizes the need for computer-readable data and for appropriate methods for integrated analysis. Data warehouses reflect the change in emphasis from browsing a database and performing simple queries to integrated analysis. Whereas maxd (The Manchester ArrayExpress Database, http://bioinf.man.ac.uk/microarray/maxd) provides a variety of tools for analyzing the transcription signals and an interface for user-defined methods, GIMS (Genome Information Management System, http://img.cs.man.ac.uk/gims) (15) focuses on integrating more information about the investigated genes, such as chromosomal location or protein-protein interactions.

Current microarray database and warehousing projects focus on making these gene annotations available in addition to integrating multiple platforms and various fields of microarray experiments by means of the flexible storage of transcription intensities and consistent, often hierarchical, sample description. However, most of the valuable information contained in an experiment annotation is currently not taken into account for analysis. This is due to the fact that the annotations are stored in a way that is not readily accessible to statistical methods. The frequencies of annotation values (e.g., within a set of experiments clustered by their expression patterns) should be countable.

Figure 2 shows an example of how to determine the association of transcription patterns with experimental parameters. The figure is based on a simple, hy-

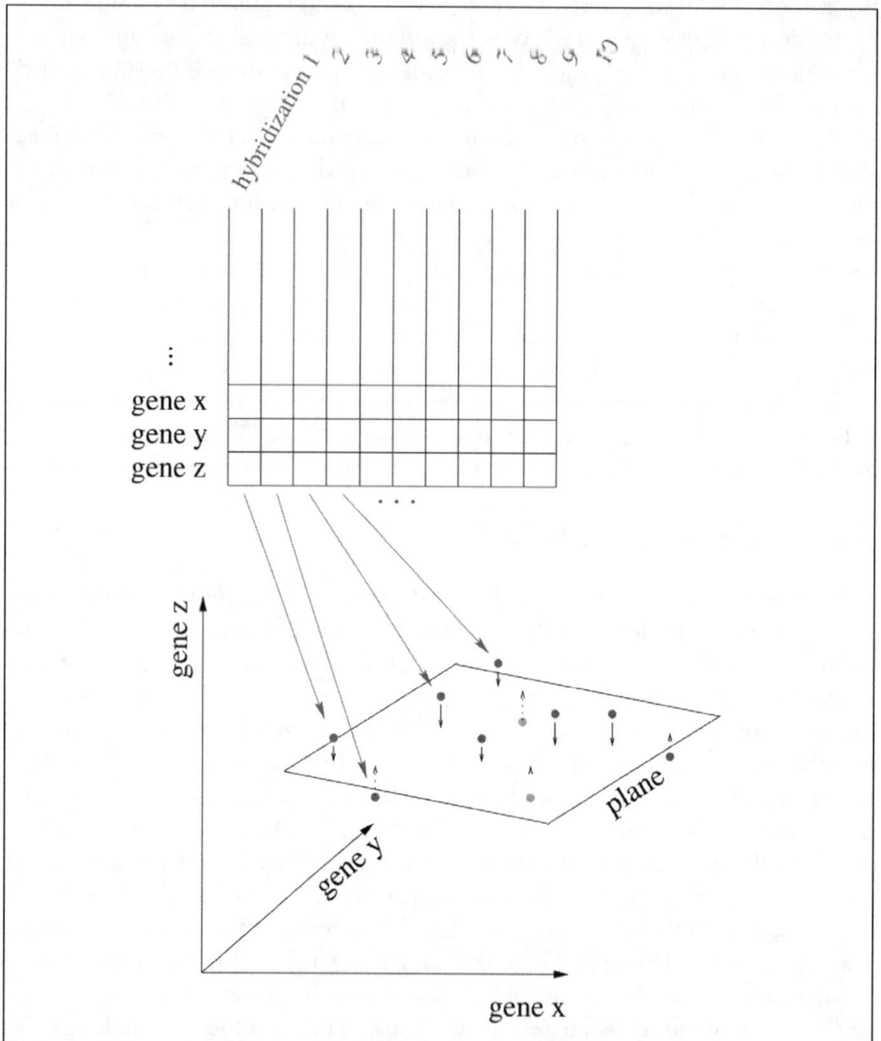

Figure 1. Planar embedding. The m columns of a table of n genes × m hybridizations are represented in n-dimensional gene space (three dimensions are shown). n ranges from a few hundred to tenths of thousands. Most microarrays comprise several thousand elements. A plane is selected such that the distance of the hybridization vectors to the plane is minimal, thus conserving point-to-point distances among these vector points as well as possible. (**See color plate A27.**)

pothetical data set of 48 microarray hybridizations. Data are depicted in a planar projection such as a MDS or PCA plot. Are there experimental parameters that are characteristic for the cluster of hybridizations in the upper left-hand corner?

Instead of using the χ^2 test statistic to determine which annotation values are characteristic, let's consider a simple way to access these associations. Consider the yeast specific enumeration-type annotation "growth phase," which can take three different values, namely "exponential," "stationary," or "pseudo-hyphal." The corresponding hybridization data points are drawn as rectangles, hexagons, and triangles, respectively. Focusing on the triangles, one can count their frequency in the encircled hybridization cluster, which is 1/2 (5 of 10), as well as in the entire set (8/48 = 1/6). Dividing the first by the second frequency suggests a 3-fold over-representation of the pseudo-hyphal value in the selected cluster. In the same manner, all values of all annotations can be scanned for being charac-

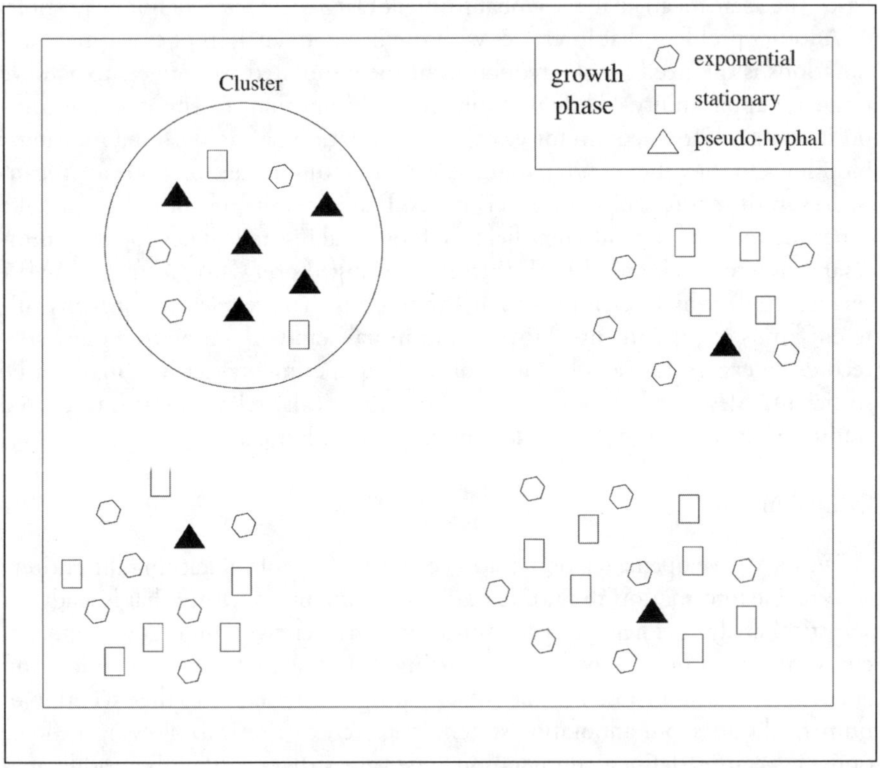

Figure 2. Correlation between transcription pattern and experimental parameter. The scheme—a planar projection (Figure 1) of virtual data—illustrates a simple method to determine the frequencies of annotation values for a cluster of hybridizations under study. The dimensions of the axes correspond to the dimensions of the axes of the original space. Triangles are filled to facilitate counting. All values of all annotations can be scanned for being characteristic; that is, over-represented or underrepresented in a hybridization cluster, thus enabling computer-based analysis of large and complex data sets. However, these values have to be countable to determine their frequencies inside and outside the cluster.

teristic, i.e., over-represented, or underrepresented in a hybridization cluster, thus enabling the automated analysis of large and complex data sets. The resulting (characteristic) experimental parameters are candidates for explaining the cluster formation; that is, they are candidates for being the active players that drive the cells to the observed transcriptional state.

Simple as it may be, this method already provides good analytical access to long lists of annotations and huge sets of hybridizations, which could hardly be evaluated by visual inspection. While this method is simple and easy to explain, statistical tests would certainly be better suited to this task. However, any statistical analysis will require that the variables (annotations) are of categorical range and that the instances of occurrence can be counted for any annotated value. Misspellings, different textual representations of semantically identical items, and, vice versa, ambiguous words whose meanings depend on the context interfere with counting such values. With these limitations to access for computer-based (i.e., statistical) analysis, global studies of large data sets will not be possible.

Another problem that interferes with the in-depth analysis of experiment annotations is the need for the reduction of the annotated parameters to achieve universal applicability. While it is suitable and important to annotate the composition of culture medium for yeast, this annotation has no meaning for cancer biopsies. On the other hand, parameters such as tumor stage or grade are meaningless in the context of yeast experiments. Public data repositories have to take experiments from very different fields of biological research. They focus on minimal standards such as MIAME (http://www.mged.org/Workgroups/MIAME/miame.html), which resemble sets of annotations that are relevant for many different kinds of experiments. However, minimal-standard annotation inevitably records an experiment at a low level of detail. Experimental settings may not be sufficiently described to detect artifacts or biologically relevant but unexpected transcriptional responses to subtle experimental variations.

Contributions

We have developed a relational storage concept capable of keeping the entirety of available information about microarray experiments in a form that is ready for statistical analysis. The concept ensures the consistency of annotations and circumvents the difficulties of free-text parsing. It is designed to permit annotation at any level of detail chosen by the collaborating biologists. With the exception of numerical values, our annotation system is entirely categorical, allowing a choice only between predefined enumeration-type values that can then be readily analyzed in an automated fashion.

Annotation storage should be flexible enough to allow the easy inclusion of new attributes and new values. We store the definitions for the annotations and their allowed values as separate tables in the database, thus avoiding a fixed, "hard-wired" structure that would be difficult to extend. Storing these definitions as table content rather than attributes (table columns) enables them to be extended without changing the database structure or adjusting the analyzing algorithms.

We have developed a storage and analysis concept called the multi-conditional hybridization intensity processing system (M-CHIPS) (24). It has been implemented as a set of organism-specific databases, such as for *Saccharomyces cerevisiae, Arabidopsis thaliana, Trypanosoma bruceï, Neurospora crassa,* and human tumor samples. While differing in the annotations used to describe the samples, these databases share a common structure and are thus accessed by the very same analysis algorithms. The concept is able to integrate all types of intensity data obtained from cDNA microarrays. It has been tailored to the needs of the collaborating groups that use cDNA microarrays with either single-channel radioactive or multichannel fluorescence read-outs.

We have also incorporated a method particularly suitable for the detection of interdependencies not only among one set of variables, such as the genes, but also between any gene and hybridization under study. CA is an exploratory technique that allows for the visualization of structures within the data and thus reveals which questions could be asked or which hypotheses could be asserted. Unlike many other methods, CA does not require any prior choice of parameters.

Similar to other projection methods, CA represents variables, such as transcription intensities of genes as vectors in a high-dimensional space. In our case, the dimensionality of the space would be the number of hybridizations involved. Both PCA and CA reveal the principal axes of this high-dimensional space, enabling the projection into a subspace of low dimensionality that accounts for the main variance in the data. Unlike PCA, CA is able to account for the genes in hybridization-dimensional space and the hybridizations in gene-dimensional space at the same time. Both representations of the data matrix will be projected into the same low-dimensional subspace, such as a plane (yielding a so-called biplot), revealing associations both within and between these two variables. We will show the general applicability of CA to microarray data analysis in examples, both with radioactive and multichannel data.

TRANSCRIPTIONAL PROFILING WITH CA

Preprocessing of Hybridization Intensities

Before high-level analysis, data have to be normalized and filtered. In M-CHIPS, preprocessing starts with the normalization of raw signal intensities. Different levels of background may result in additive offsets, or different amounts of mRNA or different label incorporation rates may lead to multiplicative distortions among the hybridizations. The normalization is based on robust affine-linear regression (4,25) and is carried out in two different ways, depending on whether the multi-conditional experiment comprises mono-channel or multichannel data (24). While this approach does not rely on any model assumptions, an example for a model-based normalization is given by Huber et al. (37). CA relies on the proportions among the data rather than on absolute values. Thus, it is equally important to tackle the problem of unduly high ratios in the low-intensity region. To reduce the impact of low intensities on the intensity ratios, the normalization method previously described (4) has been modified, additively shifting the normalized matrix back to its original expression level. To exemplify the benefit of simply adding a certain number to all of the values, consider that a shift from 0.02 to 0.04 resembles up regulation by a factor of two, whereas a change from 1000.02 to 1000.04 does not.

It is also advisable to select genes that fulfill the following criteria: (*i*) considerable absolute expression level in at least one of the conditions; (*ii*) substantial change relative to the control condition in at least one of the other conditions; and (*iii*) reproducibility in the separation from the control condition in at least one of the other conditions (4,24,25). For the analysis of the preprocessed data, we implemented hierarchical clustering (22) and CA (25). While the former method yields easy-to-interpret output and is widely used for microarray data analysis, it poorly visualizes important details. Moreover, it is of limited use for the integrated analysis of more than one variable and, therefore, is not useful for the investigation of interdependencies among genes and measurements intended here. In contrast, CA provides the capability to simultaneously visualize more than one variable.

Correspondence Analysis

Here we provide a concise summary of the technique. See References 30 and 32 for a thorough exposition. An informal, intuitive description will be given below. The aim is to embed both rows (genes) and columns (hybridizations) of a matrix in the same space, the first two or three coordinates of which contain most of the information. Let I genes and J hybridizations be collected into the $I \times J$ matrix N with elements n_{ij}. Let n_{i+} and n_{+j} denote the sum of the ith row and jth column, respectively. By n_{++}, we denote the grand total of N. The mass of the jth column is defined as $c_j = n_{+j}/n_{++}$, and, similarly, the mass of the ith row is $r_i = n_{i+}/n_{++}$. The basis for the calculation is the correspondence matrix P with elements $p_{ij} = n_{ij}/n_{++}$, from which the matrix S with elements $s_{ij} = (p_{ij}-r_i c_j)/\sqrt{r_i c_j}$ is derived. S is submitted to singular value decomposition (28); that is, it is decomposed into the product of three matrices: $S = U\Lambda V^T$. Λ is a diagonal matrix, and its diagonal elements are referred to as the singular values of S. We think of them as sorted from the largest to the smallest and denote them by λ_k. The coordinates for gene i in the new space are then given by $f_{ik} = \lambda_k u_{ik}/\sqrt{r_i}$ for k = 1,...,J. Hybridizations are viewed in the same space, with hybridization j given coordinates $g_{jk} = \lambda_k v_{jk}/\sqrt{c_j}$ for k = 1,...,J. These coordinates are called principal coordinates.

To reduce dimensionality, only the first two or three coordinates of the new space are plotted. The loss of information associated with this dimension reduction is quantified in terms of the proportion of the so-called total inertia $\Sigma_k \lambda_k^2$ that is explained by the axis displayed. Total inertia is proportional to the value of the χ^2 statistic, and thus the amount of information represented in, for example, a planar embedding $(\lambda_1^2 + \lambda_2^2)/\Sigma_k \lambda_k^2$, corresponds to the proportion of the χ^2 statistic explained by the embedding.

The above summary aims to provide all the information needed to implement a simple CA algorithm. This can be easily done by using nested for loops. A much shorter implementation without loops can be achieved in any programming language that supports matrix multiplication and provides a routine for singular value decomposition (e.g., in MATLAB™; see Appendix).

Standard Coordinates as an Aid in Visualization

CA attempts to separate dissimilar objects (genes or hybridizations) from each other; similar objects are clustered together, resulting in small distances. In contrast, the distance between a gene and a hybridization cannot be directly interpreted. For visualization of between-variable association in the plot, one includes virtual genes that have all their intensity focused in one hybridization (32). The coordinates of such a gene are called the standard coordinates of the hybridization where the gene is expressed. Similarly, one could introduce standard coordinates for genes. The standard coordinates for the genes are computed as $u_{ik}/\sqrt{r_i}$ and for the hybridizations as $v_{jk}/\sqrt{c_j}$. In practice, the spread of the set of real genes and hybridizations is much smaller than the spread introduced when one includes these virtual genes and hybridizations via their standard coordinates. As a consequence, the real points would shrink to a tiny area, so we rather depict the direction from the centroid of the data to the standard coordinates instead of the standard coordinates themselves.

Medians and Replicate Hybridizations in CA

Typically, replicate hybridizations are performed for each condition under study, which leads to several values for one gene-condition pair. The number of such repeated hybridizations is often small. Therefore, we represent these values by their gene-wise median rather than their gene-wise average because the median is less sensitive to outliers. The need remains, however, to visualize the original data and not only the median because these data contain valuable information about the experimental variance and quality of individual hybridizations. In fact, CA offers the possibility to reflect both aspects. To this end, CA is first effected by using the gene-wise medians, determining the coordinate system to embed the original hybridization intensities. These data points are then referred to as supplementary points or points without mass. Thus, the share of noise belonging to an experimental condition is shown by the spread of its hybridizations around the median. Because the dimensions of the data are reduced by using the medians of hybridizations per experimental condition, we refer to this strategy as hybridization-median determined scaling (HMS).

The embedding for hybridizations without mass is computed as follows. Let the matrix N contain only the hybridization medians and let N* of elements $n^*_{ij'}$ be the original data matrix containing all the hybridizations. N is submitted to CA. Let P have elements $p^*_{ij'} = n^*_{ij'}/n^*_{++}$. The principal coordinates for the supplementary hybridizations from correspondence matrix P* are then calculated as

$$g^*_{j'k} = \frac{1}{\sum_i p^*_{ij'}} \sum_i \frac{p^*_{ij'} f_{ik}}{\lambda_k}$$

In our own data sets, a single hybridization consists of two corresponding spot sets because each cDNA had been spotted twice on the array. We refer to these spot sets as primary and secondary spots. They tend to show a higher correlation than hybridizations that belong to the same experimental condition. Plotting them separately (duplicating the number of supplementary points) provides an atomic unit of distance in the biplot, where no units are assigned to the axes. The intensity unit cancels out when calculating the correspondence matrix P.

Interpretation of a CA Biplot

CA was originally developed for contingency tables and is intimately connected with the χ^2 test for homogeneity in a contingency table. The question asked by this is whether the differences among the rows (or columns) of the table are large enough to reject the hypothesis that the rows (columns) are homogenous. In other words, it is asked whether the discrepancies between the observed rows and an average row profile expected for the homogenous case are so large that they are unlikely to occur by chance alone. The question is answered by computing a measure of discrepancy between all the observed and expected values: the difference between observed and expected value is squared and subsequently divided by the expected value. The result is referred to as the χ^2 distance between the observed and expected value. These distances, calculated for all elements of the table, sum up to the χ^2 statistic. The value of the χ^2 statistic is high when there is an association between the rows and columns of the table. In CA, points are depicted such that the sum of the distances of the points to their centroid (called total inertia) is proportional to the value of the χ^2 statistic of the data table. The farther a point is away from the centroid, the higher is its row's contribution to the value of the statistic. In this sense, CA decomposes the overall χ^2 statistic. Distances among points are not meant to approximate Euclidean distances but rather the χ^2 distance. This distance is low when the profiles of two vectors show similar shapes, regardless of their absolute values.

Together with the row points, CA displays points representing columns using the same χ^2 criterion. This also establishes the link between the row and column points. If a column determines an outstanding entry of a row (and vice versa), then the corresponding row and column points tend to lie on a common line through the centroid. For a positive association, the two points will lie on the same side of the centroid, with the distance to it larger the stronger the association is. A negative association will cause the column-point and row-point to lie on opposite sides of the centroid. The following example demonstrates the practical use for microarray data analysis. For clarity, a hypothetical example has been constructed. Figure 3 shows virtual profiles for 24 genes, with transcription intensity plotted versus experimental condition. Let each condition comprise two repeated hybridizations.

CA will decompose the profiles corresponding to their association to the hybridizations under study. Figure 4 shows the expected planar projection. The genes are depicted as black dots, and the hybridizations are depicted as boxes, which are color-coded according to the experimental condition to which they belong. To illustrate the locations of the genes, each gene cluster is shown together with the corresponding gene profiles.

The following properties of this type of plot are useful for its interpretation.

(*i*) Hybridizations showing a high similarity in expression profile; for example, if they belong to the same experimental condition, with a short distance in the 24 dimensional gene space, they will therefore be neighbors in the projection. (*ii*) Genes with high intensities in a condition are located in the direction of this condition. The two genes located in the direction of the blue condition (upper right-hand corner) are both up-regulated, particularly in the blue condition. (*iii*) Genes particularly down-regulated under this condition are located at the opposite side of the centroid. One can regard this gene (lower left-hand corner) as being down-regulated in the blue condition. Another valid interpretation is that it is located in the direction of the bisection line between the red and the green conditions because it is equally abundant in these two conditions. (*iv*) All genes with unchanged expression or those not expressed to a measurable amount in any of the conditions under study are located near the centroid. For experiments with comprehensive or complete gene sets (i.e., sets not particularly selected for high expression), the genes that are not detectable will be the majority (61). The CA plot will show a centric cloud of many genes that lack significantly changed expression throughout the experiment. The outer regions of the plot will contain the so-called differential genes. Their distance to the centroid will reflect the significance of displaying differing expression from the "average" ones in terms of χ^2 statistics, which are placed at the center of the plot.

These items are sufficient for the correct CA plot interpretation of most real data examples. However, the second item, claiming common directions for conditions and associated genes is not mathematically correct and has to be seen more as a guideline. It can be proven that the directions of a condition and its highest possible associated gene never coincide exactly. To properly visualize the associations between the conditions and genes, virtual genes that are fully concentrated on one condition will be used. They serve as representatives of the hybridizations in gene space (see standard coordinates). Because they show the highest possible association to the corresponding conditions, they are located far away from the centroid. Plotting them directly would cause all other points representing genes and hybridizations to be located in a small centric area. We will represent them as lines from the centroid to their coordinates (see Figure 6), with the latter lying outside the plot margins after zooming into the area of conditions and "real" genes.

Multichannel Data Example—Analysis of a Well-Known Data Set

To introduce the method, its performance is demonstrated on a well-known data set. This set comprises the hybridizations referred to by Spellman et al. (53), which are publicly available at http://genome-www.stanford.edu/cellcycle/data/rawdata/combined.txt. Spellman et al. arrested the *S. cerevisiae* cell cycle by

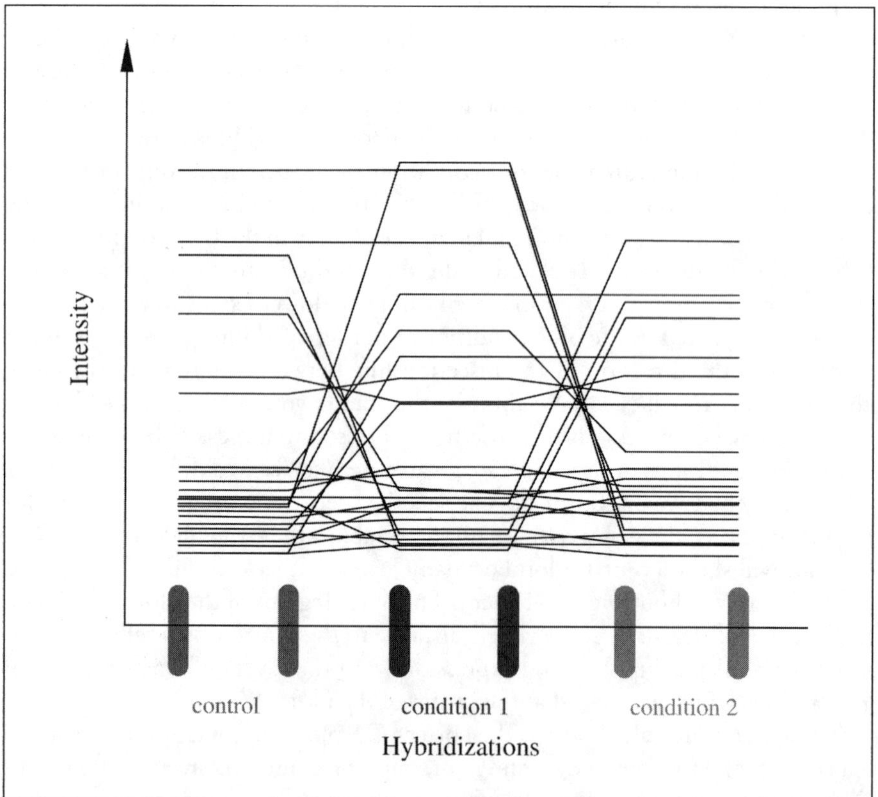

Figure 3 Virtual data example. To demonstrate the interpretation of a CA biplot, we constructed a data example. It resembles real data in that the majority of the genes are lowly or not transcribed to a measurable amount. It comprises only 24 genes and differs from the real world in perfect reproducibility among the two hybridizations of each experimental condition. **(See color plate A28.)**

four different methods, namely α factor-, *CDC15*-, and *CDC28*-, based blocking, and elutrition. We refer to these four methods as alpha, cdc15, cdc28, and elu, respectively. At certain time points after releasing the block, samples from each of the methods were drawn, their cell cycle phase was classified, and their transcriptional status was assayed by microarray hybridization.

The data consist of two-channel fluorescence signals. Following the authors, we based our analysis on the log ratio of the intensities of the two channels. To make the data analyzable by CA, they were additively shifted to a positive range. In our analysis, we gave mass to all hybridizations instead of applying HMS. The standard coordinates of the hybridization medians, on the other hand, were computed without mass and are depicted as lines emanating from the centroid. We

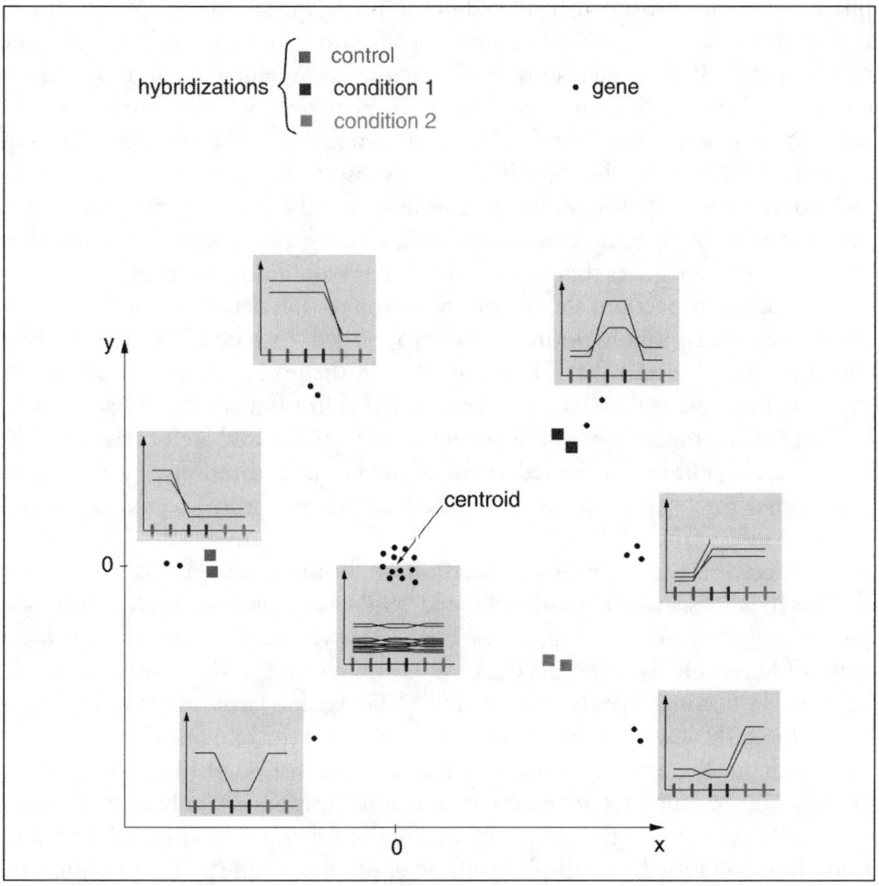

Figure 4. Planar projection by CA. The plot resembles output that can be expected for the virtual data example shown in Figure 3. It was constructed to demonstrate the properties of such projections more clearly than would be possible by showing a single plot of real data. Gene clusters are shown together with the corresponding gene profiles. The abscissa represents the first principal axis, and the ordinate represents the second. Both axes are dimensionless. (**See color plate A29.**)

analyzed the 800 cell cycle-associated genes depicted in Figure 1 in Reference 53 over all 73 hybridizations. To allow for direct comparison, part B of the original figure, which also marks the histone gene cluster, is shown here in Figure 5. The CA biplot is shown in Figure 6, with hybridizations color-coded according to their phase assignment and following the color code of the original figure.

The planar embedding produced by CA (Figure 6a) shows the hybridizations clearly separated according to their cell cycle phase. They are arranged in the circular order of correct sequence. The lines that denote the direction of the hybridization medians emphasize this arrangement. The black dots correspond to genes. Genes that show strong expression in a certain phase are located in the direction determined by the hybridizations of this phase. The farther away from the center the genes are, the more pronounced their association is with that phase. Genes that are down-regulated in this phase appear on the opposite site of the centroid. As an example of a strong association with the S-phase, the gene profiles for the histone gene cluster, also marked by Spellman et al. (53), are encircled in black. Their profiles are shown in Figure 6b, which is further subdivided according to the method of cell cycle arrest that had been used. The red-encircled genes will be discussed in the next subsection in the context of *CDC14* induction. Genes equally transcribed in most or all of the cell cycle states had been removed by Spellman and coworkers, causing a hole near the centroid of the CA plot where genes that show little change would otherwise lie.

Upon close inspection, the biplot reveals interesting details about the data. It should be noticed that hybridization cdc15_30 (cdc15-based blocking, 30-min time point) classified as M/G1 (yellow) lies in the green (classified G1) sector rather than in the yellow sector. Likewise, hybridization cdc15_70 is classified G1 but clusters together with the purple dots (S-phase), and an S-phase hybridization, cdc15_80, lies in the red sector of the G2 hybridizations. All these outliers come from the series of hybridizations where the cell cycle arrest was achieved using *CDC15*-based blocking. This arrangement of cdc15 hybridizations suggests an improper phase classification for these samples.

This hypothesis can be validated based on the gene profiles. For the histones, the shift toward an earlier stage in the cell cycle is visible in the upper right-hand panel of Figure 6b. Time points cdc15_30 through cdc15_90 show the up-regulation of the histones already at the end of M/G1 (yellow) instead of G1 (green) as well as too early down-regulation: the curves intersect the zero line (identity to the control channel) at cdc15_90, classified as G2 (red) instead of M (brown), as, for example, in the elutrition experiment. The nine histones are only a small subset of the 800 cell cycle-regulated genes. Profiles of other genes, although different from the ones plotted, also display shifting of the above time points to expression patterns associated with an earlier state in cell cycle by the remaining time points (data not shown). CA computes the projection for time points cdc15_30 to cdc15_90 according to their expression patterns in the entirety of the gene set, independent of their phase classification. Figure 6a displays them displaced in a

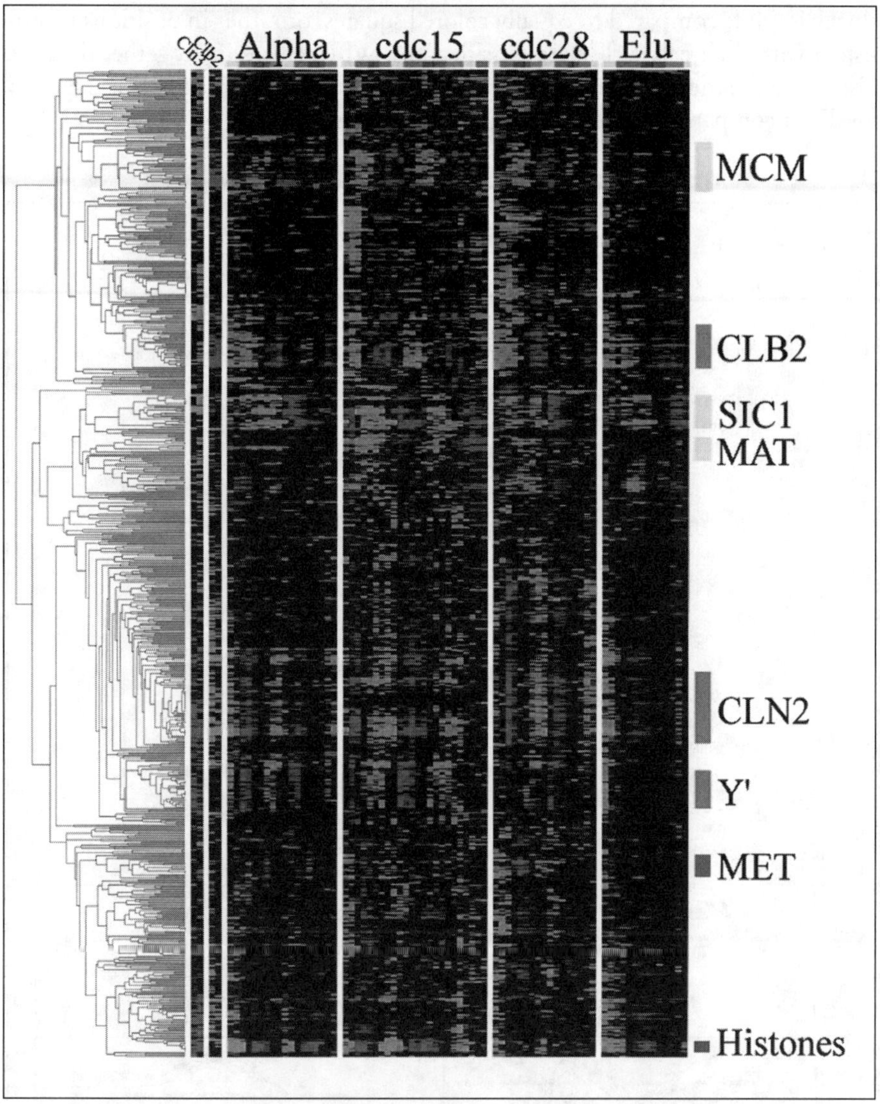

Figure 5. Gene expression during the yeast cell cycle. This figure has been reproduced from Reference 53. Genes correspond to rows, and the time points of each experiment are represented by columns. The ratio of induction/repression is shown for each gene so that the magnitude is indicated by the intensity of the colors displayed. If the color is black, then the ratio of control to experimental cDNA is equal to one, whereas the brightest colors (red and green) represent a ratio of 2.8:1. Ratios greater than 2.8 are also displayed as the brightest color. In all cases, red indicates an increase in mRNA abundance, whereas green indicates a decrease in abundance compared to the control samples (stemming from asynchronous cultures of the same cells growing exponentially at the same temperature in the same medium). Gray areas (when visible) indicate absent data or data of low quality. Color bars on the right indicate the phase group to which a gene belongs (M/G1, yellow; G1, green; S, purple; G2, red; and M, brown). These same colors indicate cell cycle phase along the top of the figure. Genes that share similar expression profiles are grouped. The dendrogram on the left shows the structure of the cluster. Reprinted from Molecular Biology of the Cell, (1988 Dec; 9(12):3273-3297), with permission by the American Society for Cell Biology. **(See color plate A30.)**

clockwise shift compared to equally colored squares, or rather in positions inconsistent with their cell cycle state classification. While clustering together the nine histone genes, the original figure by Spellman et al. (53) (printed here as Figure 5 for direct comparison) does not properly show this shift.

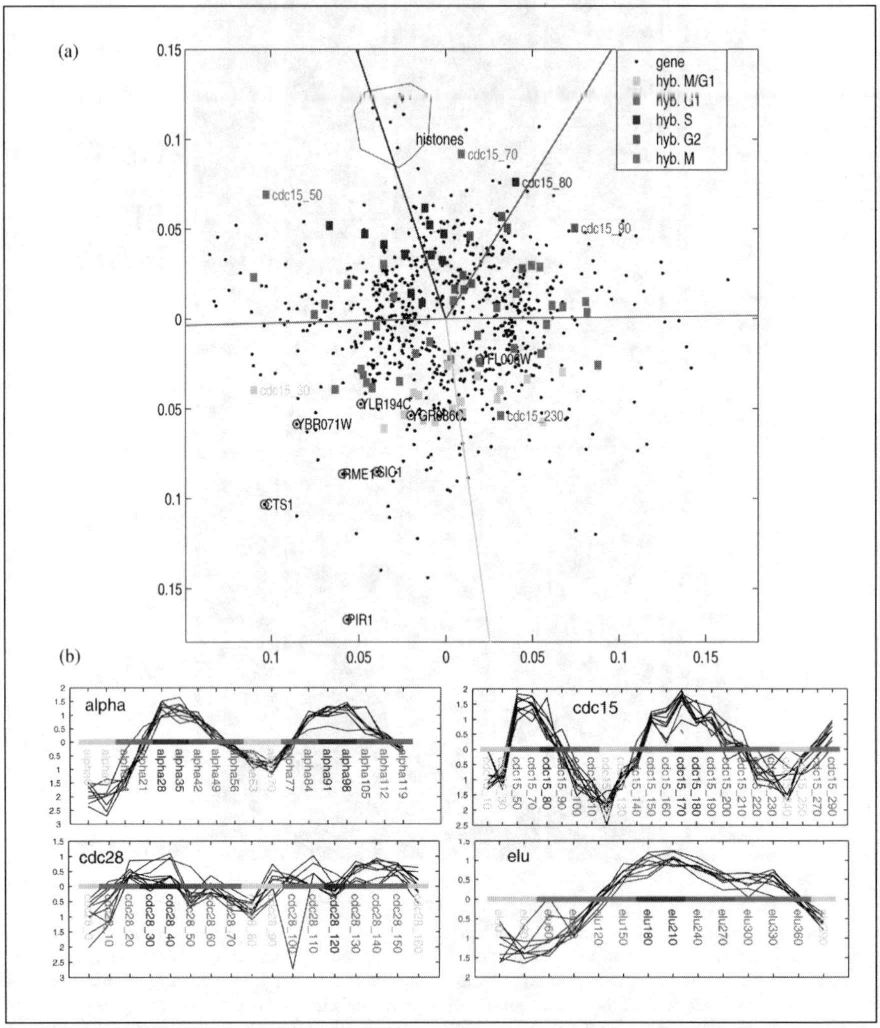

Figure 6. Cell cycle synchronization data by Spellman et al. The data set composed of 800 cell cycle-associated genes has been projected by CA as is. No HMS has been employed so as not to bias the resulting plot in terms of the separation of the cell cycle phases. The outlying hybridizations have been identified to be caused by a slight phase shift of the cdc15-based synchronization visible in the upper right-hand panel of Figure 6b, which shows the profiles of the nine histone genes *HHF1, HHF2, HHT1, HTB2, HHT2, HTB1, HTA1, HTA2,* and *HHO1* that are encircled in black in Figure 6a (see text for details). The abscissa represents the first principal axis, and the ordinate represents the second. Both axes are dimensionless. Reprinted from Fellenberg et al. (25), Copyright (2001) National Academy of Sciences, USA. **(See color plate A31.)**

Monochannel Data Example—Overexpression of *CDC14*

Instead of following the cell cycle through S-phase, G2, mitosis, and G1, this experiment focuses on the transition from mitosis to G1. In late mitosis, mitotic cyclin-dependent protein kinases have to be inactivated to exit mitosis. Cdc14p plays a major role in this transition to G1 because it is a dual specific phosphatase. In this experiment, cells were arrested in the mitotic meta phase, and *CDC14* was overexpressed by the induction of the controlling *GAL1*-promoter. Thus, one cannot observe directly the effect of *CDC14* overexpression because it will be overlaid with the gene expression changes due to the presence of galactose. To subtract for these effects, wild-type and transgenic strains were grown under repressing and inducing conditions, which lead to four samples that were subjected to array hybridization: wild-type without galactose; wild-type with galactose; transgenic yeast without galactose (no induction of *CDC14*); and transgenic yeast with galactose.

In contrast to the previous example, CA was used to analyze transcription intensities instead of two-channel ratios. The data, which consisted of three to four hybridizations per condition, were normalized. Genes (1400 of 6100) were extracted for being reproducibly differential (positive minmax separation) from the control condition (wild-type strain without induction) in at least one of the other conditions. The measurement noise was further reduced by HMS. Planar embedding by HMS (Figure 7) explains 80.8% of the total inertia, compared to 52.5% in the case of embedding all the hybridizations separately (data not shown). Hybridization medians are represented both in principal coordinates and as lines to their standard coordinates. The actual hybridizations, each separated into primary and secondary spot sets, are drawn as supplementary points. They are represented only in principal coordinates, as are the genes.

The biplot (Figure 7) clearly shows four directions corresponding to the four conditions. Genes in the direction of galactose-induced transgenic yeast are those specifically up regulated upon *CDC14* induction, as opposed to genes activated by galactose that are also in the wild-type strain, like *GAL1* and *GAL7*. This subtraction was achieved purely computationally and was based on the provision of galactose-activated genes in wild-type as a separate condition. The set of genes associated specifically to the Cdc14p overproducing condition comprises *CDC14* itself in addition to *SIC1*, which is known to accumulate in a Cdc14p-dependent fashion (44), and *CTS1*, which belongs to the cluster of *SIC1* co-regulated genes (53). *RME1*, *CRH1*, and *PST1* are known to be cell cycle-regulated with peaks in the mitosis/G1 transition, G1, or late G1, respectively, but have not yet been described in association with Cdc14p activity. *YBR071W*, *PIR1*, *YGR086C*, *YLR194C*, and *YFL006W* have not been annotated to be cell cycle-regulated, but these results show that they are. This is in agreement with the data of Spellman et al. (53) (see Figure 6, genes marked by red circles) that show these genes to be transcribed during the mitosis/G1 transition. The role of the nuclear pore protein GLE2 in a Cdc14p activation context remains unclear.

CA versus Other Methods

CA is an exploratory computational method for the study of associations between variables. For many other methods, unequal choice of parameters, such as the number of clusters, will focus on different properties of the transcription profiles and may therefore produce unequal results for the same data. CA does not require any choice of parameters. Similar to PCA, it displays a low-dimensional projection of the data (e.g., into a plane). Thus, it visualizes clusters of data points,

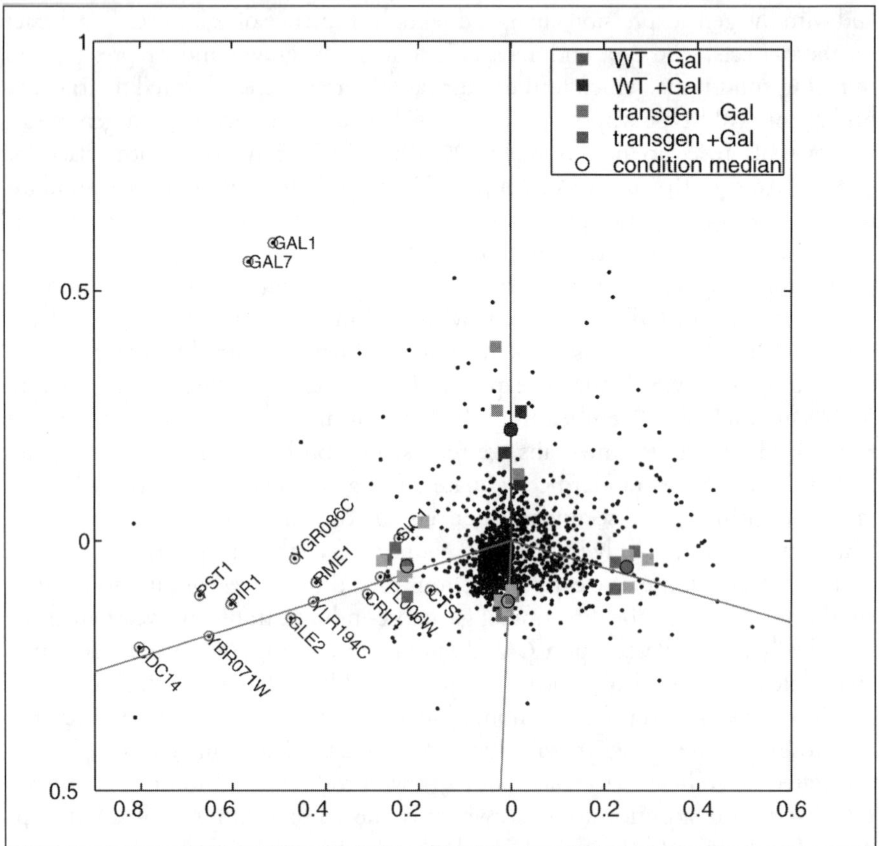

Figure 7. Overexpression of Cdc14p. In arrested yeast cultures, Cdc14p expression was induced under control of the *GAL1* promoter and investigated in comparison with uninduced transgenic, uninduced wild-type, and inductor-exposed wild-type cells. From three to four hybridizations were performed for each experimental condition. Both spot sets of the array are drawn separately for each hybridization, primary and secondary spot sets are depicted in light and dark colors, respectively. The conditions are color-coded, and their hybridization medians are marked according to the legend in the upper right-hand corner. Lines are drawn in the direction of the standard coordinates of the condition medians in appropriate colors: genes such as *GAL7* and *GAL1* are associated with both the wild type and the transgenic strain grown in the presence of galactose to an equal share. *CDC14* is associated with the induced transgenic strain only. The abscissa represents the first principal axis, and the ordinate represents the second. Both axes are dimensionless. Reprinted from Fellenberg et al. (25), Copyright (2001) National Academy of Sciences, USA. WT, wild-type; transgen, transgenic. **(See color plate A32.)**

which also shows how fuzzy or defined cluster borders are. However, it does this for two variables simultaneously, thus revealing associations between them. Visualization using CA is based on plotting the χ^2 distance among genes and among hybridizations, representing a decomposition of the value of the χ^2 statistic. Emphasis is placed on the genes and hybridizations that contribute to this value through their association. In this respect, it resembles the doubly sorted hierarchical clusterings (45), although our examples demonstrate that CA is capable of revealing intricate detail, such as subtle discrepancies between the phase classification and transcription pattern of the hybridizations. While hierarchical clustering reflects associations among the genes and also among hybridizations, CA is, in addition, more suited for visualizing interdependencies between the two variables.

The emphasis on the association between genes and hybridizations also distinguishes CA from other embedding methods such as PCA or MDS, although these methods share the idea of representing objects in a 2- or 3-D space that can be visualized. While CA and PCA use the same mathematical machinery for dimension reduction and visualization; namely, singular value decomposition, their difference stems from the different distance measures used. This also discriminates CA from Gabriel's biplot, which, while also visualizing both the genes and hybridizations, aims to approximate the values of the original data matrix by the scalar products of their plotted column and row vector points.

Alter et al. (2) successfully applied singular value decomposition to the analysis of the same data set that we used as our first example. In our plots, the distance of a given gene from the centroid represents the strength of its association with a hybridization lying in the same direction and vice versa. A direct comparison between the phase and radius in the visualization of Alter et al. (e.g., http://genome-www.stanford.edu/SVD/PNAS/Datasets/Sort_Elutriation.txt) shows that this is not necessarily the case in the singular value decomposition alone (data not shown). Moreover, CA does not depend on model assumptions, as demonstrated in our second example. There genes were identified whose transcripts are specifically up-regulated in an overexpressing mutant yeast strain that is induced by galactose, whereas they are at normal levels (or even undetectable) in this mutant strain without galactose and similarly in wild-type, regardless of the additive. Thus, we find CA to be generally applicable to and particularly well-suited to gene expression data because of its ability to simultaneously display genes and hybridizations as well as the strength of their association.

INTEGRATION OF GENE AND EXPERIMENT ANNOTATIONS WITH CA

As discussed earlier, it is possible to visualize associations both among and between genes and hybridizations with CA. However, these associations extracted from gene × measurement tables need biological interpretation to become meaningful results. Both genes and measurements can be annotated by biologically relevant information. Much like the intensity or ratio data themselves, these annotations are not suited to visual inspection due to their high number. While they are usually reduced to a few putatively important parameters, at the risk of overlooking something unexpected, we want to keep all of them. Efforts to store complex and detailed experiment annotations in a statistically accessible form—as described in the next section—have to be seen as prerequisites for integrating as much information into the analysis as possible.

General Interconnectivity among Different Visualization Plots

Analysis techniques provided by M-CHIPS include hierarchical clustering (22), CA (25), and statistical analysis of experiment annotations for arbitrary sets of hybridizations, such as those clustered by similar expression profiles. The comparison of different visualizations of a data set is facilitated by highlighting data points that have been selected in another plot. It is also possible to mark all genes bearing a certain keyword like "cell cycle" in their gene annotation or to import multiple sets of gene tags from text lists. In the CA plot, several disjoint gene sets can be visualized by different colors to highlight different functional categories or to mark interesting clusters of genes. For the latter, gene sets can be selected by encircling them by mouse clicks. Expression profiles of marked genes can be displayed in a parallel coordinate plot.

In the same manner, clusters of measurements can be selected. They are automatically scanned for significant experiment annotation values. For each value of every annotation, instances of occurrence are counted. For a particular value, its frequency in the cluster is determined as the number of its occurrences in the cluster divided by the number of measurements in the cluster. Comparison to its frequency in the whole set of measurements under study reveals whether it is over-represented or underrepresented in the cluster, as illustrated in Figure 2. An example is presented in the following subsection.

Characterization of Measurement Clusters by Experiment Annotation Scan

A time course has been recorded for wild-type *S. cerevisiae* cells under oxidative stress by 0.2 M hydrogen peroxide. Data have been preprocessed and visualized by CA (Figure 8). Experimental and computational details are given in Figure 9 and in the supplemental material to Reference 24 (http://www.dkfz-heidelberg.de/tbi/services/mchips/supplement_bioinformatics.html).

The plot comprises both genes and measurements. The genes are depicted as black dots. Measurements are shown as squares, color-coded according to the experimental condition to which they belong. There is one outlying cluster of measurements that belongs to the 30-min time point (pink), whereas other measurements of the very same condition are located in a distant area, clustering with other time points. Selecting these outliers, searching for at least 2-fold over-represented or underrepresented annotation values results in values that belong to only eight of 111 annotations (Table 1).

These annotations are possible candidates to explain the cluster formation. Some can be excluded when one considers their meaning in the experimental context. The annotation "incubation period" records the time points, and "temporary additive" describes whether or not hydrogen peroxide was present in the growth medium, both only reflecting that the selected measurements belong to the 30-min time point. "Label incorporation rate" and "total activity" of the incorporated label can also be disregarded for the characterization of the cluster because the values annotated for the measurements in the cluster show up in the mid-range for both annotations in Table 1.

The absence in the cluster of a particular researcher (annotation"experi-

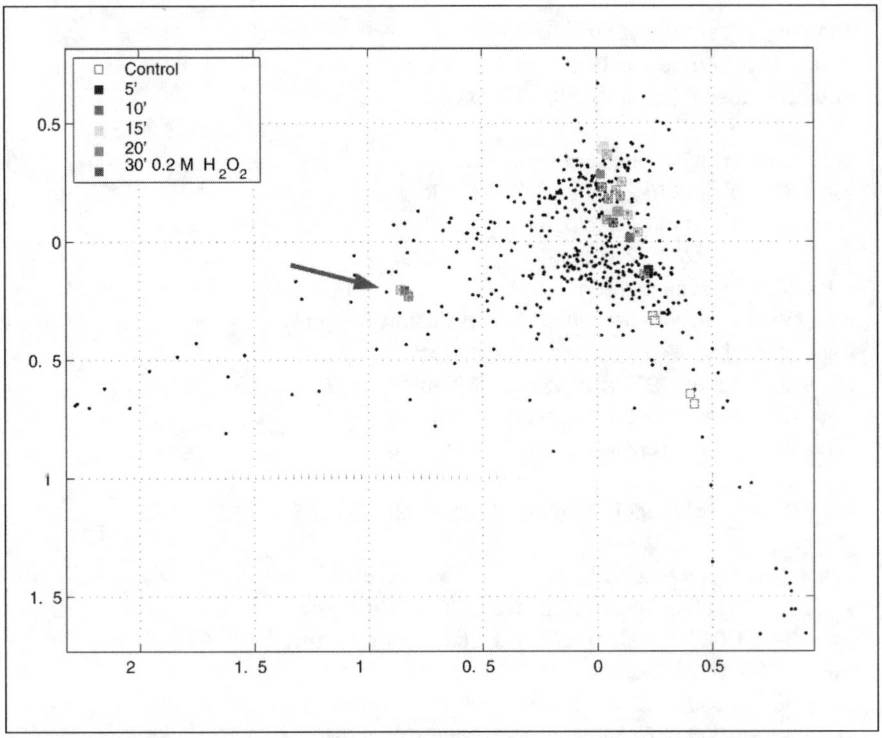

Figure 8. Oxidative stress. The CA plot shows a data set recorded from wild-type yeast cells responding to 0.2 M hydrogen peroxide in the medium. Genes are depicted as black dots, and measurements (in this case mono-channel hybridizations) are shown as squares that are color-coded according to the experimental condition (here time point) to which they belong (see text for details). The outlying cluster of pink-labeled measurements (red arrow) is further characterized by experiment annotation values that are over-represented or underrepresented in this cluster, as shown in Table 1. The abscissa represents the first principal axis, and the ordinate represents the second. Both axes are dimensionless. Modified from Fellenberg et al. (24), with permission by Oxford University Press. (**See color plate A33.**)

Table 1. Frequencies of Characteristic Annotation Values[a]

More than or exactly 2× over-represented or underrepresented:

annotation 2 array_series
value 59 is 7× over-represented (2/2 in cluster; 2/14 in total)
value 61 is absent (0/2 in cluster; 12/14 in total)

annotation 3: array_individual
value 1 is absent (0/2 in cluster; 2/14 in total)
value 2 is absent (0/2 in cluster; 2/14 in total)
value 3 is absent (0/2 in cluster; 2/14 in total)
value 4 is absent (0/2 in cluster; 2/14 in total)
value 5 is absent (0/2 in cluster; 4/14 in total)
value 6 is 7× over-represented (2/2 in cluster; 2/14 in total)

annotation 7: array_hybridization
value 5 is absent (0/2 in cluster; 1/14 in total)
value 6 is absent (0/2 in cluster; 1/14 in total)

annotation 16: label_incorporation_rate
value 44 is absent (0/2 in cluster; 1/14 in total)
value 46 is absent (0/2 in cluster; 1/14 in total)
value 51 is absent (0/2 in cluster; 2/14 in total)
value 52 is 7× over-represented (1/2 in cluster; 1/14 in total)
value 56 is 7× over-represented (1/2 in cluster; 1/14 in total)
value 59 is absent (0/2 in cluster; 1/14 in total)
value 68 is absent (0/2 in cluster; 1/14 in total)
value 84 is absent (0/2 in cluster; 1/14 in total)
value 87 is absent (0/2 in cluster; 2/14 in total)
value 88 is absent (0/2 in cluster; 2/14 in total)
value 93 is absent (0/2 in cluster; 1/14 in total)

annotation 17: total_activity
value 26 000 000 is absent (0/2 in cluster; 1/14 in total)
value 34 000 000 is absent (0/2 in cluster; 1/14 in total)
value 35 000 000 is absent (0/2 in cluster; 1/14 in total)
value 36 000 000 is absent (0/2 in cluster; 1/14 in total)
value 38 000 000 is 7× over-represented (1/2 in cluster; 1/14 in total)
value 39 000 000 is absent (0/2 in cluster; 1/14 in total)
value 43 000 000 is 7× over-represented (1/2 in cluster; 1/14 in total)
value 46 000 000 is absent (0/2 in cluster; 1/14 in total)
value 50 000 000 is absent (0/2 in cluster; 1/14 in total)
value 61 000 000 is absent (0/2 in cluster; 2/14 in total)

Table 1. Continued

value 65 000 000 is absent (0/2 in cluster; 1/14 in total)
value 71 000 000 is absent (0/2 in cluster; 1/14 in total)
value 80 000 000 is absent (0/2 in cluster; 1/14 in total)

annotation 39: experimentator
value 104: bastuk is absent (0/2 in cluster; 2/14 in total)

annotation 1053: temporary_additive
value 1123: none are absent (0/2 in cluster; 2/14 in total)

annotation 1055: incubation_period
value 5 is absent (0/2 in cluster; 4/14 in total)
value 10 is absent (0/2 in cluster; 2/14 in total)
value 15 is absent (0/2 in cluster; 2/14 in total)
value 20 is absent (0/2 in cluster; 2/14 in total)
value 30 is 3.5× over-represented (2/2 in cluster; 4/14 in total)

[a]From Reference 24, with permission by Oxford University Press.

menter"), who performed two of the 12 measurements outside the cluster, is unlikely to explain the difference between the cluster and the other measurements. The same applies to not rehybridizing the array for the 5th or 6th time (annotation "array_hybridization").

The first two annotations listed mean that the entire cluster was hybridized on "array individual" 6, which is the only one stemming from "array series" (i.e., production batch) 59, whereas all other arrays were from series 61. From other experiments, sufficient comparability among arrays of the same production series was observed, whereas arrays of different batches could not be directly compared. The differential array batch used for the hybridization in the selected measurements causes their profiles to be different. The CA plot in Figure 8 shows them clearly separated not only from the remaining measurements of the 30-min time point but also from all other measurements. This artifact distorts the projection of an otherwise sound and revealing data set. Omitting the two outlying measurements for the analysis results in the CA plot shown in Figure 9.

By disregarding the measurements hybridized on a different array series, it is possible to differentiate response phases and to identify the genes involved. Figure 9 reveals a leap in transcriptional status of the cells between 15 and 20 min after the start of treatment, which is consistent with the work of Godon and coworkers (27). This includes both genes with increased and decreased transcription levels. As an example of a marked expression pattern, we selected a cluster of four genes that were activated in the initial phase of the response but downregulated after 20 min (Figure 10).

Scanning Annotations of Continuous Range

For the earlier example, all annotations were treated as categorical. Sometimes, especially with higher numbers of measurements, it is desirable to aggregate values for annotations of continuous range (Table 1, nos. 16 and 17). "Label incorporation rate" may thus be discretized into low, medium, and high values. M-CHIPS provides methods that enable the discretization of annotation ranges into a chosen number of bins due to their particular distribution or by expert knowledge. After selecting a cluster of measurements in a CA plot, a loop is entered for the repeated extraction of characteristic annotation values. At the beginning of each loop step, a menu enables the alteration of the representation factor threshold and discretization of an annotation. The discretization routine displays a histogram of all values annotated for the selected annotation in the entire database. The number of intervals to display and linear or logarithmic scale

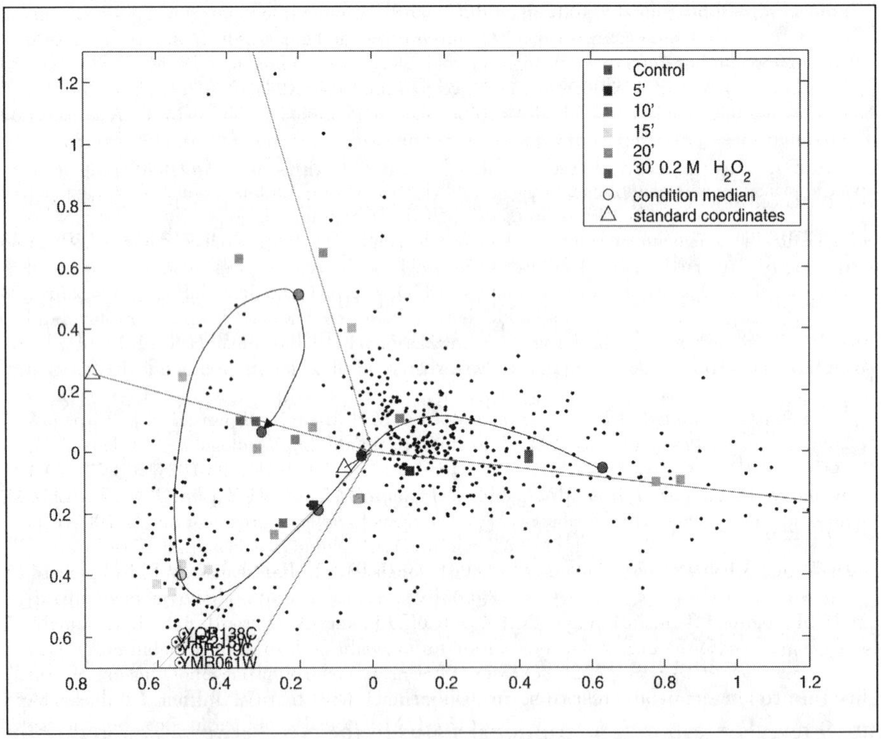

Figure 9. Oxidative stress. Wild-type yeast was grown in the presence of 0.2 M hydrogen peroxide and sampled in 5-min intervals (10 min at end). The experiment was performed twice, which yielded two hybridizations per experimental condition (time point). These conditions are color-coded according to the legend, double-spotting in dark and light colors. A black arrow shows the course of the experiment. Four genes are tagged that are almost exclusively associated with the 15-min time point. Their profiles are given in Figure 10. The abscissa represents the first principal axis, and the ordinate represents the second. Both axes are dimensionless. Modified from the supplemental material to Reference 24. (**See color plate A34.**)

can be selected for the histogram. The routine allows for discretization into equally spaced intervals. Alternatively, arbitrary interval centers can be manually selected by a mouse click. When the histogram shows the desired segmentation of the value range, the values can be replaced by the displayed interval centers. The effect of this measure in conjunction with the chosen representation threshold is immediately visible in the displayed list of characteristic values (Table 1) and can be revoked, revised, improved, or augmented by the discretization of other annotations in the next round.

EMBEDDING INTO A DATA WAREHOUSE

We applied CA as an exploratory and powerful statistical tool to study interdependencies both between and among sets of variables, such as genes and hybridizations obtained by expression profiling (25). It is necessary to carefully preprocess the data and to perform CA in a way that is capable of dealing with the measurement noise and outliers, thus adapting the method to the particular requirements of the microarray data. Furthermore, the data to be analyzed should

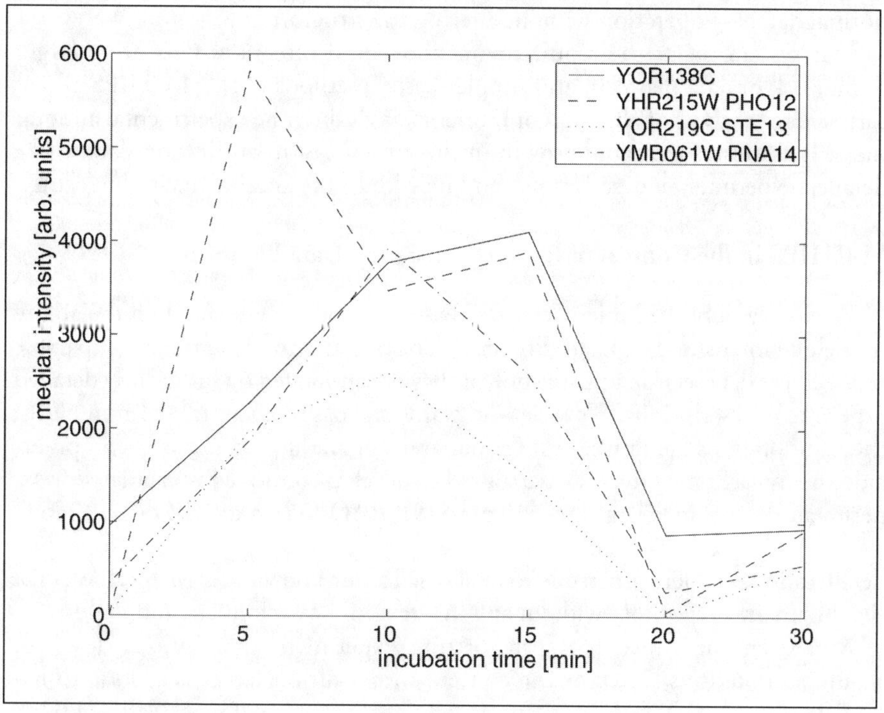

Figure 10. Profile of the four genes tagged in Figure 9. For the four genes tagged in Figure 9, the median transcription intensity is plotted against the response time. Their ORF identifiers and names are provided in the legend. The ordinate shows the arbitrary (machine-dependent) intensity units.

be accessible in a standardized format. To achieve a wide analytic scope, it is not sufficient to put the hybridization intensities into a common format as a platform for preprocessing and analysis. With more and more hybridizations, it is also necessary to provide this platform for the experiment annotations for biological interpretation. For the statistical analysis of numerous and large data sets stemming from multi-conditional microarray experiments, the data need to be properly annotated and stored in a manner so that both the data sets and their annotations are accessible to statistical analysis. Including annotated experimental parameters into statistical analysis offers the opportunity to identify the global players behind the transcription patterns.

Free-text annotation of recent microarray databases impairs direct statistical access to it. Parameter sets used for experiment descriptions have not yet reached their final shape, and standards are reduced to minimal conventions that do not yet allow for extensive description. Complex and highly diverse experimental settings cause a high complexity and diversity in experiment descriptions, also requiring a higher flexibility in data storage than that achieved by standard database solutions. This is especially true when the data are stored in a statistically accessible format restricted to defined values. A structure that is independent of the particular parameter set enables updates of annotation hierarchies during normal database operation without altering the structure.

Our system meets the requirements above and integrates CA into a larger framework of data platform and supplemental methods (24). M-CHIPS allows for the data analysis of all of its components, including the experimental annotations. It addresses the rapid growth in the amount of hybridization data, more detailed experimental descriptions, and new kinds of experiments in the future.

M-CHIPS in the Context of Recent Microarray Data Platforms

Previously published microarray database concepts have focused on the ability to include intensity data from different platforms and to make these comparable (1,9). These data sets are valuable only if they are annotated by sufficiently detailed experiment descriptions. However, in many databases, a substantial number of these annotations are in free-text format and not readily accessible to computer-aided analysis. Some projects have started to develop controlled vocabulary for experiment descriptions [e.g., ArrayExpress (9), RAD (54), and GEO (21)]. However, so far little effort has been made to categorize the descriptions down to minute detail and make them amenable to analysis. To our knowledge, M-CHIPS is the first microarray repository without any free-text in the experiment annotation.

Moreover, the concept of categorizing experiment descriptions down to atomic annotations stored as the content of definition tables is unique to M-CHIPS. GeneX (NCGR, http://www.ncgr.org/doc/genex/GeneXSchema.pdf); ArrayExpress (EBI, http://www.ebi.ac.uk/arrayexpress/Schema/); SMD (Stanford, http://genome-www4.stanford.edu/MicroArray/SMD/doc/db_specifications.

html) (52); ArrayDB (NHGRI, http://research.nhgri.nih.gov/arraydb/schema.shtml) (23); ExpressDB (Harvard University, http://arep.med.harvard.edu/ExpressDB/ExpressDB.v200.help.htm) (1); EpoDB (UPenn, http://www.cbil.upenn.edu/EpoDB/release/schema/schema.shtml) (48,55); RAD (UPenn, http://www.cbil.upenn.edu/cgi-bin/RAD2/schemaBrowserRAD.pl); and GEO (NCBI, http://www.ncbi.nlm.nih.gov/geo/info/geo.tbl) all show several biological or protocol-related keywords in their UML structure representations or attribute listings, indicating a more or less static, hard-wired implementation.

Because our concept is not meant to be implemented in a large public gene expression database, we have not dealt with the inclusion of additional platforms such as oligonucleotide chips or SAGE but have concentrated on mining the wealth of information contained in the experiment annotations. However, we have been able to serve several collaborating groups by providing databases and analysis tools for data from different areas of research (i.e., experiments with *S. cerevisiae, A. thaliana, T. bruceï, N. crassa*, and human cancer samples), obtained by different platforms (radioactive hybridization to nylon or polypropylene membranes and fluorescent hybridization to glass slides) and by different imaging software.

Similar to other data warehousing solutions, M-CHIPS focuses on analysis rather than database browsing and data retrieval by simple queries. However, instead of dealing with an extract from the data, M-CHIPS makes available all of the information gained by microarray experiments to statistical analysis. This includes complex experiment descriptions at arbitrary levels of detail. Instead of reducing the information by predefined extraction rules before analysis at the data warehouse level, it is reduced at a later stage, according to statistical significance only. Also, in contrast to providing a stock of different methods, the output of which needs to be compared, M-CHIPS provides a homogenous view with CA, which is capable of visualizing multiple variables.

Analytical Scope

The storage system provides an unprecedented level of detail for experiment descriptions captured in categorical and continuous variables. For data entry, this ensures the completeness of experiment annotation (i.e., a level of completeness that exceeds minimal standards). For analysis, it provides the opportunity to include the complete experiment information as additional variables and to study it by means of multivariate statistics. Additional attributes or additional allowed values for existing attributes can easily be added without changing the database structure or any algorithm. Previously annotated experiments can be revisited to annotate newly defined annotations. However, this is unnecessary. In a set of hybridizations that lack an annotation for some cases, this annotation will be ignored, enabling the correct analysis of the others. Thus, no pressure is put on the defining and annotating experimenter to revisit hundreds of already stored experiments when he or she wishes to define a useful new annotation.

Reliability and Universal Applicability

The M-CHIPS concept allows for information from heterogeneous experiments to be stored in databases of similar structure so that the same algorithms for analysis can be applied. The system has been used by collaborating groups since June 1999. Thus, all algorithms described earlier have been extensively tested. Currently, we have 33 yeast-specific (MIAME compliant; http://www.mged.org/Workgroups/MIAME/miame.html); 54 human tumor-specific; 71 *Arabidopsis*-specific (MIAME compliant); 41 *Trypanosome*-specific, 20 *Neurospora*-specific, and 78 common (technical, MIAME compliant) experiment annotations. Compliance with standards such as those proposed by EBI (MIAME) is independent from our storage schema. The experimenter who defines the annotations decides on the standard compliance and level of detail. The sets of hierarchically ordered annotations are listed on the associated Web page (http://www.dkfz.de/tbi/services/mchips). The descriptions of all hybridizations stored in M-CHIPS databases can be analyzed statistically. Apart from working reliably for two-and-a-half years, the system has proven stable and effective with vast amounts of data. We currently perform the administration of 2230 hybridizations on 18 databases. They belong to the above five fields of research and comprise both radioactive-labeled and multichannel experiments. Although these databases may contain different parameter sets used for experiment descriptions, they share the same structure concept [(M-CHIPS databases do not share exactly the same database structure, as illustrated in Figure 1) (24)] and therefore can be accessed by the very same algorithms for statistical analysis.

DISCUSSION

Adaption of CA

For microarray data, a thorough preprocessing step is essential. Different normalization algorithms have to be applied to single and multichannel data for the different meaning of the particular raw intensities (24). Genes with generally low reproducibility for most of the conditions under study are filtered out. However, with increasing numbers of conditions, after eliminating all genes with low reproducibility in one of the conditions, few or no genes will remain. The same is true for the intensity filter. Therefore, it is reasonable to use these filters to discard only genes with low abundance or low reproducibility (often coinciding) in all of the conditions under study. Thus, outliers also have to be handled by other measures. Otherwise, they would seriously interfere with CA, which, in contrast to other methods, is not similarity driven but aims to display variance. Any difference in the default state (expected value), such as an outlier, will be regarded as important for the projection. The larger the difference, the more the data point will attract the projection plane. We prevented this by choosing the principal axes according to the condition medians only (HMS). Furthermore, our HMS technique allows for the original data to be visible in the plot, thus combining the noise reduction capability of HMS with the quality-control aspect of retaining the original data. Projection methods are generally used to explain the major trends in the data while simultaneously ignoring minor fluctuations. HMS has been demonstrated to further enhance this effect (25).

Given a sufficient number of repeated hybridizations, these precautions should ensure that the variance explained by a CA plot reflects biological changes rather than measurement fluctuations. For the biological changes observed, the significance of differences rather than the absolute differences are plotted both among the genes and among the hybridizations in terms of the χ^2 statistic. However, the power of the CA technique is that it is able to show associations between genes and hybridizations. To fully exploit this property, it is necessary to examine the exact directions of gene association with the experimental conditions. These are given by the standard coordinates of the corresponding condition medians rather than by their principal coordinates. Plotting the standard coordinates directly would cause all principal coordinates to shrink into a small area in the middle of the plot. The introduction of the lines representing the standard coordinates is great help to the interpretation of the plots by relating genes and conditions to each other and by circumventing direct plotting.

Applicability to Microarray Data

Traditionally, CA has been predominantly used on categorical data in the social sciences (6,35), but its application has also been extended to (positive) physical quantities (30) and to proteomics (46,47). Recently, it has been applied to microarray data (14,25,39). It has been shown that CA provides an informative and concise means of visualizing these data and is capable of uncovering relationships both among genes or hybridizations and between genes and hybridizations. The method has proved to be generally applicable to microarray data, regardless of whether the data have been obtained by radioactive labeling or by two-channel fluorescent-tag labeling. Normalization procedures lead to intensity values that can be interpreted as being proportional to a cell's content of mRNA molecules per gene and per condition. For two-channel intensities, the log ratios of the red channel versus the green channel appear to work just as well. To introduce the method, we have shown its application to the well-known *S. cerevisiae* cell cycle synchronization data by Spellman et al. (53), which allowed for comparison with their visualization of this data set. Furthermore, we have applied CA to a non-time series data set of our own, thus supporting its general applicability to microarray data of different complexity, underlying structure, and experimental strategy (both two-channel fluorescence-tag and radioactive labeling). This example also shows that CA is capable of subtracting particular effects, such as the influence of galactose in the medium.

CA is most powerful if it is used in combination with a standardized data platform (e.g., a data warehouse) and supplementary algorithms that enable preprocessing as well as the interaction of the CA routines with other algorithms such as hierarchical clustering (24).

Moreover, our platform enables the integrated analysis of both transcription intensities and complex experiment annotations. Statistical analysis of experiment annotations can be applied for arbitrary sets of hybridizations in any CA plot by mouse click (e.g., for those clustered by similar expression profiles). This provides a means to reveal both experimental artifacts and biologically meaningful correlations from huge sets of experiment descriptions in an automated way. The resulting experimental parameters are candidates for being the active players that drive the cells to the expression pattern observed in the corresponding hybridization cluster. While this is a fairly simple method, it already provides good analytical access to long lists of annotations and huge sets of hybridizations that could not be thoroughly evaluated by visual inspection. More sophisticated statistical methods can also be directly applied because, unlike free-text annotation, the instances of occurrence are readily countable for all annotation values. Future plans include the integrated visualization of both transcription intensities and experiment annotations by CA.

Integration with a Data Warehouse

In microarray research and other fields, biologists and bioinformaticians have to work together. A way to share the work would be for the biologist to do the practical work up to obtaining the data, which are then given to the bioinformatician, who then performs the analysis. Statistical implications may lead to a low value of the resulting data if the bioinformatician has not been consulted for the experimental design. Analysis carried out by the bioinformatician alone might be statistically sound but has its weaknesses in biological interpretation. We consider M-CHIPS to be a communication tool as well. It supplies the complete experiment information to the bioinformatician. A technical parameter in the protocol, such as the concentration of a particular buffer component, may be identified to be associated with the down-regulation of a set of genes—even by someone not knowing its purpose. On the other hand, we implemented analysis algorithms in a way that is accessible to the biologist. A graphical user interface completely replaces the input of function names and any command line parameter, which enables one to perform the entire analysis by mouse clicks. Sharing both the complete data and all analysis methods facilitates the communication between experimenter and bioinformatician, resulting in analysis methods that are tailor-made for the special requirements of microarray data. The modular structure of the system and a MATLAB environment enable the quick implementation of new tools. Likewise, biologists are able to analyze the data themselves instantly after upload for direct and immediate feedback into follow-up experiments.

Furthermore, M-CHIPS has already served successfully as a communication tool between collaborating workgroups. The EUROFAN II, B2 data set (http://mips.gsf.de/proj/eurofan/eurofan_2/b2) was acquired by five European groups. The raw intensities were collected and uploaded by Nicole Hauser. However, only the experimenters were able to provide detailed information about the technical protocols and experimental conditions. In M-CHIPS, experiment annotation is Web-based to ensure that any experiment can be annotated remotely by the experimenters themselves. The effort involved in annotating experiments is minimized. After preprocessing, raw and preprocessed data as well as gene and experiment annotations were given to the Munich Information Center for Protein Sequences (MIPS, http://mips.gsf.de) for Web publication.

PERSPECTIVES

CA is an intuitive means of exploratory microarray data analysis. It shows which questions could be asked or which hypotheses could be asserted. It decomposes the overall χ^2 statistic; distances among points in the plots approximate χ^2 distances. Thus, they resemble the statistical significance of an observed Euclidean distance with respect to all other distances measured, rather than the absolute distance itself. However, statistical significance is not a crucial requirement for justifying an inspection of the maps. CA can be regarded as a way of re-expressing the data in pictorial form for the ease of interpretation (33). In this work, we used CA to transform data tables of genes versus measurements into 2-D maps. Alternatively, one could cluster genes or measurements into discrete groups of similar profiles (34). Moreover, CA can be used in discriminant analysis and classification (31). As mentioned earlier, CA has been predominantly applied to survey data in social sciences. To grasp the full potential of the technique, one should consider that using CA to simultaneously display genes and hybridizations is analogous to investigating a questionnaire comprising two questions. More than two variables can be integrated by applicative arrangement in a two-way table or by multiple or joint CA of multi-way tables. For microarray data analysis, this enables the straightforward inclusion of additional information such as gene and experiment annotations if they are stored in a format accessible to statistical analysis. Integration of gene and experiment annotation data may enable the identification of active players among these parameters that drive the cells to the observed expression patterns. Sequence patterns in regulatory elements also offer great promise for linking with microarray data (8,59,62). Discovery of complex interdependencies between experimental parameters, gene properties, promotor sequences, mRNA abundance, and protein levels might be possible by the integrated analysis of all these variables.

ACKNOWLEDGMENTS

First and foremost, we would like to thank Judith Boer, Marcel Scheideler, Frank Diehl, Susanne Diehl, Verena Aign, Albert Neutzner, Helene Tournu, Arno Meijer, Luis Lombardia, Manuel Beccera, Andy Hayes, Nikolaus Schlaich, José Pérez-Ortín, Arnaud Lagorce, Melanie Bier, Sonja Bastuck, Tamara Korica, Andrea Bauer, Matthias Nees, Diana Stjepandic, Kerstin Hasenpusch-Theil, Julia Klopp, Marcus Frohme, Olaf Witt, Boris Beckmann, Melanie Bey, Claudia Hartmann, Fabian Kruse, Kerstin Böhm, and Marc Valls, who entrusted the outcome of their hard work to us for data administration and analysis. Many thanks also for suggestions for improvement, critical discussions, and reported bugs. We are also grateful to Tim Beißbarth, Dieter Finkenzeller, and Christian Busold for contributing algorithms. The work relied on permanently running computer systems, which were administered by Tobias Reber, Heiko Schmidt, and Karlheinz Groß.

REFERENCES

1.**Aach, J., W. Rindone, and G.M. Church.** 2000. Systematic management and analysis of yeast gene expression data. Genome Res. *10*:431-445.

2.**Alter, O., P.O. Brown, and D. Botstein.** 2000. Singular value decomposition for genome-wide expression data processing and modeling. Proc. Natl. Acad. Sci. USA *97*:10101-10106.

3.**Basset, D.E., Jr., M.B. Eisen, and M.S. Boguski.** 1999. Gene expression informatics—it's all in your mine. Nat. Genet. *21(Suppl)*:51-55.

4.**Beißbarth, T., K. Fellenberg, B. Brors, R. Arribas-Prat, J.M. Boer, N.C. Hauser, M. Scheideler, J.D. Hoheisel, et al.** 2000. Processing and quality control of DNA array hybridization data. Bioinformatics *16*:1014-1022.

5.**Bittner, M., P. Meltzer, Y. Chen, Y. Jiang, E. Seftor, M. Hendrix, M. Radmacher, R. Simon, et al.** 2000. Molecular classification of cutaneous malignant melanoma by gene expression profiling. Nature *406*:536-540.

6.**Blasius, J. and M.J. Greenacre.** 1998. Visualization of Categorical Data, 1st ed. Academic Press, London.

7.**Braxton, S. and T. Bedilion.** 1998. The integration of microarray information in the drug development process. Curr. Opin. Biotechnol. *9*:643-649.

8.**Brazma, A., I. Jonassen, J. Vilo, and E. Ukkonen.** 1998. Predicting gene regulatory elements in silico on a genomic scale. Genome Res. *8*:1202-1215.

9.**Brazma, A., A. Robinson, G. Cameron, and M. Ashburner.** 2000. One-stop shop for microarray data. Nature *403*:699-700.

10.**Breiman, L., J.H. Friedman, R.A. Olshen, and C.J. Stone.** 1984. Classification and Regression Trees. Wadsworth and Brooks/Cole, Monterey, CA.

11.**Brown, P.O. and D. Botstein.** 1999. Exploring the new world of the genome with DAN microarrays. Nat. Genet. *21(Suppl)*:33-37.

12.**Brown, M.P.S., W.N. Grundy, D. Lin, N. Cristianini, C.W. Sugnet, and T.S. Furey.** 2000. Knowledge-based analysis of microarray gene expression data by using support vector machines. Proc. Natl. Acad. Sci. USA *97*:262-267.

13.**Chapman, S., P. Schenk, K. Kazan, and J. Manners.** 2002. Using biplots to interpret gene expression patterns in plants. Bioinformatics *18*:202-204.

14.**Charnomordic, B. and S. Holmes.** 2001. Correspondence analysis with R. Statistical Computing and Graphics *12*:19-25.

15.**Cornell, M., N. Paton, S. Wu, C. Goble, C. Miller, P. Kirby, K. Eilbeck, A. Brass, et al.** 2001. GIMS—a data warehouse for storage and analysis of genome sequence and functional data. Proc. 2nd

IEEE International Symposium on Bioinformatics and Bioengineering (BIBE). Piscataway, NJ: IEEE Press. March 04-06, p.15-22.

16. **Debouck, C. and P.N. Goodfellow.** 1999. DNA microarrays in drug discovery and development. Nat. Genet. *21 (1 Suppl)*:48-50.

17. **DeRisi, J., L. Penland, P.O. Brown, M. Bittner, P.S. Meltzer, M. Ray, Y. Chen, Y.A. Su, et al.** 1996. Use of a cDNA microarray to analyse gene expression patterns in human cancer. Nat. Genet. *14*:457-460.

18. **Dobrowolski, S.F., R.A. Banas, E.W. Naylor, T. Powdrill, and D. Thakkar.** 1999. DNA microarray technology for neonatal screening. Acta Paediatr. Suppl. *88*:61-64.

19. **Dudoit, S., Y.H. Yang, M.J. Callow, and T.P. Speed.** 2000. Statistical methods for identifying differentially expressed genes in replicated (cDNA) microarray experiments. Dept. of Statistics, University of California Berkeley, Berkeley, CA. Technical Report 578. Available online at (http://stat-www.berkeley.edu/tech-reports/)

20. **Duggan, D.J., M. Bittner, Y. Chen, P. Metzler, and J.M. Trent.** 1999. Expression profiling using cDNA microarrays. Nat. Genet. *21(Suppl)*:10-14.

21. **Edgar, R., M. Domrachev, and A.E. Lash.** 2002. Gene expression omnibus: NCBI gene expression and hybridization array data repository. Nucleic Acids Res. *30*:207-210.

22. **Eisen, M.B., P.T. Spellman, P.O. Brown, and D. Botstein.** 1998. Cluster analysis and display of genome-wide expression patterns. Proc. Natl. Acad. Sci. USA *95*:14863-14868.

23. **Ermolaeva, O., M. Rastogi, K.D. Pruitt, G.D. Schuler, M.L. Bittner, Y. Chen, R. Simon, P. Meltzer, et al.** 1998. Data management and analysis for gene expression arrays. Nat. Genet. *20*:19-23.

24. **Fellenberg, K., N.C. Hauser, B. Brors, J.D. Hoheisel, and M. Vingron.** 2002. Microarray data warehouse allowing for inclusion of experiment annotations in statistical analysis. Bioinformatics *18*:423-433.

25. **Fellenberg, K., N.C. Hauser, B. Brors, A. Neutzner, J.D. Hoheisel, and M. Vingron.** 2001. Correspondence analysis applied to microarray data. Proc. Natl. Acad. Sci. USA *98*:10781-10786.

26. **Fisher, R.A.** 1936. The use of multiple measurements in taxonomic problems. Ann. Eugenics *7*:300-308.

27. **Godon, C., G. Lagniel, J. Lee, J.M. Buhler, S. Kieffer, M. Perrot, H. Boucherie, M.B. Toledano, et al.** 1998. The H2O2 stimulon in *Saccharomyces cerevisiae.* J. Biol. Chem. *273*:22480-22489.

28. **Golub, G.H. and C. Reinsch.** 1970. Singular value decomposition and least squares solutions. Numer. Math *14*:403-420.

29. **Golub, T.R., D.K. Slonim, P. Tamayo, C. Huard, M. Gaasenbeek, J.P. Mesirov, H. Holler, M.L. Loh, et al.** 1999. Molecular classification of cancer: class discovery and class prediction by gene expression monitoring. Science *286*:531-537.

30. **Greenacre, M.J.** 1984. Special topics, p. 223. *In* Theory and Applications of Correspondence Analysis, 1st ed. Academic Press, London.

31. **Greenacre, M.J.** 1984. Use of CA in discriminant analysis, classification, regression, and cluster analysis, p. 185-206. *In* Theory and Applications of Correspondence Analysis, 1st ed. Academic Press, London.

32. **Greenacre, M.J.** 1993. Appendix 2: summary of theory, p. 181-183 and 36. *In* Correspondence Analysis in Practice, 1st ed. Academic Press, London.

33. **Greenacre, M.J.** 1993. More examples, p. 74-85. *In* Correspondence Analysis in Practice, 1st ed. Academic Press, London.

34. **Greenacre, M.J.** 1993. Clustering rows and columns, p. 111-118. *In* Correspondence Analysis in Practice, 1st ed. Academic Press, London.

35. **Greenacre, M.J. and J. Blasius.** 1994. Correspondence Analysis in the Social Sciences, 1st ed. Academic Press, London.

36. **Hilsenbeck, S.G., W.E. Friedrichs, R. Schiff, P. O'Conell, R.K. Hansen, C.K. Osborne, and S.A.W. Fuqua.** 1999. Statistical analysis of array expression data as applied to the problem of tamoxifen resistance. J. Natl. Cancer Inst. *91*:453-459.

37. **Huber, W., A. von Heydebreck, H. Sültmann, A. Poustka, and M. Vingron.** 2002. Variance stabilization applied to microarray data calibration and to the quantification of differential expression. Bioinformatics *18(Suppl 1)*:S96-S104.

38. **Khan, J., M. Bittner, Y. Chen, P.S. Meltzer, and J.M. Trent.** 1999. DNA microarray technology: the anticipated impact on the study of human disease. Biochim. Biophys. Acta *1423*:M17-M28.

39. **Kishino, H. and P.J. Waddell.** 2000. Correspondence analysis of genes and tissue types and finding genetic links from microarray data. Genome Inform. Ser. Workshop. Genome Inform. *11*:83-95.

40. **Lefkovits, I., L. Kuhn, O. Valiron, A. Merle, and J. Kettman.** 1988. Toward an objective classification of cells in the immune system. Proc. Natl. Acad. Sci. USA *85*:3565-3569.

41. **Lennon, G.G. and H. Lehrach.** 1991. Hybridization analyses of arrayed cDNA libraries. Trends Genet. *7*:314-317.

42. **Lockhart, D.J., M. Dong, M.C. Byrne, M.T. Folletie, M.V. Gallo, M.S. Chee, M. Mittmann, C. Wang, et al.** 1996. Expression monitoring by hybridization to high-density oligonucleotide arrays. Nat. Biotechnol. *14*:1675-1680.

43. **Lockhart, D.J. and E.A. Winzeler.** 2000. Genomics, gene expression and DNA arrays. Nature *405*:827-836.

44. **Morgan, D.O.** 1999. Regulation of the apc and the exit from mitosis. Nat. Cell Biol. *1*:E47-E53.

45. **Perou, C.M., S.S. Jeffrey, M. van de Rijn, C.A. Rees, M.B. Eisen, D.T. Ross, A. Pergamenschikov, C.F. Williams, et al.** 1999. Distinctive gene expression patterns in human mammary epithelial cells and breast cancers. Proc. Natl. Acad. Sci. USA *96*:9212-9217.

46. **Pleissner, K.P., V. Regitz-Zagrosek, B. Krudewagen, J. Trenkner, B. Hocher, and E. Fleck.** 1998. Effects of renovascular hypertension on myocardial protein patterns: analysis by computer-assisted two-dimensional gel electrophoresis. Electrophoresis *19*:2043-2050.

47. **Pun, T., D.F. Hochstrasser, R.D. Appel, M. Funk, V. Villars-Augsburger, and C. Pellegrini.** 1988. Computerized classification of two-dimensional gel electrophoretograms by correspondence analysis and ascendant hierarchical clustering. Appl. Theoret. Electrophoresis *1*:3-9.

48. **Salas, F., J. Haas, B. Brunk, C.J. Stoeckert, Jr., and G.C. Overton.** 1998. EpoDB: a database of genes expressed during vertebrate erythropoiesis. Nucleic Acids Res. *26*:288-289.

49. **Schena, M.** 1996. Genome analysis with gene expression microarrays. BioEssays *18*:427-431.

50. **Schena, M., D. Shalon, R.W. Davis, and P.O. Brown.** 1995. Quantitative monitoring of gene expression patterns with a complementary DNA microarray. Science *270*:467-470.

51. **Shalon, D., S.J. Smith, and P.O. Brown.** 1996. A DNA microarray system for analyzing complex DNA samples using two-color fluorescent probe hybridization. Genome Res. *6*:639-645.

52. **Sherlock, G., T. Hernandez-Boussard, A. Kasarskis, G. Binkley, J.C. Matese, S.S. Dwight, M. Kaloper, S. Weng, et al.** 2001. The stanford microarray database. Nucleic Acids Res. *29*:152-155.

53. **Spellman, P.T., G. Sherlock, M.Q. Zhang, V.R. Iyer, K. Anders, M.B. Eisen, P.O. Brown, D. Botstein, et al.** 1998. Comprehensive identification of cell cycle-regulated genes of the yeast *Saccharomyces cerevisiae* by microarray hybridization. Mol. Biol. Cell *9*:3273-3297.

54. **Stoeckert, C.J., A. Pizarro, E. Manduchi, M. Gibson, B. Brunk, J. Crabtree, J. Schug, S. Shen-Orr, et al.** 2001. A relational schema for both array-based and SAGE gene expression experiments. Bioinformatics *17*:300-308.

55. **Stoeckert, C.J., Jr., F. Salas, B. Brunk, and G.C. Overton.** 1999. EpoDB: a prototype database for the analysis of genes expressed during vertebrate erythropoiesis. Nucleic Acids Res. *27*:200-203.

56. **Tamayo, P., D. Slonim, J. Mesirov, Q. Zhu, S. Kitareewan, E. Dmitrovsky, E.S. Lander, and T.R. Golub.** 1999. Interpreting patterns of gene expression with self-organizing maps: methods and application to hematopoietic differentiation. Proc. Natl. Acad. Sci. USA *96*:2907-2912.

57. **Tang, K., D.J. Fu, D. Julien, A. Braun, C.R. Cantor, and H. Koster.** 1999. Chip-based genotyping by mass spectrometry. Proc. Natl. Acad. Sci. USA *96*:10016-10020.

58. **Tavazoie, S., J.D. Hughes, M.J. Campbell, R.J. Cho, and G.M. Church.** 1999. Systematic determination of genetic network architecture. Nat. Genet. *22*:281-285.

59. **van Helden, J., B. Andre, and J. Collado-Vides.** 1998. Extracting regulatory sites from the upstream region of yeast genes by computational analysis of oligonucleotide frequencies. J. Mol. Biol. *281*:827-842.

60. **van't Veer, L.J., H. Dai, M.J. van de Vijver, Y.D. He, A.A. Hart, M. Mao, H.L. Peterse, K. van der Kooy, et al.** 2002. Gene expression profiling predicts clinical outcome of breast cancer. Nature *415*:530-536.

61. **Velculescu, V.E., S.L. Madden, L. Zhang, A.E. Lash, J. Yu, C. Rago, A. Lal, C.J. Wang, et al.** 1999. Analysis of human transcriptomes. Nat. Genet. *23*:387-388.

62. **Vilo, J., A. Brazma, I. Jonassen, A. Robinson, and E. Ukkonen.** 2000. Mining for putative regulatory elements in the yeast genome using gene expression data. Proc. Int. Conf. Intell. Syst. Mol. Biol. *8*:384-394.

63.**Vingron, M.** 2001. Bioinformatics needs to adopt statistical thinking. Bioinformatics *17*:389-390.
64.**Yang, Y.H., M.J. Buckley, S. Dudoit, and T.P. Speed.** 2000. Comparison of methods for image analysis on cDNA microarray data. Dept. of Statistics, University of California Berkeley, Berkeley, CA. Technical Report 584. Available online at (http://stat-www.berkeley.edu/tech-reports/).

Appendix: MATLAB Implementation

Algorithm 1 lists a short implementation of simple correspondence analysis. More complex ones, adapted to the requirements of microarray data analysis, are at the core of M-CHIPS.

Algorithm 1. A simple correspondence analysis program in MATLAB

```
P=N./sum(sum(N));                   % correspondence matrix P
q=sum(P')'*sum(P);                  % q=r_i c_j
[U,D,V]=svd((P-q)./sqrt(q),0);      % ",0" produces economy-size
                                    % decomposition,
                                    % can be omitted
F=diag(1./sqrt(sum(P')))*U*D;       % gene coordinates
G=diag(1./sqrt(sum(P)))*V*D;        % hybridization coordinates

% The following statements plot out the above computed coordinates:
plot(F(: ,1) , F(: ,2),'.k'); hold on;              % plots ...
text(F(: ,1) , F(: ,2),num2str([1:size(F ,1)]'));   % & annotates genes
plot(G(: ,1) , G(: ,2),'s');                        % and hybridizations
text(G(: ,1) , G(: ,2),num2str([1:size(G ,1)]'));

% To correctly display χ2 distances, the two axes have to be scaled
% identically:
from=min([get(gca,'xlim') get(gca,'ylim')]);             % determines ...
to=max([get(gca,'xlim') get(gca,'ylim')]);               % ... max. range
set(gca,'xlim',[from to]); set(gca,'ylim',[from to]);    % fixes range
```

17 High-Throughput Functional Inference in the Post-Genomic Era

Avrum Spira[1,3], Itai Yanai[1], Rakesh Chaterjee[2], Jie Wu[2], Rhonda Harrison-Tang[1], and Charles DeLisi[1,2]

[1]Bioinformatics Graduate Program, [2]Department of Biomedical Engineering, and [3]Division of Pulmonary Medicine, Boston University, Boston, MA, USA

INTRODUCTION

Among the many concerns that were voiced when genome projects were first debated 17 years ago was the possibility that rapid sequencing would yield little more than warehouses of unevaluated data. The rate at which function was determined experimentally already lagged the rate of sequence generation, and the standard computational method of the day—the Smith-Waterman string matching algorithm—required homologs of known function to classify new sequences, severely limiting its utility for functional inference.

Ironically, the seeds to the solution of the problem of analyzing genomic data were to be found in the preponderance of the data itself. String matching algorithms find correlations between a target sequence and a sequence of known function and assume that if the correlation is significant enough, the functions will be related. But many properties, in addition to sequence, can be correlated, and these became evident as whole genomes were sequenced and new experimental technologies for analyzing whole genome transcript profiles were developed. The computational methods include domain fusion (9,23), chromosomal proximity (25), phylogenetic profiling (10,16,26,36), and various experimental technologies that include high-density arrays of probes for detecting expressed genes (20,30,34). Moreover, unlike sequence homology methods that typically provide no direct information about context (targets), each of these methods can be used for pathway and network inference. This paper briefly reviews high-throughput inferential methods as they are being developed and applied in our laboratory and by others.

Perspectives in Gene Expression
Edited by K. Appasani

METHODS

Domain Fusion Analysis

The domain fusion method is based on the observation that genes occurring separately in a given organism are likely to be functionally related if they are found fused in another organism. Thus, if a new organism is sequenced, and the gene boundaries are reasonably well delineated, functional information can be obtained by identifying all pairs of genes that are fused in some other organism. If the function of one member of such a pair is known, then the function of the other is related. A good example is provided by the genes for phosphoglycerate kinase, triphosphate isomerase, and GAPDH, which are independent in *Mycobacterium genitalium.* The first two are fused in *Thermotoga maritima*, and the second two are fused in *Phytophthora infestans.* The fusion hypothesis predicts that the three should be linked, either functionally or by direct interaction between the proteins they encode; in fact, the encoded enzymes act sequentially in glycolysis (41).

ONTOLOGIES

The main difficulty in testing the fusion hypothesis systematically resides in the limitations of currently available functional ontologies. The development of ontologies is an active field that currently focuses on controlled vocabularies to describe XML-based hierarchies (http://img.cs.man.ac.uk/stevens/ontology.html and the citations found therein). An example is the Gene Ontology (GO) (2), which uses a controlled vocabulary organized around function, cellular environment, and biological processes associated with eukaryotic gene products. These draw from the recent development of Web-based retrieval methods and will become increasingly useful for comparing functions across phyla, as more organisms are included in the hierarchy.

However, even as such efforts proceed, ongoing research in the construction of pathways and networks [(http://www.bind.ca) (http://www.genome.ad.jp/kegg/kegg2/html) and (http://www-unix.mcs.anl.gov/compbio/)] is beginning to provide the foundation for the next stage in the development of ontologies, moving them beyond simple hierarchies. Indeed, the addition of interesting and complex semantics (19) will be a step toward the development of a predictive ontology.

Because the large majority of completed genomes are microbial, in the short term, we can make use of well-curated ontologies for microbial systems, such as those associated with the COG[1] database (www.ncbi.nlm.nih.gov/COG/) pro-

[1]Clusters of orthologous genes. An ortholog is the same gene in different organisms, the difference arising by speculation. This is to be distinguished from paralogs, which are variants of the same gene in the same oganism, that have arisen by duplication and mutation.

vided by the National Center for Biotechnology Information. The ontology partitions genes into 15 broad functional categories. They include genetic functions such as translation, ribosomal structure, and biogenesis; transcription; cell division and chromosome partitioning; and signal transduction mechanisms, etc., with a 16th category for genes of unknown function.

These categories are not entirely mutually exclusive in that genes of one may subserve a common function with genes of another. Nevertheless, an analysis of the 30 fully sequenced genomes available in the year 2000 indicated that in 72% of the cases, fused genes were in the same COG functional category. In principle, the other 28% could be false positives, but there is no way to know from the ontology, which, by definition, can only provide us with the percentage of true positives.

In all, approximately 4500 functional linkages were identified (41), a number that subsequently increased to 10 000. Genes linked by this method form branched networks since a particular gene will sometimes be fused to several different genes (in different organisms). One of the limitations of domain fusion is that fusions tend to increase with cellular complexity, which severely limits the information that can be obtained about human genes.

CONSERVED CHROMOSOMAL PROXIMITY

Genes are classified as proximate if they are on the same strand and the 5′-end of one is within 300 bases of the 3′-end of the other (25). The conserved proximity hypothesis says that genes are functionally related if their proximity is conserved across a set of genomes. The hypothesis was evaluated using the COG ontology, and it was also evaluated using the 138 metabolic and regulatory pathways in the Kyoto encyclopedia of genes and genomes (www.genome.ad.jp/kegg/kegg2.html). A careful analysis using 42 microbial genomes in 25 phylogenetic lineages (41) indicates that, as one might expect, the percentage of functional agreement (i.e., common pathways) depends on the fraction of organisms in which proximity is conserved. When a pair is observed to be proximate in a given organism, then the two genes will share at least one pathway 32% of the time if they are proximate in at least one other organism. If they are proximate in at least four other organisms, then they will share at least one pathway more than 80% of the time. If they are proximate in at least six other organisms, then they will have at least one pathway in common more than 90% of the time. Thus, specificity increases as the extent of conservation increases. On the other hand, sensitivity must decrease because the number of genes proximate in, for example, at least six organisms must, by definition, be less than the number of genes proximate in at least two. When the criterion used is at least four organisms, the number of proximate genes is more than 11 000.

It is noteworthy that this result is dependent on genes being observed proximate in a given organism. However, if a genome is newly sequenced, then we would be omitting important information if we examined only those genes that are proximate in that organism since it is evident that at least some non-proximate genes will be neighbors in other organisms. When we include these "inferred" links, the number of linked genes nearly doubles, with only a slight decrease in specificity. Conserved proximity, as it has been used to date, typically links genes in the same operon. The linkages therefore are not commonly promiscuous (i.e., neighbors are relatively invariant), and the networks thus formed—in contrast to domain fusion networks—show almost no branching (42).

CORRELATED PHYLOGENETIC PROFILES

A phylogenetic profile refers to the pattern of occurrence of a gene across a set of genomes (organisms). Therefore, it is most easily represented as a string of binary digits, ones and zeroes, denoting the presence and absence of a gene across the genomes. If two profiles are correlated beyond chance expectation or if their probability of chance occurrence is unusually low, then they are hypothesized to be functionally related. The relation between a pair of profiles can be scored numerically using any number of methods (Wu, Kasif, and DeLisi, manuscript in preparation), such as a Pearson correlation coefficient, mutual information, or direct probability.

The COG database makes this method easy to apply to microbes, simply by constructing a profile for each of the 3166 COGs and then computing the similarities between the 5 010 195 pairs by any of a variety of methods. Higher order correlates (triplets of genes) can of course also be calculated. When this method is applied to yeast and *Escherichia coli* orthologs, one finds that the degree of functional correspondence at a pathway level increases as a more stringent threshold is applied for any of the three methods. Although the percentage of functional correspondence can reach as high as 90%, the number of genes they cover decreases. These methods uncover not only those pairs with identical profiles but also pairs with profiles similar in a statistical sense.

A fourth method for obtaining information related to pathway and network discovery does not involve the analysis of genomic sequences but rather the analysis of expressed genes from homogenous collections of cells that have been systematically perturbed (e.g., gene suppression and knockouts, variations in growth media, etc.). Sets of genes with correlated expression patterns under different conditions are expected to be functionally related.

DATABASE CREATION AND MINING

The earlier methods are most useful when the information they provide can be integrated, linked to related data from other sources [e.g., from yeast two hybrid (17) experiments and co-precipitation (31) data], and visualized.

A sample output from a homemade visualization system is shown below (http://structure.bu.edu/cgi-bin/project1.cgi). Here we started with CDC28, a yeast cell cycle enzyme, and a database of pairwise protein interactions that is used to add links, just as multifunctional units form branched polymers in solution. In this way, we find a 65-protein network. It can made Web accessible by several techniques, such as an html file containing a Java® applet code, using a PERL-CGI scripting language. A more flexible approach uses a CGI program that calls the database of pair interactions and constructs a network in response to user input (e.g., gene name and network size). A graphical interface can be created using a Java applet. Annotation can be added in various ways, such as by using the information-dense conventions introduced by K.W. Kohn (19).

Although much of the current information in the literature is based on yeast, the genes in the constructed pathways and networks often have homologs in other organisms, suggesting that the same network is present. Such mining approaches can be used with proteins or their yeast homologs to provide preliminary networks that would be refined using inference methods and high-throughput perturbation experiments. As network generation methods are developed, refined, and implemented, the knowledge obtained based on a few organisms can be amplified by software that constructs homologous networks in other organisms and assembles them into the database of predicted functions.

To form more reliable network models, we generally screen yeast two hybrid data, using all available microarray databases to assign quality weights to the links. This could be done, for example, by computing the correlation between co-expressed genes when such expression data are available. Microarray data are subject to error, and we confine ourselves to genes that are highly expressed and whose change in expression exceeds some threshold (e.g., 2-fold). In the future, as we move to more accurate and less expensive array technologies, we expect to obtain sufficient replicates to make decisions based on rigorous statistical analysis.

The accuracy of the constructed pathways and networks is always an important consideration. Yeast two hybrid data assembled by whole genome screens miss the majority of interactions and, as can be expected, also contain some false positives (15). Nevertheless, it is noteworthy that the CDC28 network recovered based on these data agrees well with models developed by an integration of years of effort on cdc28 cyclin-dependent protein kinase activity (24). In addition, networks constructed by others (31) predict correct functional category with an accuracy exceeding 70% and display the intuitively expected clustering of proteins with similar cellular location.

Although the quality of data on interactions, whether based on experimental yeast two hybrid determinations, co-precipitation, or computation, is imperfect and varies with method, using this information is nonetheless important since it generates hypotheses that can then be pursued in more detail. The problem of imperfect data is of course common, even in widely used methods such as microarrays, in which the number of false positives can be high. When these methods are used in conjunction with one another, the reliability of the conclusions increases, and they generate plausible hypotheses that focus and thereby substantially increase the efficiency of subsequent experimental studies. In the next section, we consider microarray technology and its applications in greater detail.

OLIGONUCLEOTIDE ARRAYS

Arrays can be manufactured by a number of methods, including in situ synthesis of oligonucleotides that are the reverse complements of the target genes and deposition of PCR-amplified gene fragments, either from plasmid DNA or clones in culture, and using them to probe complementary sequences from target cells.

In situ synthesis has the advantage of requiring only the sequence of the genes to be probed, and it can therefore be used to probe for a much larger number of transcripts. In situ methods also offer additional advantages in specificity. However, their use is costly, and overriding economic considerations make deposition methods, which are themselves both useful and versatile, the method of choice in many academic laboratories.

The studies of interest in our laboratory require the high specificity of in situ methods, but, because we are interested in diagnostic and analytical applications to large human populations, they must also be cost effective. We have therefore developed an array fabricator using in situ synthesis, but with a procedure made less expensive by the replacement of physical masks with an adaptable lens technology (33) and an optical system that is good enough in principle to obtain 15-μm pixels. We have also developed probe optimization strategies (Bhanot et al., submitted) and signal-processing algorithms to increase specificity.

Our interest in DNA microarray technology arises in part because it augments other methods for allocating newly sequenced genes to known pathways or for discovering new pathways, only more broadly because of its potential as a tool for understanding the molecular basis of diseases, including the identification and modulation of aberrant pathways. Until recently, studies of the molecular basis of pathogenesis relied on the limited one-gene-at-a-time approach. The technological advances underlying DNA chips coupled with draft human and mouse genome sequences and the complete sequences of many other organisms, including *Drosophila* and more than 50 microbes, will not only accelerate this research but will fundamentally change how we think about disease etiology. Discrete molecular changes to cellular pathways can ultimately be placed in context of the topology and stability of complex cell circuits, their haplotype dependence, and their robustness and plasticity in the face of endogenous and exogenous perturbations. Disease classification, currently based on morphology/histology, will soon be replaced by gene expression profiles and new molecular classification schemes that have pathogenetic and therapeutic implications.

Two of the most exciting applications of DNA microarray technology to the study of human disease have been in the areas of cancer biology and host response to pathogens. The potential for the development of the array-based molecular classification of cancer and infectious disease will likely have an enormous impact on the clinical practice of medicine within the next decade. In addition, the identification of key cellular pathways involved in oncogenesis and the immune response to infection will enable the development of novel targeted therapies for these diseases.

CANCER BIOLOGY

Introduction

Cancer arises from an accumulation of mutations over time in a series of genes. Gene expression profiling with DNA microarrays is a powerful tool for describing cancer cells (5). In principle, it can identify therapeutic targets based on transcript-level differences between normal and malignant tissue and is very useful in developing molecular classification schemes that will inform therapeutic choices. Until recently, cancer classification has been based primarily on histological appearance, yet tumors with similar histology may have different pathogeneses, clinical courses, and responses to therapy. In addition, expression profiling can identify genes silenced by methylation or other epigenetic mechanisms.

Class Prediction/Class Discovery

A large segment of the published DNA array cancer research has focused on diagnostic classifications, allocating tumors either to known classes or to previously unrecognized classes or subclasses. Several computational methods have been devised in an attempt to select a relatively small number of informative genes from the extremely large amount of data produced by array experiments. In the hallmark paper by Golub et al. (13), a supervised learning procedure was used to obtain a group of 50 genes that could classify an unknown sample as either acute myelogenous or acute lymphoblastic leukemia, two closely related hematological malignancies that can be difficult to distinguish histologically. MacDonald et al. (21) adapted this procedure to classify medulloblastoma as metastatic or non-metastatic, based on a committee of 85 genes.

We have modified the Golub algorithm and applied it to gene expression profiling of lung adenocarcinomas in smokers and nonsmokers (manuscript in preparation). We have established three sets of gene committees (100 genes per committee) capable of classifying a lung tissue sample as normal or cancerous in a smoker or nonsmoker. In all of these studies, the accuracy of the class prediction model was first tested by cross-validation on the initial data set and then on an independent set of samples.

Recently, a number of additional, supervised methods have been applied to class prediction, including artificial neural networks (ANN) and support vector machines. Khan et al. (18) used ANN to classify small, round blue-cell tumors into their four distinct diagnostic categories. In a study by Selaru et al. (32), ANNs were shown to have the potential to discriminate among the subtly different clinical entities of sporadic colorectal adenomas and cancers (SAC) and inflammatory bowel disease (IBD)-associated dysplasias and cancers. Van't Veer et al. (38) applied supervised classification to identify a breast cancer gene expression signature strongly predictive of a short interval to distant metastases ("poor prognosis" signature) in patients without tumor cells in local lymph nodes at diagnosis. Ramaswamy et al. (27) used a support vector machine algorithm to correctly classify 78% of 218 tumor samples (spanning 14 common tumor types) and 90 normal tissue samples. This same data set was used in a comparison study of various multi-class prediction algorithms that found that support vector machines (SVMs) outperformed the weighted committee method (43).

Allocation of an unknown sample to known categories is readily carried out using supervised learning methods in which algorithms are trained to recognize known distinctions between classes. When discovery is the goal, distinctions by definition are not known in advance, and unsupervised methods must be used to uncover novel subcategories of malignancy that may have prognostic and therapeutic significance. Alizadeh et al. (1) used hierarchical clustering to uncover two molecularly distinct forms of diffuse large-B-cell lymphoma that correlated with survival after treatment with multi-agent chemotherapy regimens. Similarly, hierarchical clustering was used by Rosenwald et al. (29) to formulate a molecular predictor of survival after chemotherapy for diffuse large-B-cell lymphoma. Hierarchical clustering has also been used to sub-classify lung adenocarcinoma into subgroups that correlate with patient survival (11).

BIOLOGICAL INSIGHT INTO CANCER PATHOGENESIS

Aside from cancer classification and prognostication, the other main goal of cancer microarray research is to define the molecular pathways associated with oncogenesis and translate these discoveries into effective treatment and prevention regimens. At its simplest level, microarrays can identify therapeutic targets based on transcript-level differences between normal and malignant tissue. Several statistical techniques have been developed to facilitate the identification of differentially expressed genes in microarray experiments (21,37). Once differentially expressed genes are identified, in vitro and in vivo experiments can be performed to assess the functional significance of these genes. Clark et al. (3) used microarrays to identify a group of extracellular matrix assembly and actin-based cytoskeleton genes that are associated with melanoma metastasis. These investigators subsequently demonstrated that overexpression of one of those genes

(RhoC) enhanced metastasis, while inhibition of this gene markedly suppressed metastases in a melanoma cell line. MacDonald et al. (21) found a group of genes downstream of the RAS/mitogen-activated protein kinase signal transduction pathway that are up-regulated in metastatic medulloblastoma. Using in vitro assays, neutralizing antibodies to a number of the protein products of these genes were shown to inhibit cell migration. By comparing gene expression profiles between the normal lung and lung adenocarcinoma from the same individuals, we found that up-regulation of Ras-related signaling pathways occurs in the transition from normal lung to cancer in smokers, while a decrease in TGFβ pathway-related genes occurs in nonsmokers (submitted).

While identification of differentially expressed genes is clearly important, the integration of the DNA microarray data with data from other platforms is central to placing gene expression alterations in biological context. Several novel strategies to integrate genomic and proteomic data from different platforms are currently in development. By coupling gene expression profiles from our lung adenocarcinoma study with interactome data from *Saccharomyces cerevisiae* found in public databases, we were able to strengthen functional inferences regarding differentially expressed genes and to help suggest where in a pathway the genes interact and exact points at which cellular pathways intersect (manuscript in preparation). Furthermore, Ge et al. (12) have shown that genes with similar expression profiles are more likely to encode interacting proteins.

Potential Problems

While the application of DNA microarrays to cancer research holds great promise, there are a number of important limitations to the current technology. Due to the relatively high cost of commercial microarrays, the sample sizes in most studies have been relatively small, resulting in difficulties with the analysis caused by the multidimensional nature of the data. Academic institutions have begun to produce inexpensive arrays that will allow researchers to design experiments with a larger number of samples to achieve better results. However, to generate high-throughput cDNA microarrays, thousands of cDNA clones have to be processed through many error-prone steps. In fact, most high-density cDNA arrays contain an error rate in excess of 10% (44). Thus, it is essential that vigorous quality-control measures are instituted, including sequencing all cDNA before the printing and placement of replicate spots on the array.

There are many potential sources of biological variability involved in cancer microarray data acquisition and analysis. Tumor tissues that are surgically removed from patients can differ in stage, grade, extent to which necrosis is present, and contain different compositions of cell type. In addition, samples taken from within the same tumor from the same individual may show marked heterogeneity. As a result, it is essential for pathologists to characterize tissue samples taken for microarray research and attempt to match patients for their stage and grade of tumor. Another approach to address the issue of tissue heterogeneity is to isolate specific cell types using laser capture microdissection (LCM) (8). While LCM can isolate pure populations of cancer cells from tissue, several amplification procedures are needed to analyze the mRNA population from a few hundred living cells (39). Given that amplification methods can yield false positives, confirmatory studies are a necessity. In addition, targeted analysis of purified cancer cells using LCM may fail to reflect the whole picture of the cancer. Cancer cells do not exist in a vacuum, and the combination of stromal and malignant cells and their interactions most accurately defines the genotype of a tumor (14). The surrounding stromal cells, inflammatory cells, and endothelial cells are clearly distinct from normal cells in the nontumor region, as confirmed by the finding that tumor endothelium cells have a unique gene expression profile (35).

HOST RESPONSE TO PATHOGENS

Introduction

The complex interaction between a microbial pathogen and host is the underlying basis of infectious disease. Each pathogen evolves its own particular strategy for manipulation of the host that optimizes microbial survival and transmission. DNA microarrays have the potential to offer an unparalleled view of the molecular events that underlie the host response to microbial pathogens. By understanding the transcriptional events of this interaction, we can identify host defense strategies, mechanisms by which they are regulated, and potentially identify microbial virulence factors. In addition, class prediction and class discovery algorithms used in cancer research can be applied to infectious disease to identify host diagnostic signature expression profiles and predictors of the clinical course of infection.

Insight into Microbial Pathogenesis

Microarrays have the potential to accelerate our understanding of the host response to microbial infection. By studying the entire host genome response simultaneously, DNA arrays may identify groups of genes activated or repressed in response to microbial infection. By performing time point experiments after infection, microarrays may allow us to follow the temporal sequence of gene activation and repression, enabling key regulatory pathways to be identified.

The first application of microarray technology to the study of host infection involved human fibroblasts infected with cytomegalovirus (CMV) (45). The authors measured fibroblast gene expression levels at three time points after infection and found several interesting and previously unrecognized features of CMV infection including up-regulation of the gene encoding HLA-E as a means to avoid natural killer cell surveillance. A similar experimental design has been used to study the effects of HIV infection on cultured CD4+ T cells. Corbeil et al. (6) demonstrated substantial HIV-induced transcriptional changes occur very early after infection with the activation of a number of cytokines.

Recently, DNA microarrays have been used to study host response to a variety of bacterial pathogens. Eckmann et al. (7) have demonstrated increased expression of many proinflammatory cytokines and chemokines in epithelial cells infected by *Salmonella.* A similar study (28) demonstrated that exposure of macrophages to purified *Salmonella* lipopolysaccharide resulted in a very similar profile to whole cells. Cohen et al. (4) examined the expression profiles of human promyelocytic cells to *Listeria monocytogenes* infection. Comparison of the data from the *Salmonella* and *L. monocytogenes* experiments suggests similarities in the proinflammatory response to both organisms but differences in the apoptotic response (induction of apoptotic genes by *Salmonella*).

While the experiments described earlier are primarily exploratory, hypothesis-driven microarray research of the differences between closely related strains of the same pathogen has attempted to uncover novel virulence factors. Comparison of expression profiles in human monocytes infected with two strains of *Ebola* virus, one infectious and one not, revealed differences in transcription response (40). We found a subset of human alveolar macrophage (AM) apoptosis-related genes to be differentially regulated by mycobacteria of differing virulence (submitted). In addition, the effect of TNF antibody was shown to change the AM gene expression profile of the avirulent *Mycobacterium tuberculosis* (Mtb) infection to resemble the AM gene expression profile of the virulent infection. Our data suggest that TNF activity is a key component in the elaboration of the virulence-associated phenotype of AM apoptosis after infection with Mtb.

Diagnostic Gene Expression Profiles

While many studies have applied class prediction and discovery algorithms to microarray studies in cancer, few studies have attempted to classify infectious disease based on host signature expression profile. To apply this paradigm to infectious disease, we must demonstrate that specific pathogens elicit a unique transcription response in the host through a unique constellation of virulence factors. By measuring gene expression profiles in peripheral blood mononuclear cells (PBMCs), diagnostic signature expression profiles for specific pathogens can be found. However, we must first determine whether exposure to a pathogen leads to a persistent and unique expression profile in PBMCs from patients with different genetic backgrounds.

There is some preliminary data to suggest that this is feasible. Manger et al. (22) have studied gene expression in a monocytic cell line after exposure to several members of the same Gram-negative bacterial genus and multiple strains of the same species. Hierarchical clustering revealed that large groups of genes respond in a similar fashion, yet a small subset of genes responded in a more discriminating manner. However, in vivo studies that address the importance of host-host variability are still necessary. Our study with Mtb found significant differences in gene expression response in human AMs from different individuals infected with the same strain of Mtb. We were, however, able to uncover a small number of genes that changed consistently between different hosts.

There are many potential advantages of using host gene expression patterns as diagnostic markers for infection. A single, easily accessible sample (i.e., PBMCs) could be used to diagnose exposure to a wide variety of pathogens. Pathogens that are difficult to culture or are found in poorly accessible anatomic sites may still be detected. In addition, we may be able to infer the amount of time that has elapsed since exposure based on variation in the signature expression profile. Finally, novel subclasses of a number of pathogens may be uncovered with prognostic and therapeutic implications.

REFERENCES

1. **Alizadeh, A.A., M.B. Eisen, R.E. Davis, C. Ma, I.S. Lossos, A. Rosenwald, J.C. Boldrick, H. Sabet, et al.** 2000. Distinct types of diffuse large B-cell lymphoma identified by gene expression profiling. Nature *403*:503-511.
2. **Ashburner, M., C.A. Ball, J.A. Blake, D. Botstein, H. Butler, J.M. Cherry, A.P. Davis, K. Dolinski, et al.** 2000. Gene ontology: tool for the unification of biology. The Gene Ontology Consortium. Nat. Genet. *25*:25-29.
3. **Clark, E.A., T.R. Golub, E.S. Lander, and R.O. Hynes.** 2000. Genomic analysis of metastasis reveals an essential role for RhoC. Nature *406*:532-535.
4. **Cohen, P., M. Bouaboula, M. Bellis, V. Baron, O. Jbilo, C. Poinot-Chazel, S. Galiegue, E.H. Hadibi, and P. Casellas.** 2000. Monitoring cellular responses to Listeria monocytogenes with oligonucleotide arrays. J. Biol. Chem. *275*:11181-11190.
5. **Cole, K.A., D.B. Krizman, and M.R. Emmert-Buck.** 1999. The genetics of cancer—a 3D model. Nat. Genet. *21(1 Suppl)*:38-41.
6. **Corbeil, J., D. Sheeter, S. Rought, P. Du, M. Ferguson, D.R. Masys, et al.** 1999. Magnitude and specificity of temporal gene expression during HIV-1 infection of a CD4+ T cell. Nat. Genet. *23*:39-40.
7. **Eckmann, L., J.R. Smith, M.P. Housley, M.B. Dwinell, and M.F. Kagnoff.** 2000. Analysis by high density cDNA arrays of altered gene expression in human intestinal epithelial cells in response to infection with the invasive enteric bacteria Salmonella. J. Biol. Chem. *275*:14084-14094.
8. **Emmert-Buck, M.R., R.F. Bonner, P.D. Smith, R.F. Chuaqui, Z. Zhuang, S.R. Goldstein, R.A. Weiss, and L.A. Liotta.** 1996. Laser capture microdissection. Science *274*:998-1001.
9. **Enright, A.J., I. Iliopoulos, N.C. Kyrpides, and C.A. Ouzounis.** 1999. Protein interaction maps for complete genomes based on gene fusion events. Nature *402*:86-90.
10. **Gaasterland, T. and M.A. Ragan.** 1998. Constructing multigenome views of whole microbial genomes. Microb. Comp. Genomics *3*:177-192
11. **Garber, M.E., O.G. Troyanskaya, K. Schluens, S. Petersen, Z. Thaesler, M. Pacyna-Gengelbach, M. van de Rijn, G.D. Rosen, et al.** 2001. Diversity of gene expression in adenocarcinoma of the lung. Proc. Natl. Acad. Sci. USA *98*:13784-13789.
12. **Ge, H., Z. Liu, G.M. Church, and M. Vidal.** 2001. Correlation between transcriptome and interactome mapping data from *Saccharomyces cerevisiae*. Nat. Genet. *29*:482-486.
13. **Golub, T.R., D.K. Slonim, P. Tamayo, C. Huard, M. Gaasenbeek, J.P. Mesirov, H. Coller, M.L. Loh, et al.** 1999. Molecular classification of cancer: class discovery and class prediction by gene expression monitoring. Science *286*:531-537.
14. **Hanahan, D. and R.A. Weinberg.** 2000. The hallmarks of cancer. Cell *100*:57-70.

15.**Hazbun, T.R. and S. Fields.** 1998. Networking proteins in yeast. Proc. Natl. Acad. Sci. USA *98*:4277-4278.
16.**Huynen, M.A. and P. Bork.** 1998. Measuring genome evolution. Proc. Natl. Acad. Sci. USA *95*:5849-5856.
17.**Ito, T., K. Tashiro, S. Muta, R. Ozawa, T. Chiba, M. Nishizawa, K. Yamamoto, S. Kuhara, and Y. Sakaki.** 2000. Toward a protein-protein interaction map of the budding yeast: a comprehensive system to examine two-hybrid interactions in all possible combinations between the yeast proteins. Proc. Natl. Acad. Sci. USA *97*:1143-1147.
18.**Khan, J., J.S. Wei, M. Ringner, L.H. Saal, M. Ladanyi, F. Westermann, F. Berthold, M. Schwab, et al.** 2001. Classification and diagnostic prediction of cancers using gene expression profiling and artificial neural networks. Nat. Med. *7*:673-679.
19.**Kohn, K.W.** 1999. Molecular interaction map of the mammalian cell cycle control and DNA repair systems. Mol. Biol. Cell *10*:2703-2734.
20.**Lipshutz, R.J., S.P. Fodor, T.R. Gingeras, and D.J. Lockhart.** 1999. High density synthetic oligonucleotide arrays. Nat. Genet. *21*:20-24.
21.**MacDonald, T.J., K.M. Brown, B. LaFleur, K. Peterson, C. Lawlor, Y. Chen, R.J. Packer, P. Cogen, and D.A. Stephan.** 2001. Expression profiling of medulloblastoma: PDGFRA and the RAS/MAPK pathway as therapeutic targets for metastatic disease. Nat. Genet. *29*:143-152.
22.**Manger, I. and D. Relman.** 2000. How the host "sees" pathogens: global gene expression responses to infection. Curr. Opin. Immunol. *12*:215-218.
23.**Marcotte, E.M., M. Pellegrini, H.L. Ng, D.W. Rice, T.O. Yeates, and D. Eisenberg.** 1999. Detecting protein function and protein-protein interactions from genome sequences. Science *285*:751-753.
24.**Mendenhall, M.D. and A.E. Hodge.** 1998. Regulation of Cdc28 cyclin-dependent protein kinase activity during the cell cycle of the yeast *Saccharomyces cerevisiae.* Microbiol. Mol. Biol. Rev. *62*:1191-1193.
25.**Overbeek, R., M. Fonstein, M. D'Souza, G.D. Pusch, and N. Maltsev.** 1999. The use of gene clusters to infer functional coupling. Proc. Natl. Acad. Sci. USA *96*:2896-2901.
26.**Pellegrini, M., E.M. Marcotte, M.J. Thompson, D. Eisenberg, and T.O. Yeates.** 1999. Assigning protein functions by comparative genome analysis: protein phylogenetic profiles. Proc. Natl. Acad. Sci. USA *96*:4285-4288.
27.**Ramaswamy, S., P. Tamayo, R. Rifkin, S. Mukherjee, C.H. Yeang, M. Angelo, C. Ladd, M. Reich, et al.** 2001. Multiclass cancer diagnosis using tumor gene expression signatures. Proc. Natl. Acad. Sci. USA *98*:15149-15154.
28.**Rosenberger, C.M., M.G. Scott, M.R. Gold, R.E. Hancock, and B.B. Finlay.** 2000. Salmonella typhimurium infection and lipopolysaccharide stimulation induce similar changes in macrophage gene expression. J. Immunol. *164*:5894-5904.
29.**Rosenwald, A., G. Wright, W.C. Chan, J.M. Connors, E. Campo, R.I. Fisher, R.D. Gascoyne, H.K. Muller-Hermelink, et al.** 2001. The use of molecular profiling to predict survival after chemotherapy for diffuse large-B-cell lymphoma. N. Engl. J. Med. *346*:1937-1947.
30.**Schena, M., R.A. Heller, T.P. Theriault, K. Konrad, E. Lachenmeier, and R.W. Davis.** 1998. Microarrays: biotechnology's discovery platform for functional genomics. Trends Biotechnol. *16*:301-306.
31.**Schwikowski, B., P. Uetz, and S. Fields.** 2001. A network of protein-protein interactions in yeast. Nat. Biotechnol. *18*:12571261.
32.**Selaru, F.M., Y. Xu, J. Yin, T. Zou, T.C. Liu, Y. Mori, J.M. Abraham, F. Sato, et al.** 2002. Artificial neural networks distinguish among subtypes of neoplastic colorectal lesions. Gastroenterology *122*:606-613.
33.**Singh-Gasson, S., R.D. Green, Y. Yue, C. Nelson, F. Blattner, M.R. Sussman, and F. Cerrina.** 1999. Maskless fabrication of light-directed oligonucleotide microarrays using a digital micromirror array. Nat. Biotechnol. *17*:974-978.
34.**Southern, E., K. Mir, and M. Shchepinov.** 1999. Molecular interactions on microarrays. Nat. Genet. *21*:5-9.
35.**St. Croix, B., C. Rago, V. Velculescu, G. Traverso, K.E. Romans, E. Montgomery, A. Lal, G.J. Riggins, et al.** 2000. Genes expressed in human tumor endothelium. Science *289*:1197-1202.
36.**Tatusov, R.L., E.V. Koonin, and D.J. Lipman.** 1997. A genomic perspective on protein families. Science *278*:631-637.
37.**Tusher, V.G., R. Tibshirani, and G. Chu.** 2001. Significance analysis of microarrays applied to the ionizing radiation response. Proc. Natl. Acad. Sci. USA *98*:5116-5121.

38. **Van't Veer, L.J., H. Dai, M.J. van de Vijver, Y.D. He, A.A. Hart, M. Mao, H.L. Peterse, K. van der Kooy, et al.** 2002. Gene expression profiling predicts clinical outcome of breast cancer. Nature *415*:530-536.

39. **Wang, E., L.D. Miller, G.A. Ohnmacht, E.T. Liu, and F.M. Marincola.** 2000. High-fidelity mRNA amplification for gene profiling. Nat Biotechnol *18*:457-459.

40. **Xiang, C., H. Young, H. Alterson, D. Reynolds, M. Bittner, Y. Chen, et al.** 1999. Comparison of cellular gene expression in Ebola-Zaire and Ebola-Reston virus-infected primary human monocytes. Nat. Genet. *23*:82-84.

41. **Yanai, I., A. Derti, and C. DeLisi.** 2001. Genes linked by fusion events are generally of the same functional category: a systematic analysis of 30 microbial genomes. Proc. Natl. Acad. Sci. USA *98*:7940-7945.

42. **Yanai, I., J.C. Mellor, and C. DeLisi.** 2002. Identifying functional links between genes using conserved chromosomal proximity. Trends Genet. *18*:176-179.

43. **Yeang, C.H., S. Ramaswamy, P. Tamayo, S. Mukherjee, R.M. Rifkin, M. Angelo, M. Reich, E. Lander, et al.** 2001. Molecular classification of multiple tumor types. Bioinformatics *17(Suppl. 1)*:S316-S322.

44. **Zhang, W., P.M. Laborde, K.R. Coombes, D.A. Berry, and S.R. Hamilton.** 2001. Cancer genomics: promises and complexities. Clin. Cancer Res. *7*:2159-2167.

45. **Zhu, H., J.P. Cong, G. Mamtora, T. Gingeras, and T. Shenk.** 1998. Cellular gene expression altered by human cytomegalovirus: global monitoring with oligonucleotide arrays. Proc. Natl. Acad. Sci. USA *95*:14470-14475.

Section VI

Emerging Technologies

18 Expression of siRNA from a pol III Promoter in Mammalian Cells

Makoto Miyagishi[1,2] and Kazunari Taira[1,2]

[1]Department of Chemistry and Biotechnology, School of Engineering, The University of Tokyo, Tokyo, and [2]Gene Function Research Laboratory, National Institute of Advanced Industrial Science and Technology (AIST), Tsukuba Science City, Japan

BACKGROUND

RNA interference (RNAi) refers to the sequence-specific silencing of gene expression induced by double-stranded RNA (dsRNA), and it has been demonstrated in various organisms, including plants, a nematode, *Drosophila*, and a protozoan. In mammalian cells, it has recently been reported that 21- or 22-nucleotide RNAs with 2-nucleotide 3′ overhangs [small interfering RNA (siRNA)] induce RNAi without the induction of the dsRNA-dependent inhibition of protein synthesis, which is known as the host defense system against viral infections. However, the introduction of siRNA into cells by lipofection appears to restrict the range of applications because of low transfection efficiencies in some types of cell and/or short-term persistence of silencing effects. Here we report a vector-based siRNA system for the expression of siRNAs that has the potential to induce RNAi in mammalian cells.

TECHNIQUES

We introduced the promoter of a human gene for U6 small nuclear RNA (snRNA) for the transcription of both strands of siRNA in vivo. This expression system is suitable for the transcription of small RNAs. Moreover, remarkably high levels of transcripts were observed when both strands were transcribed in

Perspectives in Gene Expression
Edited by K. Appasani

the same cell. Because RNAs transcribed from a U6 promoter include a stretch of approximately four uridines at the 3′-end (13), the transcribed siRNAs have an approximately 4-nucleotide overhang at each 3′-end, with the same activity as endogenous siRNAs with 2- or 3-nucleotide overhangs at the 3′-end.

RESULTS

To generate siRNAs in vivo, we designed an siRNA expression vector, in which the transcription of sense and antisense RNAs was driven independently by U6 promoters. Using a dual-luciferase reporter system, we demonstrated that this construct can efficiently suppress the expression of specific genes. Moreover, to demonstrate the power of the siRNA expression vector in suppressing the expression of endogenous genes, we constructed siRNA expression vectors targeted to transcripts of the gene for β-catenin, a protein involved in cadherin-mediated intercellular adhesion. The β-catenin-targeting siRNA was expressed from a plasmid that contained the origin of DNA replication of the Epstein-Barr virus (Ori-P), which is maintained stably and extrachromosomally in cells that express the Epstein-Barr virus nuclear antigen 1 (EBNA-1). Transformants with an siRNA expression plasmid targeted to β-catenin exhibited specific reduction in the expression of β-catenin.

CONCLUSION

We successfully engineered an siRNA expression vector with the ability to repress the expression of specific genes. Utilizing an EBNA-1/Ori-P system, we created transformants in which the expression of β-catenin was repressed. Thus, our siRNA expression system should be useful for "knock-down" by the RNAi of endogenous genes and has the potential to suppress gene expression through the generation of stable transformants. This technical advance in the silencing of gene expression should not only facilitate a wide range of functional analyses of mammalian genes but also might allow therapeutic applications by means of vector-mediated RNAi.

INTRODUCTION

The first evidence for gene disruption by dsRNA came from careful and elegant analysis in *Caenorhabditis elegans* (9). This phenomenon, called RNAi, was also observed subsequently in various organisms including plants, *C. elegans*, *Drosophila*, and a protozoan (8,11,19,23). Although the mechanism of RNAi is not fully understood, recent genetic and biochemical studies have unveiled the mechanism at the molecular level. In cells, dsRNAs are digested into fragments of 21–23 nucleotides with 2-nucleotide 3′ overhangs by an RNase III-related enzyme that has been recently identified as Dicer in *Drosophila* (2). Subsequently, the resultant siRNAs are incorporated into protein/RNA complexes (15) or RNA-induced silencing complexes (RISCs) (10) and appear to be unwound to single-stranded RNAs (15) that act as guide sequences. The active complexes containing the guide RNA recognize and cleave the target RNA.

Initially, RNAi was used as a tool to define the functions of unknown genes in various organisms, but the silencing of specific genes by RNAi was not easily detected in mammalian systems. The introduction of dsRNA into mammalian cells triggers the dsRNA-dependent nonspecific inhibition of protein synthesis; namely, the host defense system against viral infections. However, Tuschl and colleagues (4,6) made important progress toward overcoming this problem, and soon it was discovered that 21- or 22-nucleotide RNAs with 2-nucleotide 3′ overhangs, called siRNAs, induce gene silencing without inducing the nonspecific inhibition of translation in cultured mammalian cells. However, the efficiency of transfection for the expression of siRNAs in cells depends on cell type and when lipofection is performed, and the RNAi effect seems to be sustained only for a limited period of time (12,22). Thus, the application to the analysis of gene function and human gene therapy has been problematic. Therefore, a vector-based expression system for generating siRNA in vivo is required. We have developed a generally applicable method for inducing RNAi in mammalian cells using an siRNA expression vector.

METHODOLOGY

Constuction of U6-Driven siRNA Expression Vectors

The pU6icassette vector contained a human U6 promoter and two *Bsp*MI sites. *Bsp*MI is a member of Class IIS restriction enzymes that cleave outside their recognition sequences and allow the direct insertion of a DNA fragment immediately downstream of a U6 promoter. (See Figures 1 and 2 for depictions of plasmid construction.) DNA fragments harboring the sequences of the sense and antisense regions and the U6 promoter were amplified by PCR with the pU6icassette vector as template and Primer 1 and Primer 2, which included the sequences of the sense or antisense sequence plus the terminator (see Figure 2). After the digestion of the PCR products with *Bsp*MI, each fragment was ligated into the *Bsp*MI sites of the pU6icassette vector to yield a series of siRNA expression vectors. Sequences inserted in each primer (corresponding to NNNNNNNNNNNNNNNNNNN in Figure 2) were as follows (only the sense sequence is shown): hygromycin/EGFP, 5′-GTTCGACAGCGTCTCC-

GAC-3′; firefly luciferase, Site O, 5′-GCTATGAAACGATATGGG C-3′; Site A, 5′-GTTCGTCACATCTCATCTAC-3′; Site B, 5′-GTGCGCTGCTGGTGC-CAAC-3′; and Site C, 5′-ATGTACACGTTCGTCACAT-3′; *Renilla* luciferase, 5′-GTAGCGCGGTGTATTATAC-3′; and β-catenin, Site A, 5′-GTCCTGTA-TGAGTGGGAAC-3′; and Site B, 5′-GCAACAGTCTTACCTGGAC-3′. The pcOriP vector was made by PCR-based subcloning of the Ori-P sequence into pcDNA3 (Invitrogen, Carlsbad, CA, USA). The pcOriP/si β-catenin vector was generated by the PCR subcloning of siRNA expression cassette targeted against β-catenin into the pcOriP vector.

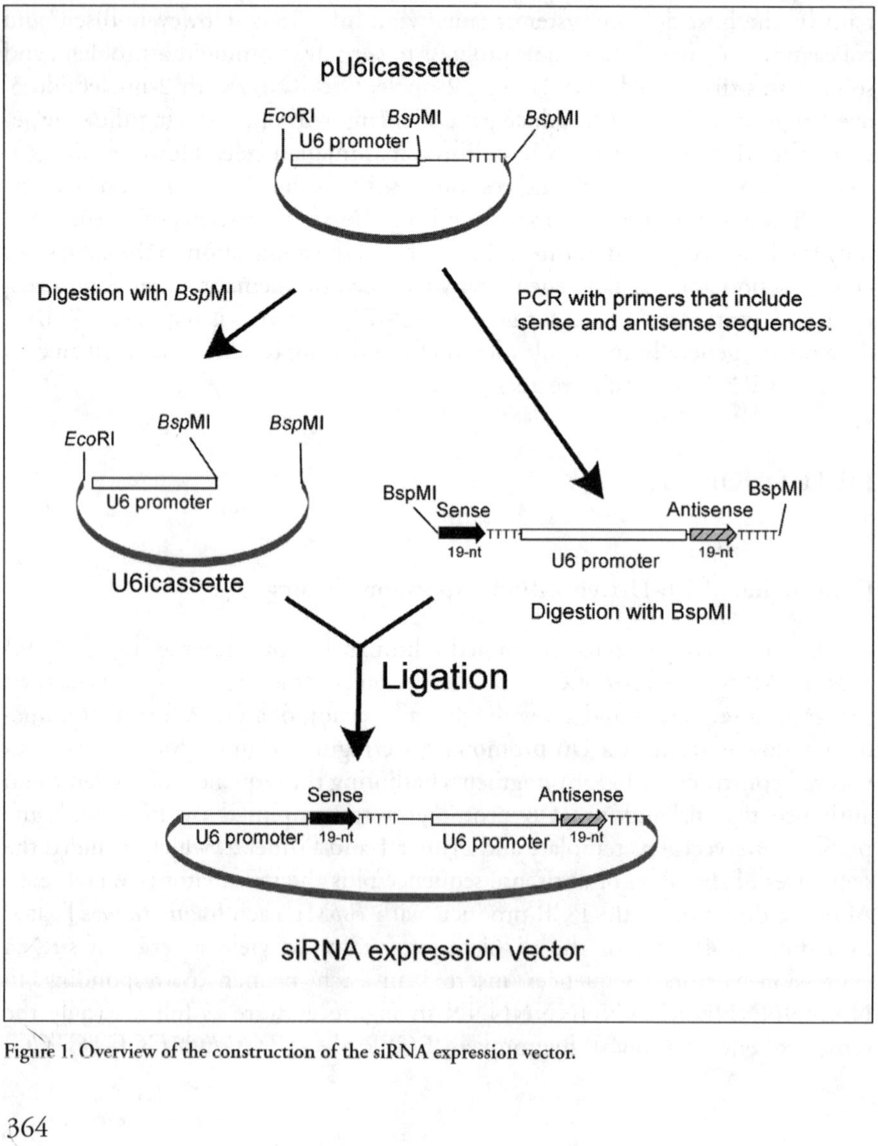

Figure 1. Overview of the construction of the siRNA expression vector.

Culture and Transfection

HeLa S3 cells and COS7 cells were cultured in DMEM supplemented with 10% FBS. HeLa S3/EBNA-1 cells were generated by stable transfection with a linearized form of pHygEGFP and hygromycin selection of a single clone, based on the level of expression of the hygromycin/EGFP fusion protein. Transfections were performed with Lipofectamine™ 2000 reagent (Invitrogen), as described by the manufacturer. In brief, 3×10^5 cells were plated in six-well plates one day before transfection. After the cells had been washed twice with PBS, 2 mL growth medium were added to each well. A solution of plasmid DNA (2 µg plasmid DNA in 100 µL optiMEM™ 1 medium) and a solution of Lipofectamine 2000 (4 µL Lipofectamine 2000 reagent in 100 µL of optiMEM 1 medium, preincubated for 5 min at room temperature) were mixed, and the mixture was kept at room temperature for 15 min. The mixed solution was added to the cells in 2 mL of the growth medium. The cultures were incubated for an additional 24 h and analyzed subsequently by fluorescence microscopy (see section entitled Fluorescence Analysis).

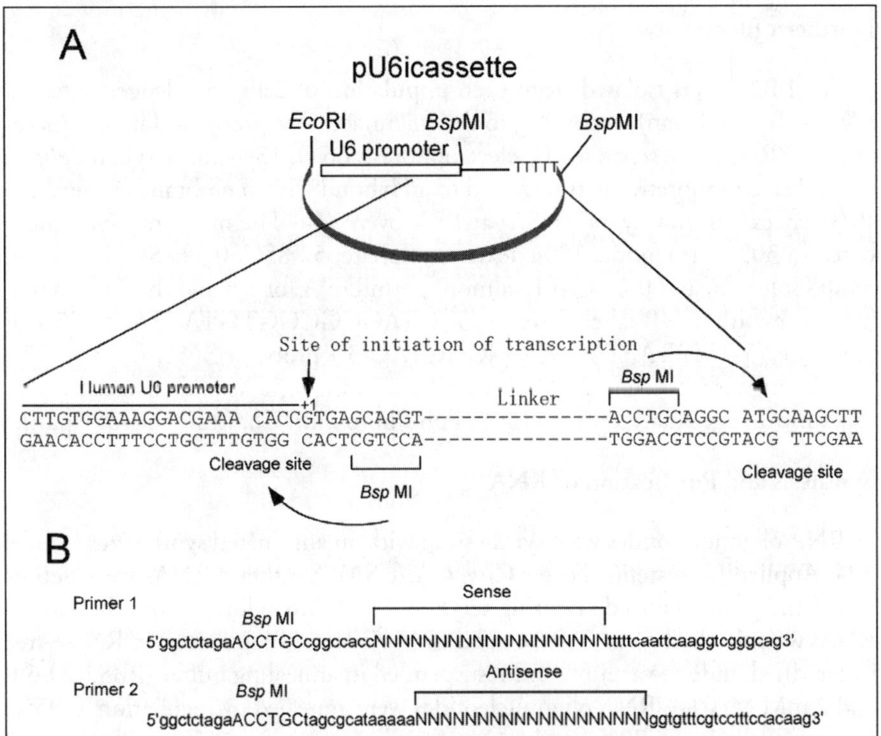

Figure 2. Cloning site of the pU6icassette vector. (A) Oligonucleotide sequences used for PCR (B) and the amplified product (C). NNNNNNNNNNNNNNNNNNNNN in each primer represents the sequence transcribed under the control of the U6 promoter. See text for specific sequences.

Fluorescence Analysis

HeLa S3 cells were co-transfected with 1 μg siRNA expression vector against hygromycin/EGFP transcript (psiHygEGFP), 1 μg hygromycin/EGFP expression vector (pHygEGFP; BD Biosciences Clontech, Palo Alto, CA, USA), 0.05 μg DsRed expression vector (pDsRed2; BD Biosciences Clontech), and 1 μg siRNA expression vector against hygromycin/EGFP (psiHygEGFP) or pU6. The cells were examined with a Leica TCS NT confocal microscope (Leica Microsystems, Mannheim, Germany).

Luciferase Assay

HeLa S3 cells were transfected with 30 ng of a *Renilla* luciferase plasmid (pRL-RSV) (14), 30 ng of a firefly luciferase expression vector (pGL3; Promega, Madison, WI, USA), and indicated amounts of siRNA duplex or siRNA expression vector in each well of 48-well plates using Lipofectamine 2000 reagent. Luciferase activities were analyzed after 24 h with a Dual Luciferase System (Promega).

Northern Blot Analysis

Total RNA was isolated from each population of cells with Isogen Reagent (Wako, Tokyo, Japan), according to the manufacturer's protocol. Fifteen micrograms of RNA were separated by electrophoresis on an 18% polyacrylamide/urea gel and electrophoretically transferred to a Hybond® N+ membrane (Amersham Biosciences, Piscataway, NJ, USA) at 15 V overnight. The membrane was incubated in 30% formamide, 10% dextran sulphate, 5× SSC, 0.5% SDS, 1× Denhardt's solution, and 0.2 mg/mL salmon sperm DNA for 1 h and then allowed to hybridize with a ^{32}P-labeled sense (5′-GTAGCGCGGTGTATTATAC-3′) or antisense (5′-GTATAATACACCGCGCTAC-3′) probe at 36°C for 3 h. The membrane was washed in 2× SSC twice for 15 min at 36°C and analyzed with a Fujix Bio-Image Analyzer (BAS1000; Fuji Photo Film Co. Ltd., Tokyo, Japan).

Synthesis and Purification of RNA

RNA oligonucleotides were synthesized with an automated synthesizer (model 394; Applied Biosystems, Foster City, CA, USA). Synthetic RNAs were deprotected and purified on a denaturing acrylamide gel. After elution from the gel, the RNAs were desalted on a NAP-10 column (Amersham Biosciences) in RNase-free water, dried under a vacuum, and resuspended in annealing buffer (PBS, pH 6.8 and 2 mM $MgCl_2$). RNA oligonucleotides were annealed by incubation at 95°C for 1 min, cooling to 70°C, and subsequent slow cooling to 4°C for 2 h at a concentration of 10 μM. The sequences of RNA oligonucleotides were as follows: firefly luciferase, Site O, 5′-GCUAUGAAACGAUAUGGGCUU-3′ (sense) and 5′-GCCCAUAUCGUUUCAUAGCUU-3′ (antisense); Site A, 5′-GUUCGU-

CACAUCUCAUCUACUU-3′ (sense) and 5′-GUAGAUGAGAUGUGAC-GAAUU-3′ (antisense); Site B, 5′-GUGCGCUGCUGGUGCCAACUU-3′ (sense) and 5′-GUUGGCACCAGCAGCGCACUU-3′ (antisense); Site C, 5′-AUGUACACGUUCGUCACAUUU-3′ (sense) and 5′-AUGUGACGAA-CGUGUACAUUU-3′ (antisense); and Renilla luciferase, 5′-GUAGCGCGGU-GUAUUAUACUU-3′ (sense; 2-nucleotide overhang) and 5′-GUAUAAUA-CACCGCGCUACUU-3′ (antisense; 2-nucleotide overhang).

Western Blot Analysis

Western blot analysis was performed essentially as described previously (14). In brief, cells were lysed in a Triton® X-100-based buffer (1% Triton X-100, 100 mM NaCl, 20 mM Tris-HCl, pH 7.6) with protease inhibitors (Complete™; Roche Applied Science, Mannheim, Germany), and the resulting lysates were cleared by centrifugation at 20 000× *g* for 15 min. Proteins (30 μg/well) were resolved by SDS-PAGE (7.5% polyacrylamide) and transferred to a PVDF membrane. The membrane was blocked in 3% skim milk for 2 h and then probed with primary antibody for 1 h at room temperature. After incubation with HRP-conjugated goat antibodies against mouse IgG or against rabbit IgG as secondary antibody (Amersham Biosciences) in blocking buffer, the immunoblots were visualized with an enhanced chemiluminescence system (Amersham Biosciences).

PROJECTS AND SAMPLE RESULTS

Design of an siRNA Expression Vector

We chose a U6 promoter for construction of an siRNA expression vector for use in mammalian cells. Because, unlike other pol III promoters (17), the U6 promoter does not contain internal promoter sequences, such as the A Box and B Box, we considered that a U6 promoter might allow direct transcription of small RNAs that can subsequently form siRNA duplexes in cells. We constructed an siRNA expression vector in which 19-nucleotide sense and antisense sequences directed against a target gene were placed under the control of U6 promoters, as shown in Figure 3. Since RNAs transcribed by a U6 promoter include approximately four uridines at the 3′-end (3), the resultant siRNAs should have an approximately 4-nucleotide overhang at each 3′-end.

Suppression of Expression of an Exogenously Introduced Gene for a Hygromycin/EGFP Fusion Protein

To assess the effect of the U6 promoter-driven siRNA, we used reporter plasmids that harbored the gene for a hygromycin/EGFP fusion protein or DsRed. Expression of these reporters is easily visualized by fluorescence microscopy in living cells. We constructed an siRNA-generating vector (psiHygEFGP) to target hygromycin/EGFP transcripts and co-transfected HeLa S3 cells with this vector plus the hygromycin/EGFP expression vector (pHygEGFP) and the DsRed expression vector (pDsRed2) into HeLa S3 cells. Figure 4 shows cells co-transfected with psiHygEGFP exhibited a significant reduction in hygromycin/EGFP expression compared to cells transfected with the pU6 control vector. By contrast, there was no difference in the expression of DsRed, which was used as an internal control. Likewise, co-transfection of COS7 cells with the psiHygEGFP and pHygEGFP plasmid plus the pDsRed2 plasmid resulted in the reduced expression of hygromycin/EGFP, without the loss of expression of DsRed (data not shown).

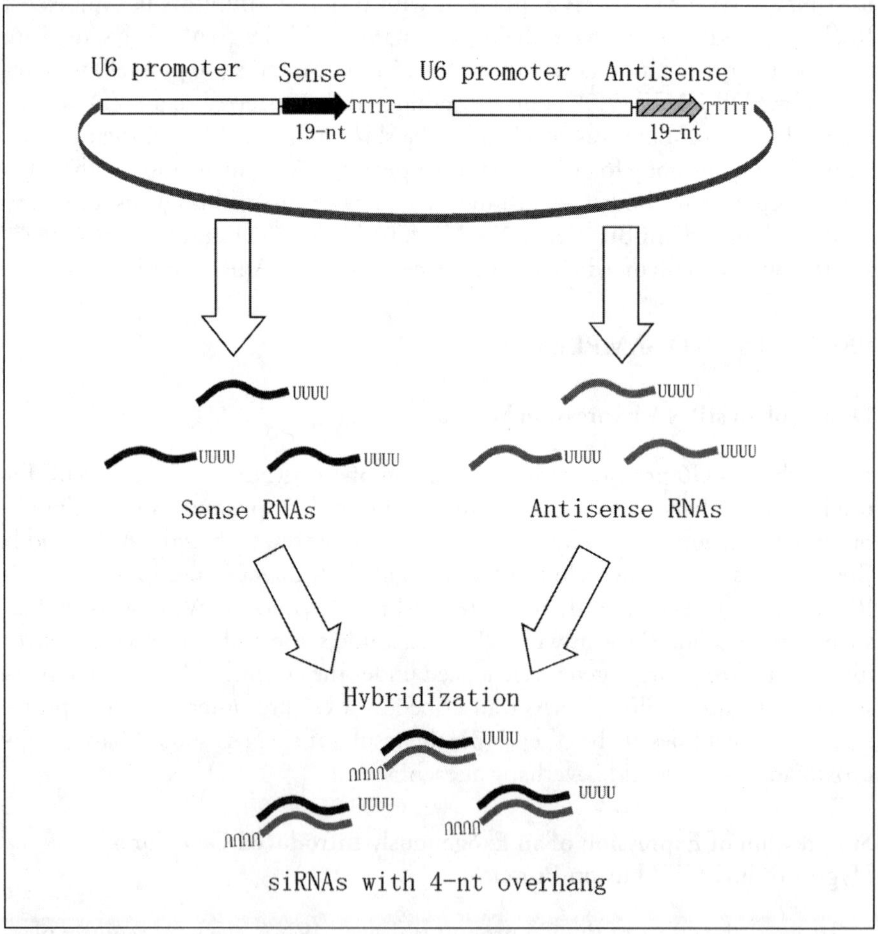

Figure 3. Strategy for generating siRNA using the U6 promoter. Each U6 promoter directs the transcription of a sense or antisense small RNA with approximately four uridines at the 3′-end. Subsequently, sense and antisense RNAs are annealed and form siRNA duplexes with a 4-nucleotide overhang at each 3′-end. nt, nucleotide.

Quantitative Analysis of Effects of siRNA Using Reporter Genes for Luciferase

To confirm the RNAi effect of the U6 promoter-driven siRNAs quantitatively, we tested for the ability of the siRNA to repress the expression of reporter genes for firefly luciferase and *Renilla* luciferase. Co-transfection experiments with HeLa S3 cells revealed that an siRNA expression vector directed against firefly luciferase disrupted the expression of the firefly gene for luciferase specifically without altering the activity of *Renilla* luciferase, and the effect was dose-dependent. The converse was true for an siRNA expression vector directed against *Re-*

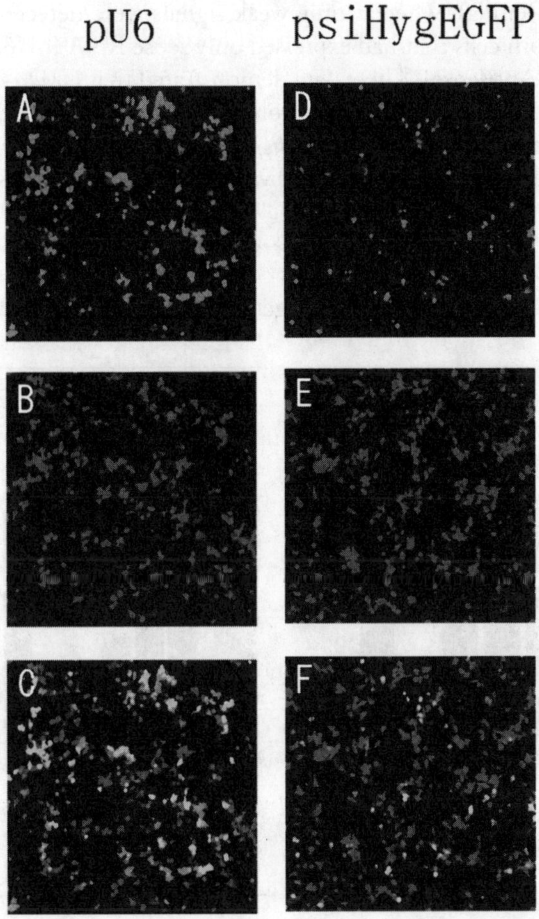

Figure 4. Effects of an siRNA expression vector targeted against the gene for a hygromycin/EGFP fusion gene. The top panels (A and D) show the expression of hygromycin/EGFP, and the middle panels (B and E) show the expression of DsRed. The lower panels (C and F) show overlay images caused by hygromycin/EGFP and DsRed. Cells that co-expressed hygromycin/EGFP and DsRed appear as yellow cells. (**See color plate A35.**)

nilla luciferase (Figure 5). Plasmids encoding only sense or antisense RNA affected expression of neither firefly nor *Renilla* luciferase. These results demonstrated that the U6 promoter-driven siRNA system was able to disrupt efficiently and specifically the expression of a gene of interest.

Northern Blot Analysis of U6-Driven siRNAs

Transcripts expressed from each plasmid were identified by Northern blot analysis (Figure 6). Fragments of expected length (with respect to 5′-phosphorylated 21- and 23-mer RNAs) were clearly discernible when RNAs were extracted from cells that had expressed both sense and antisense strands (Figure 6; the lane indicated by psiRenilla), whereas only weak signals were detected when RNAs were extracted from cells that had expressed only sense RNA (pU6Sense) or antisense RNA (pU6Antisense). These data demonstrated that (*i*) U6-driven siRNAs can be produced in cells, and (*ii*) they probably form duplexes in cells and might be protected so that (*iii*) their lifetimes might be significantly longer than those of sense and antisense strands by themselves when the latter are expressed separately.

Dependence on Target Sites of the Effects of siRNA

The efficacy of antisense molecules is generally strongly dependent on the cho-

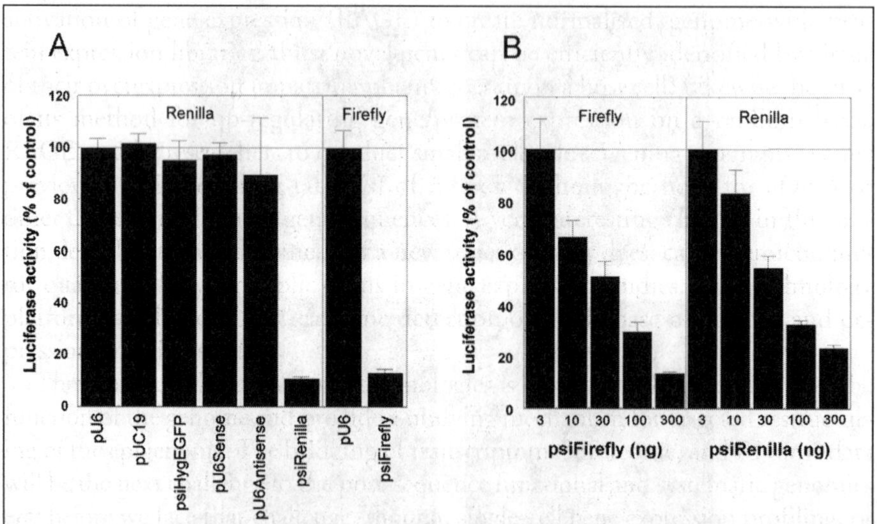

Figure 5. Effects of siRNA expression vectors directed against *Renilla* or firefly luciferase. (A) HeLa S3 cells were co-transfected with 30 ng of an RSV-*Renilla* luciferase plasmid (pRL-RSV), 30 ng of a firefly luciferase plasmid (pGL3), and 300 ng sense RNA expression vector, antisense RNA expression vector, or siRNA expression vector against firefly luciferase or *Renilla* luciferase. (B) Dose dependence of the effects of the siRNA expression vectors against *Renilla* or firefly luciferase. To ensure that equal amounts (300 ng) of DNA were used, empty plasmids (pU6) were included in each transfection.

sen target site (21) and, thus, it is possible that siRNA-generating vectors that directed different target sites in the same transcript might have suppressive effects in cells. Figure 7A shows the activities of siRNA expression vectors directed against five different sites of firefly luciferase transcript. The siRNA expression vector against Site B suppressed luciferase activity by 86%, whereas corresponding values for Site O, Site A, Site C, and Site D were approximately 0%, 56%, 62%, and 64%, respectively. To examine whether this target site-dependent suppression of luciferase expression might depend solely on the target site or is attributable to the efficiency of the transcription of individual constructs, we synthesized siRNA oligonucleotides targeted to the same sites and examined their activities.

The profile of the suppression of luciferase activity by siRNA oligonucleotides at each site (Figure 7B) was similar to that obtained with the siRNA expression vectors (Figure 7A), which suggested that the differences among the suppressive activities of the siRNA expression vectors were caused by the respective target site. The effects were most likely due to secondary structure or to RNA-binding proteins, although we cannot exclude other possibilities. The sequence of the target RNA might not be the only determinant of siRNA function because (*i*) the product release of products of a cleaved mRNA might be rate limiting, and the sequence and the GC content of the siRNA may be important factors; (*ii*) the sequence of the siRNA might affect the rate of 5′-phosphorylation, a requirement for siRNA function, of the antisense siRNA; and (*iii*) since siRNAs target sense, as well as antisense, single-stranded RNAs, it is possible that ratios of sense to antisense targeting endonuclease complexes are dependent on the sequence of the siRNA. Clearly, siRNA effects are target site- and/or siRNA-dependent, and those phenomena require clarification.

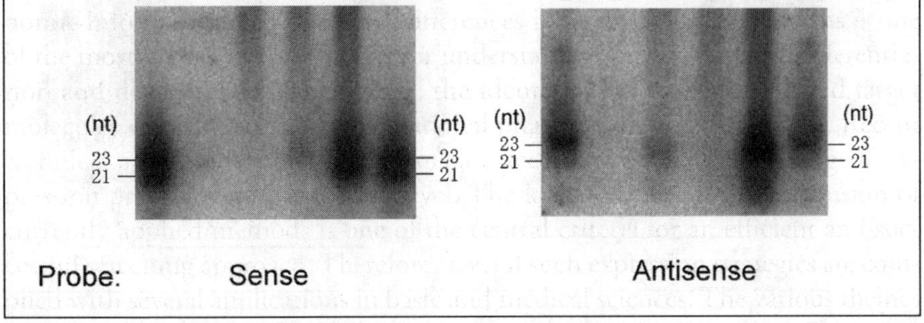

Figure 6. Northern blot analysis of U6 promoter-driven transcripts. HeLa S3 cells were transfected with 5 µg siRNA expression vector, sense RNA expression vector, antisense RNA expression vector, or pU6 empty vector. Fifteen micrograms of RNAs were loaded on each lane. RNA oligonucleotides phosphorylated at the 5′-end were used as markers. (nt), nucleotide.

Suppressive Effects in Cultured Cells of siRNA Oligonucleotides with Four-Uridine Overhangs

As mentioned earlier, siRNAs generated from the U6 promoter a four-uridine overhang at each 3′-end. Tuschl and co-workers (3) found that siRNA oligonucleotides with a longer overhang at the 3′-end are less effective than those with 2- or 3-nucleotide overhangs in experiments with *Drosophila* lysates in vitro (7). Therefore, we next examined whether the four-uridine overhang at the 3′-ends of siRNAs generated from the U6 promoter might interfere with the RNAi effect in cultured cells. We synthesized RNA oligonucleotides with 2-, 3-, or 4-nucleotide overhangs, directed to the same target site in *Renilla* luciferase and compared their suppressive activities. Co-transfection experiments using expression vectors for RNA oligonucleotides with 2-, 3-, or 4-nucleotide overhangs, *Renilla* luciferase, and firefly luciferase revealed no significant differences among siRNAs with 2- to 4-nucleotide overhangs (Figure 8). Thus, it appeared that siRNAs with a 4-nucleotide overhang produced from by our U6 promoter had suppressive activity equivalent to those of siRNAs with 2- or 3-nucleotide overhangs in cultured cells.

Simultaneous Repression of Two Target Genes by a Multi-Cassette System

We then examined whether it might be possible to suppress the expression of two different genes simultaneously. We constructed a plasmid that included two siRNA expression cassettes, directed against firefly luciferase and *Renilla* lu-

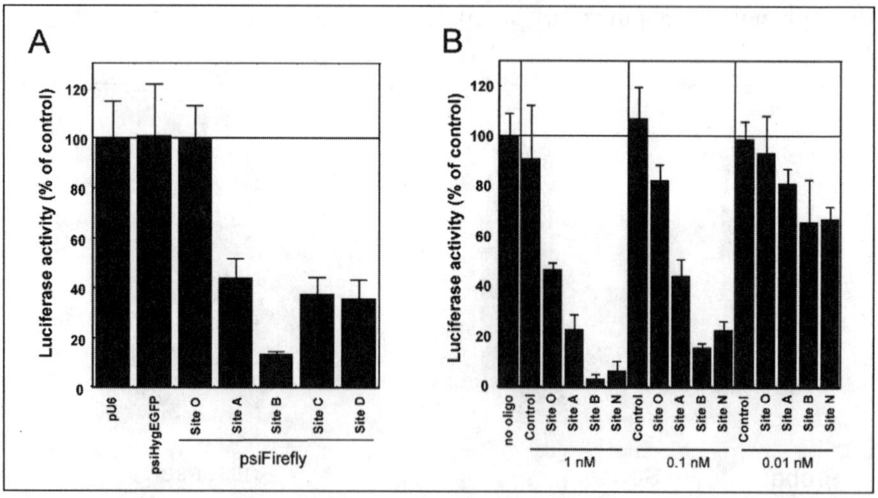

Figure 7. Dependence on target site of RNAi. (A) Dependence on the target site of the suppressive activity of siRNA expression vectors. (B) Effect of siRNA oligonucleotides targeted against the same sites as the siRNA expression vectors. HeLa S3 cells were transfected with 30 ng pRL-RSV, 30 ng pGL3, and siRNA oligonucleotides at the indicated concentrations. The siRNA oligonucleotides for site N have the same sequence as those used by Elbashir et al. (6). As controls, we used siRNA oligonucleotides against hygromycin/EGFP.

ciferase, respectively, and we co-transfected cells with this plasmid plus pGL3, pRL-RSV, and a β-galactosidase expression vector, which was used as an internal control. As shown in Figure 9, transfection with the siRNAs expression vector directed against the two luciferase genes resulted in reductions in expression of firefly and *Renilla* luciferases simultaneously and to the same extent as that observed with the siRNA expression vectors against each independent gene. Thus, each of the two cassettes in the same plasmid was capable of disrupting the expression of the corresponding gene without any interference between the two promoters or the two different kinds of siRNA.

Silencing of an Endogenous Gene by the U6-Driven siRNA Expression System

Finally, we attempted to disrupt the expression of an endogenous gene. We choose β-catenin as the target. It is a scaffold protein in cadherin-mediated intercellular adhesion and the product of a potent oncogene (18). To establish stable transformants, we used the EBNA-1/Ori-P system (5), in which plasmids harboring the sequence for the origin of replication of Epstein-Barr virus DNA (Ori-P sequence) are maintained stably and extrachromosomally in cells that express the EBNA-1 antigen. We constructed Ori-P-containing siRNA-expression plasmids targeted against two sites in the β-catenin transcript (pcOriP/si β-catenin) and transfected HeLa S3/EBNA-1 cells. After selection for one week in medium that contained G418, we examined the expression of β-catenin by Western blot analysis. As shown in Figure 10, both siRNA that harbored pcOriP/si β-catenin plasmids targeted to one of two sites in the β-catenin transcript expressed substantially lower levels of β-catenin than mock- transfected cells. This result demonstrated that the siRNA expression system can be applied to knockdown by the siRNA of endogenous genes and has the potential to suppress gene expression through the generation of stable transformants.

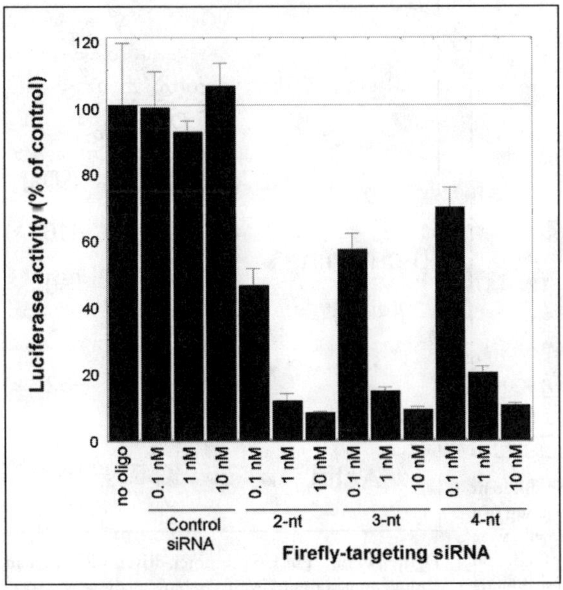

Figure 8. Dependence on the length of the 3′ overhang. HeLa S3 cells were transfected with 30 ng pRL-RSV, 30 ng pGL3, and firefly luciferase-targeted siRNA nucleotides with 2-, 3-, or 4-uridine overhangs at the 3′-end.

FUTURE TRENDS

We demonstrated that our siRNA expression vector was able to repress efficiently the expression of specific genes of interest, and it should be useful for the functional analysis of genes whose function have not been identified through the generation of stable cell lines and transgenic mice. Our siRNA expression system has several notable features, in particular, the inducible regulation of RNAi. We have already established a tetracycline-regulatable system with a U6 promoter (16). This system allows strict control of RNAi (unpublished data). Moreover, our siRNA expression cassette can be maintained in vectors or in genomic DNA with higher stability than stem-loop type or hairpin-type dsRNA-producing cassettes because the latter include unstable palindromic sequences (1). It should also be possible to generate siRNA using opposing U6 promoters, similarly to previously reported opposing T7 promoters (20). In preliminary experiments, we developed working opposing U6 promoters. Such an opposing promoter system might allow the production of randomized siRNA libraries or cDNA/genome DNA fragment-expression libraries and eventually reverse genetic screening by RNAi, achieved previously using hybrid ribozyme libraries (13).

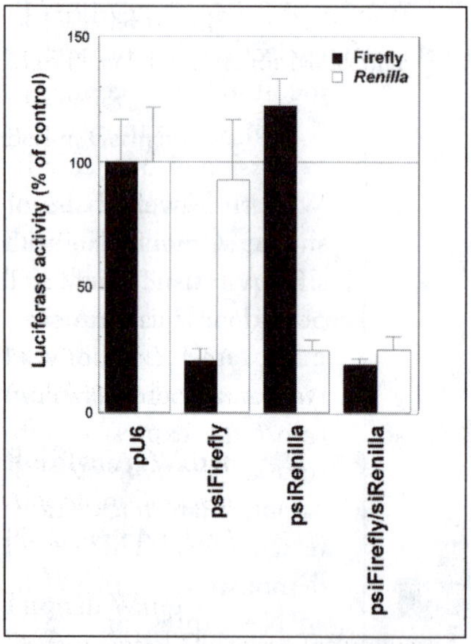

Figure 9. Simultaneous repression of two different genes by a siRNA expression plasmid harboring two siRNA expression cassettes. HeLa S3 cells were transfected with 30 ng pRL-RSV, 30 ng pGL3, 100 ng β-galactosidase expression vector, and 300 ng siRNA expression vector targeted against both *Renilla* and firefly luciferases or 300 ng siRNA expression vector against each separate luciferase. Values were normalized by reference to β-galactosidase activity.

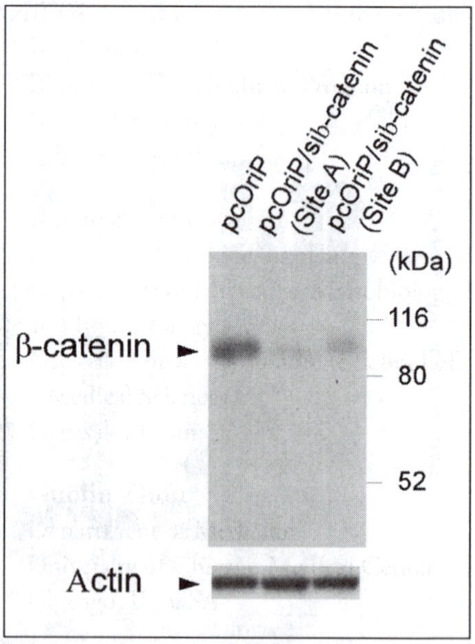

Figure 10. Effect of U6 promoter-driven siRNA on an endogenous target. Western blot analysis was performed with β-catenin-specific antibody (1:1000 dilution; Sigma, St. Louis, MO, USA) or actin-specific antibody (1:400 dilution; Sigma). Approximately 30 μg protein were loaded into each lane.

REFERENCES

1. **Akashi, H., M. Miyagishi, and K. Taira.** 2001. Suppression of gene expression by RNA interference in cultured plant cells. Antisense Nucleic Acid Drug Dev. *11*:359-367.
2. **Bernstein, E., A.A. Caudy, S.M. Hammond, and G.J. Hannon.** 2001. Role for a bidentate ribonuclease in the initiation step of RNA interference. Nature *409*:363-366.
3. **Bogenhagen, D.F. and D.D. Brown.** 1981. Nucleotide sequences in *Xenopus* 5S DNA required for transcription termination. Cell *24*:261-270.
4. **Caplen, N.J., S. Parrish, F. Imani, A. Fire, and R.A. Morgan.** 2001. Specific inhibition of gene expression by small double-stranded RNAs in invertebrate and vertebrate systems. Proc. Natl. Acad. Sci. USA *98*:9742-9747.
5. **Chittenden, T., S. Lupton, and A.J. Levine.** 1989. Functional limits of oriP, the Epstein-Barr virus plasmid origin of replication. J. Virol. *63*:3016-3025.
6. **Elbashir, S.M., J. Harborth, W. Lendeckel, A. Yalcin, K. Weber, and T. Tuschl.** 2001. Duplexes of 21-nucleotide RNAs mediate RNA interference in cultured mammalian cells. Nature *411*:494-498.
7. **Elbashir, S.M., J. Martinez, A. Patkaniowska, W. Lendeckel, and T. Tuschl.** 2001. Functional anatomy of siRNAs for mediating efficient RNAi in *Drosophila melanogaster* embryo lysate. EMBO J. *20*:6877-6888.
8. **Fire, A.** 1999. RNA-triggered gene silencing. Trends Genet. *15*:358-363.
9. **Fire, A., S. Xu, M.K. Montgomery, S.A. Kostas, S.E. Driver, and C.C. Mello.** 1998. Potent and specific genetic interference by double-stranded RNA in *Caenorhabditis elegans*. Nature *391*:806-811.
10. **Hammond, S.M., S. Boettcher, A.A.Caudy, R. Kobayashi, and G.J. Hannon.** 2001. Argonaute2, a link between genetic and biochemical analyses of RNAi. Science *293*:1146-1150.
11. **Hammond, S.M., A.A. Caudy, and G.J. Hannon.** 2001. Post-transcriptional gene silencing by double-stranded RNA. Nat. Rev. Genet. *2*:110-119.
12. **Holen, T., M. Amarzguioui, M.T. Wiiger, E. Babaie, and H. Prydz.** 2002. Positional effects of short interfering RNAs targeting the human coagulation trigger tissue factor. Nucleic Acids Res. *30*:1757-1766.
13. **Kawasaki, H., R. Onuki, E. Suyama, and K. Taira.** 2002. Identification of genes that function in the TNF-α-mediated apoptotic pathway using randomized hybrid ribozyme libraries. Nat. Biotechnol. *20*:376-380.
14. **Miyagishi, M., R. Fujii, M. Hatta, E. Yoshida, N. Araya, A. Nagafuchi, S. Ishihara, T. Nakajima, et al.** 2000. Regulation of Lef-mediated transcription and p53-dependent pathway by associating β-catenin with CBP/p300. J. Biol. Chem. *275*:35170-35175.
15. **Nykanen, A., B. Haley, and P.D. Zamore.** 2001. ATP requirements and small interfering RNA structure in the RNA interference pathway. Cell *107*:309-321.
16. **Ohkawa, J. and K. Taira.** 2000. Control of the functional activity of an antisense RNA by a tetracycline-responsive derivative of the human U6 snRNA promoter. Hum. Gene Ther. *11*:577-585.
17. **Paule, M.R. and R.J. White.** 2000. Survey and summary: transcription by RNA polymerases I and III. Nucleic Acids Res. *28*:1283-1298.
18. **Peifer, M. and P. Polakis.** 2000. Wnt signaling in oncogenesis and embryogenesis—a look outside the nucleus. Science *287*:1606-1609.
19. **Sharp, P.A.** 2001 RNA interference. Genes Dev. *15*:485-490.
20. **Wang, Z., J.C. Morris, M.E. Drew, and P.T. Englund.** 2000. Inhibition of *Trypanosoma brucei* gene expression by RNA interference using an integratable vector with opposing T7 promoters. J. Biol. Chem. *275*:40174-40179.
21. **Warashina, M., T. Kuwabara, Y. Kato, M. Sano, and K. Taira.** 2001. RNA-protein hybrid ribozymes that efficiently cleave any mRNA independently of the structure of the target RNA. Proc. Natl. Acad. Sci. USA *98*:5572-5577.
22. **Yang, S., S. Tutton, E. Pierce, and K. Yoon.** 2001. Specific double-stranded RNA interference in undifferentiated mouse embryonic stem cells. Mol. Cell Biol. *21*:7807-7816.
23. **Zamore, P.D.** 2001. RNA interference: listening to the sound of silence. Nat. Struct. Biol. *8*:746-750.

19 Short Hairpin-Type dsRNAs Transcribed by pol III Polymerase Induce Efficient Sequence-Specific Silencing in Human Cells

Hiroaki Kawasaki and Kazunari Taira
Department of Chemistry and Biotechnology, School of Engineering, The University of Tokyo, Tokyo, and Gene Function Research Center, National Institute of Advanced Industrial Science and Technology (AIST), Tsukuba Science City, Japan

BACKGROUND

Double-stranded RNA (dsRNA)-mediated interference (RNAi) has recently emerged as a powerful reverse-genetic tool for silencing gene expression in organisms that include plants, *Caenorhabditis elegans*, and *Drosophila.* The discovery that synthetic double-stranded small interfering RNAs (siRNAs) of 21 nucleotides can trigger gene-specific silencing in mammalian cells has extended its utility to mammalian systems. Specific inhibition of gene expression can also be achieved by the stable and inducible expression of dsRNA in animals and plants. In this chapter, we introduce two kinds of dsRNA expression vectors, which are controlled by a $tRNA^{Val}$ and a U6 promoter, respectively.

TECHNIQUES

We introduced the pol III system and the promoter of a human gene for $tRNA^{Val}$ for the efficient transcription of dsRNA in vivo. This expression system is suitable for the transcription of small RNAs. Moreover, levels of expression were several orders of magnitude higher than those obtained with a pol II system. Because

Perspectives in Gene Expression
Edited by K. Appasani

our tRNA^{Val}-dsRNA was exported to the cytoplasm and, thus, both the dsRNAs and RNA-induced silencing complexes (RISCs) were present within the same cellular compartment, our tRNA^{Val}-dsRNA could exert high levels of activity in vivo.

RESULTS

To examine the relationship between the localization of dsRNA and the target specificity of RNAi in mammalian cells, we constructed two kinds of dsRNA expression vectors; one was controlled by the promoter of a human gene for tRNA^{Val} and the other by a U6 promoter. Transcripts of tRNA-dsRNA were localized in the cytoplasm and efficiently processed by a ribonuclease III complex. By contrast, transcripts of U6-dsRNA were localized in the nucleus and not processed to any significant extent. In addition, tRNA-dsRNA directed against a mutant k-*ras* transcript cleaved its target mRNA efficiently in assays of RNAi not only in vitro in a cytoplasmic extract but also in vivo. Moreover, the same tRNA-dsRNA had no effect on the degradation of the wild-type k-*ras* mRNA in vitro and in vivo. In addition, tRNA-dsRNA was significantly more effective than U6-dsRNA when these constructs were examined under identical conditions.

CONCLUSIONS

We have successfully developed a tRNA^{Val}-dsRNA expression system. The tRNA^{Val}-dsRNA expressed in this system is exported to the cytoplasm for colocalization with the RISC. Because the tRNA-dsRNA was significantly more effective than the U6-dsRNA when both were examined under identical conditions, our tRNA-dsRNA technology should be a powerful tool for studies of the RNAi mechanism and the functions of various genes in mammalian cells, with potential utility as a therapeutic agent.

INTRODUCTION

RNAi is a phenomenon whereby dsRNA induces the sequence-dependent degradation of a cognate mRNA (10,27). The natural roles of RNAi might include defense against viral infection (9,15,22,26,32) and regulation of the expression of cellular genes (1,34). Genetic and biological studies have revealed that RNAi is a very complex process that involves many different proteins with, for the most part, mostly unidentified functions (1,3,14,32,34). It has been demonstrated that dsRNA targets mRNA for cleavage in lysates of early *Drosophila* embryos and in extracts of cultured *Drosophila* S2 cells (12,39,41) and that such reactions in vitro require ATP (41). The molecular basis for the requirement for ATP is due in part to a requirement for ATP in the initial processing of long dsRNA into the 21- to 25-nucleotide siRNAs that serve as guides for targeted cleavage (2,7,11,38,41).

Recent studies with synthetic RNA duplexes demonstrated that the siRNA duplex must have 2- or 3-nucleotide 3′-end overhangs for the efficient cleavage of its target (7). Such 3′ overhangs are characteristic of the products of cleavage reactions catalyzed by RNase III, and, in cultured *Drosophila* S2 cells, cleavage of dsRNA into siRNAs requires a multidomain RNase III, known as Dicer (2). Subsequently, siRNAs seem to associate with a multicomponent nuclease, identified in *Drosophila* and designated RISC, and then they guide this enzyme so that it catalyzes the sequence-specific degradation of mRNA (2,12,13).

RNAi provides a method for inactivating genes of interest and, thus, provides a powerful tool for studies of gene function in *C. elegans, Drosophila melanogaster*, and plants. Specific inhibition of gene expression can also be achieved by the stable or inducible expression of dsRNA in animals and plants (14). The inactivation of genes by dsRNA has been achieved in mouse embryonic carcinoma (EC) cells and embryonic stem (ES) cells (4,29), but the elicitation of RNAi using long dsRNAs has generally been less successful in cultured mammalian cells. Such failures can be explained most readily by the action of two latent enzymes that form part of the interferon (IFN) defense pathways that are activated by long dsRNAs (>30 base pairs) (35).

It was recently reported that 21-nucleotide synthetic siRNAs (6) and siRNAs that were transcribed from a U6 or H1 promoter (5,21,25,28,30,26,40) specifically suppressed the expression of endogenous genes in several lines of mammalian cells. These 21-nucleotide siRNA duplexes were able to evade the IFN defense system, which suggested that RNAi or RNAi-related systems might exist in mammals. Some mammalian homologs of RNAi-associated proteins, such as RED-1, MUT-7, and Dicer have, indeed, been identified (2,18,37). However, the details of the characteristics and mechanisms of RNAi in mammalian somatic cells remain undetermined.

This chapter describes the construction of two kinds of dsRNA expression vectors, which were controlled by the promoter of a human gene for $tRNA^{Val}$ and the widely used U6 promoter, respectively, to examine the relationship between the localization and target specificity of RNAi in mammalian cells. Transcripts of tRNA-dsRNA were localized in the cytoplasm and processed efficiently by the ribonuclease III complex. By contrast, transcripts of U6-dsRNA were localized in the nucleus and not processed to any significant extent. In addition, tRNA-dsRNA directed to a mutant k-*ras* gene cleaved its target mRNA efficiently both in vitro and in vivo. However, the tRNA-dsRNA did not affect the expression of wild-type k-Ras in HeLa cells. Our results indicate that RNAi in mammalian cells occurs in the cytoplasm and that our tRNA-dsRNA expression system should be a powerful tool for studies of the RNAi mechanism and other gene functions in mammalian cells, with possible potential utility as a therapeutic agent.

METHODOLOGY

Construction of dsRNA Expression Plasmids

To construct vectors for expression of tRNA-dsRNA, we used the pPUR-tRNA plasmid that includes the chemically synthesized promoter for a human gene for $tRNA^{Val}$ between the *Eco*RI and *Bam*HI sites of pPUR (BD Biosciences Clontech, Palo Alto, CA, USA). Chemically synthesized oligonucleotides encoding k-*ras*-directed dsRNA consist of sense sequence 5′-GGTAGTTGGAGCTGTTGGCGTAGGCAAGA-3′, antisense sequence 5′-TCTTGCCTACGCCAACAGCTCCAACTACC-3′, and loop sequences (17). These nucleotides were amplified as double-stranded sequences by PCR using a forward primer that contained *Csp*45I linker and a reverse primer that contained *Kpn*I linker (17). After digestion with *Sac*I and *Kpn*I, the fragments were cloned downstream of the promoter of the tRNA gene in pPUR-tRNA. The construction of vectors for expression of dsRNA from the U6 promoter has been described elsewhere (16). The *Eco*RI and *Xho*I sites were used for the insertion of the sequence of k-*ras*-directed dsRNA (17). The constructed vectors for tRNA- and U6-promoter driven dsRNAs are designated piGENE™ tRNA and piGENE U6, respectively.

Culture and Transfection of Cells

SW480 human colon cancer cells were cultured in L-15 medium (ICN Biomedicals, Costa Mesa, CA, USA) supplemented with 10% FBS, and HeLa cells were cultured in DMEM supplemented with 10% FBS. Transfections were performed with the Effectin™ reagent (Qiagen, Hilden, Germany), according to the manufacturer's instructions. dsRNA-expressing SW480 and dsRNA-expressing HeLa cells were selected by incubation with puromycin for two weeks.

Preparation of the Nuclear Fraction and the Cytoplasmic Fraction of Cells

SW480 cells or HeLa cells were grown to approximately 1×10^7 cells and then transfected with a tRNA-dsRNA or U6-dsRNA expression vector with the Effectin reagent. Thirty-six hours after transfection, the cells were harvested. For the preparation of the cytoplasmic fraction, collected cells were washed twice with PBS and then resuspended in digitonin lysis buffer (50 mM HEPES/KOH, pH 7.5, 50 mM potassium acetate, 8 mM $MgCl_2$, 2 mM EGTA, and digitonin) on ice for 10 min. The lysate was centrifuged at 1000× *g*, and the supernatant was collected as the cytoplasmic fraction. The pellet was resuspended in Nonidet™ P-40 (NP-40) lysis buffer (20 mM Tris-HCl, pH 7.5, 50 mM KCl, 10 mM NaCl, 1 mM EDTA, and 0.5% NP-40) and held on ice for 10 min, and the resultant lysate was used as the nuclear fraction.

Northern Blot Analysis

Cytoplasmic RNA and nuclear RNA (nRNA) were extracted and purified from the cytoplasmic fraction and the nuclear fraction, respectively, with ISOGEN™ reagent (Wako BioProducts, Osaka, Japan). Thirty micrograms of total RNA per lane were loaded on a 15% polyacrylamide gel. After electrophoresis, the bands of RNA were transferred to a Hybond-N™ nylon membrane (Amersham Biosciences, Buckinghamshire, UK). The membrane was probed with synthetic oligonucleotides that were complementary to the sequences of the k-*ras* gene. The synthetic probe was labeled with ^{32}P by T4 polynucleotide kinase (Takara Shuzo, Kyoto, Japan).

Western Blot Analysis

SW480 or HeLa cells that had been transfected with individual dsRNA-expression vectors (piGENE tRNA and piGENE U6 for tRNA- and U6-promoter based vectors, respectively) were harvested. Proteins were resolved by SDS-PAGE (10% polyacrylamide) and transferred to a PVDF membrane (Funakoshi, Tokyo, Japan) by electroblotting. Immune complexes were visualized with an ECL™ kit (Amersham Biosciences) with specific polyclonal antibodies against K-Ras (UBI, Charlottesville, VA, USA) and actin (Santa Cruz, CA, USA).

Assay of RNAi In Vitro

For the cleavage of a target RNA, we synthesized a mutant k-*ras* template DNA and a wild-type k-*ras* template DNA (70 nucleotides each) using an automated DNA synthesizer (Model 394; Applied Biosystems, Foster City, CA, USA). For the preparation of k-*ras* mRNA substrates, we amplified template DNAs by PCR using a k-*ras*-specific forward primer that included the T7 promoter sequence and a k-*ras*-specific reverse primer. The amplified k-*ras* DNA templates were transcribed by T7 polymerase, and transcribed wild-type and mutant k-*ras* mRNA substrates were purified by PAGE. These mRNAs were labeled with ^{32}P by T4 polynucleotide kinase. For the detection of the cleavage of the target mRNA, target mRNA (5–10 nM) was incubated with a lysate of SW480 cells that expressed either tRNA-dsRNA or U6-dsRNA under standard conditions (41) for 2 h at 25°C. In the case of siRNA targeted to k-*ras* mRNA, 100 nM siRNA were incubated with 5–10 nM target RNA in a lysate of SW480 cells under standard conditions (41) for 2 h at 25°C. The rate of proliferation of each line of cells was measured with a Cell Proliferation Kit II (Roche Applied Science, Basel, Switzerland), according to the manufacturer's instructions.

PROJECTS AND SAMPLE RESULTS

Construction of Two Kinds of dsRNA-Expression Plasmid with pol III Promoters

It has not yet been determined whether dsRNA acts in the nucleus or the cytoplasm. Therefore, we used two kinds of pol III promoter (the promoter of a human gene for $tRNA^{Val}$ and the U6 promoter) for the expression of dsRNAs in mammalian cells. If designed appropriately, then transcripts generated with the promoter of a human gene for $tRNA^{Val}$ are transported efficiently to the cytoplasm. We demonstrated previously that localization in the cytoplasm is important in efforts to improve ribozyme activities in vivo (16,19,20). However, small

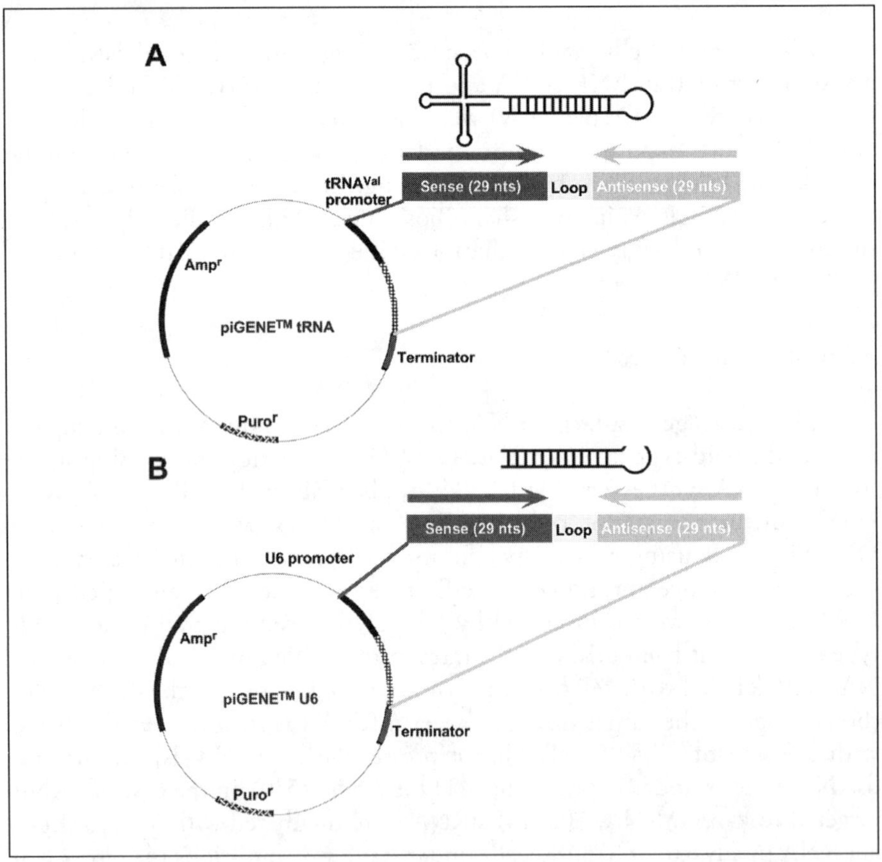

Figure 1. Construction of dsRNA expression plasmids. (A) The plasmid for the expression of tRNA-dsRNA included the promoter sequence from a human gene for $tRNA^{Val}$ and a terminator sequence. The loop sequence was inserted between a k-*ras* sense-strand sequence (29 bases) and an antisense strand sequence (29 bases). (B) The plasmid for the expression of U6-dsRNA included a mouse U6 promoter and terminator. The dsRNA and loop sequence in U6-dsRNA were the same as those in tRNA-dsRNA. nts, nucleotides.

RNAs transcribed under the control of the U6 promoter remain localized in the nucleus (16). Thus, we attached the gene for dsRNA (an extended stem-loop RNA; Figure 1, panels A and B) to the 3′-end of the promoter of a gene for tRNA^{Val} (tRNA-dsRNA) or to the 3′-end of the U6 promoter (U6-dsRNA). We constructed dsRNA expression plasmids targeted to the mRNA for a mutant form of k-Ras with a point mutation in codon 12 of the k-*ras* gene (Figure 2). The length of the double-stranded region within the dsRNA was kept at 29 bp because long dsRNAs (>30 nucleotides) activate the IFN defense pathway (35).

Processing of piGENE tRNA-Driven Transcripts by a Putative RNase III Complex in the Cytoplasm

In RNAi, long dsRNAs are first processed to short RNA duplexes of predominantly 21 and 22 nucleotides in length and with staggered 3′-ends in an RNase III-like reaction (2,41). To examine whether tRNA-dsRNA and U6-dsRNA might be processed by an RNase III complex in mammalian cells, we performed Northen blot analysis with a k-*ras* mRNA-specific probe. SW480 cells were transfected with plasmids that encoded dsRNA under the control of the tRNA^{Val} or U6 promoter. Forty-eight hours after transfection, the cells were collected and separated into cytoplasmic and nuclear fractions. Total RNA in each fraction was isolated and fractionated on a 15% polyacrylamide gel. As shown in Figure 3, in

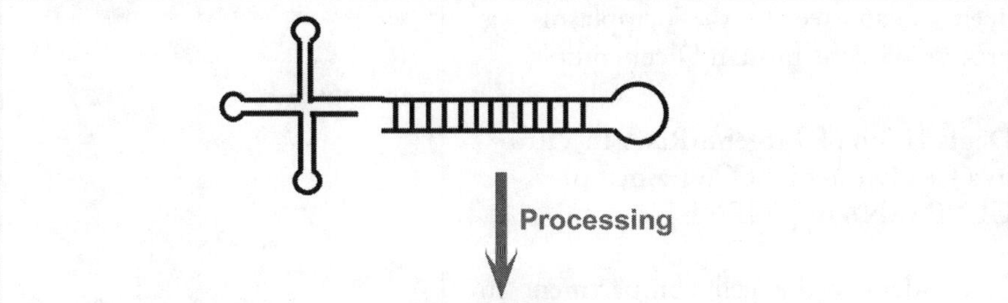

Figure 2. The target k-*ras* mutant gene that was the target of tRNA-dsRNA and U6-dsRNA. Sequences of mutant k-*ras* and wild-type k-*ras* mRNAs that were targeted by dsRNA. The mutant k-*ras* gene had a point mutation (codon 12 GGT→GTT).

cells that expressed tRNA-dsRNA, processed siRNAs were detected in the cytoplasmic fraction but not in the nuclear fraction. Moreover, the sequences of processed siRNAs were confirmed by the cloning and sequencing of the siRNAs (data not shown). However, in cells that expressed U6-dsRNA, predominantly unprocessed precursor dsRNAs were detected in the nucleus, and the nuclear fraction contained very few siRNAs in cells that expressed U6-dsRNA. Our U6-dsRNA with a hairpin structure was not transported to the cytoplasm (Figure 3). Under different conditions, other U6-dsRNAs prepared by several independent groups (the lengths of dsRNAs and the sizes and sequences of hairpin loops in these constructs were different from those of our U6-dsRNA) could support RNAi in mammalian cells (21,25,28,30,36,40). In one case, a microRNA motif was used as the loop motif in dsRNA, and the efficacy is likely enhanced by a microRNA pathway (5,17). Nevertheless, as in this specific example when a microRNA-derived loop was not used, tRNA-dsRNA supported RNAi in mammalian cells significantly more effectively than U6-dsRNA (17). In addition, a mammalian analog of Dicer has been detected predominantly in the cytoplasm by immunostaining in situ (4). Thus, it seems likely that the tRNA-dsRNA that had been transported to the cytoplasm was processed by the human Dicer complex.

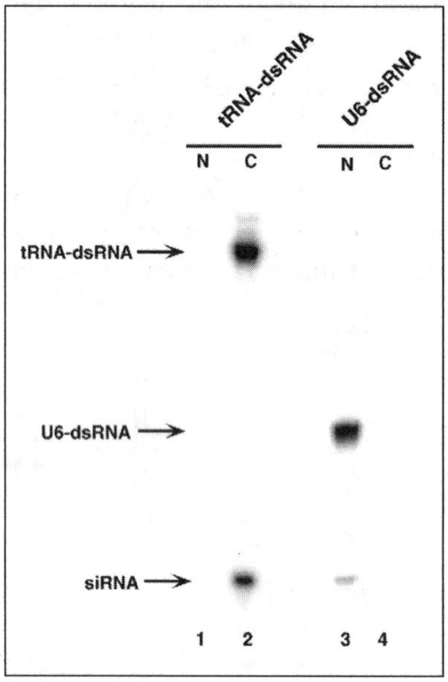

Figure 3. Detection of precursor dsRNAs and siRNAs. The presence of precursor dsRNAs and siRNAs was analyzed by Northern blot analysis. Plasmids encoding piGENE tRNA or piGENE U6 were introduced into SW480 cells. After 48 h, the cells were collected and divided into cytoplasmic (C) and nuclear (N) fractions. Total RNA in each fraction was isolated and fractionated on a 15% polyacrylamide gel. Northern blot analysis was performed as described in the text. Lane 1, nuclear fraction of cells that expressed tRNA-dsRNA; lane 2, cytoplasmic fraction of cells that expressed tRNA-dsRNA; lane 3, nuclear fraction of cells that expressed U6-dsRNA; and lane 4, cytoplasmic fraction of cells that expressed U6-dsRNA.

Degradation of Target mRNA In Vitro in a Cell Extract that Contained piGENE tRNA or piGENE U6

To identify the cell compartment in which the degradation of a target mRNA by dsRNA-mediated gene silencing occurs, we performed assays of RNAi in vitro using cell extracts that contained transcripts of piGENE tRNA or piGENE U6. In these assays, we used mutant and wild-type k-*ras* partial mRNAs that had been transcribed in vitro by T7 polymerase as substrates. For cleavage of the target mRNA, each substrate was incubated for 2 h at 25°C with an extract of SW480 cells that had been transfected with the piGENE tRNA or piGENE U6 expression vector.

The 5′-cleavage products were resolved on sequencing gels. Figure 4 shows that the mutant k-*ras* mRNA substrate was cleaved in the cytoplasmic fraction of cell extracts that contained tRNA-dsRNA (lane 4). In the nuclear fraction of the cell extracts, the substrate was not cleaved (lane 3). These results support previous reports that the RISC is included in a ribosomal fraction (2,13,41). In the case of the wild-type k-*ras* mRNA transcript, significantly smaller amounts of 5′-cleavage products were detected, and these were found only in the cytoplasmic fraction of cell extracts that contained transcripts of tRNA-dsRNA (Figure 4, lane 2). These results are in accord with the report that the efficiency of cleavage of a target mRNA by synthesized siRNAs, with one mismatched base pair in the middle of the siRNA, was reduced in an assay of RNAi in vitro with lysates of *D. melanogaster* (8). By sequencing analysis, we confirmed that the siRNA generated from the piGENE tRNA formed a mismatched base pair with wild-type k-*ras* mRNA in the middle of the siRNA (data not shown). We obtained similar results with synthesized siRNAs targeted against the mutant k-*ras* mRNA (Figure 5). Thus, our results suggested that the degradation of a target mRNA by dsRNA-mediated gene silencing occurs in the cytoplasm and that siRNAs are capable of recognizing one mismatched base pair in a middle position.

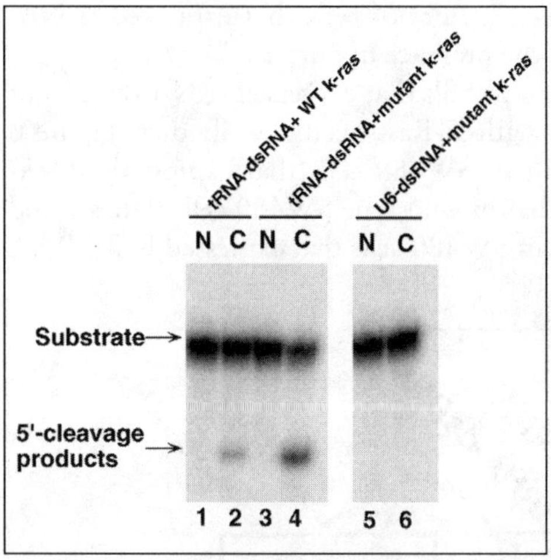

Figure 4. Detection of dsRNA- and siRNA-mediated cleavage of mutant k-*ras* mRNA. Cleavage of mutant k-*ras* mRNA by cell extracts that contained tRNA-dsRNA or U6-dsRNA. The assay of RNAi in vitro was performed as described in the text. Lane 1, nuclear fraction (N) of cells that expressed tRNA-dsRNA and wild-type (WT) k-*ras* RNA; lane 2, cytoplasmic fraction (C) of cells that expressed tRNA-dsRNA and wild-type k-*ras* mRNA; lane 3, nuclear fraction of cells that expressed tRNA-dsRNA and mutant k-*ras* mRNA; lane 4, cytoplasmic fraction of cells that expressed tRNA-dsRNA and mutant k-*ras* mRNA; lane 5, nuclear fraction of cells that expressed U6-dsRNA and mutant k-*ras* mRNA; and lane 6, cytoplasmic fraction of cells that expressed U6-dsRNA and mutant k-*ras* mRNA.

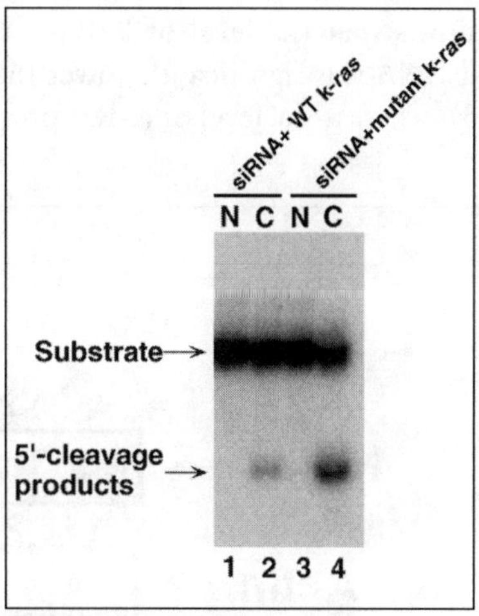

Figure 5. Cleavage of mutant k-*ras* mRNA mediated by k-*ras*-directed siRNAs and extracts of SW480 cells. Lane 1, nuclear fraction of SW480 cells and wild-type (WT) k-*ras* mRNA; lane 2, cytoplasmic fraction of SW480 cells and wild-type k-*ras* mRNA; lane 3, nuclear fraction of SW480 cells and mutant k-*ras* mRNA; and lane 4, cytoplasmic fraction of SW480 cells and mutant k-*ras* mRNA.

The step at which degradation of mRNA occurs is a critical aspect of RNAi-mediated gene silencing. Two groups have proposed that an RNA-directed RNA polymerase (called RdRP) chain reaction with siRNA amplifies the interference caused by a small amount of "trigger" dsRNA in *C. elegans* and *Drosophila* (23,33). However, the mechanism remains unclear in mammalian cells because an RdRP homolog does not appear to exist in mammalian cells. In this study, we demonstrated that a cytoplasmic fraction from cells that expressed tRNA-dsRNA had mRNA-degradation activity (Figure 4). In addition, siRNA in the cytoplasmic fraction also had mRNA-degradation activity (Figure 5). Thus, it is likely that an siRNA-associated silencing complex (SASC), including Dicer or "Slicer," is localized in the cytoplasm of mammalian cells.

Efficiency and Specificity of RNAi by tRNA-dsRNA in Colon Cancer Cells

To examine the efficiency and specificity of RNAi by tRNA-dsRNA in human colon cancer SW480 cells, we introduced piGENE tRNA or piGENE U6 expression plasmids into SW480 and HeLa cells. The mutant k-*ras* gene was expressed only in SW480 cells (24). We used HeLa cells that expressed a normal k-*ras* gene as controls, and generated stable lines of cells that expressed tRNA-dsRNA or U6-dsRNA by selection in the presence of puromycin.

We examined levels of K-Ras protein in cells that expressed tRNA-dsRNA or U6-dsRNA by Western blot analysis with K-Ras-specific antibodies. Figure 6 shows that the level of K-Ras protein in SW480 cells that expressed tRNA-dsRNA was significantly lower than that in wild-type SW480 cells (lanes 1 and 3), whereas the level of K-Ras protein in SW480 cells that expressed U6-dsRNA

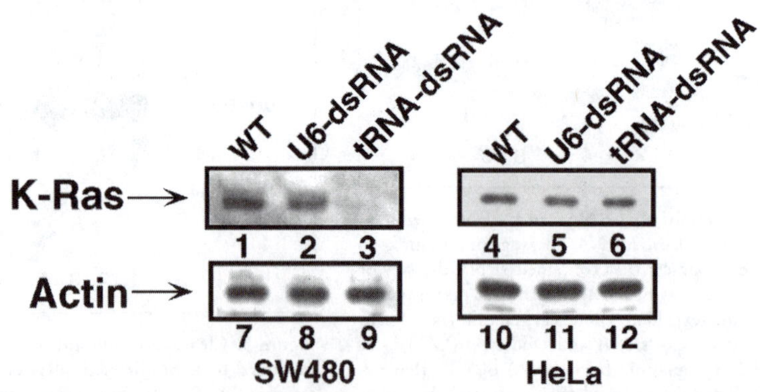

Figure 6. Efficiency and specificity of RNAi by tRNA-dsRNA in vivo. The level of K-Ras protein in cells that expressed tRNA-dsRNA or U6-dsRNA was analyzed by Western blot analysis. Lanes 1 and 7, wild-type (WT) SW480 cells; lanes 2 and 8, SW480 cells that expressed U6-dsRNA; lanes 3 and 9, SW480 cells that expressed tRNA-dsRNA; lanes 4 and 10, wild-type HeLa cells; lanes 5 and 11, HeLa cells that expressed U6-dsRNA; and lanes 6 and 12, HeLa cells that expressed tRNA-dsRNA.

was reduced only slightly compared to that in wild-type SW480 cells (lanes 1 and 2). The level of actin, chosen as an endogenous control, remained constant in these cell lines. In HeLa cells that expressed tRNA-dsRNA, the level of K-Ras protein was similar to that in wild-type cells and in cells that expressed U6-dsRNA. These results clearly demonstrated both the efficiency and the specificity of the tRNA-dsRNA in human cancer cells.

To examine the phenotype of cells that expressed tRNA-dsRNA or U6-dsRNA, we analyzed the rates of proliferation of various lines of cells. As shown in Figure 7, SW480 cells that expressed tRNA-dsRNA proliferated significantly more slowly than wild-type SW480 cells and cells that expressed U6-dsRNA. By contrast, the rate of proliferation in HeLa cells that expressed tRNA-dsRNA was the same as that of wild-type SW480 cells and cells that expressed U6-dsRNA. These results indicated that the reduced rate of proliferation of SW480 cells that expressed tRNA-dsRNA correlated with the reduction in the level of K-Ras protein in the cells. Thus, our tRNA-dsRNA targeted to the mutant k-*ras* gene appears to have potential utility as a therapeutic agent.

Future Trends

The discovery that dsRNA can induce gene silencing in animals and plants has raised the possibility that RNAi might be a nearly universal mechanism of gene silencing. An outline of the processes involved in RNAi and several components of the reaction have been identified in *C. elegans*, *D. melanogaster*, and plants, but details of the mechanism and many of the necessary participants in

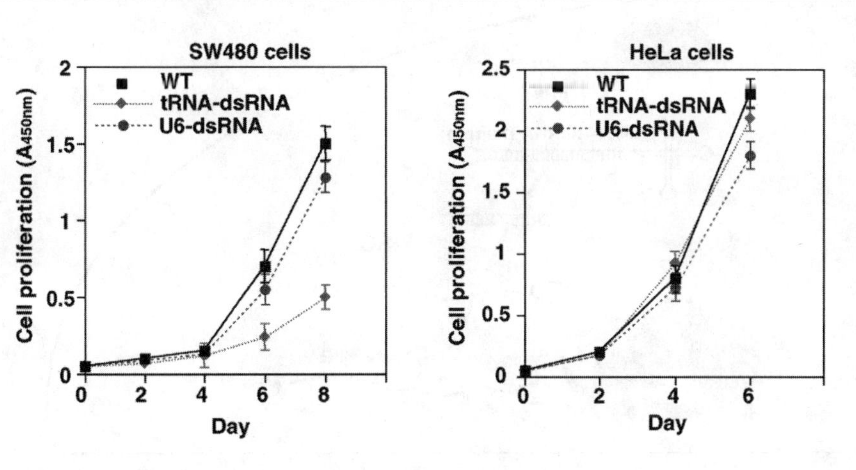

Figure 7. Effects on the proliferation of cells of tRNA-dsRNA and of U6-dsRNA. The rates of proliferation were determined as described in the text. In each case, the values are the means ± SD of the results from three replicates. WT, wild-type. (**See color plate A36.**)

RNAi in mammalian cells remain uncharacterized. In the work described here, in an attempt to clarify the relationship between the localization and the target specificity of RNAi in mammalian cells, we constructed two kinds of dsRNA expression plasmids; namely, piGENE tRNA and piGENE U6 expression plasmids. A model of the tRNA-dsRNA-induced silencing of gene expression is shown in Figure 8. We demonstrated that the tRNA-dsRNA that was localized in the cytoplasm was efficiently processed by the RNase III complex. An initial step in RNAi is the cleavage by Dicer, which is localized in the cytoplasm (4) of long dsRNAs. Although short dsRNAs are cleaved less effectively by Dicer in vitro (2,7), our tRNA-dsRNA was processed with significant efficiency in mammalian cells (Figure 3).

RNAi has been shown to be a powerful tool for studies of gene function in *C. elegans*, *D. melanogaster*, and plants. However, in mammalian cells, a long dsRNA causes a nonspecific reduction in the expression of many genes. Thus, it was initially believed that RNAi could not be used for gene inactivation in mammalian cells. However, Elbashir et al. (6) demonstrated that siRNA could suppress the expression of a target gene specifically. The exploitation of RNAi in mammalian cells requires the evasion of the IFN defense system. In addition, since the putative SASC is located in the cytoplasm, it is important that dsRNA transcripts be localized in the cytoplasm. If properly designed, then tRNA-dsRNAs can be transported to the cytoplasm and escape the IFN defense system. Indeed, we found that dsRNA-dependent protein kinase (PKR) was not acti-

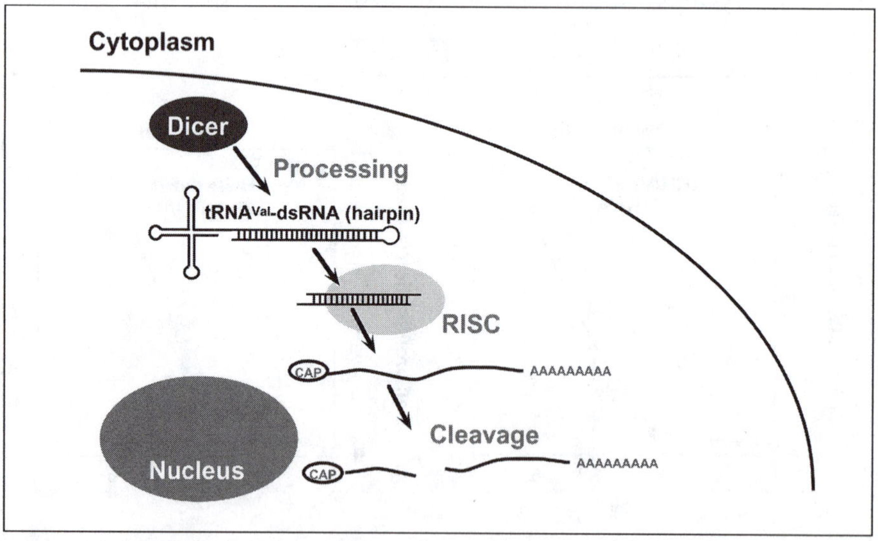

Figure 8. A model of tRNA-dsRNA-induced silencing of gene expression. tRNA-dsRNA that is localized in the cytoplasm is efficiently processed by the RNase III complex. The siRNAs that are generated become associated with RISC and then siRNAs guide this complex so that a ribonuclease can catalyze the sequence-specific degradation of an mRNA.

vated in cells that expressed tRNA-dsRNA (data not shown). Moreover, tRNA-dsRNA induced the efficient degradation of a target mRNA both in vitro and in vivo. By contrast, tRNA-dsRNA did not affect the expression of the wild-type k-*ras* mRNA in HeLa cells.

Taken together, our results indicate that RNAi in mammalian cells occurs in the cytoplasm and is very specific in mammalian cells. Our tRNA-dsRNAs should be powerful tools for studies of the mechanism of RNAi and the functions of specific genes in mammalian cells, and they might also be useful as therapeutic agents because RNAi might be observed at significantly lower concentrations of tRNA-dsRNAs than of U6-dsRNAs. However, it should be emphasized that the piGENE U6 transcript that has a microRNA-derived loop is at least as effective as the piGENE tRNA transcript (17).

REFERENCES

1. **Aravin, A.A., N.M. Naumova, A.V. Tulin, V.V. Vagin, Y.M. Rozovsky, and V.A. Gvozdev.** 2001. Double-stranded RNA-mediated silencing of genomic tandem repeats and transposable elements in the *D. melanogaster* germline. Curr. Biol. *11*:1017-1027.
2. **Bernstein, E., A.A. Caudy, S.M. Hammond, and G.J. Hannon.** 2001. Role for a bidentate ribonuclease in the initiation step of RNA interference. Nature *409*:363-366.
3. **Bernstein, E., A.M. Denli, and G.J. Hannon.** 2001. The rest is silence. RNA *7*:1509-1521.
4. **Billy, E., V. Brondani, H. Zhang, U. Müller, and W. Filipowicz.** 2001. Specific interference with gene expression induced by long, double-stranded RNA in mouse embryonal teratocarcinoma cell lines. Proc. Natl. Acad. Sci. USA *98*:14428-14433.
5. **Brummelkamp, T.R., R. Bernards, and R. Agami.** 2002. A system for stable expression of short interfering RNAs in mammalian cells. Science *296*:550-553.
6. **Elbashir, S.M., J. Harborth, W. Lendeckel, A. Yalcin, K. Weber, and T. Tuschl.** 2001. Duplexes of 21-nucleotide RNAs mediate RNA interference in mammalian cell culture. Nature *411*:494-498.
7. **Elbashir, S.M., W. Lendeckel, and T. Tuschl.** 2001. RNA interference is mediated by 21- and 22-nucleotide RNAs. Genes Dev. *15*:188-200.
8. **Elbashir, S.M., J. Martinez, A. Patkaniowska, W. Lendeckel, and T. Tuschl.** 2001. Functional anatomy of siRNAs for mediating efficient RNAi in *Drosophila melanogaster* embryo lysate. EMBO J. *20*:6877-6888.
9. **Fire, A.** 1999. RNA-triggered gene silencing. Trends Genet. *15*:358-363.
10. **Fire, A., S. Xu, M.K. Montgomery, S.A. Kostas, S.E. Driver, and C.C. Mello.** 1998. Potent and specific genetic interference by double-stranded RNA in *Caenorhabditis elegans.* Nature *391*:806-811.
11. **Hamilton, A.J. and D.C. Baulcombe.** 1999. A species of small antisense RNA in posttranscriptional gene silencing in plants. Science *286*:950-952.
12. **Hammond, S.M., E. Bernstein, D. Beach, and G.J. Hannon.** 2000. An RNA-directed nuclease mediates post-transcriptional gene silencing in *Drosophila* cells. Nature *404*:293-296.
13. **Hammond, S.M., S. Boettcher, A.A. Caudy, R. Kobayashi, and G.J. Hannon.** 2001. Argonaute 2, a link between genetic and biochemical analysis of RNAi. Science *293*:1146-1150.
14. **Hammond, S.M., A.A. Caudy, and G.J. Hannon.** 2001. Post-transcriptional gene silencing by double-stranded RNA. Nat. Rev. Genet. *2*:110-119.
15. **Hunter, C.P.** 1999. A touch of elegance with RNAi. Curr. Biol. *9*:R440-R442.
16. **Kato, Y., T. Kuwabara, M. Warashina, H. Toda, and K. Taira.** 2001. Relationships between the activities in vitro and in vivo of various kinds of ribozyme and their intracellular localization in mammalian cells. J. Biol. Chem. *276*:15378-15385.
17. **Kawasaki H. and K. Taira.** 2003. Functional anatomy of RNAi in human cells using dsRNA expression systems that are controlled by pol III promoters. Nucleic Acids Res. *31*:700-707.
18. **Ketting, R.F., T.H. Haverkamp, H.G. van Luenen, and R.H. Plasterk.** 1999. Mut-7 of *C. elegans,*

required for transposon silencing and RNA interference, is a homolog of Werner syndrome helicase and RNaseD. Cell *99*:133-141.

19.**Kuwabara, T., M. Warashina, S. Koseki, M. Sano, J. Ohkawa, K. Nakayama, and K. Taira.** 2001. Significantly higher activity of a cytoplasmic hammerhead ribozyme than a corresponding nuclear counterpart: engineered tRNAs with an extended 3′ end can be exported efficiently and specifically to the cytoplasm in mammalian cells. Nucleic Acids Res. *29*:2780-2788.

20.**Kuwabara, T., M. Warashina, M. Sano, H. Tang, F. Wong-Staal, E. Munekata, and K. Taira.** 2001. Recognition of engineered tRNAs with an extended 3′ end by Exportin-t (Xpo-t) and transport of tRNA-attached ribozymes to the cytoplasm in somatic cells. Biomacromol. *2*:1229-1242.

21.**Lee, N.S., T. Dohjima, G. Bauer, H. Li, M.J. Li, A. Ehsani, P. Salvaterra, and J. Rossi.** 2002. Expression of small interfering RNAs targeted against HIV-1 rev transcripts in human cells. Nat. Biotechnol. *20*:500-505.

22.**Li, W.X. and S.W. Ding.** 2001. Viral suppressors of RNA silencing. Curr. Opin. Biotechnol. *12*:150-154.

23.**Lipardi, C., Q. Wei, and B.M. Paterson.** 2001. RNAi as random degradative PCR: siRNA primers convert mRNA into dsRNAs that are degraded to generate new siRNAs. Cell *107*:297-307.

24.**Mitchell, C.E., S.A. Belinsky, and J.F. Lechner.** 1995. Detection and quantitation of mutant K-ras codon 12 restriction fragments by capillary electrophoresis. Anal. Biochem. *224*:148-153.

25.**Miyagishi, M. and K. Taira.** 2002. U6 promoter driven siRNAs with four uridine 3′ overhangs efficiently suppress targeted gene expression in mammalian cells. Nat. Biotechnol. *20*:497-500.

26.**Montgomery, M.K. and A. Fire.** 1998. Double-stranded RNA as a mediator in sequence-specific genetic silencing and co-suppression. Trends Genet. *14*:255-258.

27.**Montgomery, M.K., S. Xu, and A. Fire.** 1998. RNA as a target of double-stranded RNA-mediated genetic interference in *Caenorhabditis elegans.* Proc. Natl. Acad. Sci. USA *95*:15502-15507.

28.**Paddison, P.J., A.A. Caudy, E. Bernstein, G.J. Hannon, and D.S. Conklin.** 2002. Short hairpin RNAs (shRNAs) induce sequence-specific silencing in mammalian cells. Genes Dev. *16*:948-958.

29.**Paddison, P.J., A.A. Caudy, and G.J. Hannon.** 2002. Stable suppression of gene expression by RNAi in mammalian cells. Proc. Natl. Acad. Sci. USA *99*:1443-1448.

30.**Paul, C.P., P.D. Good, I. Winer, and D.R. Engelke.** 2002. Effective expression of small interfering RNA in human cells. Nat. Biotechnol. *20*:505-508.

31.**Plehn-Dujowich, D. and S. Altman.** 1998. Effective inhibition of influenza virus production in cultured cells by external guide sequences and ribonuclease P. Proc. Natl. Acad. Sci. USA *95*:7327-7332.

32.**Sharp, P.A.** 2001. RNA interference 2001. Genes Dev. *15*:485-490.

33.**Sijen, T., J. Fleenor, F. Simmer, K.L. Thijssen, S. Parrish, L. Timmons, R.H. Plasterk, and A. Fire.** 2001. On the role of RNA amplification in dsRNA-triggered gene silencing. Cell *107*:465-476.

34.**Smardon, A., J. Spoerke, S. Stacey, M. Klein, N. Mackin, and E. Maine.** 2000. EGO-1 is related to RNA-directed RNA polymerase and functions in germ-line development and RNA interference in *C. elegans.* Curr. Biol. *10*:169-178.

35.**Stark, G.R., I.M. Kerr, B.R. Williams, R.H. Silverman, and R.D. Schreiber.** 1998. How cells respond to interferons. Annu. Rev. Biochem. *67*:227-264.

36.**Sui, G., C. Soohoo, E.B. Affar, F. Gay, Y. Shi, W.C. Forrester, and Y. Shi.** 2002. A DNA vector-based RNAi technology to suppress gene expression in mammalian cells. Proc. Natl. Acad. Sci. USA *99*:5515-5520.

37.**Tabara, H., M. Sarkissian, W.G. Kelly, J. Fleenor, A. Grishok, L. Timmons, A. Fire, and C.C. Mello.** 1999. The rde-1 gene, RNA interference, and transposon silencing in *C. elegans.* Cell *99*:123-132.

38.**Tuschl, T.** 2001. RNA interference and small interfering RNAs. Chem. Biochem. 2:239-254.

39.**Tuschl, T., P.D. Zamore, R. Lehmann, D.P. Bartel, and P.A. Sharp.** 1999. Targeted mRNA degradation by double-stranded RNA in vitro. Genes Dev. *13*:3191-3197.

40.**Yu, J.Y., S.L. DeRuiter, and D.L. Turner.** 2002. RNA interference by expression of short-interfering RNAs and hairpin RNAs in mammalian cells. Proc. Natl. Acad. Sci. USA *99*:6047-6052.

41.**Zamore, P., T. Tuschl, P. Sharp, and D. Bartel.** 2000. RNAi: double-stranded RNA directs the ATP-dependent cleavage of mRNA at 21 to 23 nucleotide intervals. Cell *101*:25-33.

20 Inverse Genomics™: A Gene Discovery Platform Based on Inactivation of Gene Expression

Leigh N. Pierce, Philip K. Tan, Qi-Xiang Li, and Flossie Wong-Staal
Immusol, San Diego, CA, USA

INTRODUCTION

Gene Expression and Diseases

The study of gene expression lies at the heart of a wide variety of biomedical and biopharmaceutical research. This includes the classification of diseases, understanding of the molecular basis of disease, disease identification, and searching for cures. Ultimately, the goal of these studies is to accelerate the discovery, development, and testing of therapies that target disease.

The first draft of the human genome has already been completed, and the information will be fine-tuned over the next few years. However, identifying the majority of human genes is only the initial step to understanding what these genes do and how their expression impacts our normal physiological and pathological processes. Understanding the functions of the 50 000–100 000 genes in the human genome, their implications in disease states, and variations within the population represent a much more difficult task than the mapping and sequencing efforts that are currently underway. The challenge now is to establish a correlation between genes and function, in the hope that better therapeutic targets can be discovered more efficiently.

Loss of function and aberrant expression of key genes are the causes of many human diseases. Such defects are especially common in human cancers, which can occur when regulatory genes are deleted or inactivated (e.g., tumor suppressors) or become constitutively active (e.g., oncogenes). Unlike some other genetic

Perspectives in Gene Expression
Edited by K. Appasani

diseases, such as cystic fibrosis, in which a single defective gene is responsible, human cancer development may result from the mutation of multiple genes, sometimes accumulating as the disease progresses. For example, colorectal cancer arises from a gradual series of mutations in several genes, which may in turn cause the aberrant expression of other genes, resulting in the transition from normal colonic epithelia to dysplastic adenoma and finally to colorectal cancer (76). The key to developing effective drugs against these diseases is identifying the genes that are responsible for their initiation or progression.

Gene Discovery Methods

Traditional forward and reverse genetics. Geneticists have tried for a long time to study genes and their roles in the structure, growth, health, and disease of organisms. The traditional forward genetics technique (from phenotype to gene) involves first generating a desired phenotype by random mutagenesis (e.g., chemicals, UV irradiation, or heat) and then subsequent efforts to identify the genetic lesion in the mutated cells using gene mapping, positional cloning, or phenotypic rescue with cDNA libraries. Although this approach can lead to gene identification and functional assignment, it is tedious and prone to failure. Phenotypic rescue is generally not possible in mammals or mammalian cells, and gene mapping and positional cloning techniques are laborious and may require hundreds of intercrosses and backcrosses. In a well-known example, the cloning of the mouse *ob* gene, which is important in preventing obesity and diabetes, required eight years of concerted effort (61,90). Clearly, traditional forward genetics approaches cannot rapidly identify large numbers of genes for any given phenotype.

With reverse genetics (from gene to phenotype), the investigator focuses on a specific gene of interest and determines its function using gene knock out or knock in (transgenics) technology. This approach has been used successfully not only in cell culture but also in many organisms, including *Saccharomyces cerevisiae, Drosophila, Caenorhabditis elegans,* and zebrafish (1,41,53,67). It is preferable to perform reverse genetic screens in these and other organisms with relatively short life spans and lower genomic complexity, but such screens are more challenging in mammalian systems. Nonetheless, a great deal of research has been done using transgenic and knockout mice (24,33,55). Reverse genetics is a powerful technology and can generate a lot of useful information. It is particularly attractive in the post-genomic era, with the large number of genes awaiting scrutiny. However, reverse genetics approaches are also prone to failure due to the complexity of the genome. In some instances, the mutation or knockout of a gene of interest leads to lethality, revealing little of the gene function except for its importance to life. In many other instances, targeted gene mutations or disruptions produce no phenotype because of redundancies in the genome or other compensatory mechanisms such as the up-regulation of another separate but related gene. In one recent example, the knockout of the mouse β3 adrenergic receptor, which is presumed to be important in thermogenesis and weight control, produced no phenotype because of the compensatory activity of the related β1- and β2-adrenergic receptors. Instead, the phenotype was produced by the simultaneous triple knockout of each receptor gene (3). While reverse genetics may help to reveal the function of a gene, the primary limitation is the difficulty in adapting this approach for high-throughput gene discovery.

Differential gene expression: techniques and strategies. One strategy to identify functionally relevant genes is to correlate variations in gene expression with specific phenotypes. There are various methods to examine transcripts that are differentially expressed between two different cell populations. Differential display works by first producing a set of cDNA fragments that have been prepared in identical manners from different RNA samples, usually by restriction enzyme digestion of the cDNA or by producing PCR products with arbitrary primers. Next, the fragments are resolved on a gel that produces a characteristic pattern of bands for each sample. The bands from different samples are then compared to reveal those bands that differ in intensity between lanes. The cDNA fragment within the band can be excised for further analysis. The disadvantage of this technique is that the identity of the differentially expressed genes is only revealed after the excision of the band from the gel and sequencing, thus requiring significant labor for the gene identification step. Also, the technique can be prone to false positives that arise from various factors, including PCR-induced amplification biases. Although there are many successful variations of the differential display approach, they do not allow for the rapid and efficient identification of expression levels of large numbers of genes simultaneously (43).

Subtractive hybridization is used to produce a cDNA library that has sequences that are present in one sample of RNA but not another. Typically, tumor mRNA is subtracted from normal mRNA (or vice versa) to find transcripts that may have been suppressed or overexpressed in the process of tumor formation. Generally, the approach is to hybridize the two samples (normally cDNA) in solution. An excess of one sample, the "driver," hybridizes to most of the unwanted common sequences from the other sample, the "tracer." The driver is usually labeled so that molecules containing one or both strands in common with the driver are removed or otherwise not cloned. The remaining cDNA, which consists mostly of tracer, can be cloned to form a library for further analysis (10).

Other subtractive methods used to find differentially expressed genes include suppression subtractive hybridization and representational difference analysis (RDA). These newer techniques incorporate PCR amplification steps to enable one to work from smaller quantities of starting material. RDA is an effective way to compare two sets of DNA by hybridization and subtraction, utilizing either genomic DNA or cDNA. Overall, these subtractive techniques have been used successfully to identify many important disease-related genes, but these approaches allow only a pairwise comparison and require a time-consuming cloning step, making them unsuitable for automated high-throughput gene expression analysis (14,30).

A more powerful, genomics-based tool is gene expression profiling using DNA microarray, often referred to as DNA "chips." Two basic types of DNA microarrays are currently available: oligonucleotide arrays and cDNA arrays. Both approaches involve the immobilization of DNA sequences in a gridded array on the surface of a solid support such as a glass microscope slide or silicon wafer. For both microarray approaches, the next step involves the purification of RNA from the source of interest (e.g., different tissue types or phenotypically different cultured cells), enzymatic fluorescent labeling of the RNA, and hybridization of the fluorescently labeled material to the microarray. Hybridization events are then captured by scanning the surface of the microarray with a laser-scanning device and measuring the fluorescence intensity at each position in the microarray. The fluorescence intensity of each spot on the array is proportional to the level of expression of the gene represented by that spot. Unlike the older procedures, microarrays make it possible to simultaneously monitor the expression of thousands of genes (2).

Another method for comprehensive analysis of gene expression is the serial analysis of gene expression (SAGE). This is a sequencing-based technology that provides an in-depth quantitative assessment of gene expression. In SAGE, the investigator sequences a small and unique fragment of each expressed gene (called a SAGE tag) and quantifies the number of times it appears (called the SAGE tag number). Therefore, the SAGE tag numbers directly reflect the abundance of the corresponding transcript. SAGE works by storing digital values electronically, providing absolute gene expression levels that make historical comparisons and database compilation easy. It is useful for comparing small numbers of tissue or cellular samples derived from well-controlled experiments. On the negative side, the number of samples than can be processed by SAGE is small as compared to, for example, DNA arrays, because of the time it takes to construct a SAGE library. The potential to analyze hundreds of samples by SAGE for a single experiment is not a practical option for the technology in its current form (83).

While these global gene expression technologies are powerful tools for screening many genes simultaneously and can reveal interesting genes, the expression of which correlates with specific phenotypes, they do not imply any causal relationships. Ultimately, the differentially expressed genes must still be individually validated at the functional level.

INVERSE GENOMICS™ GENE DISCOVERY

We have developed a technology platform called Inverse Genomics (Immusol, San Diego, CA, USA) to identify therapeutic drug target genes in different disease areas. Inverse Genomics uses a combinatorial library of hairpin ribozymes with randomized substrate-binding sequences. Once introduced into a disease-based cell system, a ribozyme will recognize and cleave unique RNA transcripts, thereby decreasing the translation of various proteins. Based on the ribozyme's substrate-binding sequence (ribozyme sequence tag or RST), a biologically relevant target gene that is linked to the therapeutic phenotype can be identified. Thus, Inverse Genomics is similar to other forward genetics approaches, but with the advantage that it provides a means to specifically and quickly identify the genes that are functionally relevant.

General Properties of Ribozymes

One of the more surprising discoveries of the past 16 years was that some RNA molecules have enzymatic activity. These RNAs, called ribozymes, can cleave RNA at sequence-specific sites. Ribozymes were originally discovered in Tetrahymena and were found capable of splicing out their introns without the aid of any protein enzymes (88). Subsequently, ribozyme activities have been found to be associated with some plant viroid RNA and the hepatitis Delta virus (4,16,21). It was also determined that under certain conditions, these self-splicing RNA molecules could act *in trans* to cleave other RNA molecules (89). Since this discovery, the unique ability of genetically engineered ribozymes to cleave target mRNAs has been effectively utilized in many different systems to knockdown the expression of specific genes (25,70,78,79,86). The two most extensively used catalytically active RNAs are the hammerhead and the hairpin ribozymes, both of which originated from a satellite RNA associated with the tobacco ringspot virus (16,36). The hairpin ribozyme, which is utilized in the Inverse Genomics approach, possesses a target recognition site that contains two short sequences (four and eight bases, respectively) complementary to the target mRNA, which flank an obligatory GUC triplet (Figure 1). Cleavage occurs immediately 5′ of the GUC sequence. Most mRNA molecules contain multiple GUC sites (statistically, this should occur every 64 nucleotides). Although not all of these will be accessible to ribozyme cleavage because of bound proteins or secondary structures, most mRNAs contain one or more ribozyme-cleavable target sequences.

Targeted Gene Knockdown by Ribozymes

Due to the complex intracellular environment that leads to ribozyme cleavage site inaccessibility, the efficacy of ribozymes tested using in vitro cleavage assays does not necessarily correlate with ribozyme efficacy in living, intact cells. This means that for any given target, several ribozymes will need to be tested in cell culture regardless of successful in vitro cleavage activity (45,59). The following experiment illustrates the activity of ribozymes against two genes, interleukin-1β (IL-1β) and IL-1β converting enzyme (ICE) in cells (40). IL-1β is a potent cytokine involved in a variety of disease states. ICE is a specific protease that activates the proenzyme form of IL-1β by cleavage. IL-1β and ICE are shown to be involved in various diseases, such as pancreatitis, rheumatoid arthritis, hepatic metastasis, as well as other disorders (15,58,60,75).

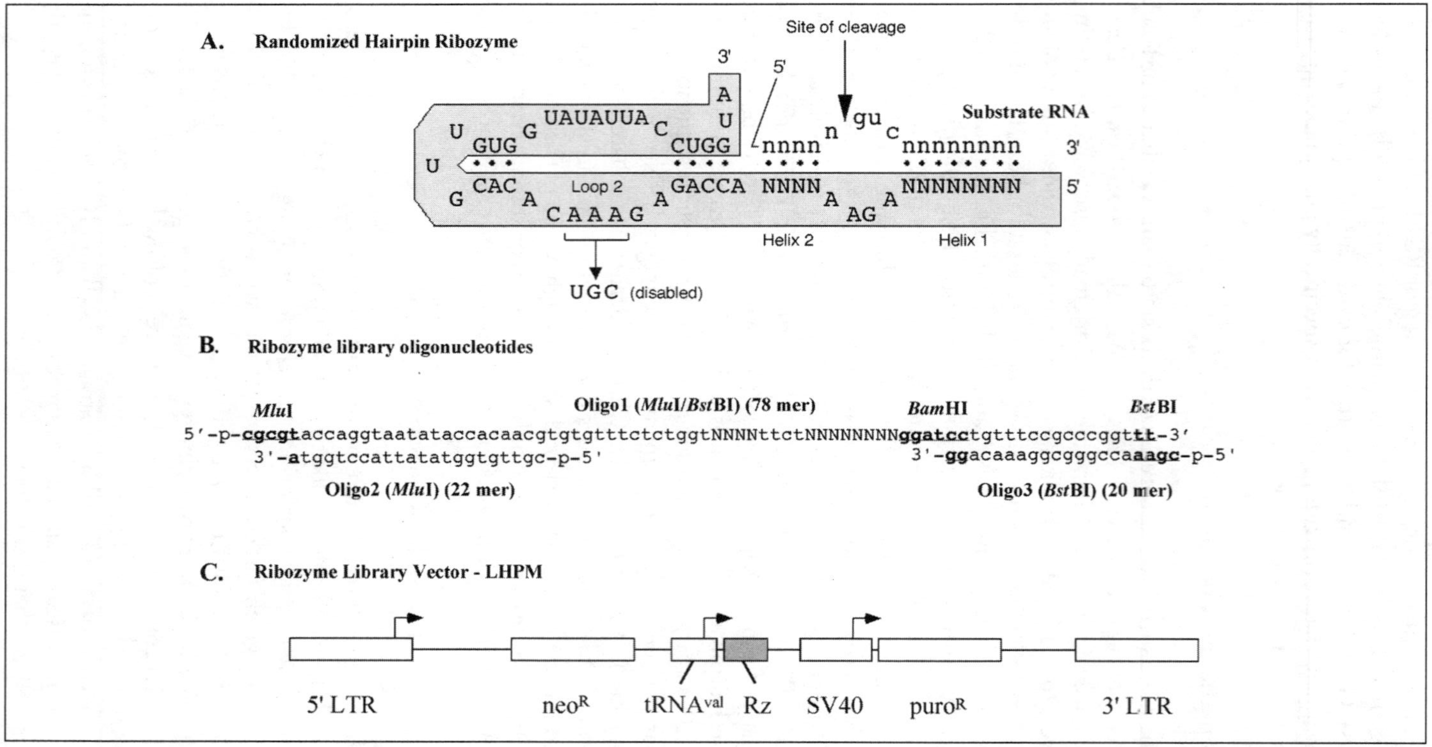

Figure 1. Randomized Ribozyme Library Components. (A) The hairpin ribozyme consists of a 52-nucleotide RNA molecule (shaded in uppercase letters) that binds to an RNA substrate (lowercase letters). Two helices, helices 1 and 2, form between the ribozyme and the substrate via Watson-Crick basepairing. The substrate RNA must contain a GUC for maximal cleavage, and cleavage occurs immediately 5′ of the G as indicated. The catalytic but not substrate-binding activity of the ribozyme can be disabled by mutating the AAA in loop 2 to CGU, as shown. (B) The construction scheme of the retroviral vector containing a $tRNA^{val}$ Pol III promoter cassette-expressing randomized ribozyme library. N can be any of the four deoxyribonucleotides. (C) The ribozyme library was cloned into the retroviral vector pLHPM. SV40, simian virus 40. Adapted with permission from Li et al. (42).

Six anti-IL-1β ribozymes were stably transduced into THP-1 cells (ATCC, Manassas, VA, USA), a human monocytic leukemia cell line. All of the ribozymes caused a decrease in mRNA, although to variable extents as measured by Northern hybridization. Even after stimulation with lipopolysaccharide (LPS), which increased IL-1β levels approximately 50-fold, significant reductions in extracellular IL-1β levels of up to 10-fold were observed in all the ribozyme transduced cells, indicating that the ribozymes are active over a wide range of mRNA copy numbers. In general, the reduction of the extracellular IL-1β protein correlated with reductions in IL-1β mRNA. These results were reproduced in a second cell line, a human fibrosarcoma cell line, HT1080 (ATCC). Furthermore, the co-transduction of multiple ribozymes into the same cells led to greater reduction of the IL-1β mRNA (up to 94%).

Eight anti-ICE ribozymes were tested; all of which caused a decrease in ICE protein of up to 55% in THP-1 cells. The three most potent ribozymes also showed significant reductions in ICE mRNA. After induction with LPS, the levels of ICE protein were greatly reduced in most of the cultures (up to 80%).

This study showed that several effective ribozymes could be identified by screening fewer than 10 hairpin ribozyme constructs in two target systems, IL-1β and ICE. Isolating transduced single cell clones or introducing a second ribozyme into the cells could further enhance the degree of knockdown. Since it is likely that most messages within living cells contain one or more accessible ribozyme cleavage sites, ribozymes can be used in genome-wide high-throughput assays to identify novel genes of therapeutic interest.

Combinatorial Hairpin Ribozyme Library

We have adapted ribozyme technology for use in genome-wide screens. The core component of the Inverse Genomics' approach is the combinatorial hairpin ribozyme library, which theoretically can knock down the expression of every gene in the genome. The ribozyme library is generated by randomizing the two short target recognition sequences (four and eight bases, respectively; Figure 1). Therefore, the theoretical complexity of the library is 4×10^{12} or 1.7×10^{7}. The ribozyme library is generated as follows.

Three oligonucleotides are synthesized: Oligo1 (5′-p-CGCGTACCAGGTA-ATATACCACAACGTGTGTTTCTCTGGTNNNNTTCTNNNNNNNN-GGATCCTGTTTCCGCCCGGTTT-3′), which encodes the ribozyme sequence; Oligo2 (5′-p-CGTTGTGGTATATTACCTGGTA-3′), which is antisense to the 5′-end of the ribozyme; and Oligo3 (5′-p-CGAAACCGGGCG-GAAACAGG-3′), which is antisense to the 3′-end of Oligo1 (Figure 1). Oligo1, Oligo2, and Oligo3 are mixed together in a 1:20:20 ratio, respectively, and are annealed together by heating to boiling for 2 min and then slowly cooled to room temperature. The annealed product is a partial dsDNA that contains an *Mlu*1 overhang on the 5′-end and a *Bst*BI overhang on the 3′-end. This product is then ligated into the *Bst*BI/*Mlu*I-prepared LHPM backbone vector (see Figure 1, B and C). As shown in Figure 1, The retroviral vector LHPM contains the puromycin resistance gene, whose expression is driven by the 5′ long terminal repeat (LTR). Ribozyme expression is directed by a $tRNA^{val}$ promoter, which is a pol III promoter that has been shown to efficiently transcribe active ribozymes (74). The prepared vector and annealed oligonucleotides are ligated together at a vector-to-insert ratio of 1:5 at 14°C overnight. Twenty micrograms of the ligated mixtures are transformed into DH10B electro-competent bacteria at a DNA-to-bacteria ratio of 1 µg:40 µL. From ten to twenty 245 mm LB ampicillin plates are then used to plate the entire library. Following incubation at 37°C overnight, the bacteria from all of the plates are pooled and stored and -80°C. The complexity of the ribozyme library is then assessed by the sequencing of random clones (42,80).

The ribozyme library and individual ribozymes are introduced into cell systems by transduction with infectious, replication-incompetent retrovirus vectors. This is produced by lipid-mediated transfection of Cf2Th cells (a canine thymus cell line) (ATCC) with pLHPM-RzLib, two helper plasmids containing gag-pol, and vesicular stomatitis virus G protein (VSV-G) in a 1:0.5:0.25 ratio. Retroviral particles are collected on days 2–5 after transfection by harvesting the supernatant from the transfected cells. The virus-containing supernatant is filtered through a 0.8-µm filter, and the titer is determined on HT1080 cells by puromycin-resistant colony formation of transduced HT1080 cells.

Once the ribozyme library is stably transduced into cells, the cells are subjected to selection

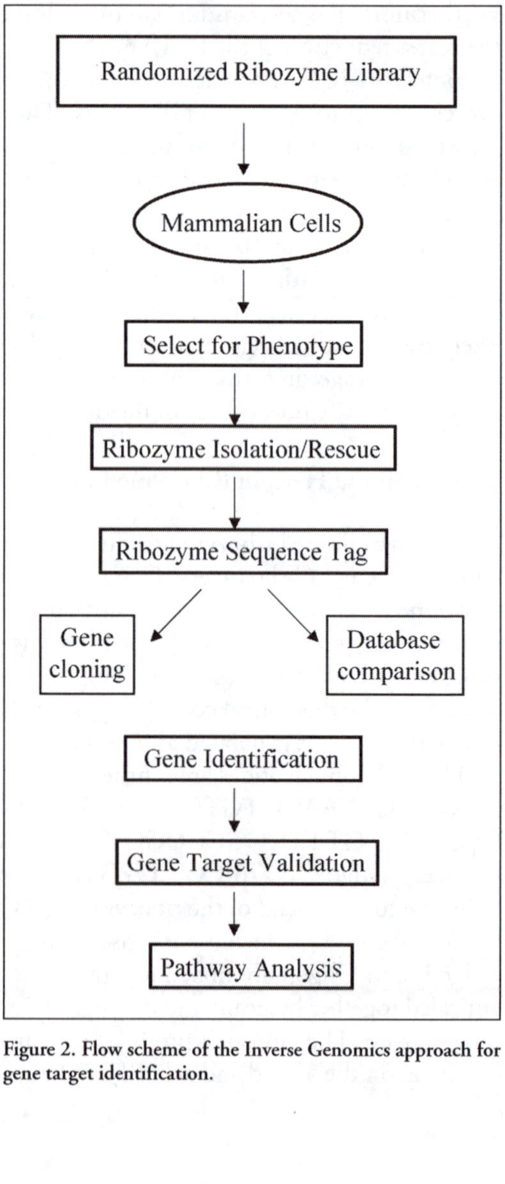

Figure 2. Flow scheme of the Inverse Genomics approach for gene target identification.

for a desired phenotype. This can be considerably varied, as shown by the examples described in the following section. Because of the high complexity of the ribozyme library, it is necessary to isolate the functional ribozymes through successive cycles of enrichment. This is accomplished by recovering the ribozymes responsible for the desired phenotypic change by helper virus rescue or PCR and re-transducing them into fresh parental cells for additional rounds of selection. A strong phenotype is usually attainable after 3–5 cycles of enrichment, and, at that point, individual ribozymes are tested for their functional activity. Once a single ribozyme is shown to produce a phenotype, the recognition sites of the confirmed ribozymes can then be used as tags (RSTs) to identify the target genes of interest based on sequence complementarity. This can be done by searching relevant databases for a sequence match, or the RST itself can be used to clone the gene of interest. Figure 2 illustrates the Inverse Genomics approach for gene identification using the randomized ribozyme library. This can be done by searching relevant databases for a sequence match, or the RST itself can be used to clone the genes of interest.

Examples of Selection Schemes

An early essential step in Inverse Genomics is the development of a powerful selection system with minimal background for the desired phenotype. The basis of selection can be quite varied: cell survival, anchorage-independent growth, flow cytometry based on fluorescent reporter genes [e.g., enhanced GFP (EGFP)], and immunostaining of cell surface or intracellular proteins. Some specific examples of Inverse Genomics applications utilizing different selection schemes have been published and are briefly described below.

Tumor suppressor genes. As a first test of the principle, we chose a HeLa cell revertant, HF, which has lost the ability to form colonies in soft agar or tumors in nude mice (9,80). Morphologically, they also appear flat and non refractile, which is more characteristic of non-transformed cells. We speculated that the change in HF cells was due to the activation of one or more tumor suppressor genes (9). The ribozyme library was introduced into HF cells via stable transfection. As a negative control, the ribozyme CNR3, which targets the negative RNA strand of the human hepatitis C virus and has no known cellular targets, was also stably transfected into HF cells. The cells were stably selected and seeded into soft agar. After three weeks of soft agar selection, the library-expressing cells only produced a modest increase (2.5-fold to 4-fold) over the control HF cells. However, when these colonies were picked and replated into soft agar for a second round of selection, the library-expressing cells showed a 300-fold increase over the control ribozyme-expressing cells. Ribozymes from these cells were then rescued from the selected cells by viral and PCR rescue, and their individual sequences were determined. One ribozyme, called Rz568, showed up repeatedly. To verify that Rz568 conferred a selective growth advantage to HF cells plated in soft agar, the ribozyme was stably transfected into HF cells and seeded into soft agar. After two rounds of selection, Rz568 significantly increased the HF colony growth in soft agar but not a disabled ribozyme control (Figure 3), suggesting that the catalytic activity of Rz568 is required for the phenotype. This also verified that Rz568 alone confers anchorage independence. As mentioned earlier, ribozymes recognize their targets by sequence complementarity; therefore, the recognition sequence of a functionally selected ribozyme predicts a corresponding 16-base target sequence tag (RST) that can be used to search the gene with BLAST and EST databases or as a primer for the 3′ and 5′ rapid amplification of cDNA ends (RACE). Using 3′ and 5′ RACE, we identified a potential cDNA that matched a recently cloned *Drosophila* gene, *peter pan* (*ppan*), which was shown to be involved in cell growth, DNA replication, and possibly cell-to-cell communication during development (49). Northern blot analysis showed that cells stably expressing Rz568 showed a 30%–35% reduction in hPPAN mRNA, and HeLa

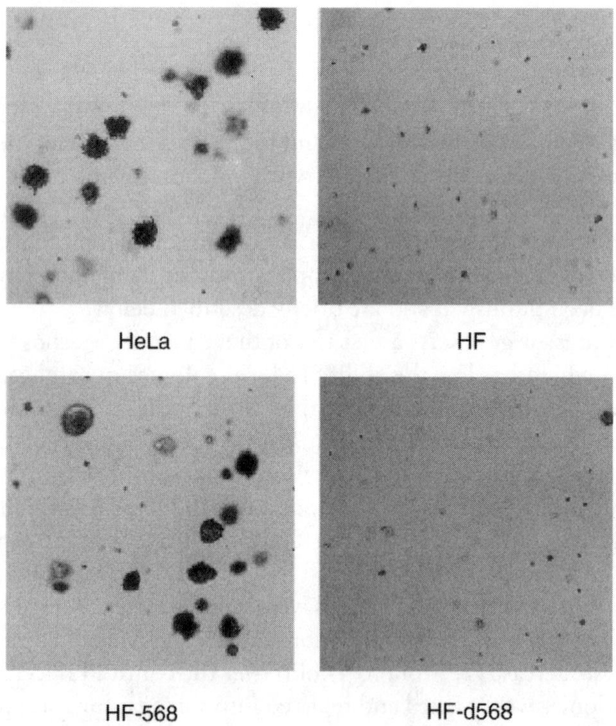

Figure 3. Verification of Rz568. Expression of catalytically active Rz568 confers anchorage independence to HF cells. HF cells were stably transfected with pLHPM vector only (HF) or pLHPM expressing either Rz568 (HF-568) or catalytically disabled 568 (HF-d568). Transfected cells were plated into soft agar, and the colonies resulting from two rounds of soft agar selection are shown. As a positive control for soft agar growth, untransfected HeLa cells were plated in soft agar in parallel. Adapted with permission from Welch et al. (80).

Table 1. Validation of PPAN in Anchorage Dependence and as the Functional Target of Rz568

	Colony Growth in Soft Agar[a]		
Ribozyme	First Round	Second Round	PPAN mRNA[b]
Control	6	15	101 ± 2
TV1	215	>50 000	82 ± 9
TV2 + TV3	75	>50 000	76 ± 13
TV4 + TV5	103	>50 000	74 ± 17

[a]Number of soft agar colonies per 10^5 cells.
[b]PPAN Northern blot analysis from three independent experiments after the first round of soft agar selection, relative to G3PDH mRNA ± standard error.
Adapted with permission from Welch et al. (80).

cells expressed only 15%–20% less hPPAN than HF cells. These observations suggest that small differences in the level of hPPAN expression may contribute to the dramatic differences in phenotype between HeLa and HF cells. We then formally validated hPPAN as a tumor suppressor gene by choosing five unique cleavage sites (GUC sequences) within the hPPAN mRNA. When stably transfected into HF cells, the designed "validation ribozymes" conferred the same anchorage-independent phenotype as Rz568 (Table 1). Furthermore, when the full-length hPPAN cDNA was stably transfected and overexpressed in HeLa cells, the ability of HeLa cells to form colonies in soft agar was significantly reduced. These experiments clearly show that hPPAN is involved in regulating anchorage-independent growth and validates Inverse Genomics as a gene discovery technology.

Through Inverse Genomics, novel functions for previously known genes are frequently discovered. An early example was the discovery of the mouse telomerase reverse transcriptase (mTERT) gene as a tumor suppressor gene (42). The original design of this selection system was to identify genes involved in fibroblast transformation. NIH 3T3 cells (ATCC) were stably transduced with the retroviral ribozyme library. After three to four weeks, there were 2-fold to 5-fold more foci in the ribozyme library transduced culture than the mock control. The foci were harvested and replated to determine the stability of the phenotype. The mock transduced cells still grew as a typical monolayer culture as did non-transduced NIH 3T3 cells. In contrast, the ribozyme library transduced cells showed a morphologically transformed phenotype and formed many foci. At this point, the ribozymes were PCR rescued from the transformed cells. They were reintroduced into the NIH 3T3 cells as a sub-library and subjected to another round of selection. After we replated the foci, the sub-library population again showed a striking transformed phenotype. The selected sub-library ribozymes were rescued by PCR, and there was enrichment of a dominant ribozyme sequence, Rz007, which was subsequently confirmed in the transformation assay. With a BLAST

Table 2. mTERT mRNA Level and Telomerase Activity in Cells Transformed by Rz007

Ribozymes	mTERT mRNA Level[a] (%)	Telomerase Activity[b] (%)
CNR3	100 ± 10	100 ± 6
Rz007	47 ± 13	16 ± 2

[a]mRNA level was measured by the TaqMan RT-PCR method. It is represented by the mean of mTERT to mGAPDH ratio of duplicates and presented as a relative quantity.

[b]Telomerse activity was measured by TRAP-based ELISA. It is represented by the mean of the activity to total protein ratio of duplicates and is presented as a relative quantity.

Adapted with permission from Li et al. (42).

search against the public database, the RST of Rz007 identified a match with the mouse gene-encoding telomerase reverse transcriptase (TERT). TaqMan® RT-PCR RNA quantitation showed that Rz007 transduced cells had an approximate 50% decrease in mTERT expression. In addition, telomerase activity was measured using a telomeric repeat amplification protocol (TRAP)-based ELISA. The telomerase activity in the Rz007 transduced cells was reduced by approximately 80% (Table 2). Target validation ribozymes generated against other ribozyme cleavage sites (GUC sequences) within mTERT also increased focus formation in the NIH 3T3 assay and reduced telomerase activity. It is clear that the down-regulation of mTERT can result in cellular transformation.

Telomerase has been implicated in many cellular processes because of its important role in cell proliferation, including development, cancer, and aging. Telomerase is a ribonucleoprotein enzyme complex that consists of RNA template and protein components, including a specialized TERT (8,26). Telomerase is responsible for maintaining the physical termini of eukaryotic chromosomes. Although little is known about the mechanism of regulation, telomerase activity in cells is apparently regulated at the level of TERT mRNA. In previous studies, the activation of human telomerase or an increased level of hTERT has been detected in many human tumors. In addition, the inhibition of telomerase activity can cause tumor cells to undergo senescence. Therefore, telomerase activity is believed to contribute to the immortalization and transformation of cells, and our discovery that mTERT actually suppresses the transformation phenotype in mouse fibroblasts was somewhat counterintuitive. However, subsequent studies from other groups showed that a double knock out of mTERT and p53 (p53 is mutated in NIH 3T3 cells) could lead to increased chromosomal instability and cancer (12). Thus, Inverse Genomics was able to uncover a novel and unexpected function of a previously known gene.

Cellular cofactors for hepatits C virus internal ribosome entry site. Antiviral therapy has largely targeted viral genes, particularly viral enzyme targets. However, due to the high mutagenic potential of some viruses, long-term treatments that act against viral targets may be ineffective. Through Inverse Genomics, we set out to identify cellular factors involved in hepatitis C viral internal ribosome entry site (IRES)-mediated function (39). Translation of cellular mRNAs usually requires a cap modification at the 5′-end of the RNA. In some viral and cellular mRNAs, a cap-independent mechanism of ribosome binding is mediated by an IRES (31). The 5′ untranslated region (UTR) of hepatitis C virus (HCV) is highly conserved and functions as an IRES to initiate the translation of HCV proteins. We determined that the use of our randomized ribozyme library would be an excellent tool to discover cellular cofactors involved in IRES-mediated translation. The reporter construct pHyg-5′*tk* was designed and expresses a bicistronic message that includes the hygromycin resistance gene upstream of an IRES that directs the translation of the HSV-*tk* suicide gene. The construct was designed so that cells that normally express HSV-*tk* through the IRES will consequently die upon the addition of the HSV-*tk* suicide gene substrate, ganciclovir (GCV). Conversely, cells that contain ribozymes that inhibit IRES-mediated translation will survive GCV treatment due to the inhibition of HSV-*tk* production. The ribozyme library was stably transduced into a HeLa cell clone, and the stable expression of the pHyg-5′*tk* reporter was ensured by hygromycin selection. The cells were then grown in GCV. Ribozymes that conferred GCV-resistant colonies were enriched through four rounds of rescue and re-transduction as described earlier. At that stage, individual ribozymes were tested for their ability to confer resistance to GCV killing as well as reduction in IRES-mediated HCV core protein translation. Figure 4 shows the results for two ribozymes, RzHCV1 and RzHCV2. Interestingly, Northern blot analysis revealed that there was an increase in the core transcript in both populations, such that the reduction in translatability of the mRNA was actually greater than the reduction in proteins. RzHCV1 caused an 83% reduction of HCV IRES-dependent translation, RzHCV2 caused a 76% reduction, while reduction in core protein synthesis was reduced by 47% and 58%, respectively. In contrast, the analysis of cap-dependent proteins, GAPDH and β-actin, and cell growth did not reveal any significant changes (Figure 4).

Searches of the National Center for Biotechnology Information (NCBI) nonredundant and EST databases failed to identify acceptable matches to potential target genes for RzHCV1. Therefore, 5′-RACE amplification was performed using HeLa mRNA or cDNA templates, with the RST as primers. A 580-bp fragment was amplified that shared significant homology to the rat eIF2Bγ subunit. Based on this sequence, the entire human eIF2Bγ was cloned. Target validation ribozymes designed against other cleavage sites (GUC sequences) in eIF2Bγ also caused an increase in GCV-resistant colonies and translatability of HCV core protein, similar to RzHCV1 and RzHCV2 (Figure 5). This confirmed that eIF2Bγ is the gene target for RzHCV1.

For RzHCV2, a search in the public databases revealed a match with a second protein in the same functional pathway, human eIF2γ. Target validation ribozymes were once again designed and tested in the HeLa/pHyg-5′*tk* cell line. The number of GCV-resistant colonies was increased, and the translatability of HCV core RNA was reduced (Figure 5).

Interestingly, eIF2Bγ and eIF2γ are subunits of eIF2B and eIF2, respectively. eIF2B and eIF2 are translation initiation factors that bind to encephalomyocarditis virus IRES with high affinity (65). eIF2B is a guanine-nucleotide exchange factor that is necessary for protein synthesis initiation in eukaryotes. Phosphorylation of eIF2 inhibits eIF2B activity in response to stress or starvation (56). The function of the γ subunits of these factors is still unclear.

Using Inverse Genomics, we showed that the HCV IRES function can be preferentially inhibited by the depletion of eIF2γ and eIF2Bγ, while general protein translation and cell growth remain unaffected. Thus, these two factors are potential therapeutic targets for inhibiting HCV translation and replication.

Regulation of BRCA1. BRCA1 is linked to genetic susceptibility to familial breast and ovarian cancers through mutational loss of expression. While mutational loss of BRCA1 does not seem to contribute to sporadic breast and ovarian cancers, in these cancers and cell lines derived from them, BRCA1 expression is often reduced or even undetectable (23,28,48,69,72,82,84). Thus, the identification of genes involved with BRCA1 regulation through Inverse Genomics could potentially lead to the development of new therapies (5).

To measure BRCA1 ex-

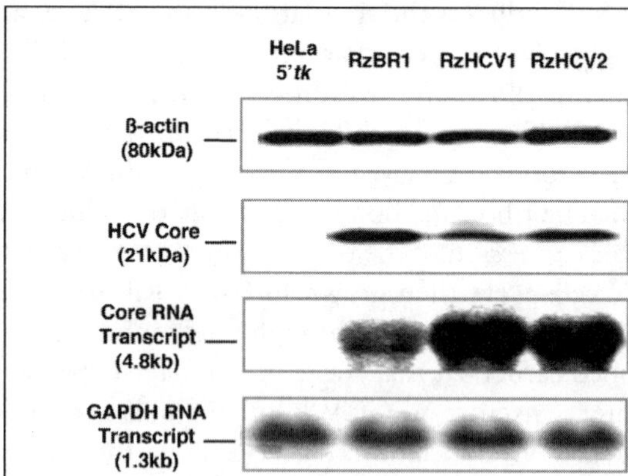

	HeLa 5′*tk*	RzBR1	RzHCV1	RzHCV2
Relative core protein expression (%)	0	100	52±5.6	41.6±4.9
Relative core RNA expression (%)	0	100	302±30	172±11
Core protein/ core RNA (% of control)	0	100	17	24

Figure 4. Immunoblot and RNA analysis of GCV-resistant HeLa-5′*tk* reporter cells after transduction with single ribozymes. Relative levels of the HCV core protein normalized to β-actin present in protein lysate (15 μg) from GCV-selected single Rz transduced reporter cells. Signals for β-actin were used to normalize the core protein signals. The activity of the RzBR1 control (lane 2) was set as 100%. Parental cells expressing the bi-cistronic vector 5′*tk* (lane 1) lack core protein expression derived from transduction with the retroviral ribozyme construct. Fifteen micrograms of total RNA isolated from GCV-resistant cells were analyzed by Northern blot analysis. HCV core RNA was normalized to GAPDH, and the ratio of core protein:core RNA transcript is expressed as a percent of the negative control. The data are presented as the mean and SEM of three experiments. Adapted with permission from Kruger et al. (39).

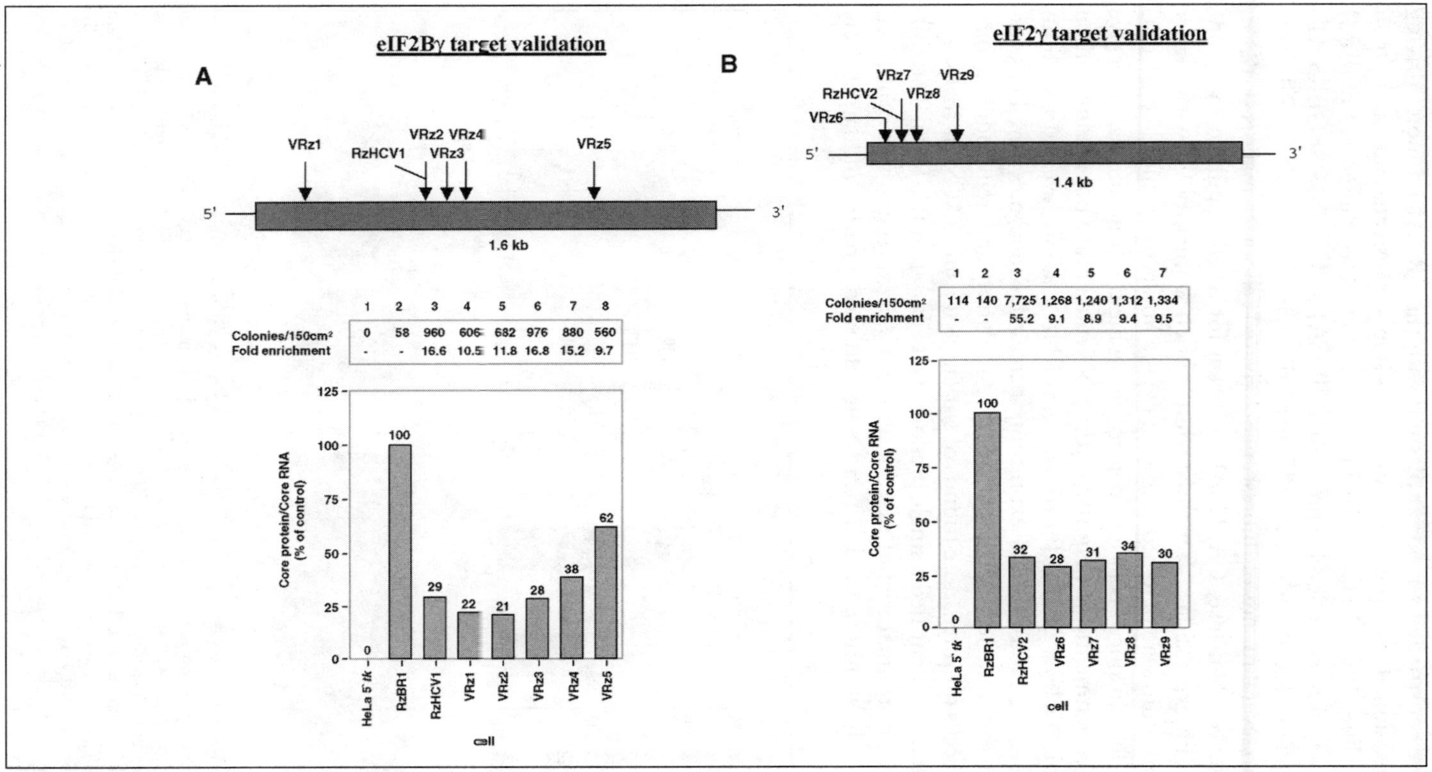

Figure 5. Ribozyme-mediated validation of eIF2Bγ and eIF2γ. (A) Five validation ribozymes (VRz1-5) were designed directed against GUC sites at positions 253 (VRz1), 643 (VRz2), 695 (VRz3), 759 (VRz4), and 1203 (VRz5), within the human eIF2Bγ mRNA. HeLa-5′*tk* cells were stably transduced with retroviral particles from individual ribozymes. GCV-resistant colonies were counted 21 days after exposure to GCV, pooled, and analyzed for HCV core and GAPDH RNA transcript levels by Northern blot analysis. (B) Validation of eIF2γ by four ribozymes directed against GUC sites at positions 143 (VRz6), 172 (VRz7), 199 (VRz8), and 370 (VRz9). Adapted with permission from Kruger et al. (39).

pression, we constructed a reporter plasmid with the BRCA1 promoter driving EGFP expression. This reporter construct was stably transfected into an ovarian cancer cell line, PA1 (ATCC). An individual clone was chosen (P8EGBR3) based on its intermediate-level expression of BRCA1 and EGFP. P8EGBR3 cells were stably transduced with the randomized ribozyme library. These cells were subjected to several rounds of fluorescence-activated cell sorter analysis (FACS®; BD Biosciences, San Jose, CA, USA) to enrich for a cell population with increased EGFP expression. After three rounds of enrichment, cells with higher EGFP expression were detected in the ribozyme library transduced cells compared to non-transduced cells and control ribozyme transduced cells. After five rounds of enrichment, the majority of the ribozyme library population shifted toward a higher level of EGFP expression. mRNA analysis revealed that both EGFP and endogenous BRCA1 expression increased. Ribozymes were PCR rescued from the enriched ribozyme library population and reintroduced into P8EGBR3 cells for three more rounds of enrichment using FACS. At that stage, sequence analysis identified a single predominant ribozyme, Rst-RH5. The ability of Rst-RH5 to increase both EGFP and BRCA1 expression was confirmed (Figure 6A). In addition, the Rst-RH5 cells exhibited a flatter and larger mor-

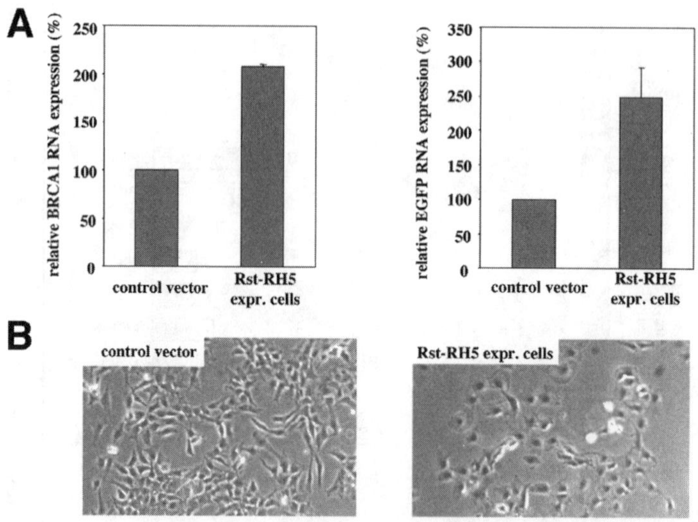

Figure 6. BRCA1 expression and morphology of selected cells expressing Rst-RH5. Cells transduced with an enriched ribozyme library were subjected to several rounds of selection for high EGFP expression and ribozyme rescue, resulting in a population expressing Rst-RH5 as the sole predominant ribozyme. These cells were examined for BRCA1 expression and cell morphology. (A) Real-time quantitative RT-PCR analysis of EGFP and endogenous BRCA1 RNA normalized to GAPDH in control vector transduced or Rst-RH5-expressing cells. Numbers are given as relative RNA levels in percent compared to control vector transduced cells (mean ± SEM). (B) Morphology in monolayer culture of Rst-RH5-expressing cells compared to control vector transduced cells. Adapted with permission from Beger et al. (5).

phology with a larger cytoplasm resembling non-transformed cells (Figure 6B) and lost the ability to form colonies in soft agar, which is consistent with the expected phenotype when BRCA1 expression was increased.

BLAST analysis of the 16 nucleotides from the Rst-RH5 target recognition sequences revealed a partial match to the *Id4* gene. *Id4* is one of four related helix-loop-helix (HLH) proteins that are known to function as dominant negative regulators of transcription (6,52). Northern blot analysis of *Id4* showed that Rst-RH5 caused a reduction in *Id4* mRNA. To validate *Id4*, target validation ribozymes were designed against other cleavage sites (GUC sequences) located within the coding region or the 3′UTR of *Id4*. The target validation ribozymes were stably transduced into P8EGBR3 cells. All of the target validation ribozymes caused an increase in BRCA1 expression as well as a concomitant decrease in *Id4* expression (Figure 7A). Each of the target validation ribozymes also decreased anchorage-independent growth of P8EGBR3 cells. When *Id4* was

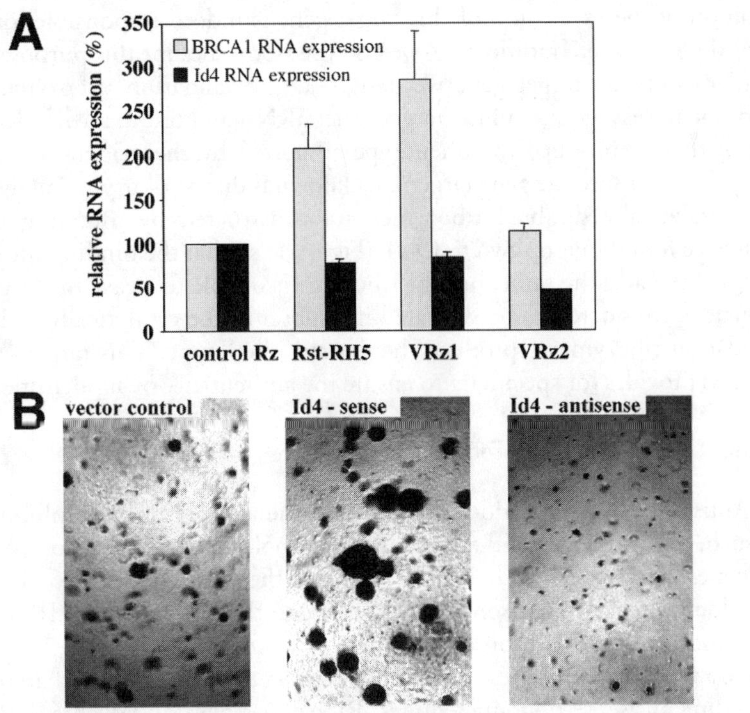

Figure 7. Validation of Id4 as a negative regulator of the BRCA1 promoter. (A) Effects of Rst-RH5 and two validation ribozymes (VRz1 and VRz2) on endogenous BRCA1 and Id4 expression as measured by real-time quantitative RT-PCR. Expression of BRCA1 or Id4 mRNA was normalized to GAPDH mRNA. Numbers are given as relative RNA levels in percent compared to the control ribozyme (mean ± SEM). (B) Anchorage-independent growth of P8EGBR3 cells transduced with control vector, Id4-sense or Id4-antisense expression vectors. Adapted with permission from Beger et al. (5). **(See color plate A37.)**

overexpressed, anchorage-independent growth increased (Figure 7B). In addition, when BRCA1 expression was increased in the breast cancer cell line MCF7 (ATCC) by estrogen stimulation, *Id4* expression was reduced more than 2-fold. These experiments confirmed the *Id4*-mediated regulation of BRCA1 expression via its promoter, which was a previously unknown function for this protein.

Identification of *Id4* as a regulator of BRCA1 illustrates the utility of Inverse Genomics as an approach to discover a new function for a previously known protein. In addition, the discovery of a new regulatory pathway for BRCA1 expression may lead to new therapies for the treatment of breast and ovarian cancer.

INVERSE GENOMICS TARGET VALIDATION

Ribozymes

As illustrated in the earlier examples, once a potential target gene is identified for a particular selection system, the next essential step is to validate that the inhibition of the expression of this target gene is indeed responsible for the observed phenotype. Traditionally, we also use ribozymes for this purpose. To validate a candidate target gene we design "target validation" ribozymes against additional cleavage sites in the target gene mRNA and test to see if at least a subset of them reproduce the phenotype conferred by the originally selected ribozyme. To ensure the gene target knockdown is due to ribozyme catalytic activity, catalytically disabled ribozymes are constructed by mutating the AAA sequence found in loop two to CGT (Figure 1) so that the binding site of the ribozyme remains the same, but the ribozyme is unable to cleave the target RNA. A gene is considered to be validated if a subset of the catalytically active target validation ribozymes reproduce the desired phenotype. This target validation process provides for specificity to ensure the authenticity of an identified gene.

Other Gene Repression Tools

Antisense oligonucleotides also confer sequence-specific gene inhibition and, as such, can also be a useful target validation tool. They are often compared to ribozymes with respect to their ability to cause the targeted inhibition of gene expression. Direct comparison has been made to the inhibition of HIV-1 replication and the expression of HIV-1 integrase, c-fos, and CAT (11,27,64,68), to name a few. In each of these cases, ribozymes were more effective than the corresponding antisense oligonucleotides. Ribozymes have an advantage over antisense because of its potential to operate both as antisense, via the complementary sequences in helices 1 and 2 (Figure 1), and as site-specific ribonucleases, to degrade target RNA catalytically. Antisense oligonucleotides are not catalytically active and thus can only stoichiometrically inactivate their targets.

RNA interference is emerging as a powerful new tool for gene silencing and target validation in mammalian cells. It has been previously shown that sequence-specific gene silencing results from the introduction of sense and antisense RNA or double-stranded RNA (dsRNA) in *C. elegans*, which was at least 10-fold more potent than sense or antisense RNAs alone (20). This blocking of gene expression by dsRNA was termed RNA interference (RNAi). RNAi has since been discovered in a diverse group of organisms including but not limited to *Drosophila*, trypanosomes, zebrafish, and mice (34,51,77,81). dsRNA induces gene silencing post-transcriptionally in an ATP-dependent manner and involves mRNA degradation (87). During the reaction, the dsRNA are first processed by the RNase III, or Dicer family of enzymes (35,37) into small (approximately 21–25 nucleotides) dsRNAs. These small dsRNAs are called short interfering RNAs (siRNAs). The introduction of siRNA has been shown to specifically repress the expression of endogenous and heterologous genes in invertebrate and vertebrate systems (18). However, until recently, RNAi was not effectively exploited in mammalian cells because of the global, nonspecific inhibition of protein synthesis by interferon, which is induced by dsRNA. In a recent exciting study, Elbashir and colleagues (18) described the use of shorter versions of siRNA, optimally 21 bp in length, to specifically suppress the expression of endogenous and heterologous genes in a variety of mammalian cells, without inducing the interferon pathway. The ability to selectively degrade virtually any desired mRNA makes it an ideal tool to study gene functions and to validate potential gene targets. It may also be possible to apply RNAi technology in genome-wide high-throughput screens in mammalian systems, but this has not yet been reported.

PATHWAY ANALYSIS AS AN INTEGRATED PART OF INVERSE GENOMICS

Inverse Genomics usually identifies several RSTs in any given selection system, suggesting that many genes in the same or separate pathways can affect a phenotypic change. To further expand the concept of pathway elucidation and our capability to identify relevant gene targets, we have incorporated some standard functional genomics approaches into the overall scheme of Inverse Genomics. These include the identification of functional, interacting protein partners of the RST gene targets using yeast two-hybrid analysis as well as the identification of co-regulated genes in the RST-expressing cells using microarray analysis.

Yeast Two-Hybrid Analysis

Protein-protein interactions play fundamental roles in many biological processes. Studying how these proteins interact may lead to the discovery of novel proteins in the same biochemical pathway and provide contextual information for novel or even known proteins. The yeast two-hybrid system is a method that investigates protein interaction by using a simple two-plasmid expression system in yeast. The basis for the test is the yeast GAL4 transcriptional activator that comprises two domains, a DNA-binding domain and an activation domain, both of which must be in close proximity to initiate transcription. A gene for one protein under investigation is spliced next to the GAL4 DNA-binding domain on one plasmid, and this protein acts as "bait." On another plasmid, a gene for another protein being tested is spliced to the activation domain, and this protein is called the "target." The target often consists of a randomized cDNA library, from which an interacting protein can be selected. The two plasmids are then introduced into the same cell. The physical interaction between the target and bait proteins leads to the expression of an engineered GAL4-dependent reported gene, which allows for phenotypic selection or screening. The plasmid-containing active target is rescued to identify the gene or genes encoding the interacting proteins. The two-hybrid system can also be automated to facilitate large-scale searches for protein interactions throughout the proteome (66,73).

The yeast two-hybrid system is a useful addition to our Inverse Genomics approach and can be utilized to identify the partners of newly discovered genes. This is particularly helpful if the gene that is discovered is not a good drug target because some of the proteins associated with the newly discovered gene may be good drug targets.

Microarray Technology

The mechanics of microarray analysis was described at the beginning of this chapter. Once a desired phenotype is achieved using our Inverse Genomics approach, the cells or tissues can be analyzed for global expression changes using microarray analysis. Several different ribozymes targeting genes that produce the same phenotype can be introduced into different cell populations (or systems). Looking for genes that are coordinately up-regulated or down-regulated can provide many additional targets that can be validated using ribozyme or siRNA technology. In this way, additional targets may be generated with this pathway analysis.

RIBOZYMES AS THERPEUTICS

A potential benefit of Inverse Genomics is that the functionally selected ribozymes may be used directly as therapeutics, either in the setting of gene therapy or as synthetic molecules to be delivered in vivo. The utility of ribozymes as therapeutic molecules has been demonstrated in many preclinical studies of different disease models. For example, many tumor associated genes are aberrantly expressed or overexpressed in tumors and may be appropriate targets for ribozyme strategies. Ribozymes directed at mutated H-ras, fos, metallothionein

(MT) II(a), and MDR genes have demonstrated antitumor efficacy by the reversion of the malignant phenotypes and by the inhibition of tumor growth both in vitro and in vivo (22,32,38,71). In one study, a hammerhead ribozyme targeting a mutated H-ras oncogene was investigated in an experimental bladder cancer model. After tumors were established in nude mice, the injection of the anti-H-ras ribozyme resulted in significant antineoplastic effects. The complete regression of the tumor was achieved in several cases without recurrence. These results suggest the efficacy of ribozymes in the therapy of human bladder cancer (32). In another study, ribozyme-mediated down-regulation of human MT II(a) was shown to induce apoptosis in human cancer cell lines. Human MTs are overexpressed in prostate and ovarian cancers. It is believed that this increased expression confers resistance to radiation and cytotoxic anticancer drugs. Ribozymes were chosen to examine the function of MT because no inhibitors of MTs are known. A hammerhead ribozyme was designed to cleave class II MT and was transiently transfected into prostate and ovarian cancer cell lines. This resulted in a complete knockdown of MT II(a) mRNA levels and cell death by apoptosis. In addition, both the proto-oncogene c-myc and the apoptosis inhibitory mediator bcl-2 were down-regulated, suggesting the involvement of these signaling pathways. These results further validated MT II(a) as an important cell-survival or anti-apoptotic factor for prostate and ovarian cancer cells (71).

Many groups have also used ribozymes in antiviral gene therapy, particularly against HIV. Most of these studies utilized ribozymes that targeted viral sequences directly (54,85). More recently, we and others have used ribozymes to down-regulate an essential host factor for HIV replication; namely, CCR5, a co-receptor for certain HIV-1 strains. CCR5 and CXCR4 are two major co-receptors used by macrophage tropic (M-tropic or R5) and T cell tropic (T-tropic or X4) HIV-1 strains, respectively (7,17,50). Apparently, R5 strains are usually involved in establishing initial infection. Thus, individuals who lack a functional CCR5 gene are relatively resistant to HIV-1 infection (13,44). This suggests that the elimination or partial reduction of CCR5 expression may protect individuals from HIV-1 infections (46,57,63). Since CCR5 appears to be dispensable for the host but important for initial HIV-1 infection, it is an excellent therapeutic target for ribozyme-mediated knockdown.

We stably transduced PM1 cells (AIDS Research and Reference Reagent Program, National Institutes of Health, Rockville, MD, USA) with anti-CCR5 hairpin ribozymes and disabled ribozyme controls. The PM1 cell is a human T cell line that expresses both CXCR4 and CCR5 and is therefore susceptible to infection by both CXCR4-specific and CCR5-specific HIV-1 strains (19). Northern blot analysis revealed that there was a 50%–70% reduction of endogenous CCR5 mRNA in the ribozyme transduced cells as compared to the disabled controls. In addition, the ribozymes were able to reduce cell surface CCR5 by up to 80% when they were stably introduced into PM1 cells. Furthermore, CD4/HIV env-mediated membrane fusion was inhibited, and the cells became resistant to

R5 HIV-1. Since this anti-CCR5 ribozyme approach is aimed at the entry level of HIV before the start of replication, it should limit the possibility of generating HIV drug-resistant mutants. It is conceivable that ribozymes that target both viral sequences and host genes can be used in combination to prevent or slow the disease progression of HIV-1 infection.

For some disease indications, ribozymes can be used directly as drugs (70). A therapeutic ribozyme can be chemically synthesized (synthetic ribozyme drug) and delivered directly to a tissue or organ of choice. An ideal application for a synthetic ribozyme drug is for treatments that can be administered locally and transiently. A particularly attractive application is in the prevention of restenosis following percutaneous transluminal coronary angioplasty (PTCA). The PCTA procedure often leads to stretching of the artery wall, abrasion of the endothelial lining, and general trauma caused by the cracking of the plaque at the lesion site and injury to the vessel. As a result of this injury and placement of a stent, restenosis or recurrent narrowing of the artery can occur (62).

Although successful in vivo application of naked ribozymes has been achieved (47), ribozymes introduced directly into an organ or tissue are susceptible to degradation by the ubiquitously present RNases. To increase resistance to degradation, chimeric DNA-RNA ribozymes that retain catalytic activity but resist RNase degradation can be chemically synthesized. We have locally administered a chimeric ribozyme drug targeting proliferating cell nuclear antigen (PCNA) at the site of angioplasty using drug delivery catheters in a pig model. A liposome formulation was used to facilitate the delivery of the ribozyme drug to the target cells (29). In comparison to stent alone or a scrambled ribozyme control, the ribozyme targeting PCNA significantly inhibited restenosis in the treated animals. This inhibition resulted in there being no clinically significant blockage, as determined by measuring the percent diameter stenosis and by histological evaluation. Our studies showed that treatment of coronary arteries with chimeric ribozymes targeting PCNA after PTCA reduces the degree of restenosis following stent placement (62).

SUMMARY

A major challenge in the post-genomics era is to link genes to function, with the ultimate goal of identifying therapeutic targets for human diseases. This is usually done at the gene expression level, either by correlating the patterns of gene expression with a disease phenotype or by determining the effects of overexpression or repression of specific genes with phenotypic changes. Most of these approaches yield only correlative information rather than a cause-and-effect relationship. Here we described a platform technology called Inverse Genomics, which begins with a therapeutic phenotype and works backward to identify the genes involved. This is a rapid way to identify genes that are specifically linked to a phenotype.

As illustrated, our Inverse Genomics approach overcomes limitations with other methods of gene discovery and has been useful for identifying genes and revealing gene function in a variety of cell-based phenotypic model systems. The complexity of the ribozyme library theoretically allows for cleavage of any mRNA within the genome. Since the ribozyme library is screened in cell-based model systems, presumably any biologically relevant gene can be identified for a given phenotype. This further overcomes a limitation of cDNA libraries and some differential gene expression techniques that are never fully representative of the genome because of the incomplete or inefficient conversion of certain mRNAs. The Inverse Genomics approach also allows for more rapid gene discovery because the ribozyme target recognition sequence can be utilized to directly identify a gene, foregoing the much slower positional cloning approaches used in traditional forward genetics. In addition, the target validation process of Inverse Genomics allows for the rapid identification of gene function, with no need for the mutagenesis, subcloning, and recombination steps performed in reverse genetics approaches. Furthermore, the ribozymes used in Inverse Genomics repress gene expression, rather than completely knocking out gene expression, thereby overcoming the cytotoxicity problems sometimes encountered with reverse genetic gene knockouts. Interestingly, the repression of gene expression more closely resembles the activity of drugs, which transiently knock down gene activity, and therefore the genes identified through Inverse Genomics may be better drug candidates. Finally, the phenotypes produced by Inverse Genomics can be integrated with other technologies such as the yeast two-hybrid system and microarray analysis to identify additional genes in disease pathways.

Inverse Genomics may also prove useful for the application of ribozymes as therapeutics. The functionally selected ribozymes can be delivered as therapeutic genes or chemically synthesized oligonucleotides, depending on the indications. This does not require the identification of the target genes themselves, provided that the ribozymes are not toxic to the cells. This allows for a direct path from Inverse Genomics to drug discovery, even without the identification of a target gene. The ability to apply Inverse Genomics as a gene identification technology as well as a direct therapeutic makes it a versatile tool for the pharmaceutical industry.

REFERENCES

1. **Ahringer, J.** 1997. Turn to the worm! Curr. Opin. Genet. Dev. *7*:410-415.
2. **Albelda, S.M. and D. Sheppard.** 2000. Functional genomics and expression profiling: be there or be square. Am. J. Respir. Cell Mol. Biol. *23*:265-269.
3. **Bachman, E.S., H. Dhillon, C.Y. Zhang, S. Cinti, A.C. Bianco, B.K. Kobilka, and B.B. Lowell.** 2002. βAR signaling required for diet-induced thermogenesis and obesity resistance. Science *297*:843-845.
4. **Been, M.D. and G.S. Wickham.** 1997. Self-cleaving ribozymes of hepatitis delta virus RNA. Eur. J. Biochem. *247*:741-753.
5. **Beger, C., L.N. Pierce, M. Kruger, E.G. Marcusson, J.M. Robbins, P. Welcsh, P.J. Welch, K. Welte,**

et al. 2001. Identification of *Id4* as a regulator of BRCA1 expression by using a ribozyme-library-based inverse genomics approach. Proc. Natl. Acad. Sci. USA *98*:130-135.

6. **Benezra, R., R.L. Davis, D. Lockshon, D.L. Turner, and H. Weintraub.** 1990. The protein Id: a negative regulator of helix-loop-helix DNA binding proteins. Cell *61*:49-59.
7. **Berger, E.A.** 1997. HIV entry and tropism: the chemokine receptor connection. AIDS *11*:Suppl. A:S3-S16.
8. **Blackburn, E.** 1999. Telomeres, p. 603-635. *In* R.F. Gesteland, T.R. Cech, and J.F. Atkins (Eds.), The RNA World. CSH Laboratory Press, Cold Spring Harbor, NY.
9. **Boylan, M.O., M. Athanassiou, B. Houle, W. Wang, and H. Zarbl.** 1996. Activation of tumor suppressor genes in nontumorigenic revertants of HeLa cervical carcinoma cell line. Cell Growth Differ. *7*:725-735.
10. **Byers, R.J., J.A. Hoyland, J. Dixon, and A.J. Freemont.** 2000. Subtractive hybridization—genetic takeaways and the search for meaning. Int. J. Exp. Pathol. *81*:391-404.
11. **Cameron, F.H. and P.A. Jennings. 1989**. Specific gene suppression by engineered ribozymes in monkey cells. Proc. Natl. Acad. Sci. USA *86*:9139-9143.
12. **Chin, L., S.E. Artandi, Q. Shen, A. Tam, S.L. Lee, G.J. Gottlieb, C.W. Greider, and R.A. DePinho.** 1999. p53 deficiency rescues the adverse effects of telomere loss and cooperates with telomere dysfunction to accelerate carcinogenesis. Cell *97*:527-538.
13. **Deng, H.K., D. Unutmaz, V.N. KewalRamani, and D.R. Littman.** 1997. Expression cloning of new receptors used by simian and human immunodeficiency viruses. Nature *388*:296-300.
14. **Diatchenko, L., Y.F. Lau, A.P. Campbell, A. Chenchik, F. Moqadam, B. Huang, S. Lukyanov, K. Lukyanov, et al.** 1996. Suppression subtractive hybridization: a method for generating differentially regulated or tissue-specific cDNA probes and libraries. Proc. Natl. Acad. Sci. USA *93*:6025-6030.
15. **Dinarello, C.A.** 1996. Biologic basis for interleukin-1 disease. Blood *87*:2095-2147.
16. **Doherty, E.A. and J.A. Doudna.** 2000. Ribozyme structures and mechanisms. Annu. Rev. Biochem. *69*:597-615.
17. **Doranz, B.J., J.F. Berson, J. Rucker, and R.W. Doms.** 1997. Chemokine receptors as fusion cofactors for human immunodeficiency virus type 1 (HIV-1). Immunol. Res. *16*:15-28.
18. **Elbashir, S.M., J. Harborth, W. Lendeckel, A. Yalcin, K. Weber, and T. Tuschl.** 2001. Duplexes of 21-nucleotide RNAs mediate RNA interference in cultured mammalian cells. Nature *411*:494-498.
19. **Feng, Y., M.C. Leavitt, R. Tritz, E. Duarte, D. Kang, M. Mamounas, P. Gilles, F. Wong-Staal, et al.** 2000. Inhibition of CCR5-dependent HIV-1 infection by hairpin ribozyme gene therapy against CC-chemokine receptor 5. Virology *276*:271-278.
20. **Fire, A., S.Q. Xu, M.K. Montgomery, S.A. Kostas, S.E. Driver, and C.C. Mello.** 1998. Potent and specific genetic interference by double-stranded RNA in *Caenorhabditis elegans.* Nature *391*:806-811.
21. **Flores, R., J.A. Navarro, M. de la Pena, B. Navarro, S. Ambros, and A. Vera.** 1999. Viroids with hammerhead ribozymes: some unique structural and functional aspects with respect to other members of the group. Biol. Chem. *380*:849-854.
22. **Funato, T., T. Ishii, M. Kanbe, K.J. Scanlon, and T. Sasaki.** 1997. Reversal of cisplatin resistance in vivo by an anti-fos ribozyme. In Vivo *11*:217-220.
23. **Futreal, P.A., Q. Liu, D. Shattuck-Eidens, C. Cochran, K. Harshman, S. Tavtigian, L.M. Bennet, A. Haugen-Strano, et al.** 1994. BRCA1 mutations in primary breast and ovarian carcinomas. Science *266*:120-122.
24. **Garcia-Bellido, A. and J.F. de Celis.** 1992. Development genetics of the venation pattern of *Drosophila.* Annu. Rev. Genet. *26*:277-304.
25. **Gilles, P., Q. Yang, G. Yu, and M. Yu.** 1996. Development of the hairpin ribozyme as a gene therapy approach to human immunodeficiency virus type-1 infection, p. 26-31. *In* P.L. Felgner, M.J. Heller, P. Lehn, J.P. Behr, and F.C. Szoka Jr. (Eds.), Artificial Self-Assembling Systems for Gene Delivery. American Chemical Society, Washington, D.C.
26. **Greider, C.W.** 1996. Telomere length regulation. Annu. Rev. Biochem. *65*:337-365.
27. **Homann, M., S. Tzortzakaki, K. Rittner, G. Sczakiel, and M. Tabler.** 1993. Incorporation of the catalytic domain of a hammerhead ribozyme into antisense RNA enhances its inhibitory effect of the replication of human immunodeficiency virus type 1. Nucleic Acids Res. *21*:2809-2814.
28. **Hosking, L., J. Trowsdale, H. Nicolai, E. Solomon, W. Foulkes, G. Stamp, E. Signer, and A. Jeffreys.** 1995. A somatic BRCA1 mutation in an ovarian tumor. Nat. Genet. *9*:343-344.
29. **Huang, L. and S. Li.** 1997. Liposomal gene delivery: a complex package. Nat. Biotechnol. *15*:620-

621.

30. **Hubank, M. and D.G. Schatz.** 1994. Identifying differences in mRNA expression by representational difference analysis of cDNA. Nucleic Acids Res. *22*:5640-5648.
31. **Iizuka, N., C. Chen, Q. Yang, G. Johannes, and P. Sarnow.** 1995. Cap-independent translation and internal initiation of translation in eukaryotic cellular mRNA molecules. Curr. Top. Microbiol. Immunol. *203*:155-177.
32. **Irie, A., B. Anderegg, M. Kashani-Sabet, T. Ohkawa, T. Suzuki, M. Halks-Miller, D.T. Curiel, and K.J. Scanlon.** 1999. Therapeutic efficacy of an adenovirus-mediated anti-H-ras ribozyme in experimental bladder cancer. Antisense Nucleic Acid Drug Dev. *9*:341-349.
33. **Joyner, A.L. and F. Guillemot.** 1994. Gene targeting and development of the nervous system. Curr. Opin. Neurobiol. *4*:37-42.
34. **Kennerdell, J.R. and R.W. Carthew.** 1998. Use of dsRNA-mediated genetic interference to demonstrate that frizzled and frizzled 2 act in the wingless pathway. Cell *95*:1017-1026.
35. **Ketting, R.F., S.E. Fischer, E. Bernstein, T. Sijen, G.J. Hannon, and R.H. Plasterk.** 2001. Dicer functions in RNA interference and in synthesis of small RNA involved in developmental timing in *C. elegans.* Genes Dev. *15*:2654-2659.
36. **Kijima, H., J. Ishida, T. Ohkawa, M. Kashani-Sabet, and K.J. Scanlon.** 1995. Therapeutic applications of riboyzmes. Pharmacol. Ther. *68*:247-267.
37. **Knight, S.W. and B.L. Bass.** 2001. A role for the RNase III enzyme DCR-1 in RNA interference and germ line development in *Caenorhabditis elegans.* Science *293*:2269-2271.
38. **Kobayashi, H., Y. Takemura, and H. Miyachi.** 2001. Novel approaches to reversing anti-cancer drug resistance using gene-specific therapeutics. Hum. Cell *14*:172-184.
39. **Kruger, M., C. Beger, Q.X. Li, P.J. Welch, R. Tritz, M.C. Leavitt, J.R. Barber, and F. Wong-Staal.** 2000. Identification of eIF2Bγ and eIF2γ as cofactors of hepatitis C virus internal ribosome entry site-mediated translation using a functional genomics approach. Proc. Natl. Acad. Sci. USA *97*:8566-8571.
40. **Leavitt, M.C., G. Yu, C. Zhou, and J.R. Barber.** 2000. Inhibition of interleukin-1β (IL-1β) production in human cells by ribozymes against IL-1β and IL-1β converting enzyme (ICE). Antisense Nucleic Acid Drug Dev. *10*:409-414.
41. **Lekven, A.C., K.A. Helde, C.J. Thorpe, R. Rooke, and R.T. Moon.** 2000. Reverse genetics in zebrafish. Physiol. Genomics *2*:37-48.
42. **Li, Q.X., J.M. Robbins, P.J. Welch, F. Wong-Staal, and J.R. Barber.** 2000. A novel functional genomics approach identifies mTERT as a suppressor of fibroblast transformation. Nucleic Acids Res *28*:2605-2612.
43. **Liang, P.** 2002. A Decade of Differential Display. BioTechniques *33*:338-346.
44. **Liao, F., G. Alkhatib, K.W.C. Peden, G. Sharma, E.A. Berger, and J.M. Farber.** 1997. STRL33, A novel chemokine receptor-like protein, functions as a fusion cofactor for both macrophage-tropic and T cell line-tropic HIV-1. J. Exp. Med. *185*:2015-2023.
45. **Lieber, A. and M. Strauss.** 1995. Selection of efficient cleavage sites in target RNAs by using a ribozyme expression library. Mol. Cell. Biol. *15*:540-551.
46. **Liu, R., W.A. Paxton, S. Choe, D. Ceradini, S.R. Martin, R. Horuk, M.E. MacDonald, H. Stuhlmann, et al.** 1996. Homozygous defect in HIV-1 coreceptor accounts for resistance of some multiply-exposed individuals to HIV-1 infection. Cell *86*:367-377.
47. **Lyngstadaas, S.P., B.S. Sproat, M. Blaschke, J. Ludwig, T. Rupp, and H.P. Prydz.** 1998. In vivo knockout of a tissue-specific gene by synthetic ribozymes, p. 371-386. *In* K.J. Scanlon (Ed.), Therapeutic Application of Ribozymes. Humana Press, Totowa, NJ.
48. **Merajver, S.D., T.M. Pham, R.F. Caduff, M. Chen, E.L. Poy, K.A. Cooney, B.L. Weber, F.S. Collins, et al.** 1995. Somatic mutations in the BRCA1 gene in sporadic ovarian tumors. Nat. Genet. *9*:439-443.
49. **Migeon, J.C., M.S. Garfinkel, and B.A. Edgar.** 1999. Cloning and characterization of *peter pan,* a novel *Drosophila* gene required for larval growth. Mol. Biol. Cell *10*:1733-1744.
50. **Moore, J.P., A. Trkola, and T. Dragic.** 1997. Co-receptors for HIV-1 entry. Curr. Opin. Immunol. *9*:551-562.
51. **Ngo, H., C. Tschudi, K. Gull, and E. Ullu.** 1998. Double-stranded RNA induces mRNA degradation in *Trypanosoma brucei.* Proc. Natl. Acad. Sci. USA *95*:14687-14692.
52. **Norton, J.D., R.W. Deed, G. Craggs, and F. Sablitzky.** 1998. Id helix-loop-helix proteins in cell

growth and differentiation. Trends Cell Biol. *8*:58-65.

53.**Nusslein-Volhard, C. and E. Wieschaus.** 1980. Mutations affecting segment number and polarity in *Drosophila*. Nature *287*:795-801.

54.**Ojwang, J.O., A. Hampel, D.J. Looney, F. Wong-Staal, and J. Rapapport.** 1992. Inhibition of human immunodeficiency virus type 1 expression by a hairpin ribozyme. Proc. Natl. Acad. Sci. USA *89*:10802-10806.

55.**Palapattu, G.S., S. Bao, T.R. Kumar, and M.M. Matzuk.** 1998. Transgenic mouse models for tumor suppressor genes. Cancer Detect. Prev. *22*:75-86.

56.**Pavitt, G.D., K.V. Ramaiah, S.R. Kimball, and A.G. Hinnebusch.** 1998. eIF2 independently binds two distinct eIF2B subcomplexes that catalyze and regulate guanine-nucleotide exhange. Genes Dev. *12*:514-526.

57.**Paxton, W.A., S.R. Martin, D. Tse, T.R. O'Brien, J. Skurnick, N.L. VanDevanter, N. Padian, J.F. Braun, et al.** 1996. Relative resistance to HIV-1 infection of CD4 lymphocytes from persons who remain uninfected despite multiple high-risk sexual exposure. Nat. Med. *2*:412-417.

58.**Ramadani, M., Q. Yang, F. Gansauge, F. Gansauge, and H.G. Beger.** 2001. Overexpression of caspase-1 (interleukin-1β converting enzyme) in chronic pancreatitis and its participation in apoptosis and proliferation. Pancreas *22*:383-387.

59.**Ramezani, A., W. Marhin, M. Weerasinghe, and S. Joshi.** 1997. A rapid and efficient system for screening HIV-1 Pol mRNA-specific ribozymes. Can. J. Microbiol. *43*:92-96.

60.**Randle, J.C., M.W. Harding, G. Ku, M. Schonharting, and R. Kurrle.** 2001. ICE/Caspase-1 inhibitors as novel anti-inflammatory drugs. Expert Opin. Investig. Drugs *10*:1207-1209.

61.**Rink, T.** 1994. Genetics. In search of a satiety factor. Nature *372*:406-407.

62.**Robbins, J.M., A. Frimerman, N. Eigler, F. Litvak, and J. Barber.** 1999. Ribozymes for the treatment of restenosis following PTCA, p. 243-253. *In* J.J. Rossi and L. Couture (Eds.), Intracellular Ribozyme Applications: Principles and Protocols. Horizon Scientific Press, Wymondham, UK.

63.**Samson, M., F. Libert, B.J. Doranz, J. Rucker, C. Liesnard, C.M. Farber, S. Saragosti, C. Lapoumeroulie, et al.** 1996. Resistance to HIV-1 infection in caucasian individuals bearing mutant alleles of the CCR-5 chemokine receptor gene. Nature *382*:722-725.

64.**Scanlon, K.J., L. Jiao, T. Funato, W. Wang, T. Tone, J.J. Rossi, and M. Kashani-Sabet.** 1991. Ribozyme-mediated cleavage of c-fos mRNA reduces gene expression of DNA synthesis enzymes and metallothionein. Proc. Natl. Acad. Sci. USA *88*:10591-10595.

65.**Scheper, G.C., A.A. Thomas, and H.O. Voorma.** 1991. The 5′ untranslated region of encephalomyocarditis virus contains a sequence for very efficient binding of eukaryotic initiation factor eIF-2/2B. Biochim. Biophys. Acta *1089*:220-226.

66.**Serebriiskii, I.G., V. Khazak, and E.A. Golemis.** 2001. Redefinition of the yeast two-hybrid system in dialogue with changing priorities in biological research. BioTechniques *30*:634-655.

67.**Shiratori, A., T. Shibata, M. Arisawa, F. Hanaoka, Y. Murakami, and T. Eki.** 1999. Systematic identification, classification, and characterization of the open reading frames which encode novel helicase-related proteins in *Saccharomyces cerevisiae* by gene disruption and Northern analysis. Yeast *15*:219-253.

68.**Sioud, M. and K. Drlica.** 1991. Prevention of human immunodeficiency virus type 1 integrase expression in *Escherichia coli* by a ribozyme. Proc. Natl. Acad. Sci. USA *88*:7303-7307.

69.**Sourvinos, G. and D.A. Spandidos.** 1998. Decreased BRCA1 expression levels may arrest the cell cycle through activation of p53 checkpoint in human sporadic breast tumors. Biochem. Biophys. Res. Commun. *245*:75-80.

70.**Sun, L.Q., M.J. Cairns, E.G. Saravolac, A. Baker, and W.L. Gerlach.** 2000. Catalytic nucleic acids: from lab to applications. Pharmacol. Rev. *52*:325-347.

71.**Tekur, S. and S.M. Ho.** 2002. Ribozyme-mediated downregulation of human metallothionein II(a) induces apoptosis in human prostate and ovarian cancer cell lines. Mol. Carcinog. *33*:44-55.

72.**Thompson, M.E., R.A. Jensen, P.S. Obermiller, D.L. Page, and J.T. Holt.** 1995. Decreased expression of BRCA1 accelerates growth and is often present during sporadic breast cancer progression. Nat. Genet. *9*:444-450.

73.**Topcu, Z. and K.L.B. Borden.** 2000. The yeast two-hybrid system and its pharmaceutical significance. Pharm. Res. *17*:1049-1055.

74.**Tritz, R., M.C. Leavitt, and J.R. Barber.** 1999. Screening promoters for optimal expression of ribozymes, p. 115-123. *In* J.J. Rossi and L. Couture (Eds.), Intracellular Ribozyme Applications: Principles and Protocols. Horizon Scientific Press, Wymondham, UK.

75. **Vidal-Vanaclocha, F., A. Alvarez, A. Asumendi, B. Urcelay, P. Tonino, and C.A. Dinarello.** 1996. Interleukin 1 (IL-1)-dependent melanoma hepatic metastasis in vivo; increased endothelial adherence by IL-1-induced mannose receptors and growth factor productin in vitro. J. Natl. Cancer Inst. *88*:198-205.

76. **Vogelstein, B., E.R. Fearon, S.R. Hamilton, S.E. Kern, A.C. Preisinger, M. Leppert, Y. Nakamura, R. White, et al.** 1988. Genetic alterations during colorectal-tumor development. N. Engl. J. Med. *319*:525-532.

77. **Wargelius, A., S. Ellingsen, and A. Fjose.** 1999. Double-stranded RNA induces specific development defects in zebrafish embryos. Biochem. Biophys. Res. Commun. *263*:156-161.

78. **Welch, P.J., J. Barber, and F. Wong-Staal.** 1998. Expression of ribozymes in gene transfer systems to modulate target RNA levels. Curr. Opin. Biotechnol. *9*:486-496.

79. **Welch, P.J., A. Hampel, J. Barber, F. Wong-Staal, and M. Yu.** 1996. Inhibition of HIV replication by the hairpin ribozyme, p. 315-327. *In* F. Eckstein and D.M.J. Lilley (Eds.), Nucleic Acids and Molecular Biology. Springer-Verlag Berlin, Heidelberg.

80. **Welch, P.J., E.G. Marcusson, Q.X. Li, C. Beger, M. Kruger, C. Zhou, M.C. Leavitt, F. Wong-Staal, et al.** 2000. Idenfication and validation of a gene involved in anchorage-independent cell growth control using a library of randomized hairpin ribozymes. Genomics *66*:274-283.

81. **Wianny, F. and M. Zernicka-Goetz.** 2000. Specific interference with gene function by double-stranded RNA in early mouse development. Nat. Cell Biol. *2*:70-75.

82. **Wilson, C.A., L. Ramos, M.R. Villasenor, K.H. Anders, M.F. Press, K. Clarke, B. Karlan, J.J. Chen, et al.** 1999. Localization of human BRCA1 and its loss in high-grade, non-inherited breast carcinomas. Nat. Genet. *21*:236-240.

83. **Ye, S.Q., T. Lavoie, D.C. Usher, and L.Q. Zhang.** 2002. Microarray, SAGE and their applications to cardiovascular diseases. Cell Res. *12*:105-115.

84. **Yoshikawa, K., K. Honda, T. Inamoto, H. Shinohara, A. Yamauchi, K. Suga, T. Okuyama, et al.** 1999. Reduction of BRCA1 protein expression in Japanese sporadic breast carcinomas and its frequent loss in *BRCA1*-associated cases. Clin. Cancer Res. *5*:1249-1261.

85. **Yu, M., J. Ojwang, O. Yamada, A. Hampel, J. Rapapport, D. Looney, and F. Wong-Staal.** 1993. A hairpin ribozyme inhibits expression of diverse strains of human immunodeficiency virus type 1. Proc. Natl. Acad. Sci. USA *90*:6340-6344.

86. **Yu, M., E. Poeschla, O. Yamada, P. Degrandis, M.C. Leavitt, M. Heusch, J.K. Yees, F. Wong-Staal, et al.** 1995. In vitro and in vivo characterization of a second functional hairpin ribozyme against HIV-1. Virology *206*:381-386.

87. **Zamore, P.D., T. Tuschl, P.A. Sharp, and D.P. Bartel.** 2000. RNAi: double-stranded RNA directs the ATP-dependent cleavage of mRNA at 21 to 23 nucleotide intervals. Cell *101*:25-33.

88. **Zaug, A.J., M.D. Been, and T.R. Cech.** 1986. The Tetrahymena ribozyme acts like an RNA restriction endonuclease. Nature *324*:429-433.

89. **Zaug, A.J. and T.R. Cech.** 1986. The intervening sequence RNA of Tetrahymena is an enzyme. Science *231*:470-475.

90. **Zhang, Y., R. Proenca, M. Maffei, M. Barone, L. Leopold, and J.M. Friedman.** 1994. Postional cloning of the mouse obesity gene and its human homologue. Nature *372*:425-432.

21 Random Activation of Gene Expression (RAGE) Technology

Evan A. Facher, Kurt Brunden, and James J. Kovach
Athersys, Cleveland, OH, USA

RANDOM ACTIVATION OF GENE EXPRESSION (RAGE)—A NOVEL APPROACH TO GENE EXPRESSION

Overview

Random activation of gene expression (RAGE) is a proprietary gene expression technology that is capable of expressing protein from virtually any gene in the human genome. RAGE is carried out without knowledge or use of cDNA molecules or cloned genomic fragments containing a gene of interest. The RAGE technology platform is built on a simple, yet elegant process that can be adapted to multiple applications. Although it has been estimated that the human genome contains somewhere between 30 000 and 100 000 genes (and perhaps a much larger set of splice variants), only a subset of these genes is active in any given cell. Each gene has its own switch that regulates gene activation and expression in response to developmental cues, disease conditions, environmental stimuli, and other factors. Consequently, some genes are active in a particular cell type and are transcribed at different frequencies, and other genes are silent. The RAGE technology consists of a collection of vectors that contain a powerful switch for controlling gene activation and, thus, do not rely on the normal signals that control mRNA production. Random insertion of RAGE vectors throughout the genome results in a library of cells that together express the entire collection of proteins encoded by the genome.

Unlike other insertion vectors (4,7,8,10,14,15,18,19,24–26), RAGE vectors are designed to activate expression from the diverse range of gene structures present in the genome. In its basic form, the RAGE vector contains a promoter linked to an exon and unpaired splice donor site (Figure 1). The process of creating a RAGE library begins by inducing a temporary chromosome break at

Perspectives in Gene Expression
Edited by K. Appasani

random points throughout a eukaryotic host cell genome. This is achieved through methods optimized to permit approximately one breakage event per cell. At about the same time the breakage events occur, the RAGE vector is introduced into the cells. The cells' normal DNA repair machinery then facilitates vector integration at the site of the chromosome break, and the vector is able to activate expression of the downstream endogenous gene. This produces a chimeric transcript in which the 5′-end is a vector-encoded exon, and the 3′-end is derived from the endogenous gene. (The effect of replacing exon I encoded by the endogenous gene is accounted for by a series of RAGE vectors that generically mimic different exon I structures.) The vector integration process, illustrated in Figure 2, results in a library in which each cell expresses, on average, a single RAGE-activated gene.

The RAGE technology has been validated extensively. Furthermore, the process has been industrialized, resulting in comprehensive protein expression libraries that collectively express virtually every gene in a host's genome, regardless of whether a given gene is normally expressed in the host cell. These genome-wide protein expression libraries can be created in various cell types that subsequently may be used in many different applications. One application is novel gene discovery. To date, RAGE has been used to build a proprietary genomics database containing thousands of rarely expressed, tightly regulated genes. Another application is high-throughput phenotypic screening in a variety of assay formats, such as biochemical assays, cellular assays, or whole organism assays (e.g., mouse, *Drosophila*, and *Caenorhabditis elegans*). RAGE libraries can be readily adapted to assays designed to identify gene products that have roles in specific biological pathways. The RAGE technology may also be used to screen for cells that express a previously known and validated biological or drug target whose isolated gene sequence may be the subject of an issued or pending patent. Cell lines expressing the desired proteins enable researchers to conduct small molecule screening programs against the validated target(s) of interest, with no need for cloned cDNA or gene sequences.

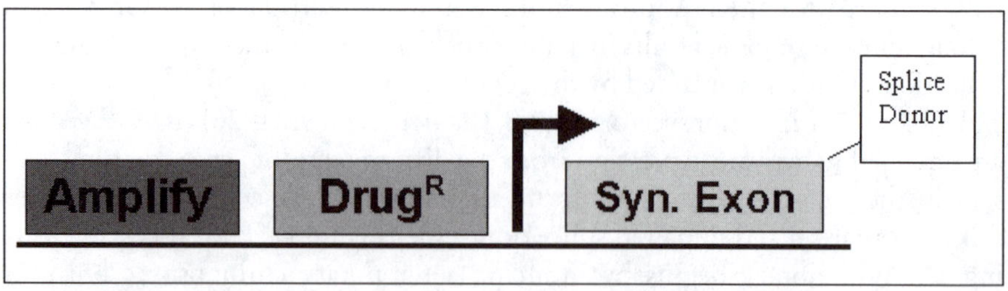

Figure 1. Structural organization of a representative RAGE vector. A representative RAGE vector is composed of a constitutive or inducible promoter (indicated by the arrow), followed by sequences that encode a synthetic exon and an unpaired splice donor sequence. The synthetic exon sequence allows any activated gene to be tagged for easy identification. It also enables the expression of chimeric proteins, if desired. The vector may also contain a selectable marker (Drug^R) and an amplifiable marker.

EXPERIMENTAL VALIDATION AND APPLICATION OF THE RAGE TECHNOLOGY—COMPREHENSIVE GENE ACTIVATION

As a first step toward validating the RAGE technology, a library was created using a prototype RAGE vector, such as the vector shown in Figure 1. Calculations based on the size of a typical human gene, frequency of induced random breakage events, and the estimated complexity of the human genome indicated that the RAGE library represents several-fold coverage of the human genome. The vector-encoded sequence tag was a tool for further validation that the RAGE tech-

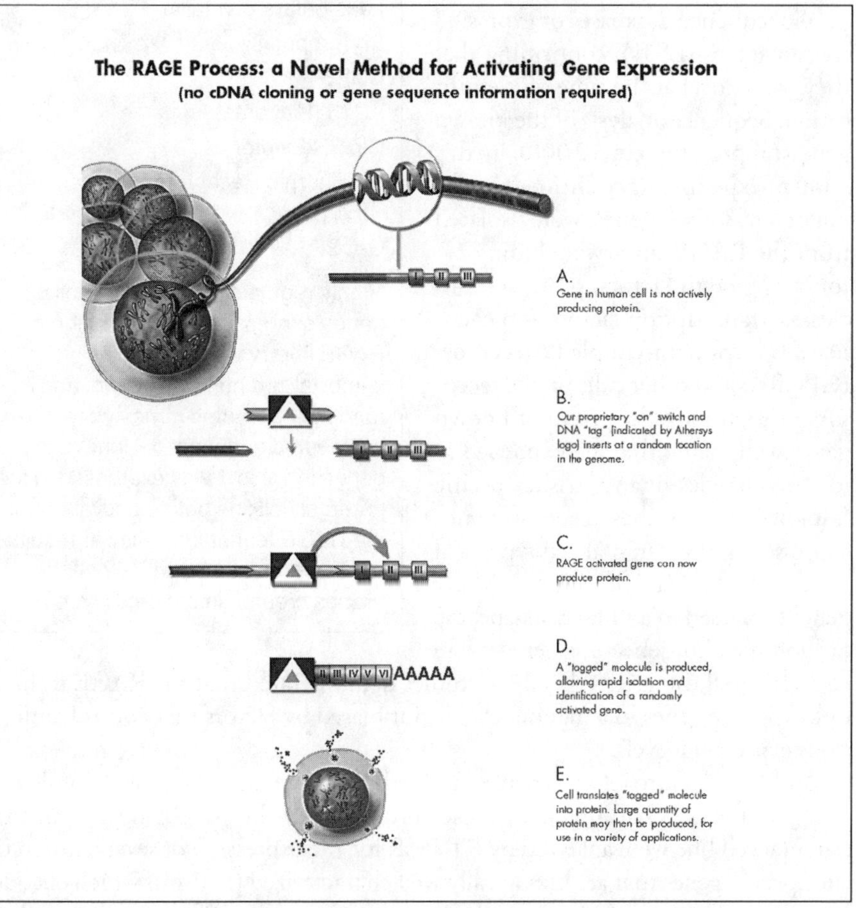

Figure 2. RAGE is a novel method for activating gene expression. By inserting proprietary vectors containing strong promoters and synthetic exon sequences randomly throughout the genome of a host, RAGE can activate previously silent genes within in a cell. These activated genes produce a tagged message that is translated into the appropriate corresponding protein, which can then be used in a variety of assays. (Figure reprinted with permission from Genetic Engineering News. Figure previously appeared in Genetic Engineering News Volume 22, No. 3, p. 28–31.)

nology activates genes in a random manner. The sequence tag facilitates RAGE transcript analysis through a variety of approaches, such as DNA sequencing, bioinformatic analysis, and expression profile determination. Using these techniques, over 20 000 unique human genes were sequenced from the RAGE library. At the time that this work was conducted, more than 50% of these were absent from public sequence databases of expressed sequence tags (ESTs), confirming that they were novel at the time of identification. Sequence analysis of the known genes (approximately 10 000) in this series of experiments confirmed that all major classes of genes were isolated from the RAGE library, including cytokines, growth factors, proteases, nucleases, transcription factors, ion channels, G protein-coupled receptors (GPCRs), and other cell surface receptors. Also among the group of known genes were many that are expressed at low frequencies or in a tissue-specific fashion; however, these genes were not expressed in the parental cell type used to create the RAGE library. Many genes expressed in a non-tissue-specific fashion (i.e., housekeeping genes) were also activated using RAGE. These studies demonstrated that the RAGE technology expresses genes in a manner that is not biased by factors that control endogenous expression levels.

Table 1. Highly Regulated Genes Activated in a RAGE Library

Human Growth Hormone
Prothrombin
Cox VIIa-L
Cyclophilin
BCNT
Oligodendrocyte Myelin Glycoprotein
Vascular Anticoagulant
Thyroglobulin
Lens beta-s-crystallin
Erythropoietin
CCR5
PP15 (Placenta)
Glp1 Receptor
Cholinergic Receptors (e.g., α3, α4, α7, β2)
G-CSF

Highly regulated or tissue-specific genes expressed in a RAGE library. A RAGE library was created using an immortalized human cell line, and randomly activated genes were harvested and sequenced. None of the genes in Table 1 was expressed in the parental cell line before activation with the RAGE technology. Similar results have been obtained with RAGE libraries created in other cell types.

Additional approaches were used to further explore the extent of genomic coverage by the RAGE library. In these studies, both the RAGE library and the parental cell line were analyzed by RT-PCR for the expression of several dozen tissue-specific genes that are biologically well characterized (many of which encoded known and validated drug targets or therapeutic proteins). While the parental cell line expressed none of these genes, the RAGE activation library expressed each of the candidate genes at comparable frequencies (Table 1). The frequencies of the activation of these genes and of the more than 20 000 genes in our original database suggest that libraries created using the RAGE technology appear to be inherently normalized. In other words, genes expressed endogenously at extremely low levels, in rare cells, or during short developmental windows are activated in RAGE libraries at the same frequency as more abundantly expressed genes, such as actin. Collectively, these studies strongly suggest that gene expression using the RAGE technology yields comprehensive genome coverage.

CHARACTERIZATION OF THE ATHERSYS GENE SEQUENCE COLLECTION

Identification of Novel Genes

Many novel genes have been identified by a variety of approaches that typically rely on expression-dependent techniques or begin with a cDNA library constructed using a cell or tissue source of interest (2,3,5,11,16,17,20,23,27). However, a fundamental limitation of these approaches is the paucity or absence of cDNA molecules that represent silenced or rarely transcribed genes in the cell or tissue source used to make the cDNA library. Consequently, many biologically interesting and therapeutically relevant genes have been overlooked using current expression-dependent approaches.

A more recent approach to identifying gene sequences, referred to as whole-genome-shotgun sequencing, has been used to generate a significant amount of raw genomic sequence information (1,6,12,13,21,22). This type of approach can be useful once the raw genomic sequences have been correctly assembled into a contiguous map, and the sequences have been analyzed using sophisticated bioinformatic algorithms to identify regions containing candidate genes. However, this approach is limited by the accuracy of the algorithms to identify potential coding sequences. While some bioinformatic algorithms have proven to be reasonably adept at this process, they are generally incapable of validating and specifying unambiguously the actual structure of an expressed gene so that validation must be confirmed by independent experiments. Furthermore, bioinformatic algorithms may not be able to identify some genes hidden in the raw genomic sequence. This particular problem was illuminated with a series of experiments, in which a database of RAGE-activated gene sequences was assembled and compared with the published sequence for human chromosome 22 (6). These analyses revealed that the RAGE approach identified a significant number of genes that were missed in the bioinformatic analysis previously conducted on the raw genomic sequence (9).

Perhaps the most significant limitation of both the cDNA-based EST approach (which typically produces gene fragments but not full-length cDNAs) and the whole-genome-shotgun sequencing approach (which produces a virtual candidate gene sequence useful for bioinformatic analysis) is that neither is well suited to facilitate the expression of proteins for subsequent analysis. This limitation impedes further studies of protein structure and function, and biological or clinical relevance. The RAGE technology can overcome all of the limitations cited for these alternative methods.

As previously mentioned, a prototype library was created and more than 20 000 unique human genes were partially sequenced. A thorough examination of the public databases, as well as some proprietary EST databases, confirmed that over 10 000 (or 50%) of the genes identified by RAGE were novel at the time of identification. Using RT-PCR to determine the expression profile, many of the novel genes were those that are expressed in a highly tissue specific manner or at extremely low levels. Expression frequencies of some of these ranged from one in 10^6 to one in 10^8 transcripts. Such transcripts would be virtually undetectable using conventional cDNA approaches, yet they were represented in the RAGE library with the same relative frequency as many

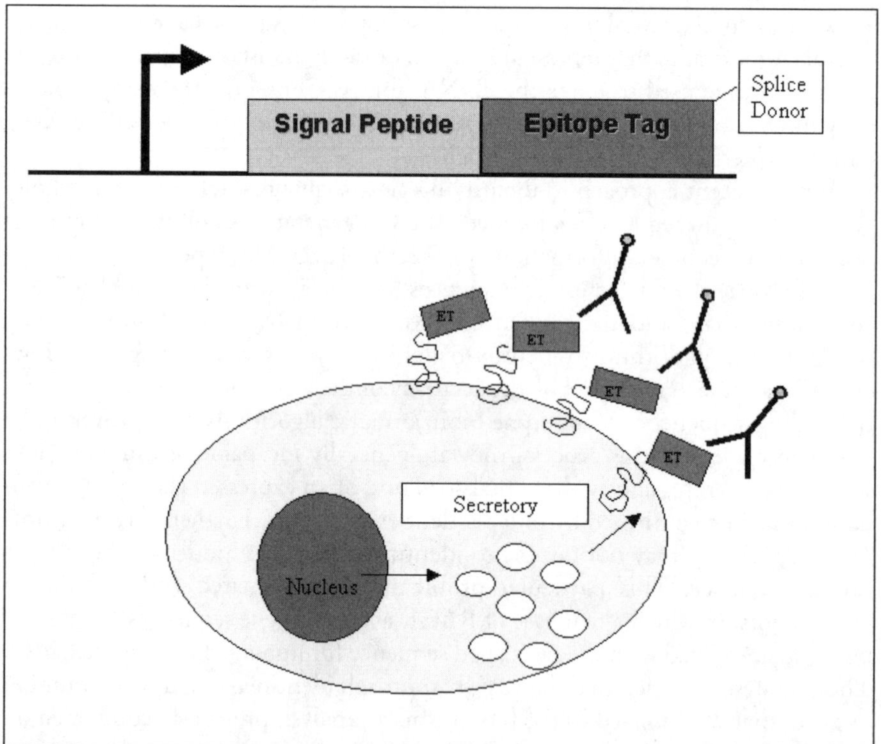

Figure 3. Creation of a specialized RAGE-TMT vector to screen for integral membrane proteins. A RAGE vector was created that contains a strong promoter (indicated by the arrow), a synthetic exon sequence encoding a signal peptide directing the protein into the secretory pathway, an epitope tag, and an unpaired splice donor sequence. Upon activation of an endogenous gene that encodes a protein containing a transmembrane domain, the chimeric protein enters the secretory pathway, and the epitope tag (ET) is displayed on the cell surface, enabling the detection and isolation of the cell using epitope-specific antibodies and FACS or alternative cell sorting technology.

broadly expressed genes. Furthermore, in contrast to conventional approaches, the RAGE technology is not significantly biased toward the isolation of more abundantly expressed genes.

Identification of Class-Specific Novel Genes—The RAGE Transmembrane Trap

It has been estimated that over 70% of validated drug targets identified and used by the pharmaceutical industry are derived from a single general class of proteins: the integral membrane proteins. Examples of these are GPCRs and ion channels. To determine whether RAGE could be used to select preferentially or screen for activated integral membrane proteins, a specialized family of RAGE vectors was created (Figure 3). These vectors are referred to as the RAGE transmembrane trap (TMT) vectors.

When a gene containing a transmembrane domain is activated by the integration of the RAGE-TMT vector, a chimeric protein is expressed that contains a strong signal peptide (directing secretion of the protein), a small synthetic epitope tag (enabling rapid detection of the protein with a monoclonal antibody specific for the epitope tag), and the transmembrane domain from the endogenously encoded protein. The chimeric protein enters the secretory pathway, which results in the cleavage of the signal peptide. Activated proteins lacking a transmembrane domain will not display the epitope on the cell surface; instead, it will be secreted from the cell. Many of the expressed membrane proteins will display the cell surface epitope tag, which will facilitate separation from cells that do not express the tag in this fashion. This can be done using either fluorescence-activated cell sorting (FACS) analysis or alternative cell-sorting techniques (Figure 4).

In one set of pilot experiments, more than 500 clones expressing randomly activated genes were harvested and analyzed using the RAGE-TMT approach. Similar to the more general gene discovery approach, over 50% of the genes identified were determined to be novel, based on bioinformatic analysis. The examination of the RAGE-TMT clones expressing previously characterized genes (e.g., genes present in the public gene sequence databases) confirmed that more than 90% of these genes encoded an integral membrane protein. Furthermore, the analysis revealed that this approach is capable of isolating adhesion molecules, single membrane-spanning cell surface receptors, multi-spanning integral membrane proteins, such as ion channels and GPCRs, and intracellular membrane proteins, such as nuclear pore proteins and mitochondrial proteins. Similarly, many of the novel genes identified display the bioinformatic characteristics of transmembrane proteins, such as structural homology to other classes of transmembrane proteins (e.g., GPCRs and ion channels). To date, several thousand confirmed or candidate integral membrane proteins have been identified using the RAGE-TMT approach and the general RAGE gene discovery approach.

PRODUCTION OF GENOME-WIDE PROTEIN EXPRESSION LIBRARIES

Genomics initiatives are providing a wealth of data on the structure and sequence of human genes. The sequence information derived from these efforts has contributed in part to understanding the specific role that genes play in health and disease, but the ability to convert sequence data into medical products requires an intimate understanding of gene and protein function. Perhaps this limitation is most evident in novel genes that have no recognizable homology to genes with known functions. The examination of the sequence and structural homology is one approach to predict biochemical activity; but alone, this approach is insufficient to yield an adequate prediction of biological function or medical relevance. For example, knowing that a protein has protease activity does not indicate where the protease is expressed, whether it is intracellular or extracellular, the identity of its specific substrate, the biological process or pathway in which it is involved, or its function in the organism. In short, this gene-to-function approach, while powerful for select genes, is generally a slow process that is not easily scaled. As a result, determining the gene or protein function is one of the major bottlenecks in the process of developing genomics-based therapeutic and diagnostic products.

The biotechnology and pharmaceutical industries have a tremendous need for methodologies that are capable of identifying, en masse, the proteins associated with medically important pathways or processes. The RAGE technology addresses this need because it can rapidly link protein function and gene structure. RAGE protein expression libraries, which comprise cells that collectively express virtually every gene and protein in the human genome, can be used in countless biological and cellular assays of interest to scientists at pharmaceutical and

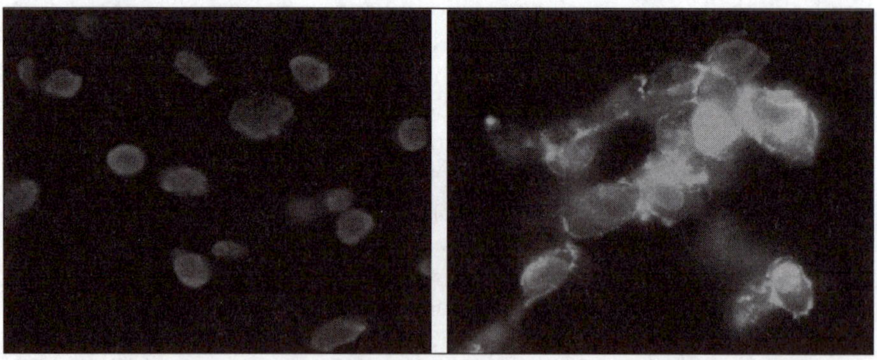

Figure 4. Creation of a RAGE library expressing known and novel integral membrane proteins. A RAGE library was created using the vector shown in Figure 3. The cells were sorted using FACS and an antibody specific for the cell surface epitope encoded by the RAGE vector. Control cells used to make the RAGE library are shown on the left, and a heterogeneous population of RAGE-activated cells expressing chimeric integral membrane proteins is shown on the right (both obtained after sequential FACS sorts). **(See color plate A37.)**

biotechnology companies and research institutes. It begins with a system that assays for a given function and identifies the gene and gene product that give rise to that function. In contrast to the gene-to-function approach, this strategy is referred to as a function-to-gene approach.

Experimental Validation of RAGE Protein Expression Libraries

A prototype RAGE protein expression library comprising an estimated 3- to 5-fold coverage of the genome was created and pools containing 5×10^3 clones were screened by ELISA for the expression of two human proteins: erythropoietin (hEpo) and growth hormone (hGH). These proteins were undetectable in the parental cells when tested by RT-PCR, the most sensitive measure of expression. Cultured supernatants from various pools of RAGE library cells were concentrated and examined by ELISA for hEpo and hGH. Of the pools tested, five had relatively high concentrations of hGH, and two had readily detectable concentrations of hEpo. These results demonstrated that culture supernatants from pooled RAGE library clones could be screened successfully for the presence of secreted proteins, such as cytokines and growth factors.

Additional work has revealed that RAGE libraries can also be tested using biological and cell-based assays. Upon successful identification of a library pool containing a factor with the desired activity, it is possible to isolate the specific clone using FACS, phenotypic identity, or "sib" selection and limiting dilution, enabling the separation of the specific RAGE clone from all other library members. The RAGE gene sequence may then be determined quickly by a modified PCR technique using a RAGE synthetic exon-specific forward primer and a reverse primer complementary to the poly(A) tail, resulting in the rapid amplification of cDNA ends (9).

The powerful capabilities afforded by the RAGE technology greatly accelerate the process of linking structural and functional information. Historically, this process has been difficult and slow. Many techniques, such as positional cloning and a variety of "reverse genetics" approaches, have been developed to move from function to gene. Nevertheless, these approaches still require a number of steps that are ultimately time and labor intensive. These steps might include: the identification of a cell or tissue with a desired biological activity; purification of the protein and verification that it is responsible for the desired activity; identification of the amino acid sequence of the protein; analysis of the potential coding sequence of the gene; and eventual identification of the actual gene sequence. This process is simplified significantly when the RAGE technology is coupled with a sensitive and robust assay. Many such assay systems are commercially available. A partial list of applications in which the RAGE technology enables the drug discovery process might include the following:

- Identification of a previously unidentified receptor for a known ligand
- Identification of a novel ligand that binds to a known receptor

- Identification of proteins with specific activities (e.g., protease, kinase, and changes in cAMP or specific ion concentrations)
- Identification of proteins that regulate differentiation along a particular pathway
- Identification of proteins modulating drug activity/potency (e.g., resistance to chemotherapeutic agents)
- Identification of proteins involved in cytotoxic processes (e.g., apoptosis)

These examples provide some insight to how versatile the RAGE system can be. The system is powerful because it couples the ability to screen the entire genome with the ability to identify rapidly the target gene and gene product responsible for the desired activity in the screening assay. These capabilities of the RAGE technology will greatly accelerate the functional characterization of the human genome.

CREATION OF CELL LINES EXPRESSING PREVIOUSLY VALIDATED DRUG TARGETS

Another important way that the RAGE technology will impact the pharmaceutical industry is through enabling drug discovery against certain validated targets covered by patent restrictions. As the discoveries of novel genes rise, the number of pending and issued patents that cover the use of cloned and isolated gene sequences is increasing considerably. Many patents have been issued each year on the use of cloned cDNAs encoding specific genes that may subsequently prove to be interesting and valuable drug targets. However, once the patent is issued, an organization interested in that target for drug discovery must either (*i*) license the commercial rights to the gene from the innovator; (*ii*) pursue difficult and/or inconvenient alternatives to obtain access to the necessary materials; or (*iii*) terminate the research program if a license to the gene is unavailable. These constraints can encumber the progress to develop potentially important and valuable drugs.

RAGE allows one to create cell lines expressing previously validated drug targets without the use of proprietary information on cloned and isolated gene sequences. By obtaining access to cell lines expressing validated targets by virtue of the RAGE technology, researchers can conduct high-throughput screening against drug targets that often do not infringe on conventional intellectual property limitations. This capability has been tested by selecting several dozen known and validated drug targets and subsequently using a variety of approaches to analyze for the expression of such genes. To date, each gene that was a target in the screen has been identified successfully. In the case where a previously constructed RAGE library was used in the screening process, the expression of the desired target was confirmed typically in less than a week. Furthermore, clonal cell lines expressing high levels of these drug targets have also been isolated rapidly. Some important biological targets that have been identified using the RAGE technology include receptors, enzymes, and certain viral targets.

RAGE cells lines that overexpress validated drug targets have been adapted for use in high-throughput screening applications by verifying appropriate second messenger responses in the presence of target specific ligands. This was demonstrated in a series of studies in which a clone expressing the Gs-linked β_3 adrenergic receptor (β_3AR) was identified in a RAGE library that was created in human embryonic kidney (HEK)293 cells. The clone was expanded, creating a cell line for subsequent analyses. Experiments to validate this cell line using a known β_3AR agonist and antagonist demonstrate how this cell line could be used in a screen for novel β_3AR-selective drugs (Figure 5). These studies demonstrated that intracellular cAMP levels in the RAGE clonal cell line increased in response to the agonist. Simultaneous addition of the agonist and antagonist blocked the cAMP response. Furthermore, no change in cAMP levels was observed when similar studies were conducted in the parental HEK293 cell line. These results indicate that cAMP increases were due to stimulation of the Gs-linked receptor encoded by the RAGE-activated gene. Studies like this one demonstrate how the RAGE technology may be useful in creating cell lines for drug screening against targets that may otherwise be encumbered by patents covering the use of the cloned and isolated gene sequences.

It is important to note that cell lines that express previously validated drug targets via the RAGE technology do not provide a solution to all intellectual property claims one would examine before drug screening. The use of the RAGE technology to provide the freedom to operate must be determined on a

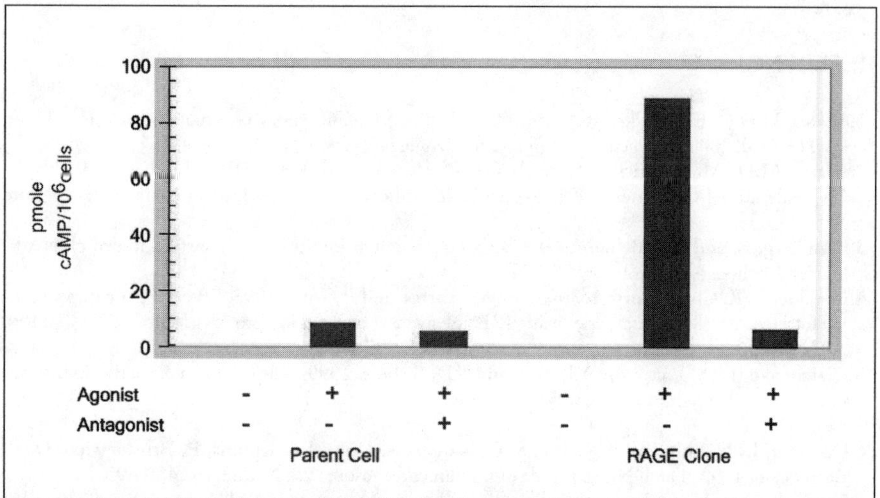

Figure 5. RAGE-activated targets exhibit full functionality. Upon the creation of a RAGE library in HEK293 cells, a clone expressing the β_3AR was identified. The addition of an appropriate agonist to the receptor caused intracellular levels of cAMP to increase. Simultaneous addition of an appropriate antagonist blocked the agonist effect. No change in calcium level was observed upon the addition of the agonist to parental HEK293 cells. (Figure reprinted with permission from Genetic Engineering News. Figure previously appeared in Genetic Engineering News Volume 22, No. 3, p. 28–31.

target-by-target basis and must account for the claims against the desired drug target as well as the screening assays to be used in conjunction with each particular target.

CONCLUSION

RAGE is a robust gene expression technology that provides a unique ability to express virtually all proteins encoded in the human genome from a single cell source. The resulting library of expressed proteins can be used to screen for desired activities and, subsequently, to identify the genes encoding the proteins responsible for those activities. Through its ability to lead to genome-wide expression, RAGE is useful for a variety of applications. The RAGE technology can be used in conjunction with bioinformatic tools as a gene discovery system, which can be tailored either to broad selection criteria or, in some cases, to class-specific criteria (i.e., transmembrane proteins). In addition, RAGE can be used for functional genomics. Combining RAGE with cellular or biochemical screens is a powerful means to identify genes that lead to particular phenotypes. Finally, RAGE is an alternative approach for creating cell lines for use in high-throughput screening. This capability will loosen the constraints that patents have imposed on the drug discovery process. In summary, the RAGE technology will help to accelerate the functional characterization of the human genome and the development of novel therapeutics.

REFERENCES

1. **Adams, M.D., S.E. Celniker, R.A. Holt, C.A. Evans, J.D. Gocayne, P.G. Amanatides, S.E. Scherer, P.W. Li, et al.** 2000. The genome sequence for *Drosophila melanogaster*. Science *287*:2185-2195.
2. **Adams, M.D., M. Dubnick, A.R. Kerlavage, R. Moreno, J.M. Kelley, T.R. Utterback, J.W. Nagle, C. Fields, and J.C. Venter.** 1992. Sequence identification of 2375 human brain genes. Nature *355*:632-634.
3. **Blumberg, B. and J.C. Belmonte.** 1999. Subtractive hybridization and construction of cDNA libraries. Methods Mol. Biol. *97*:555-574.
4. **Bonaldo, P., K. Chowdhury, A. Stoykova, M. Torres, and P. Gruss.** 1998. Efficient gene trap screening for novel developmental genes using IRES beta geo vector and in vitro preselection. Exp. Cell Res. *244*:125-136.
5. **Diatchenko, L., S. Lukyanov, Y.F. Leu, and P.D. Siebert.** 1999. Suppression subtractive hybridization: a versatile method for identifying differentially expressed genes. Methods Enzymol. *303*:349-380.
6. **Dunham, I., N. Shimizu, B.A. Roe, S. Chissoe, A.R. Hunt, J.E. Collins, R. Bruskiewich, D.M. Beare, et al.** 1999. The DNA sequence of human chromosome 22. Nature *402*:489-495.
7. **Friedrich, G. and P. Soriano.** 1991. Promoter traps in embryonic stem cells: a genetic screen to identify and mutate developmental genes in mice. Genes Dev. *5*:1513-1523.
8. **Hardouin, N. and A. Nagy.** 2000. Gene-trap-based target site for *cre*-mediated transgenic insertion. Genesis *26*:245-252.
9. **Harrington, J.J., B. Sherf, S. Rundlett, P.D. Jackson, R. Perry, S. Cain, C. Leventhal, M. Thornton, et al.** 2001. Creation of genome-wide protein expression libraries using random activation of gene expression. Nat. Biotechnol. *19*:440-445.

10.**Hay, B.A., R. Malle, and G.M. Rubin.** 1997. P element insertion-dependent gene activation in *Drosophila* eye. Proc. Natl. Acad. Sci. USA *94*:5195-5200.

11.**Hillier, L.D., G. Lennon, M. Becker, M.F. Bonaldo, B. Chiapelli, S. Chissoe, N. Dietrich, T. DuBuque, et al.** 1996. Generation and analysis of 280 000 human expressed sequence tags. Genome Res. *6*:807-828.

12.**Lander, E.S., L.M. Linton, B. Birren, C. Nusbaum, M.C. Zody, J. Baldwin, K. Devon, K. Dewar, et al.** 2001. Initial sequencing and analysis of the human genome. International Human Genome Sequencing Consortium. Nature *409*:860-921.

13.**Mewes, H.W., K. Albermann, M. Bahr, D. Frishman, A. Gleissner, J. Hani, K. Heumann, K. Kleine, et al.** 1997. Overview of the yeast genome. Nature *387*:7-65.

14.**Neilan, E.G. and G.S. Barsh.** 1999. Gene trap insertional mutagenesis in mice: new vectors and germ line mutations in two novel genes. Transgenic Res. *8*:451-458.

15.**Niwa, H., K. Araki, S. Kimura, S. Taniguchi, S. Wakasugi, and K. Yamamura.** 1993. An efficient gene-trap method using poly A trap vectors and characterization of gene-trap events. J. Biochem (Tokyo) *113*:343-349.

16.**Nomura, N., N. Miyajima, T. Sazuka, A. Tanaka, Y. Kawarabayasi, S. Sato, T. Nagase, N. Seki, et al.** 1994. Prediction of the coding sequences of unidentified human genes. I. The coding sequences of 40 new genes (KIAA0001-KIAA0040) deduced by analysis of randomly sampled cDNA clones from human immature myeloid cell line KG-1. DNA Res. *1*:27-35.

17.**Orr, S.L., T.P. Hughes, C.L. Sawyers, R.M. Kato, S.G. Quan, S.P. Williams, O.N. Witte, and L. Hood.** 1994. Isolation of unknown genes from human bone marrow by differential screening and single-pass cDNA sequence determination. Proc. Natl. Acad. Sci USA *91*:11869-11873.

18.**Ross-Macdonald, P., P.S. Coelho, T. Roemer, S. Agarwal, A. Kumar, R. Jansen, K.H. Cheung, A. Sheehan, et al.** 1999. Large-scale analysis of the yeast genome by transposon tagging and gene disruption. Nature *402*:413-418.

19.**Skarnes, W.C., B.A. Auerbach, and A.L. Joyner.** 1992. A gene trap approach in mouse embryonic stem cells: the lacZ reported is activated by splicing, reflects endogenous gene expression, and is mutagenic in mice. Genes Dev. *6*:903-918.

20.**Tanaka, T., A. Ogiwara, I. Uchiyama, T. Takagi, Y. Yazaki, and Y. Nakamura.** 1996. Construction of a normalized directionally cloned cDNA library from adult heart and analysis of 3040 clones by partial sequencing. Genomics *35*:231-235.

21.**The *C. elegans* sequencing consortium.** 1998. Genome sequence of the nematode *C. elegans*: a platform for investigating biology. Science *282*:2012-2018.

22.**Venter, J.C., M.D. Adams, E.W. Myers, P.W. Li, R.J. Mural, G.G. Sutton, H.O. Smith, M. Yandell, et al.** 2001. The sequence of the human genome. Science *291*:1304-1351.

23.**Wan, J.S. and M.G. Erlander.** 1997. Cloning differentially expressed genes by using differential display and subtractive hybridization. Methods Mol. Biol. *85*:45-66.

24.**Wiles, M.V., F. Vauti, J. Otte, E.M. Fuchtbauer, P. Ruiz, A. Fuchtbauer, H.H. Arnold, H. Lehrach, et al.** 2000. Establishment of a gene-trap sequence tag library to generate mutant mice from embryonic stem cells. Nat. Genet. *24*:13-14.

25.**Wurst, W., J. Rossant, V. Prideaux, M. Kownacka, A. Joyner, D.P. Hill, F. Guillemot, S. Gasca, et al.** 1995. A large-scale gene-trap screen for insertional mutations in developmentally regulated genes in mice. Genetics *139*:889-899.

26.**Zambrowicz, B.P., G.A. Friedrich, E.C. Buxton, S.L. Lilleberg, C. Person, and A.T. Sands.** 1998. Disruption and sequence identification of 2000 genes in mouse embryonic stem cells. Nature *392*:608-611.

27.**Zhang, H., R. Zhang, and P. Liang.** 1996. Differential screening of gene expression difference enriched by differential display. Nucleic Acids Res. *24*:2454-2455.

22 Semiconductor Quantum Dots and Optically Encoded Microbeads for Gene Expression Analysis

Xiaohu Gao and Shuming Nie
Departments of Biomedical Engineering, Chemistry, Hematology, Oncology, and the Winship Cancer Institute, Georgia Institute of Technology and Emory University School of Medicine, Atlanta, GA, USA

INTRODUCTION

A key feature in many gene analysis technologies, such as DNA chips (22,56), molecular beacons (61,62), and TaqMan® nuclease assays (33,38), is the use of fluorescent probes. These probes are prepared by covalently linking sequence-recognition oligonucleotides to organic fluorophores such as fluorescein, rhodamine, and carbocyanines. However, because of their complex molecular structures, organic fluorophores exhibit absorption and emission properties that are unfavorable to sensitive and multiplexed optical detection (28). For example, organic dyes often undergo rapid photobleaching, are sensitive to environmental quenching, and exhibit broad and asymmetric emission profiles. Additionally, it is difficult to detect more than two or three organic fluorophores at a single excitation wavelength.

The recent development of semiconductor nanocrystals, also known as quantum dots (QDs), has made available a new class of fluorescent labels with unique advantages and applications (9). In contrast to that of organic dyes, the fluorescence emission spectra of QDs can be continuously tuned by changing the particle size, and a single wavelength can be used for the simultaneous excitation of all different-sized QDs (26). Surface-passivated QDs are highly stable against photobleaching and have narrow, symmetric emission peaks (25–30 nm full

Perspectives in Gene Expression
Edited by K. Appasani

width at half maximum [FWHM]). It has been estimated that CdSe QDs are about 20 times brighter and 100–200 times more stable than single rhodamine 6G molecules (10). These properties offer new possibilities in several areas of research, including single-molecule biophysics, multiplexed medical diagnostics, high-throughput drug screening, and gene expression analysis.

In this chapter, we focus on the gene expression applications of luminescent QDs. In particular, we discuss two areas in which QDs could have a significant impact: (*i*) the use of bioconjugated QDs as fluorescent labels for ultrasensitive and multicolor imaging of gene chips or microarrays and (*ii*) the development of QD-encoded microbeads for massively parallel analysis of mRNA targets. We briefly review the synthesis of highly luminescent QDs, their optical properties, surface modifications, bioconjugation chemistry, and biological multiplexing. The chapter concludes with a brief discussion of the future challenges and prospects.

PRINCIPLES AND SYNTHESIS

Semiconductor quantum dots are composed of atoms from groups I-VII, II-VI, or III-V elements and are defined as particles with physical dimensions smaller than the exciton Bohr radius (1,7). For spherical CdSe particles, this occurs when the particle diameter is less than approximately 10 nm. This quantum-confinement effect gives rise to unique optical and electronic properties that are unavailable in either discrete atoms or in bulk solids.

Both group II-VI (e.g., CdSe, CdTe, CdS, and ZnSe) and group III-V (e.g., InP and InAs) nanocrystals have been synthesized and characterized (25,27, 35,45,55,57,60,64). In particular, CdSe QDs are highly luminescent and have been used as a model system (12,31,46,49). In one synthetic method, organometallic and chalcogenide compounds are dissolved in tri-n-butyl phosphine (TBP) or tri-n-octyl phosphine (TOP), which are injected into a hot coordinating solvent (e.g., tri-n-octylphosphine oxide or TOPO) at 340°C to 360°C. Rapid nucleation is observed by changes in the reaction mixture's color. With this procedure, technical-grade TOPO (90%–92% purity) is preferred, but the role of impurities such as alkyl phosphonics and alkylphosphonic acids is still unclear. Peng et al. (50) and Qu et al. (54) recently used phosphonic or fatty acid additives in pure TOPO to synthesize high-quality CdS, CdSe, and CdTe nanocrystals. The results indicate that CdO and $Cd(CH_3COO)_2$ can be used to replace dimethylcadmium, which is an expensive organometallic compound. By using organic solvents to synthesize QDs, the surfactant molecules TOPO and TOP reside on the particle's surface and act to stabilize the nanocrystals against aggregation. Due to the nonpolar nature of TOPO and TOP, the resulting nanocrystals are soluble in organic solvents such as chloroform and hexane.

Both the size and shape of the nanocrystal can be changed by manipulating the kinetics of particle growth. Recently, a novel method was developed to control the shape of CdSe nanoclusters (51,52). Wurtzite-type particles grow in an anisotropic fashion, in which growth is favored along the c-axis. The manipulation of solvent conditions (e.g., addition of hexyl-phosphonic acid and tetradecylphosphonic acid in pure TOPO) alters the growth rates, leading to the formation and isolation of nanoparticles with discrete aspect ratios (1:1 to 1:30). The size of nanocrystals can be controlled by several procedures. For example, nucleated CdSe QDs can be grown at high temperatures (300°C or higher) for an extended period of time (ranging from two minutes to two weeks, depending on the desired particle size) (53). In this process (also known as Ostwald ripening), smaller nanocrystals are broken down, and the dissolved atoms are transferred to larger crystals. The rate of this "ripening" process is dependent on both the temperature and the amount of the limiting reagent (15,16).

The quantum yields of these CdSe nanocrystals are usually 2%–15% but can be dramatically increased by capping the dots with an inorganic layer such as ZnS or CdS. Zinc sulfide has a wider bandgap than that of CdSe (0.9 eV greater) (40), but the Zn-S bond length is similar to that of Cd-Se (31). These conditions allow for the deposition of a thin ZnS layer on the CdSe core by epitaxial growth (12,31,49). In practice, the $Zn(CH_3)_2/S$ solution is added slowly in small aliquots to the CdSe/TOPO solution to prevent ZnS nucleation. The quantum yields of the capped CdSe nanocrystals are typically 30%–50%, but values as high as 85% at room temperature have been reported (49). Figure 1 shows a set of different-sized ZnS-capped CdSe QD excited with a handheld UV-lamp.

The observed increase in quantum yields can be attributed to the removal of shallow trap states on the surface of QDs. These states are defined as discrete electronic energy levels that arise from structural defects. For CdSe nanocrystals, these defects are mainly located on the surface, where the bonding geometries of chalcogenide atoms are different from those in the interior. These shallow traps lead to a broad emission peak at 700–800 nm (12,31,49). As the ZnS layer be-

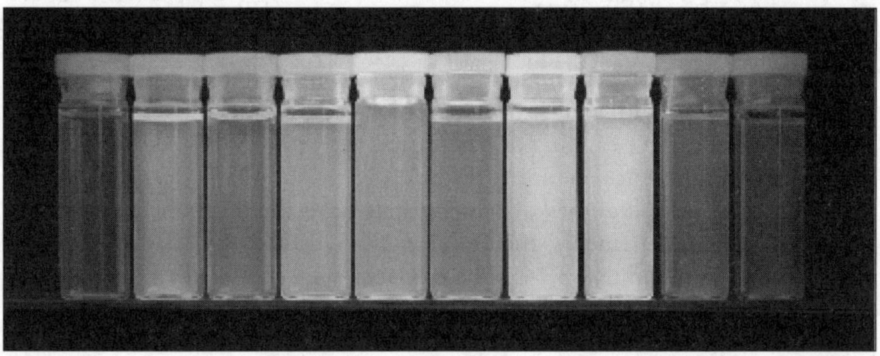

Figure 1. Ten distinguishable emission colors of ZnS-capped CdSe QDs excited with a near-UV lamp. From left to right (blue to red), the emission maxima are located at 443, 473, 481, 500, 518, 543, 565, 587, 610, and 655 nm. (Adapted with permission from Han, Gao, Su, and Nie, Nat. Biotechnol. 19:631, 2001). **(See color plate A38.)**

comes thicker, the quantum efficiency increases to approximately 50% and then levels off. Several groups have determined that the optimal capping thickness is from one to two monolayers of ZnS (12,59).

Group III-V nanocrystals can also be prepared by using the TOPO/TOP solvent system. To synthesize InP, a dehalosilylation reaction occurs when $InCl_3$ and $P(Si(CH_3)_3)$ are injected into a TOPO/TOP solution from 240°C to 265°C. The size distribution of the InP nanoparticles is narrowed by size-selective precipitation. InP nanocrystals in the size range of 2.5–6.0 nm are isolated with a relative standard deviation of 10%–15% (25,45). Similar procedures have been used to prepare InAs nanocrystals.

OPTICAL PROPERTIES

The optical properties of semiconductor nanoclusters stem from interactions between electrons, holes, and their local environments. Semiconductor QDs absorb photons when the energy of excitation exceeds the bandgap energy. During this process, electrons are promoted from the valence band to the conduction band. Measurements of UV-vis spectra reveal a large number of energy states in QDs. The lowest excited energy state is shown by the first observable peak (also known as the quantum-confinement peak), at a shorter wavelength than the fluorescence emission peak. Excitation at shorter wavelengths is possible because multiple electronic states are present at higher energy levels. In fact, the molar extinction coefficient gradually increases toward shorter wavelengths (Figure 2) (8). This is an important feature for biological applications because it allows for the simultaneous excitation of multicolor QDs with a single light source.

Light emission arises from the recombination of mobile or trapped charge carriers. The emission from mobile carriers is called excitonic fluorescence and is observed as a sharp peak. The emission spectra of single ZnS-capped CdSe QDs are as narrow as 13 nm (FWHM) at room temperature (10). Defect states in the crystal interior or on its surface can trap the mobile charge carriers (electrons or holes), leading to a broad emission peak that is red-shifted from the excitonic peak. Nanocrystals with a large number of trap states generally have low quantum yields, but surface capping or passivation can remove these defect sites and improve the fluorescence quantum yields.

The excited-state lifetimes of nanocrystals have three exponential components. In bulk measurements, the lifetimes are 5, 20–30, and 80–200 ns, with the 20–30 ns dominating (13,14). These excited-state decay rates are slower than those of organic dyes (1–5 ns) but much faster than those of lanthanide probes (1 μs–1 ms). Single-dot measurement reveals that the excited-state lifetimes are dependent on the emission intensity, but the exact origins of this multi-exponential behavior remain unclear (14).

The excitonic fluorescence is dependent on the size of the nanocrystal. Research in several groups has demonstrated a close relationship between the particle size and the bandgap energy (12,53). This quantum-size effect is similar to that observed for a "particle in a box." Outside of the box, the potential energy is considered to be infinitely high. Thus, mobile carriers (similar to the particle) are confined within the dimensions of the nanocrystal (similar to the box), with discrete wavefunctions and energy levels. As the physical dimensions of the box become smaller, the bandgap energy becomes higher. For CdSe nanocrystals, the sizes of 2.5 and 5.5 nm correspond to fluorescence emission at 500 and 620 nm, respectively. In addition to size, the emission wavelength can be varied by changing the semiconductor material. For example, InP and InAs QDs usually emit in the far-red and near-infrared (25,45), while CdS and ZnSe dots often emit in the blue or near-UV (32). It is also interesting to note that elongated QDs (called

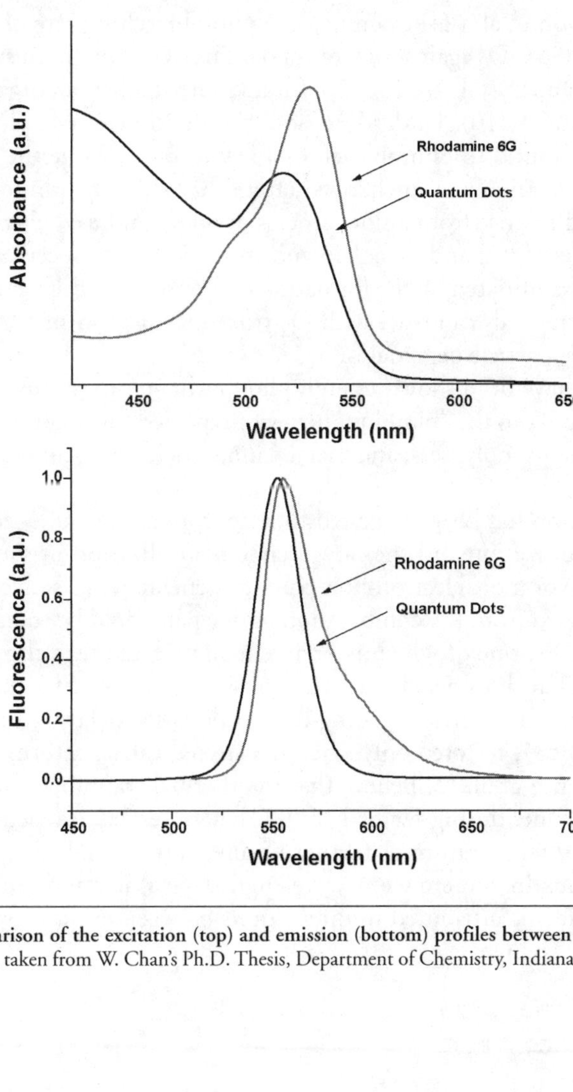

Figure 2. Comparison of the excitation (top) and emission (bottom) profiles between rhodamine 6G and CdSe QDs. Data taken from W. Chan's Ph.D. Thesis, Department of Chemistry, Indiana University (8).

quantum rods) show linearly polarized emission (34) whereas the fluorescence emission from spherical CdSe dots is circularly polarized or not polarized (17,19).

In comparison with organic dyes such as rhodamine 6G and fluorescein, CdSe nanocrystals show similar or slightly lower quantum yields at room temperature. The lower quantum yields of nanocrystals are compensated by their larger absorption cross sections and much reduced photobleaching rates. Dabbousi et al. (12) and Murray et al. (46) estimated that the molar extinction coefficients of CdSe QDs are from 10^5 to 10^6 M^{-1} cm^{-1}, depending on the particle size and the excitation wavelength. These values are 10–100 times larger than those of organic dyes but are similar to the absorption cross sections of phycoerytherin, a multi-chromophore fluorescent protein. Chan and Nie (10) estimated that single ZnS-capped CdSe QDs are approximately 20 times brighter than single rhodamine 6G molecules.

Another attractive feature of using QDs as biological labels is their high photostability. Gerion et al. (24) examined the photobleaching rate of silica-coated ZnS-capped CdSe QDs against that of rhodamine 6G. It has been suggested that capped CdSe nanocrystals are 100–200 times more stable than organic dyes and fluorescent proteins (10). Under intense UV excitation, single phycoerytherin molecules were found to photobleach in a few seconds, while the fluorescence emission of QDs remained unchanged after 600 s (8). The photobleaching of QDs is believed to arise from a slow process of photo-induced chemical decomposition. Baral et al. (3) and Henglein and co-workers (29) speculated that CdS decomposition is initiated by the formation of S or SH radicals upon optical excitation. These radicals can react with O_2 from the air to form a SO_2 complex, resulting in slow particle degradation.

Single QDs have been shown to emit photons in an intermittent on-off fashion (18,47), similar to the "blinking" behavior reported for single fluorescent dye molecules, proteins, polymers, and metal nanoparticles. The fluorescence of single QDs turns on and off at a rate that is dependent on the excitation power. This phenomenon has been suggested to arise from a light-induced process involving photoionization and the slow charge neutralization of the nanocrystal (47). When two or more electron-hole pairs are generated in a single nanocrystal, the energy released from the combination of one pair could be transferred to the remaining carriers, one of which is preferentially ejected into the surrounding matrix. Subsequent photogenerated electron-hole pairs transfer their energy to the resident, unpaired carrier, leading to non-radiative delay and dark periods. The luminescence is restored only when the ejected carrier returns to neutralize the particle. Banin et al. (2) believe that the thermal trapping of electrons and holes is also a contributing factor because they observed a dependence of the blinking rate on temperature. A further finding is that single dots exhibit random fluctuations in the emission wavelength (spectral wandering) over time (5,18). This effect is attributed to interactions between excitons with optically induced surface changes.

SURFACE CHEMISTRY

Colloidal metal nanoparticles have found widespread use in biology and medicine in the past 20 years, but semiconductor nanocrystals have only recently been used for biological labeling. The reason is that high-quality QDs are often protected with hydrophobic ligands and are not compatible with biomolecules. Mahtab and co-workers (41,42) were the first to demonstrate the use of semiconductor nanoparticles as biosensors, in which oligonucleotide molecules were detected by nonspecific interactions with the nanocrystal surface. For other applications, there is a great need to develop simple methods for conjugating biomolecules to QDs. TOPO-synthesized nanocrystals are preferred over aqueous-synthesized dots due to their superior optical properties. The surface chemistry is a key factor in determining QD solubility, biocompatibility, and molecular conjugation.

The complex surface of nanocrystals has been studied by NMR and X-ray photoelectron spectroscopies (4,36). Morphologically, QDs are not smooth spherical particles but faceted with many planes and edges. They are generally considered to be negatively charged because of molecules adsorbed on the surface (48). TOPO strongly coordinates to the surface metal atoms, while TOP or TBP is weakly bound and can easily desorb. The surface properties (such as TOPO molecules interacting with each other and their dimerization at the phosphine oxide end) improve the long-term solubility of QDs in organic solvents (39).

Two methods have been developed to prepare water-soluble QDs. Bruchez and co-workers (6) reported the use of a silica/siloxane coating for creating water-soluble ZnS-capped CdSe QDs. In this procedure, 3-(mercaptopropyl)trimethoxysilane (MPS) is directly adsorbed onto the nanocrystals in which TOPO molecules are displaced. A silica/siloxane shell is formed on the surface by the introduction of a base and hydrolysis of the silanol groups (from MPS). Polymerizing silanol groups help stabilize the nanocrystals against flocculation. The QDs become soluble in intermediate polar solvents, such as methanol and dimethyl sulfoxide. Further reaction with bifunctional methoxy compounds such as aminopropyl trimethoxysilane or trimethoxysilyl propyl urea renders the particles soluble in aqueous buffer. The second method involves the direct adsorption of bifunctional ligands such as mercaptoacetic acid or dithiothrietol to the QD surface (10). Here a mercapto compound and an organic base are added together to TOPO QDs dissolved in chloroform. The base deprotonates both the thiol and the carboxylic groups, leading to favorable electrostatic binding between negatively charge sulfur atoms and surface Cd^{2+} or Zn^{2+} ions. The semiconductor QDs precipitate out of the solution but can be redissolved in an aqueous solution (pH >5.0).

Polymerized siloxane-coated QDs are highly stable against flocculation, but only small amounts (in the mg range) can be prepared per batch. Also, residual silanol groups on the nanocrystal surface often lead to precipitation and gel formation at neutral pHs. In comparison, the direct adsorption procedure yield gram quantities of water-soluble dots, but the mercapto ligands are not completely stable. Slow desorption of mercaptoacetic acid molecules often results in the aggregation and precipitation of the solubilized QDs. The stability of the mercapto-derivatized dots is dependent on the density of mercapto-molecules on the surface, its size, and the number of thiol functional groups per mercapto-molecule (8). When solubilized with mercaptoacetic acid, the QDs are stable for 3–7 days in neutral PBS buffer. The conjugation of the complex proteins onto the surface further stabilizes the particles, making them stable from three months to a year (depending on the protein).

It should be noted that the quantum efficiencies of ZnS-capped CdSe nanocrystals are generally lower in water than in chloroform. The highest quantum yield reported for water-soluble QDs is 20%–30%. Traditionally, it is believed that capping a second semiconductor layer protects and isolates the core from the outside environment (31). Recent research in our group indicates that the ZnS layer is not uniform and that surface-adsorbed molecules can still influence the properties of the nanocrystal core (8). In fact, both large and small molecules, such as polymers, proteins, and ions, have been found to influence the optical properties of ZnS-capped CdSe QDs (23).

BIOCONJUGATION

For specific biomolecular binding or recognition, it is important to develop stable and functional QD bioconjugates. Reactive functional groups that can be placed on the nanocrystal surface include primary amines, carboxylic acids, alcohols, and thiols. Primary amines react with carboxylic acids, catalyzed with a carbodiimide or sulfo N-hydroxysuccinimide ester, to form a stable amide bond. A thioether bond is formed when sulfhydryl groups react with maleimides. Other methods for linking fluorescent labels to biomolecules are described in a book by Hermanson (30). Another approach for linking biomolecules to nanocrystals is to use a thiol-exchange reaction. Here mercapto-coated QDs (by direct adsorption) are mixed with thiolated biomolecules (such as oligonucleotides and proteins) (44). After overnight incubation at room temperature, a chemical equilibrium is reached between the adsorbed thiols and the free thiols. A similar approach has recently been used by Mattoussi and co-workers (43), in which engineered proteins with a linear poly-lysine chain are directly adsorbed onto negatively charge nanocrystals through electrostatic interactions. Figure 3 shows several possible conjugation methods.

The surface area of the QDs is large enough for linking to multiple biomolecules. Two to five protein molecules and 50 or more small molecules (such as oligonucleotides and peptides) may be conjugated to a single 4-nm dot (23). We envision that multifunctional or smart nanostructures could be developed by conjugating both DNA and proteins to QDs. The attached proteins or peptides could guide the nanoparticles to specific cells (e.g., cancer cells) or to specific locations inside a cell where mRNA molecules are detected.

Several applications have been demonstrated for QDs, including DNA hybridization, immunoassays, and receptor-mediated endocytosis. In addition, Lacoste et al. (37) and Weiss (63) showed the ability to measure distances between ZnS-capped CdSe QDs and fluorescent beads beyond the diffraction limit. Lifetime measurement was also used for time-gated fluorescence imaging of tissue sections (13). The background fluorescence lifetimes are similar to those of organic dyes and, thus, interfere with the target signals. For lanthanide-chelating probes, the excited-state lifetimes are very long, but their quantum efficiencies are relatively low. In contrast, semiconductor nanocrystals have excellent emission efficiencies and relatively long decay times, both of which are attractive features for imaging tissue sections.

APPLICATIONS IN GENE EXPRESSION ANALYSIS

One application is simply to replace fluorescent dyes with luminescent QDs for ultrasensitive and multicolor imaging of gene expression chips. Due to their brightness and photostability, the use of QDs should yield more sensitive and

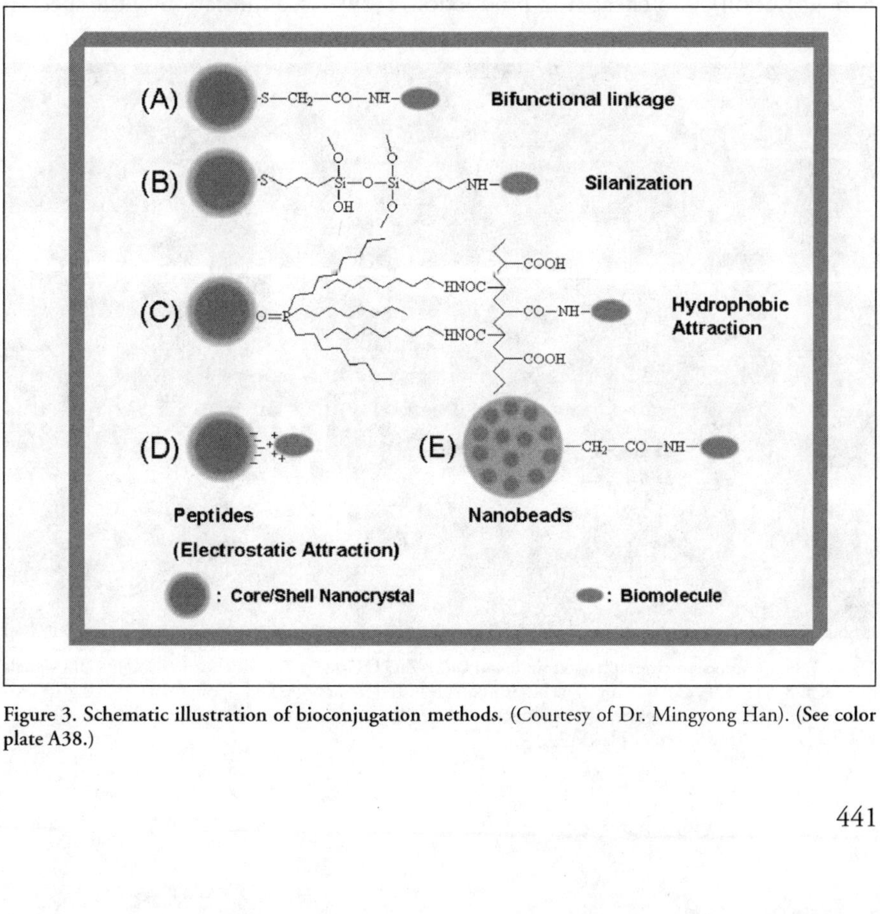

Figure 3. Schematic illustration of bioconjugation methods. (Courtesy of Dr. Mingyong Han). **(See color plate A38.)**

quantitative results. In particular, by repetitive scanning and signal averaging, many low-abundance transcripts could be detected in complex mixtures, thus overcoming a major limitation in current gene chip technologies. A potential problem with QDs is their larger size, which could present steric hindrance problems and slow binding kinetics.

An additional application of QDs is in mulitplexed optical-encoding and high-throughput analysis of genes and proteins, as reported by Han and coworkers (26). Polystyrene beads are embedded with multicolor CdSe QDs at various color and intensity combinations (Figure 4). The use of six colors and 10 in tensity levels can theoretically encode one million protein or nucleic acid sequences. Specific capturing molecules such as peptides, proteins, and oligonucleotides are covalently linked to the beads and are encoded by the bead's spectroscopic signature. A single light source is sufficient for reading all the QD-encoded beads. To determine whether an unknown analyte is captured or not, conventional assay methodologies (similar to direct or sandwich immunoassay) can be applied. This "bar-coding" technology can be used for gene profiling and high-throughput drug and disease screening.

In comparison with planar DNA chips, this encoded-bead technology is expected to be more flexible in target selection (e.g., adding new genes or single nucleotide mutations) (11), faster in binding kinetics (similar to that in homogeneous solution), and cheaper in production. However, a limitation of the spectral

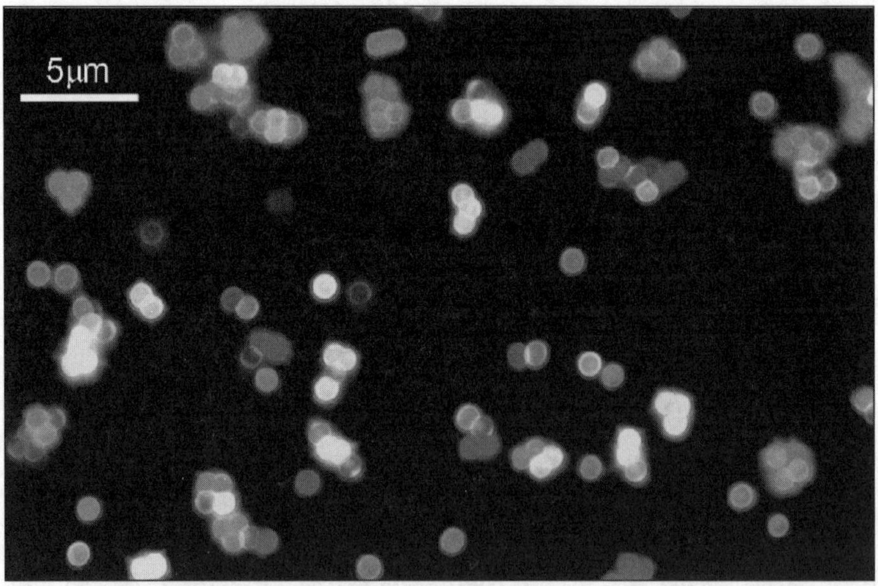

Figure 4. Fluorescence micrograph of a mixture of CdSe/ZnS QD-tagged beads emitting single-color signals at 484, 508, 547, 575, and 611 nm. The beads were spread and immobilized on a poly-lysine-coated glass slide that caused a slight clustering effect. (Reproduced with permission from Han, Gao, Su, and Nie, Nat. Biotechnol. 19, 631, 2001). **(See color plate A39.)**

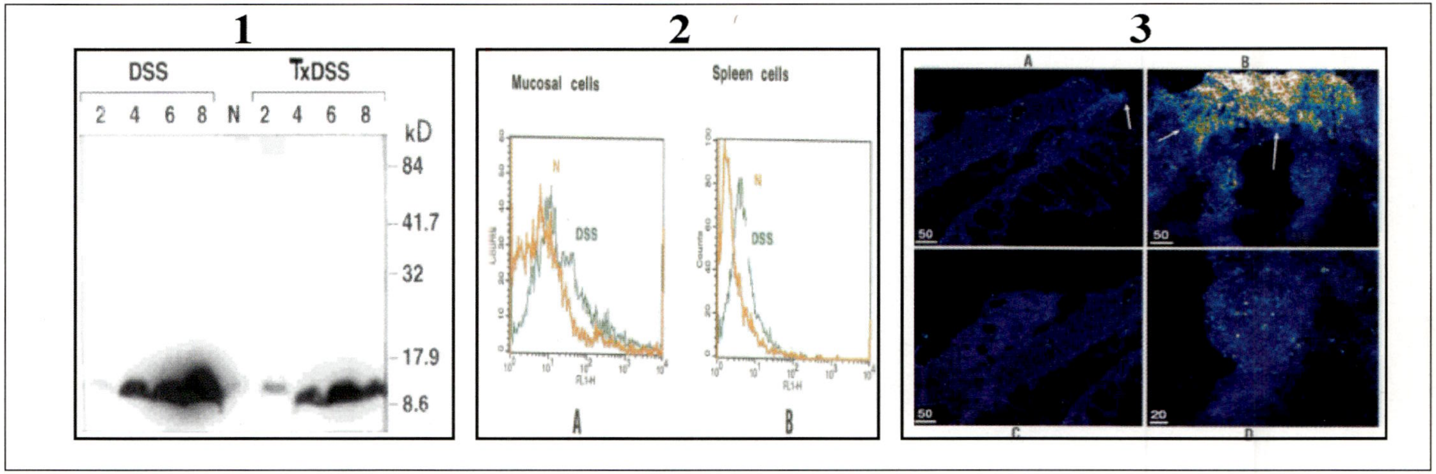

Chapter 4, Figure 5. Protein and mRNA localization of the CRG9/Ly-6A/E gene. 1. Western blot analysis. Protein extracts from mice experimental colitis tissue samples were separated on 12% PAGE, blotted to a PVDF membrane, probed with Ly-6A/E antibody, and the blot was developed by chemiluminescent method. 2. FACS® analysis. DSS-induced colitis samples: N, normal; A, mucosal cells; B, spleen cells. 3. Fluorescence in situ hybridization (FISH) analysis. (A) Normal tissue with antisense probe; (B) colitis tissue with antisense probe; (C) normal tissue with sense probe; and (D) colitis tissue with sense probe. The localized signals are shown with arrows.

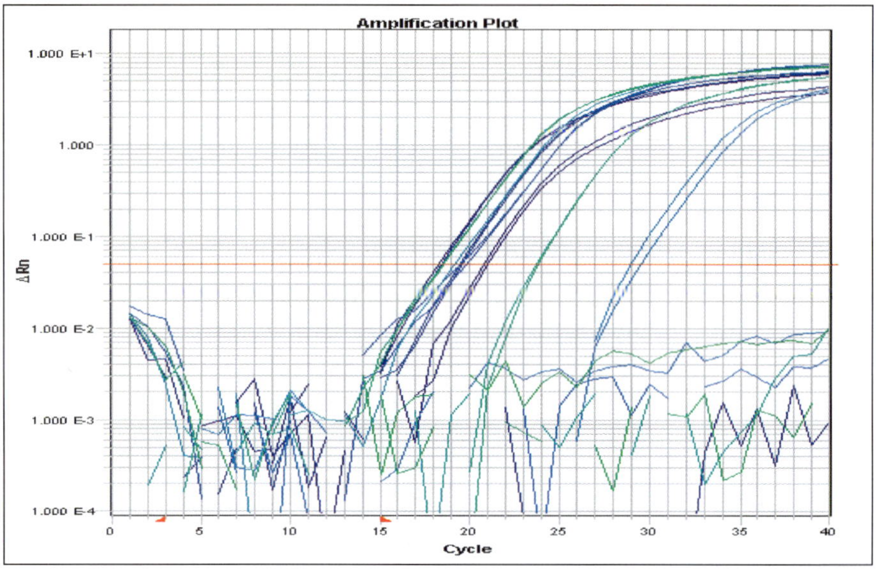

Chapter 6, Figure 1. Amplification plot generated by a real-time quantitative PCR assay using the ABI PRISM 7900. The ΔRn is calculated by dividing both the background emission intensity and the experimental emission intensity by the passive reference emission intensity and then subtracting the background value from the experimental value. The result is a logarithmic plot that depicts the amount of product at each PCR cycle.

	1A	1B	2A	2B	3A	3B	4A	4B	probe
1A	7333 100.0% in 1700bp	381 80.4% in 1340bp							
1B	629 80.4% in 1340bp	5631 100.0% in 1600bp							
2A			11621 100.0% in 1800bp	19 74.9% in 447bp					
2B			0 74.9% in 447bp	5165 100.0% in 1200bp					
3A					9522 100.0% in 900bp	30 68.7% in 560bp			
3B					0 68.7% in 560bp	10024 100.0% in 1800bp			
4A							7817 100.0% in 900bp	151 65.3% in 739bp	
4B							26 65.3% in 739bp	4449 100.0% in 900bp	
target									

Chapter 7, Figure 5. Cross-hybridization experiment. Clones 1A and 1B have 80.4% sequence similarity; 2A and 2B have 74.9%; 3A and 3B have 68.7%; and 4A and 4B have 65.3%. The length of each clone is shown in each box. The signal intensity (red) shows the intensity of self-hybridization.

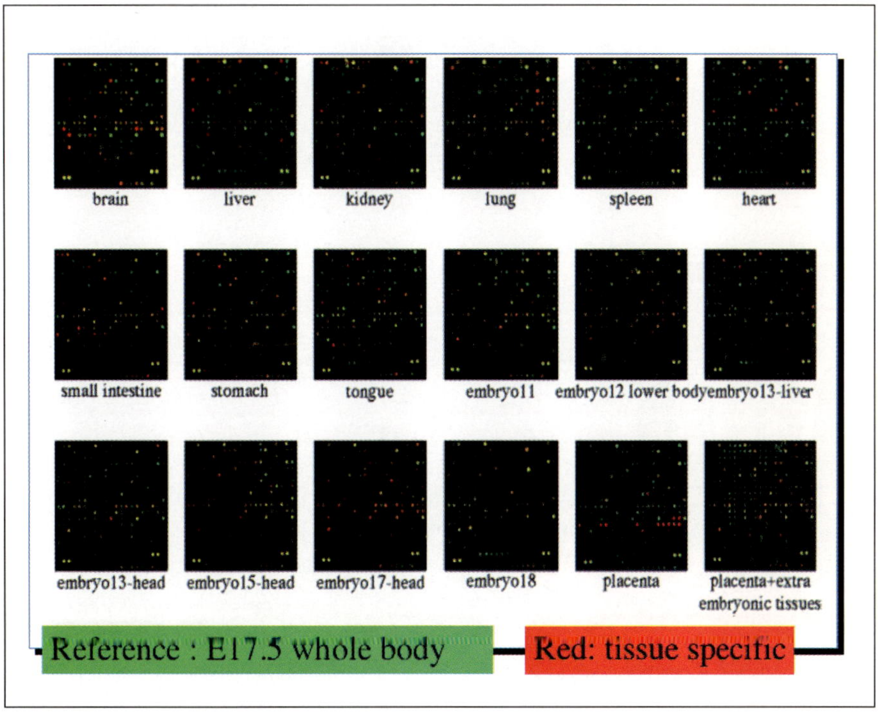

Chapter 7, Figure 6. False-color overlaid images of various tissues using RIKEN 19K mouse cDNA microarrays. The reference RNA for each hybridization was from a 17.5-day-old embryo. Only a part of the microarray (21 × 21 = 441 spots) is shown. A red spot shows that the genes are highly expressed in each tissue, and a green spot shows that the gene was expressed less in that tissue.

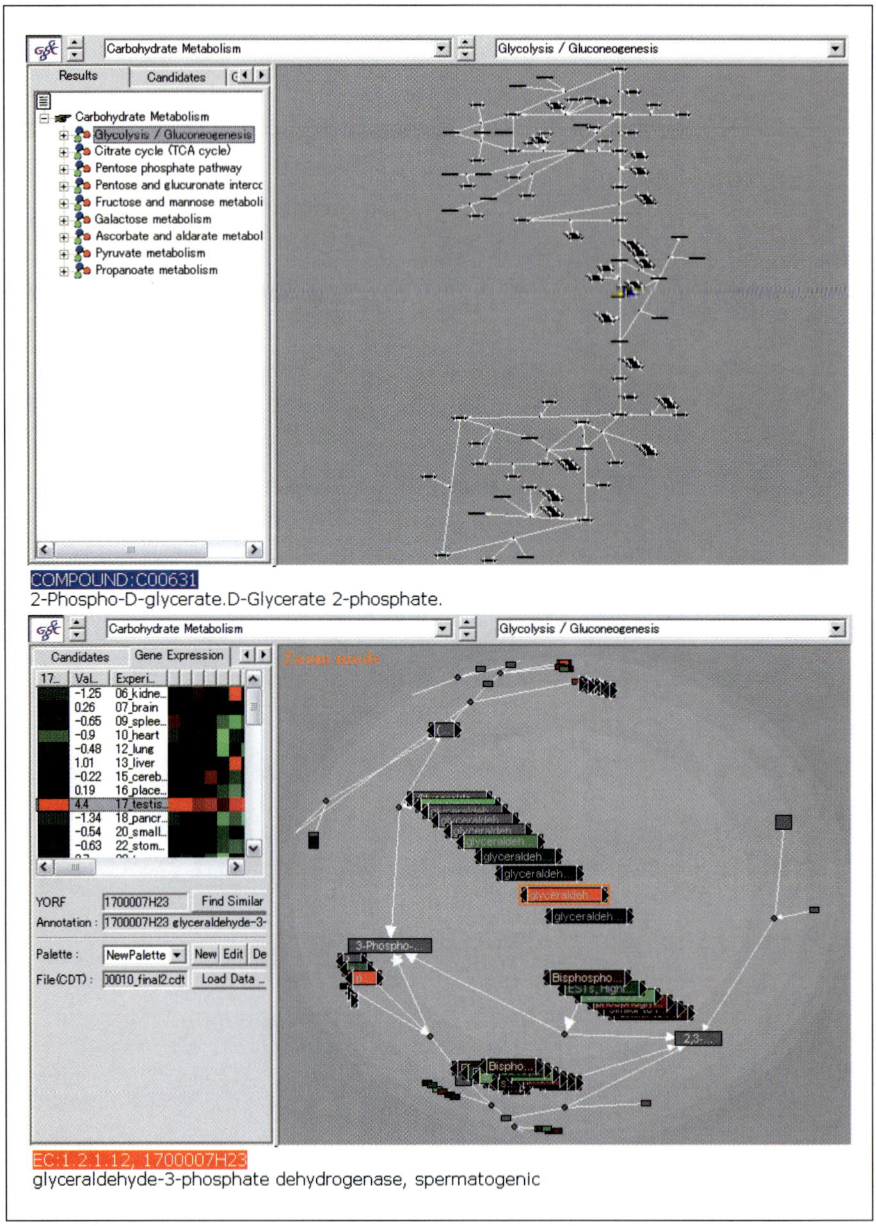

Chapter 7, Figure 8. Screen shots of GSCope interface. Upper panel is a bird's-eye view of GSCope. In this view, pathways are listed in the left column, and the metabolic cascade from glucose (glycolysis and TCA cycle) is shown on the right. Lower panel is a wide-angle view of GSCope. A part of the upper metabolic pathway is magnified, and boxes representing enzymes are colored. In the right window, each node represents an enzyme, and a testis-specific G3PDH is colored and surrounded with an orange frame. The corresponding expression profile in each tissue is shown in the left window. The second column shows the base-2 log ratio relative to the reference (E17.5 whole-embryo RNA). The first column shows the color code based on this log ratio. The third column shows the tissue used for the experiment. The other genes are represented in the fourth and subsequent columns.

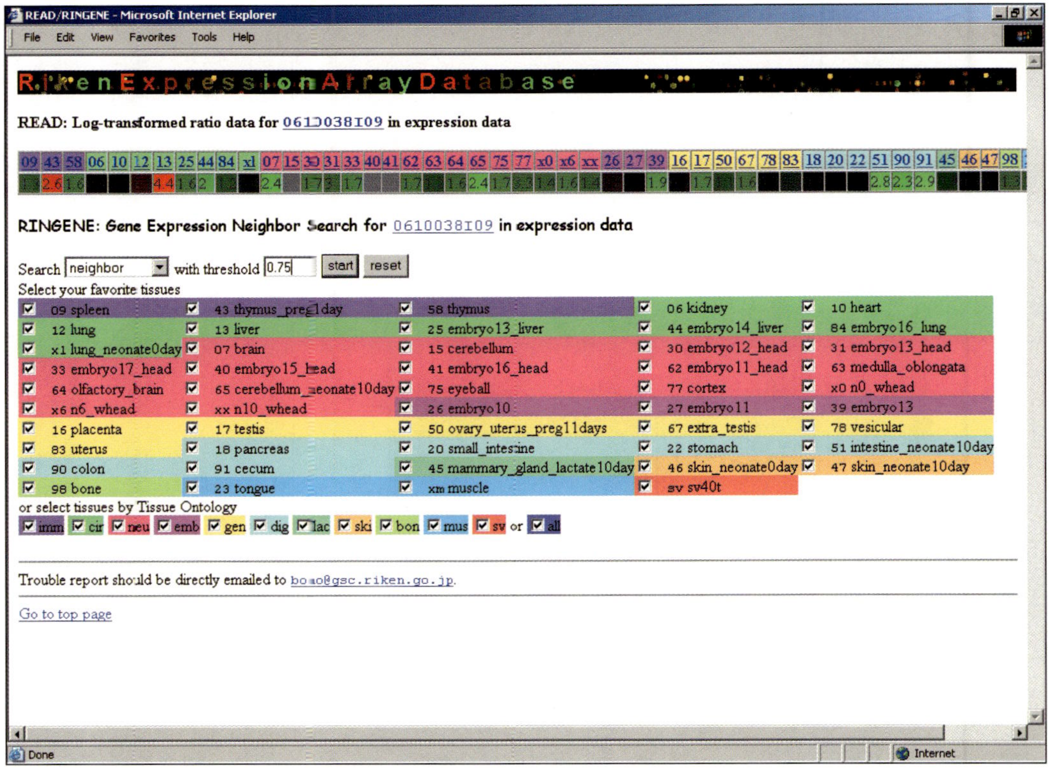

Chapter 7, Figure 10. A screenshot of ID search results from READ. The gene expression profiles for 49 adult and embryonic tissues are listed at the top of the figure. Anatomically similar tissues are clustered and colored the same (e.g., digestive organs such as the pancreas, stomach, etc., are blue and on the right side of the table). The lower part of the figure is a Web form for searching RINGENE. Users can check boxes to indicate which tissues they want to search for a similar expression pattern.

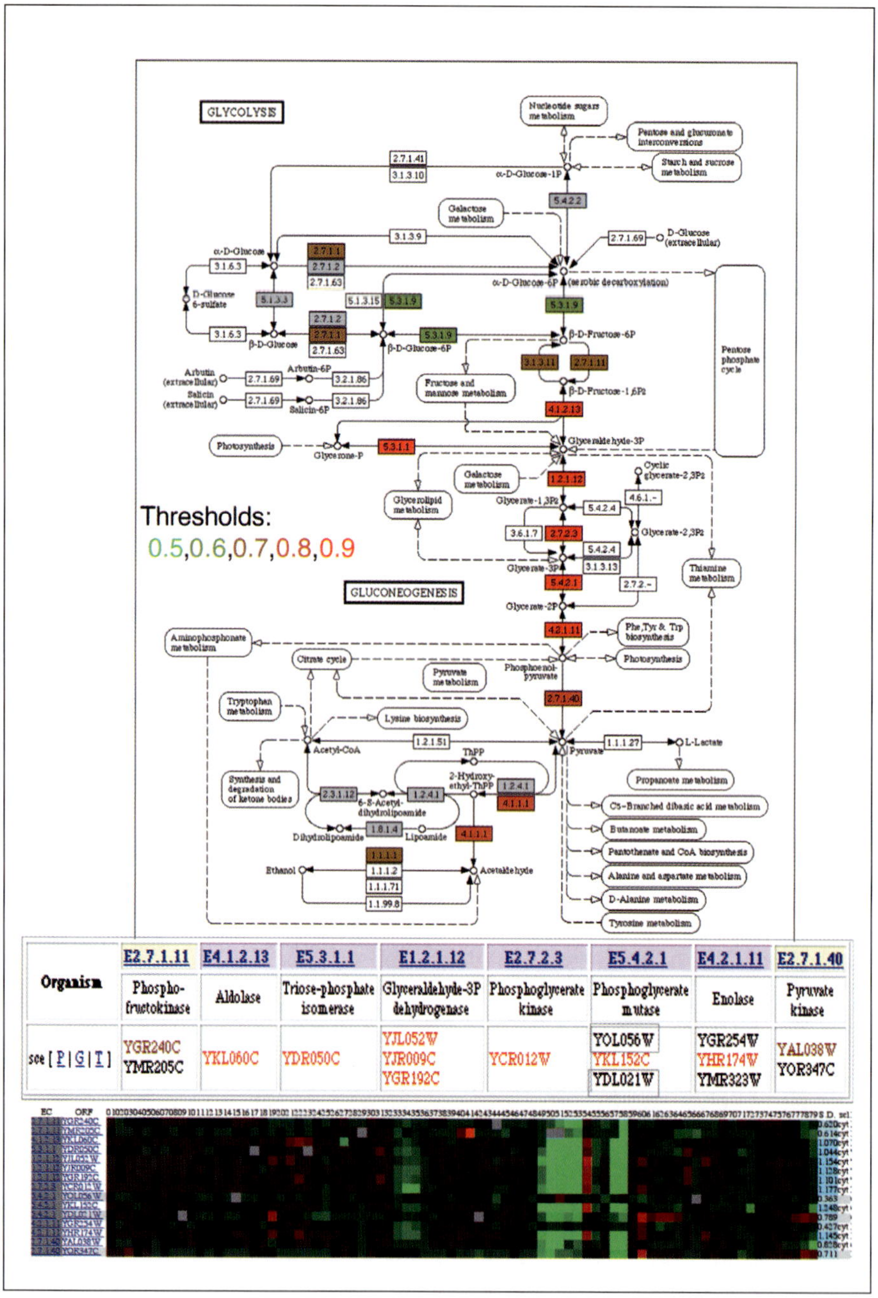

Organism	E2.7.1.11	E4.1.2.13	E5.3.1.1	E1.2.1.12	E2.7.2.3	E5.4.2.1	E4.2.1.11	E2.7.1.40
	Phospho-fructokinase	Aldolase	Triose-phosphate isomerase	Glyceraldehyde-3P dehydrogenase	Phosphoglycerate kinase	Phosphoglycerate mutase	Enolase	Pyruvate kinase
sce [P | G | T]	YGR240C YMR205C	YKL060C	YDR050C	YJL052W YJR009C YGR192C	YCR012W	YOL056W YKL152C YDL021W	YGR254W YHR174W YMR323W	YAL038W YOR347C

Chapter 7, Figure 12. The second largest cluster from UPGMA cluster analysis of the yeast data mapped to the glycolysis pathway (top). At the top of the figure and in the middle table, the relatedness in the gene expression pattern is represented in color (red for closely related genes). The gene expression pattern at the bottom (red for up-regulated and green for down-regulated) was drawn using a tool in the EPTable software (http://eptable.sourceforge.net/).

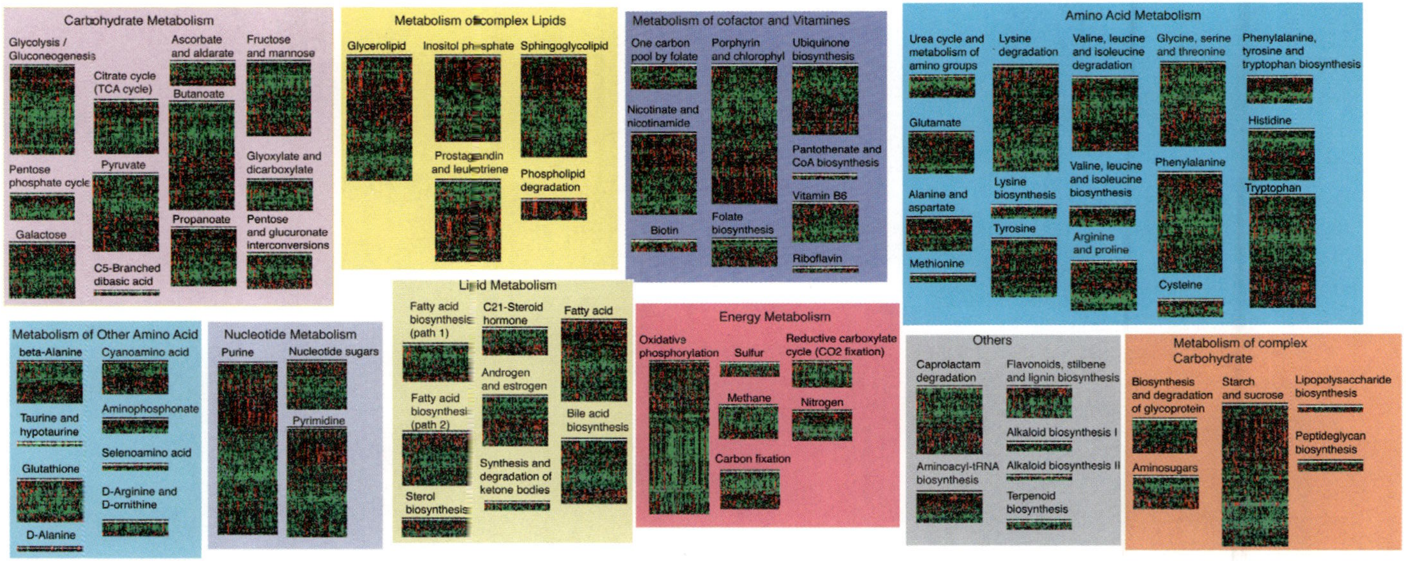

Chapter 7, Figure 16. Thumbnail images of expression profiles (in hierarchical tree view) of the clustered genes for all metabolic pathways (metabolome). The y-axis of each thumbnail represents clones, and the x-axis of each figure shows 49 tissues. Groups of functionally similar metabolic pathways are shaded with the same colors (metabolism of carbohydrates, red; amino acids, blue; and lipids, yellow). All the genes included in each pathway are listed, together with links to gene expression information, at (http://genome.gsc.riken.go.jp/READ/supplementary/allpathway/).

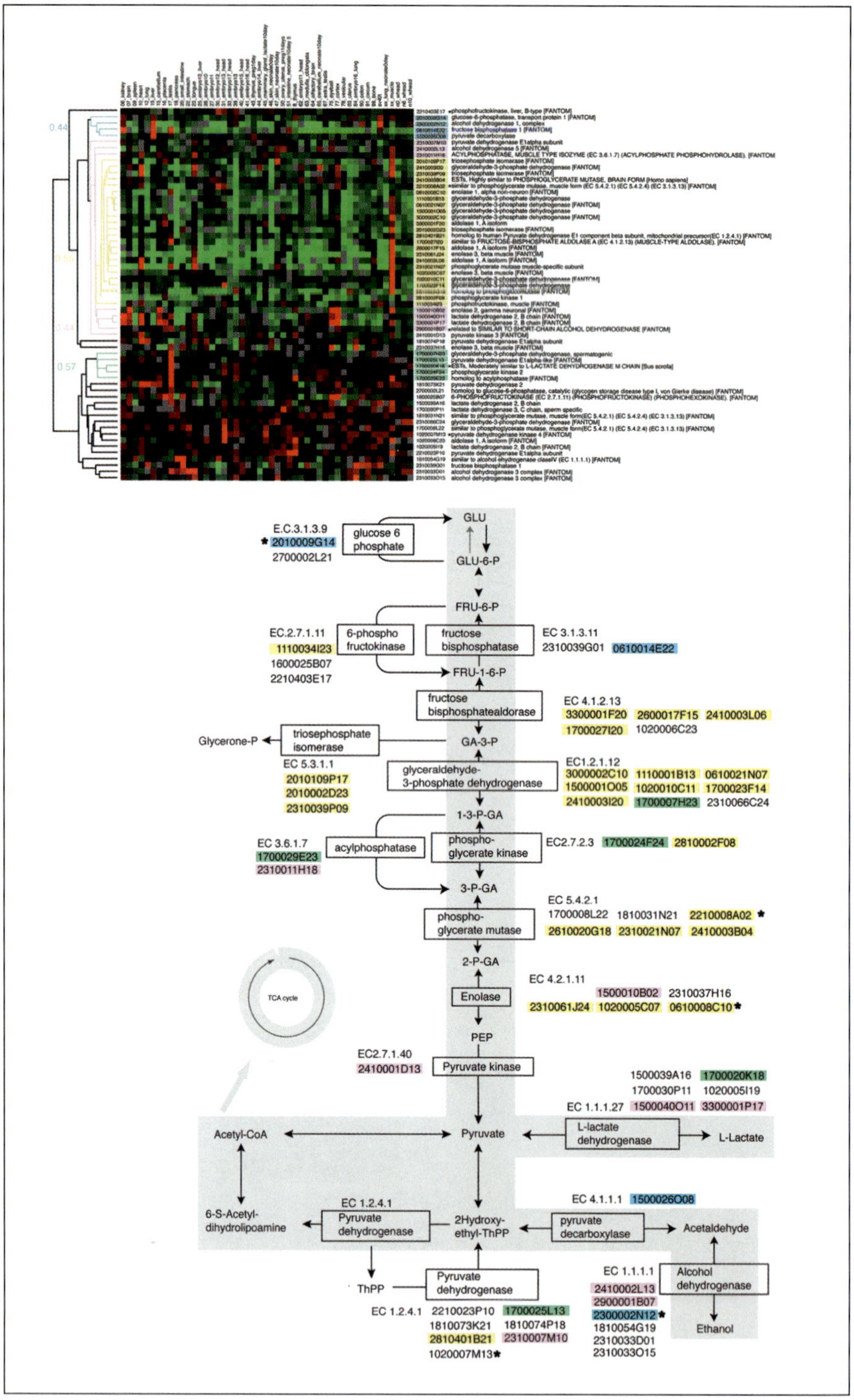

E.C.3.1.3.9
2010009G14
2700002L21
glucose 6 phosphate
GLU
GLU-6-P
FRU-6-P
EC.2.7.1.11
1110034I23
1600025B07
2210403E17
6-phospho fructokinase
fructose bisphosphatase
EC 3.1.3.11
2310039G01
0610014E22
FRU-1-6-P
fructose bisphosphatealdorase
EC 4.1.2.13
3300001F20
2600017F15
2410003L06
1700027I20
1020006C23
Glycerone-P
triosephosphate isomerase
GA-3-P
EC 5.3.1.1
2010109P17
2010002D23
2310039P09
glyceraldehyde-3-phosphate dehydrogenase
EC1.2.1.12
3000002C10
1110001B13
0610021N07
1500001O05
1020010C11
1700023F14
2410003I20
1700007H23
2310066C24
1-3-P-GA
EC 3.6.1.7
1700029E23
2310011H18
acylphosphatase
phospho-glycerate kinase
EC2.7.2.3
1700024F24
2810002F08
3-P-GA
EC 5.4.2.1
phospho-glycerate mutase
1700008L22
1810031N21
2210008A02
2610020G18
2310021N07
2410003B04
2-P-GA
TCA cycle
EC 4.2.1.11
Enolase
1500010B02
2310037H16
2310061J24
1020005C07
0610008C10
PEP
EC2.7.1.40
2410001D13
Pyruvate kinase
1500039A16
1700020K18
1700030P11
1020005I19
EC 1.1.1.27
1500040O11
3300001P17
Acetyl-CoA
Pyruvate
L-lactate dehydrogenase
L-Lactate
EC 4.1.1.1
1500026O08
6-S-Acetyl-dihydrolipoamine
EC 1.2.4.1
Pyruvate dehydrogenase
2Hydroxy-ethyl-ThPP
pyruvate decarboxylase
Acetaldehyde
ThPP
Pyruvate dehydrogenase
EC 1.1.1.1
Alcohol dehydrogenase
2410002L13
2900001B07
2300002N12
1810054G19
2310033D01
2310033O15
Ethanol
EC 1.2.4.1
2210023P10
1700025L13
1810073K21
1810074P18
2810401B21
2310007M10
1020007M13

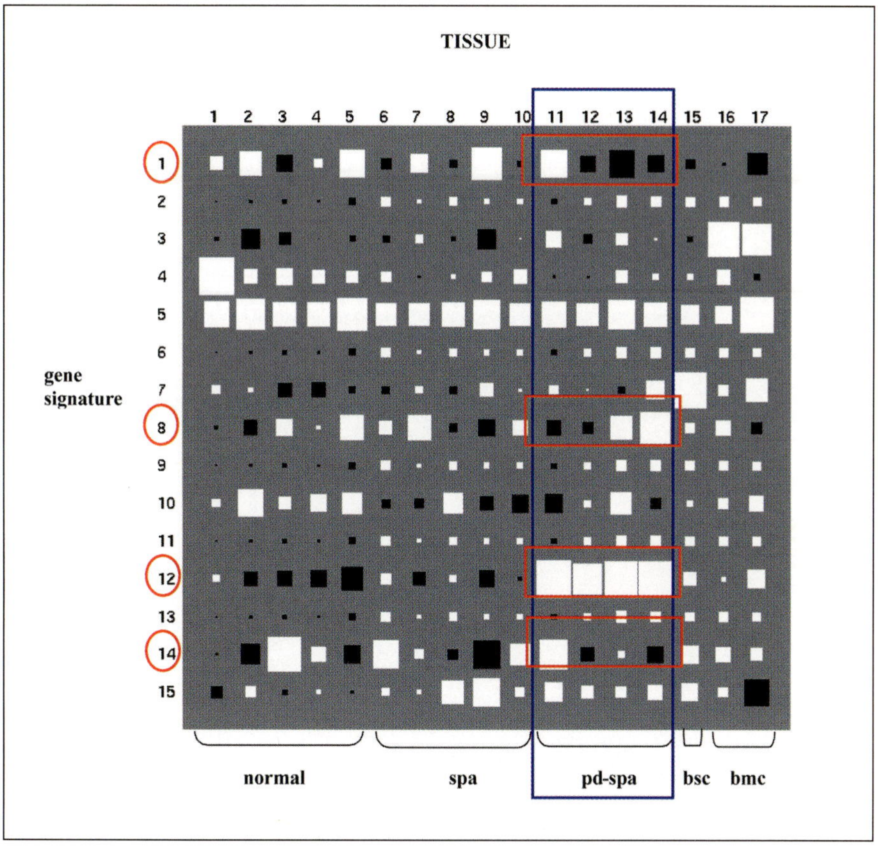

Chapter 8, Figure 3. ICA of tailored cDNA array data: unveiling gene expression patterns in ovarian hybridization profiles. Tissues 1–5 correspond to normal ovary; tissues 6–10 to spa; tissues 11–14 to pd-spa; tissue 15 to bsc; and tissues 16 and 17 correspond to samples of ovarian bmc. Several gene expression profiles (gene signatures) were unveiled by the ica run and are represented at each row. Gene signatures 3, 7, and 12 are seen to be distinctively prominent in bmc, bsc, and pd-spa samples, respectively. Intra-tissue type differences could then be assessed, based on a number of further defined gene signatures (see Figure 5) (101).

Chapter 7, Figure 17. Hierarchical tree view of the clustered genes involved in the glycolysis pathway (upper panel). The shaded colors in cloneids correspond to the color labels in the bottom pathway map and correspond to genes that have high correlations with each other. Genes colored yellow, blue, and green have correlation coefficient values of 0.55, 0.44, and 0.57, respectively. The yellow genes (high correlation coefficient) were mapped close to each other in the pathway. The pink genes clustered with a lower coefficient of correlation and mapped outside the core pathway are shown in yellow. The asterisk indicates that the functional annotations for these genes were inherited from the FANTOM database.

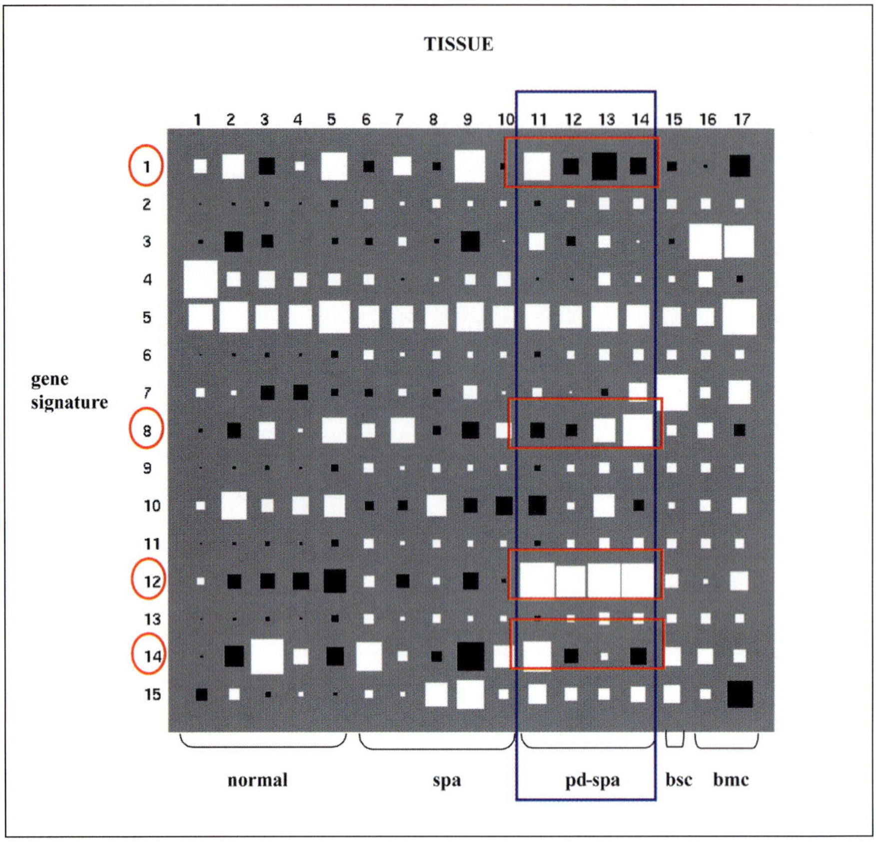

Chapter 8, Figure 5. Assessing the significance of gene signatures: tissue-type-specific profiles versus other gene expression patterns. ICA of ovarian tailored cDNA array results revealed tissue-type-specific gene signatures and tissue type-independent gene signatures present in varying amounts among the ovarian samples surveyed (red circles). For example, gene signature 12 can be observed consistently predominant in the pd-spa sample group (delineated in blue), while inherent variations within the histologically similar samples (i.e., pd-spa) can be observed at gene signatures 1, 8, and 14. Molecular analysis of tissue samples in similar multi-parameter schemes may help to classify tissues based on several underlying disease processes. This could help to identify patients who would benefit from different treatment strategies (101).

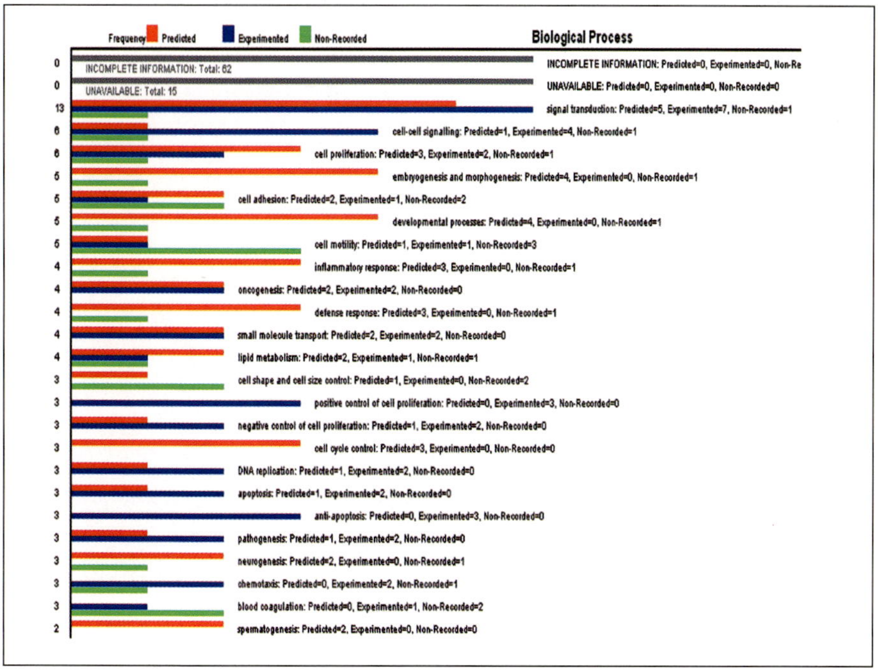

Chapter 9, Figure 1. Graphical representation of the biological role of differentially expressed genes in ovarian cancer cells. The figure is generated with Onto-Express v. 1.

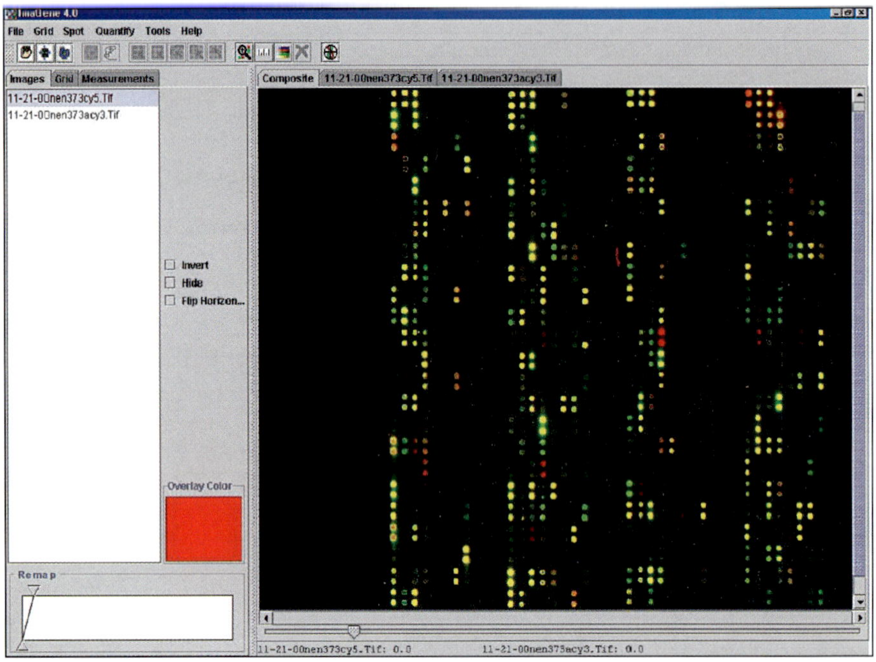

Chapter 9, Figure 2. Quantitation of differential hybridization signals between normal ovarian surface epithelial cells (green) and ovarian cancer cells (red) from a homemade ovarian-specific cDNA microarray with 768 genes printed in duplicate. Analysis was performed with software Imagene v. 4.1. Yellow indicates almost equal signals from both samples. In this analysis, we identified over 40 differentially expressed genes. One microgram of total RNA from tumor or HOSE cells was labeled with the MICROMAX™ TSA system.

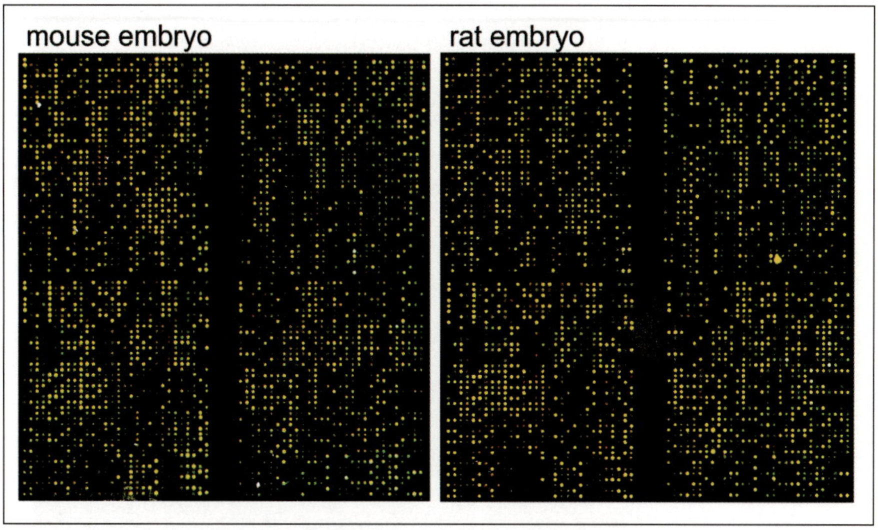

Chapter 10. Figure 2. Quality-control assessment by hybridization of mouse and rat embryonic RNA to human probe. RNA procured from the headfold of normal mouse and rat embryos at the 4–8 somite stage of gestation was labeled for two-channel competitive hybridization to the 4800-element MICROMAX MPS621 probe, which has duplicated spots for 2400 genes, including positive (spiked RNA) and negative (unspiked) plant control genes. Each image represents a red-green channel overlay for the intraspecies comparison of normal embryos. High-fidelity hybridization is clearly reflected in: (*i*) sensitivity, shown in the high percentage of spots detected in early embryos; (*ii*) precision, reflected in the degree of similarity between overlay images of mouse-mouse to human probe (left panel) and rat-rat to human probe (right panel); (*iii*) dynamic range, reflected in the signals for duplicate elements in the array; and (*iv*) the consistency between laboratories. The RNA samples from mice were collected and processed at Thomas Jefferson University (Philadelphia, PA), whereas the RNA samples from rats were collected by Dr. C. Lau and Dr. S. Hunter at the National Health and Environmental Effects Research Laboratory, EPA (Research Triangle Park, NC), and shipped to Philadelphia for subsequent processing.

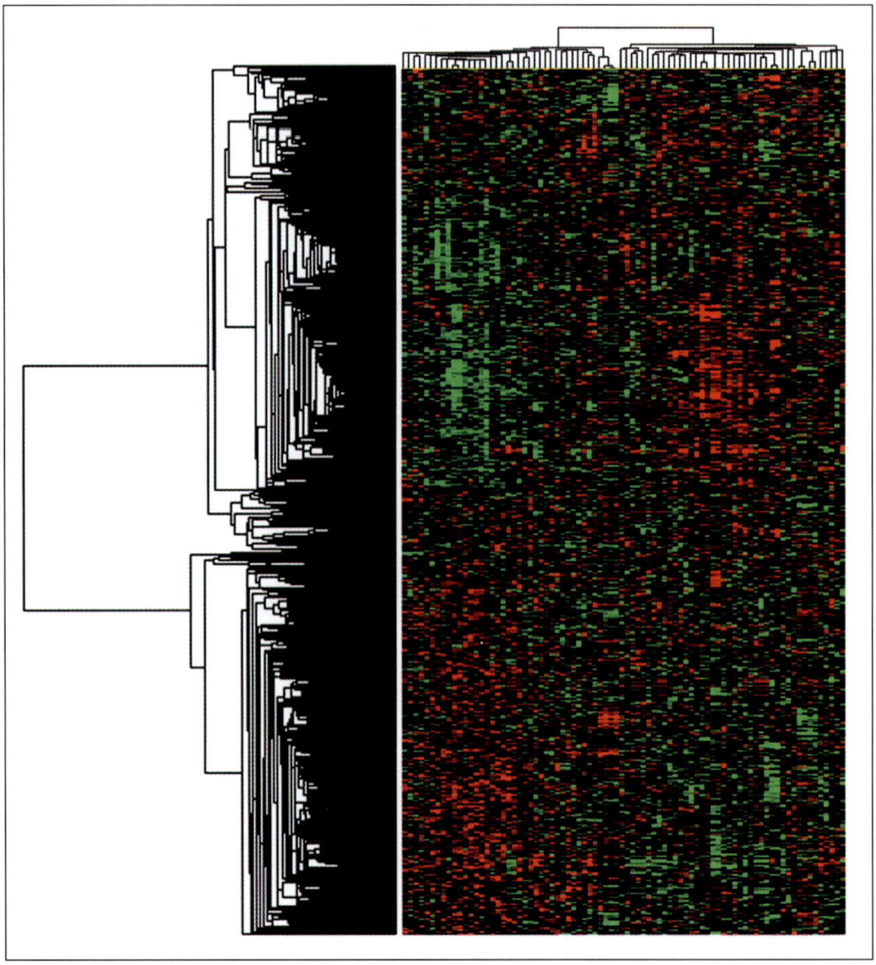

Chapter 10. Figure 5. Hierarchical clustering of genes filtered from the DTME data set. Filtering the 2400 genes on the MPS621 microchip by the concordance error model across 40 replicated experiments in the early mouse embryo produced a list of 941 genes that passed at 1.5-fold change in 15% of the replicated samples. Ratiometric gene expression data were subjected to lowess smoothing, scaling, hierarchical clustering, and visualization. Horizontal taxa in the TreeView dendritogram are samples (k = 80), and the vertical relationships are genes (p = 2400). Color-coding: red, up-regulated; green, down-regulated; and black for no change. The intensity of the color indicates the amplitude of change.

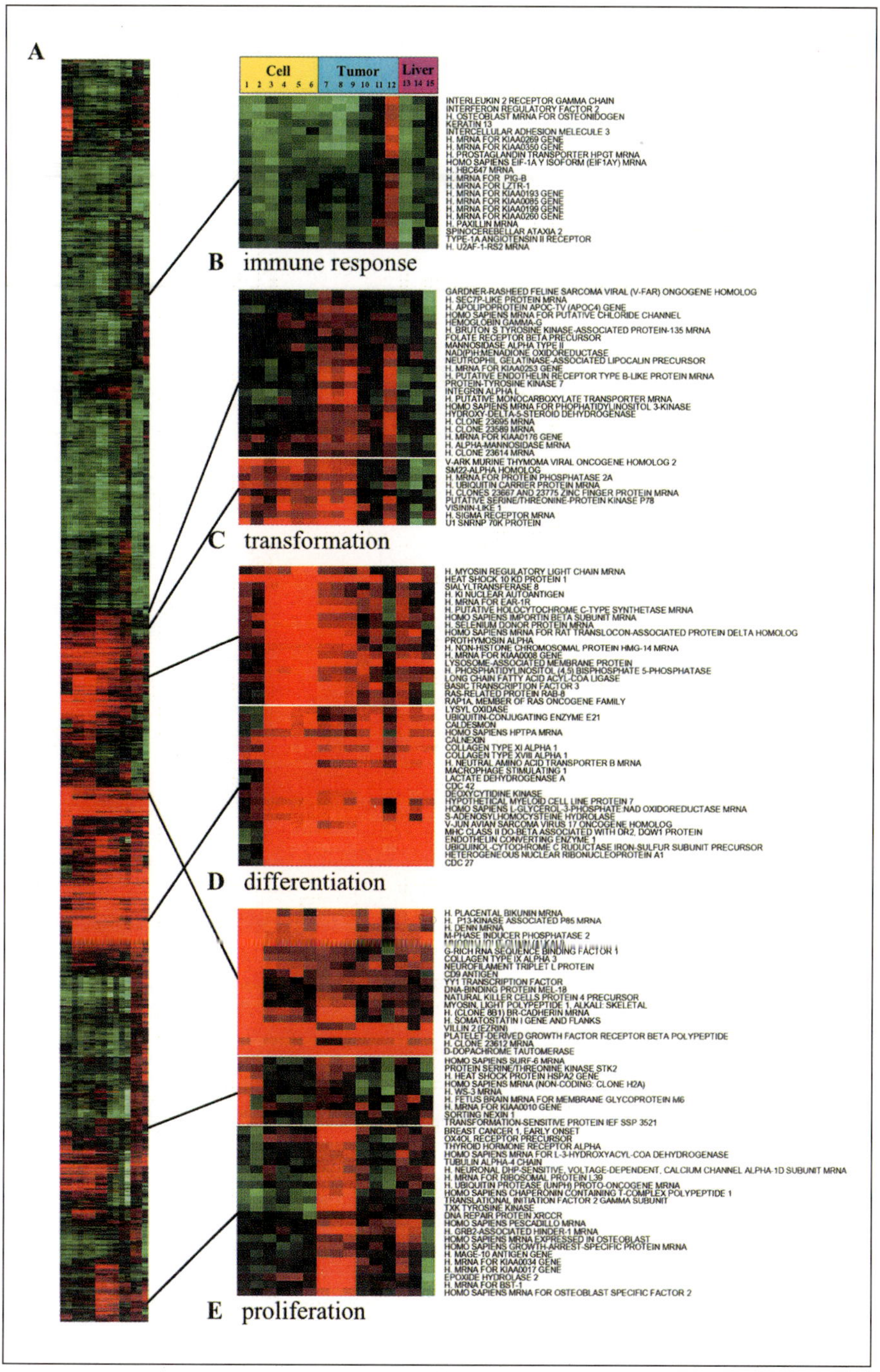

Chapter 11, Figure 8. Molecular portrait of the reverted phenotype of PC3 M cells and tumors treated with antisense RIα. See Chapter for details.

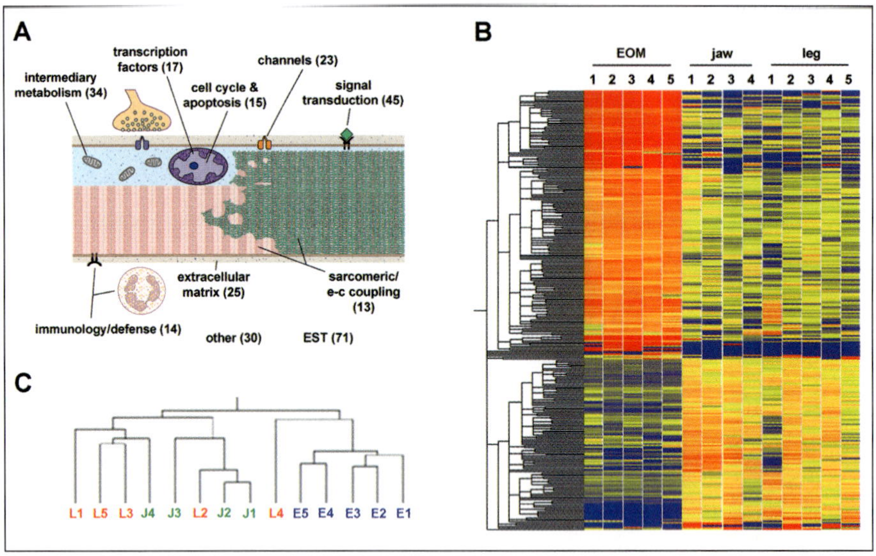

Chapter 12, Figure 4. Gene expression profiling of extraocular muscle in comparison with jaw (masseter) and hindlimb (combined gastrocnemius/soleus) muscles in mouse. [Data replotted from Porter et al. (45)] (A) Muscle fiber schematic showing the number of transcripts by gene class that are differentially regulated in extraocular muscles versus other skeletal muscles. Note the breadth of the functional class differences. (B) Hierarchical cluster analysis showing normalized replicate signal values for the 287 genes that are differentially regulated in extraocular muscle (EOM) versus jaw or leg muscles (n = 4–5 replicates/muscle group; red denotes high expression and blue denotes low expression). Note the similarities between the jaw and leg muscles and the differences in both from the extraocular muscle. Also note that the major differences between extraocular muscles and other muscle groups lie predominately in the level of gene expression, as opposed to genes that are uniquely expressed in eye muscle. (C) Dendogram showing the distribution of skeletal muscle samples when compared across all differentially regulated genes shown in panel B. Extraocular muscle replicates cluster together and separate from jaw and leg replicates, while the jaw and leg replicates are intermixed. E, extraocular muscle; J, jaw; and L, leg.

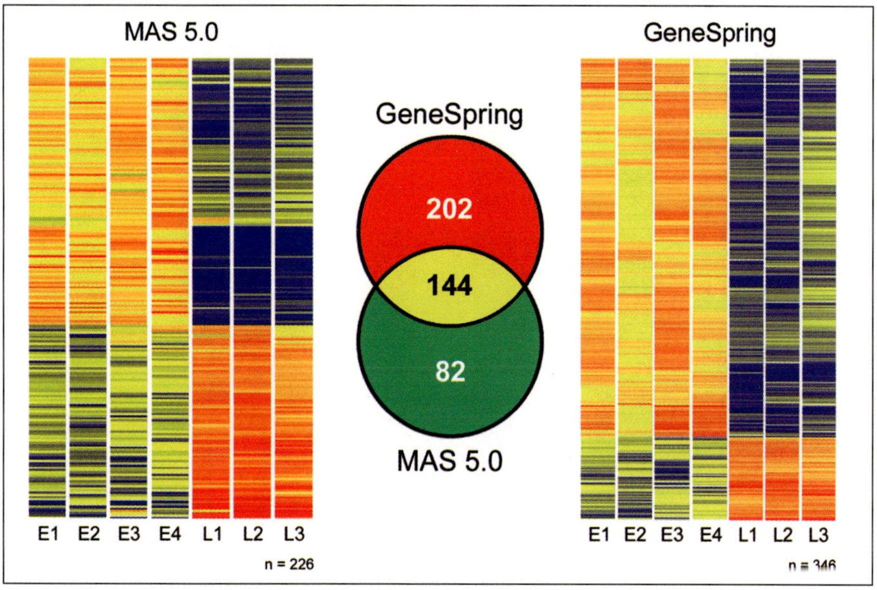

Chapter 12, Figure 6. Comparison of differential results obtained from the same sample set (rat extraocular and leg muscles run on U34A chip) with Affymetrix Microarray Suite 5.0 and GeneSpring 4.2.1. Selection criteria used with both programs are given in text. Note the higher number of differentially regulated genes determined by GeneSpring (right panel) versus Microarray Suite (left panel). The Venn diagram in the center indicates that, while there is substantial overlap between techniques, a large number of genes are detected as differentially regulated by only one of the two programs. MAS, Microarray Suite; E, extraocular; and L, leg.

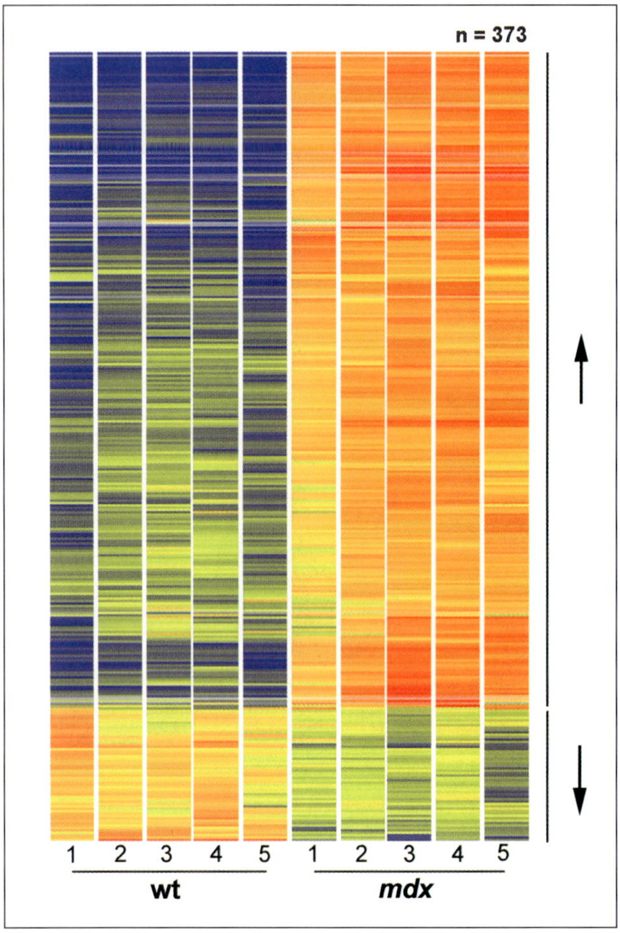

Chapter 12, Figure 7. Hierarchical cluster analysis showing differentially regulated genes in comparisons between wild-type and dystrophic mouse hindlimb musculature at eight weeks of age. [Data replotted from Porter et al. (46)]. Data were initially processed with Affymetrix Microarray Suite 5.0, filtering for those genes with both consistent I or D calls across all five replicate comparisons and a ≥2-fold average change. Cluster analysis was then performed on this subset using GeneSpring. The majority of differential regulated genes showed increased expression (indicated by the up arrow; down-regulated transcripts are denoted by the down arrow) in dystrophic leg muscle. wt, wild-type; *mdx*, dystrophic mouse.

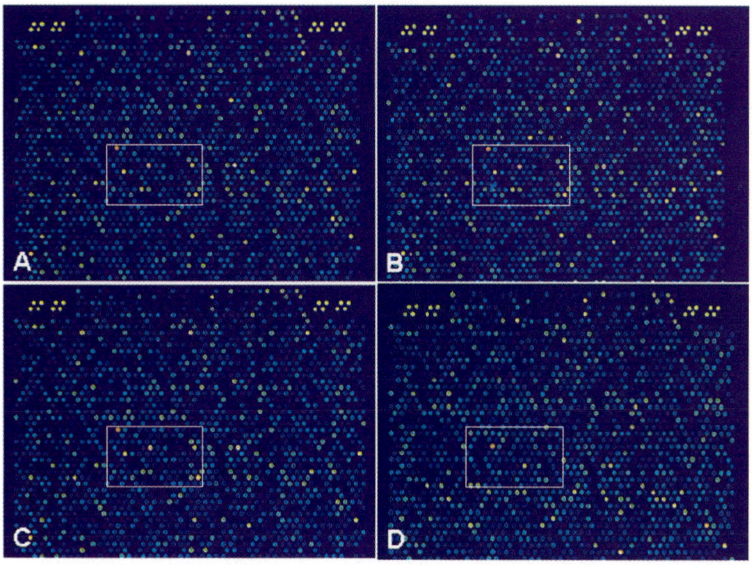

Chapter 14, Figure 3. CodeLink UniSet Human I Bioarray Images. Each bioarray image results from an individual hybridization. Each sample was run on triplicate arrays, and all the images were scanned on a Series A scanner. The images shown are the top section of each array (approximately 20% of the total array). The images (A–C) represent each of the triplicate arrays from a nontumor sample, and panel D represents one of the triplicate arrays from the corresponding patient's tumor sample. The white boxes in each panel indicate identical regions of the four arrays, where remarkable similarities in gene expression are seen in panels A–C, and distinct differences are noted between the tumor (D) and the three nontumor samples.

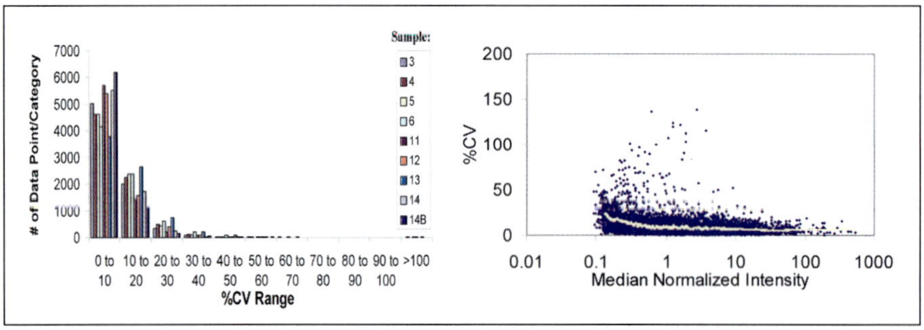

Chapter 14, Figure 4. Array-to-array reproducibility. (A) The CV distribution chart displays the frequency of the CV values for all samples. The distribution is skewed toward the lower range of numbers where the majority of the CVs are located. (B) The CV was plotted against the average median normalized intensity for sample 13. The yellow trendline represents a 100-point moving average of the CVs. The 100-point moving average demonstrates the consistency of the platform throughout all ranges of intensities.

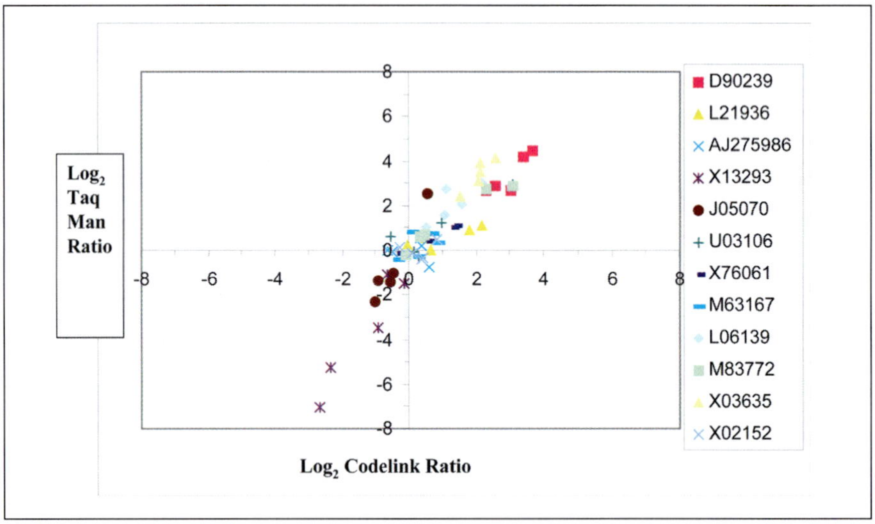

Chapter 14, Figure 5. Log$_2$ ratio scatter plot comparing expression results from the CodeLink platform to results obtained by real-time quantitative PCR (TaqMan). Twelve genes were compared on both platforms, with a strong correlation between the two, with an r^2 value of 0.8957. The GenBank accession numers are shown. Reverse transcription and PCR were conducted in one step using the TaqMan PCR Reagent Kit and gene-specific primers. Data were collected and analyzed using the TaqMan 7700 Sequence Detector (Applied Biosystems). Cyclophilin served as the endogenous control in these sets of experiments for the normalization of target mRNA results.

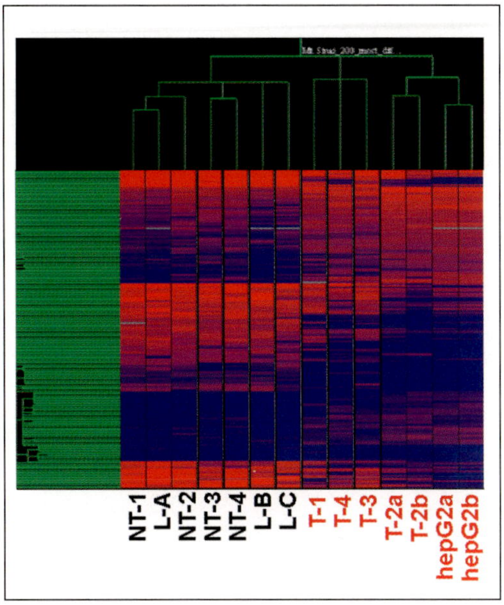

Chapter 14, Figure 6. Clustering of the 200 most differentially expressed genes. Each sample is represented by the average of the replicates. Arrays were clustered with Smooth Pearson correlation coefficients, and the genes were clustered with distance algorithms by GeneSpring software. Up-regulated genes are shown in red, and down-regulated genes are shown in blue. In addition, normal liver and HepG2 samples from a previously performed experiment were included in this analysis. The normal liver samples clustered with the nontumor samples (black text), and the HepG2 samples clustered with the tumor samples (red text).

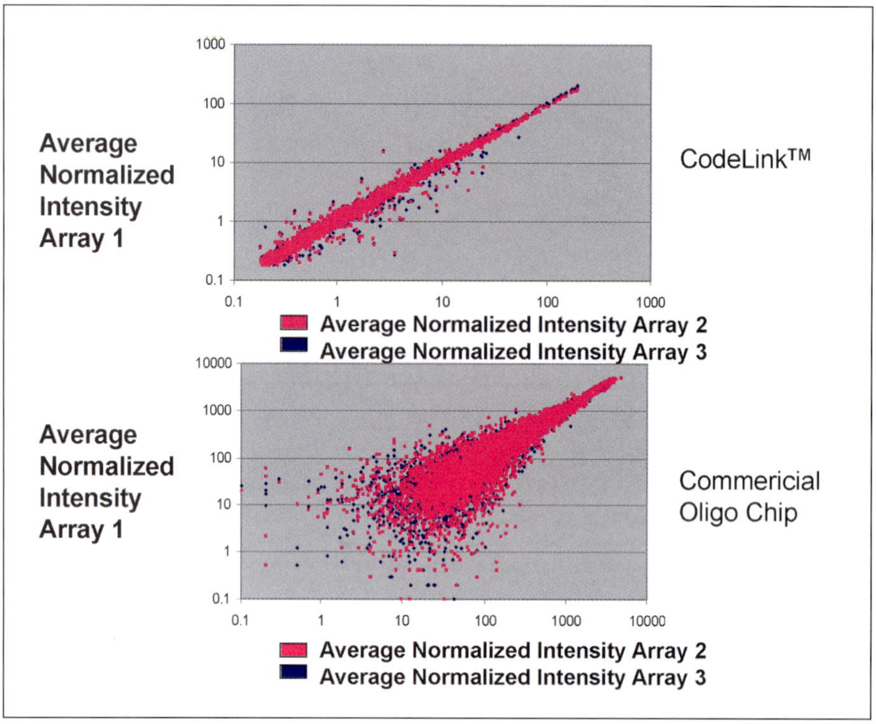

Chapter 14, Figure 7. Reproducibility of replicate arrays, CodeLink versus commercial oligonucleotide chip. cRNA synthesized from Universal Human Reference RNA (Stratagene) was labeled with biotin, according to each chip manufacturer's instructions, in three separate reactions for each platform. Each target preparation was hybridized to a single chip, the data were collected with OEM software, and the array-to-array reproducibility was assessed by comparing the scatter in the data. The closer the expression data points are to the line of identity, the more reproducible the platform becomes.

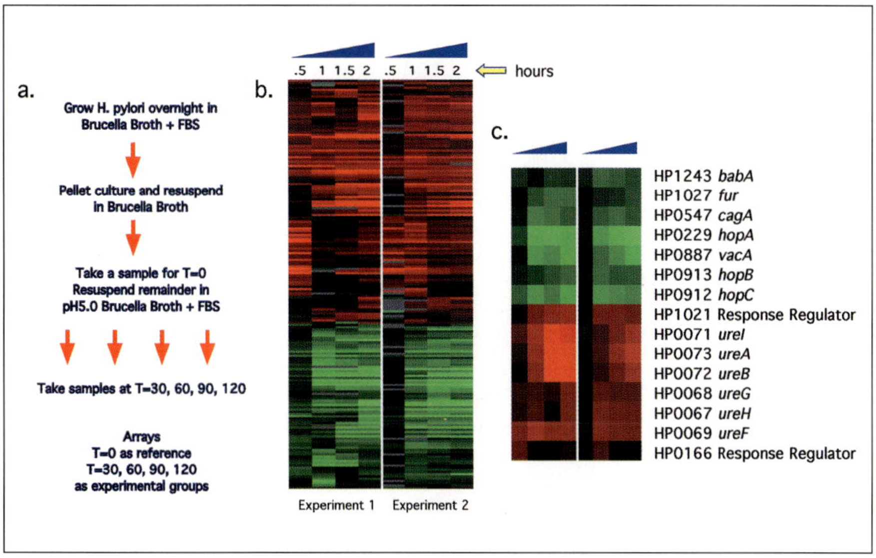

Chapter 15, Figure 2. Low pH regulated genes of *H. pylori.* (a) Schematic depiction of experimental design by which genes affected by a shift to low pH were determined. (b) Hierarchical cluster of all 183 genes from two independent experiments that passed the filtering criteria as explained in the text. Red indicates increased expression, green represents decreased expression, and gray indicates spots for which data were not reliable. The progression of time is indicated by the arrow located above each experimental set. (c) Enlargement of a subset of the genes contained within panel b. Gene designations and names are indicated to the right of the corresponding cluster diagram.

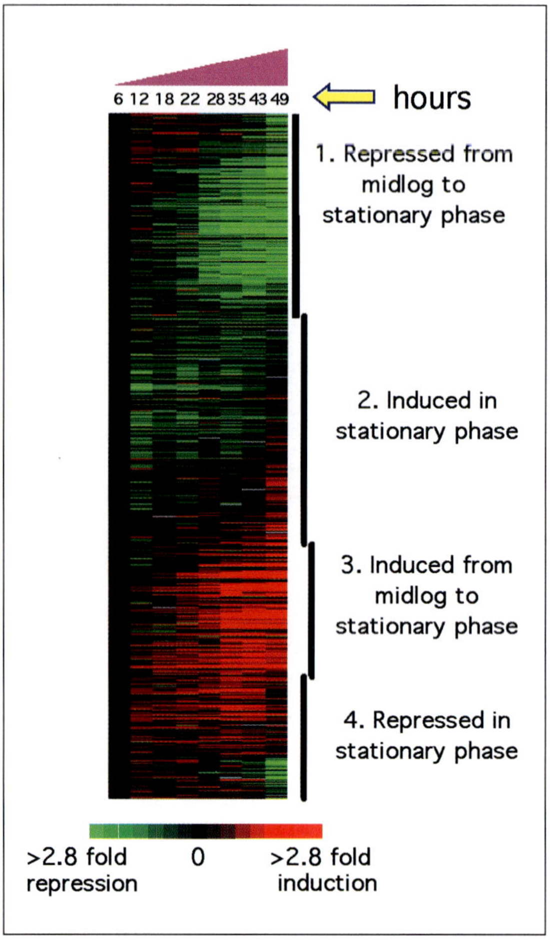

Chapter 15, Figure 3. Growth-phase-dependent gene expression of *H. pylori* strain SS1. Hierarchical cluster of self organizing maps that show the temporal dependence of gene expression changes during the time course for all genes that passed the filtering criteria described within the text. The major classes of expression patterns are indicated, and the progression of time is indicated by the triangle above the figure.

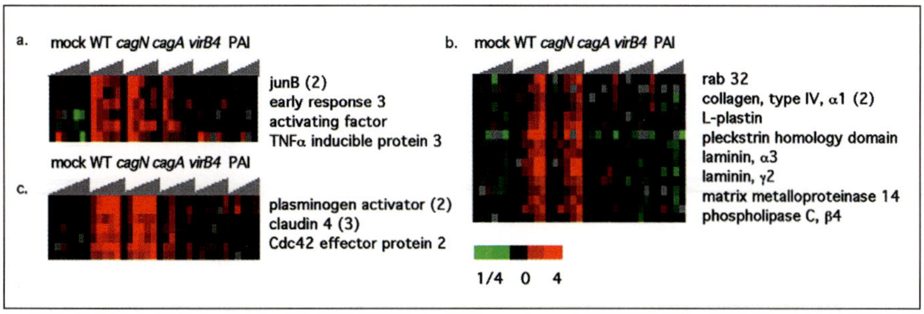

Chapter 15, Figure 4. AGS transcriptome analysis. Gene expression during infection time courses with wild-type *H. pylori* and isogenic mutants lacking individual PAI genes or the entire PAI as indicated. Each row represents an individual gene, and each column represents an individual condition, with increasing time during the infection (0, 1, 3, 6, 12, and 24 h) represented by the width of the gray triangle. The fold induction relative to the zero hour time point is indicated colorimetrically according to the scale bar. Three groups of co-regulated genes are shown (panels a-c), with representative gene names indicated.

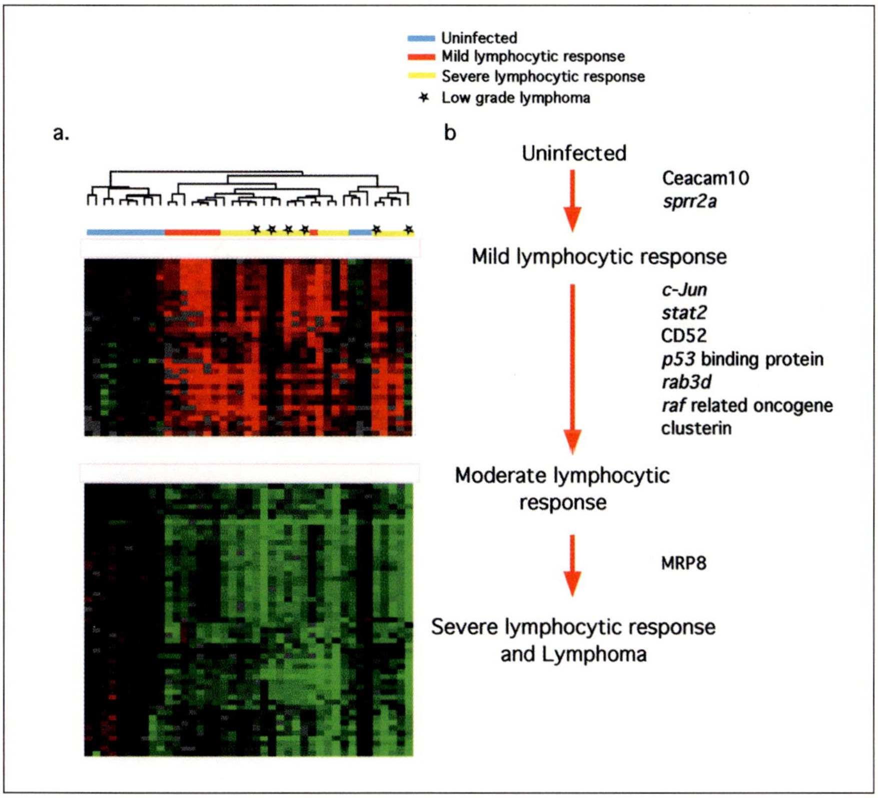

Chapter 15, Figure 5. Unsupervised hierarchical clustering of 41 *H. heilmannii* infected and uninfected murine arrays. Infection and histopathology grading are indicated below the dendogram. Distinct nodes of infected and uninfected are seen as, with few exceptions, arrays from uninfected animals cluster tightly together and in a separate node from arrays from infected animals. (b) Schematic depiction of basis for SAM of the arrays shown in panel a. A selection of representative induced genes is shown to the right.

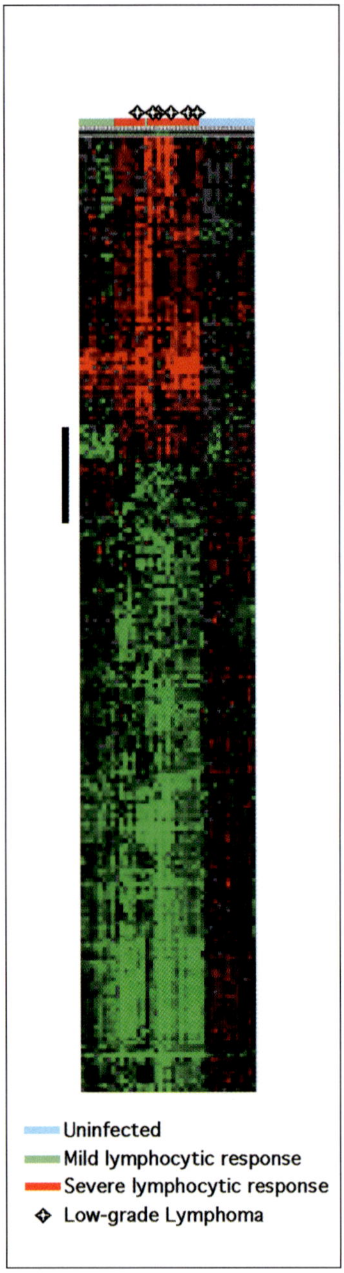

Chapter 15, Figure 6. Supervised hierarchical clustering of 41 *H. heilmannii* infected and uninfected murine arrays. A paired Student's *t* test was used to calculate p values for every gene contained on the murine array. Comparisons were made only between infected animals, and p values were determined for genes between animals showing mild lymphocytic responses and animals showing severe lymphocytic responses. The 280 genes with p values less than 0.001 were carried forward for cluster analysis. The resulting dendogram tightly clusters uninfected animals into their own node and further segregates the low-grade and high-grade pathology animals into distinct nodes. Potentially predictive clusters where animals exhibiting low-grade and high-grade pathology differ significantly from one another are marked by a vertical black bar on the left side of the figure.

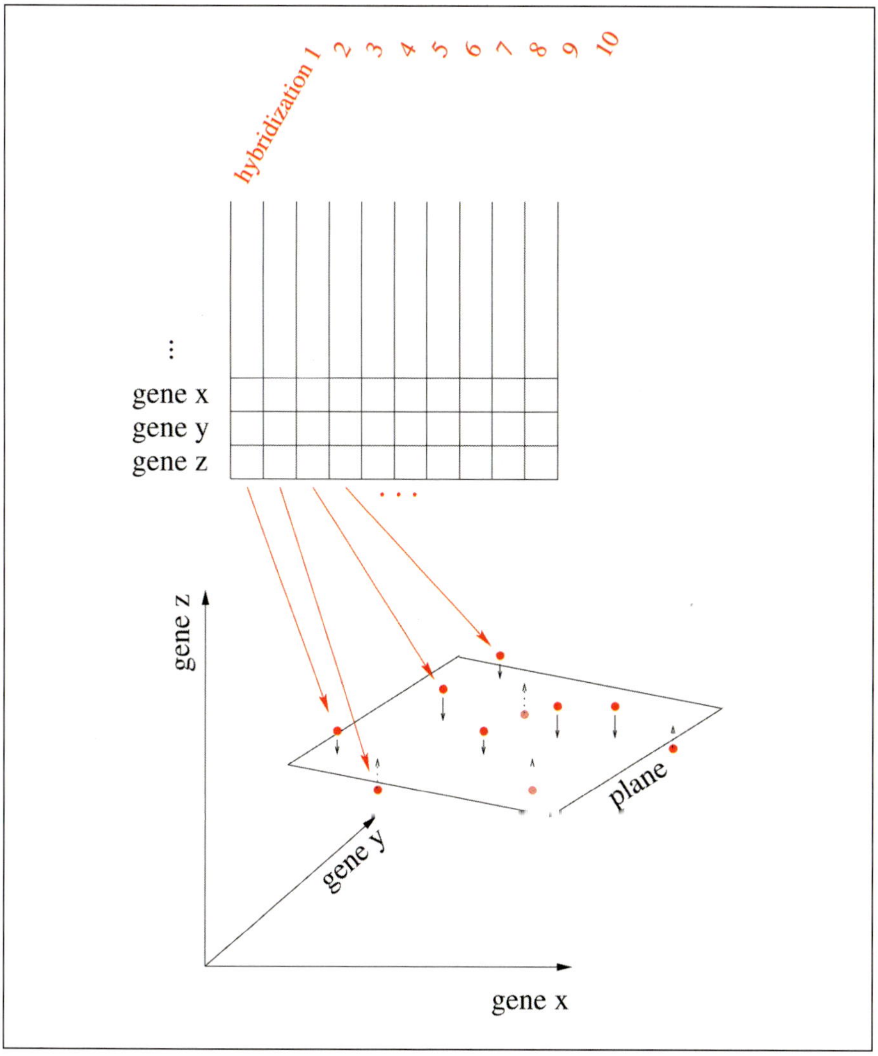

Chapter 16, Figure 1. Planar embedding. The *m* columns of a table of *n* genes × *m* hybridizations are represented in *n*-dimensional gene space (three dimensions are shown). *n* ranges from a few hundred to tenths of thousands. Most microarrays comprise several thousand elements. A plane is selected such that the distance of the hybridization vectors to the plane is minimal, thus conserving point-to-point distances among these vector points as well as possible.

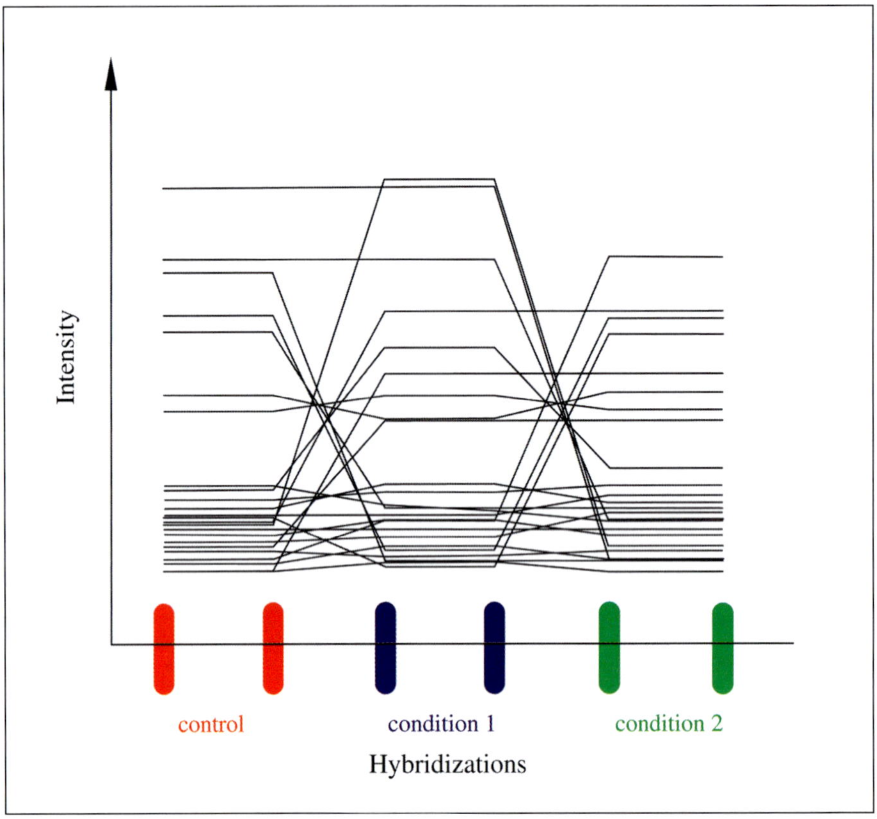

Chapter 16, Figure 3. Virtual data example. To demonstrate the interpretation of a CA biplot, we constructed a data example. It resembles real data in that the majority of the genes are lowly or not transcribed to a measurable amount. It comprises only 24 genes and differs from the real world in perfect reproducibility among the two hybridizations of each experimental condition.

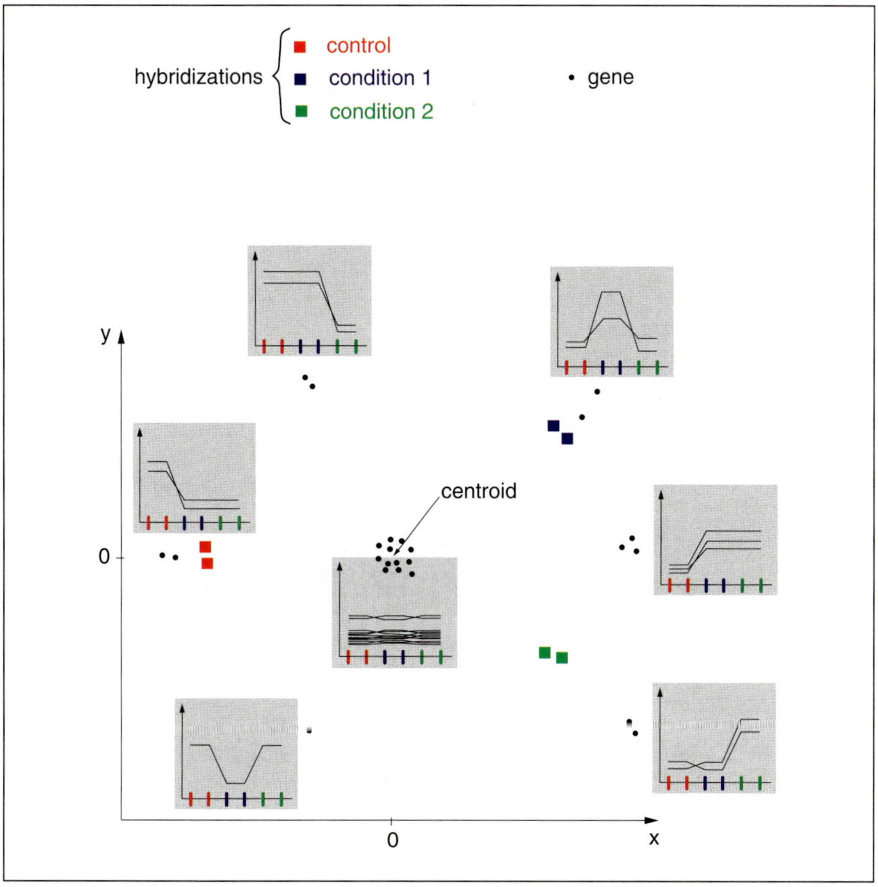

Chapter 16, Figure 4. Planar projection by CA. The plot resembles output that can be expected for the virtual data example shown in Figure 3. It was constructed to demonstrate the properties of such projections more clearly than would be possible by showing a single plot of real data. Gene clusters are shown together with the corresponding gene profiles. The abscissa represents the first principal axis, and the ordinate represents the second. Both axes are dimensionless.

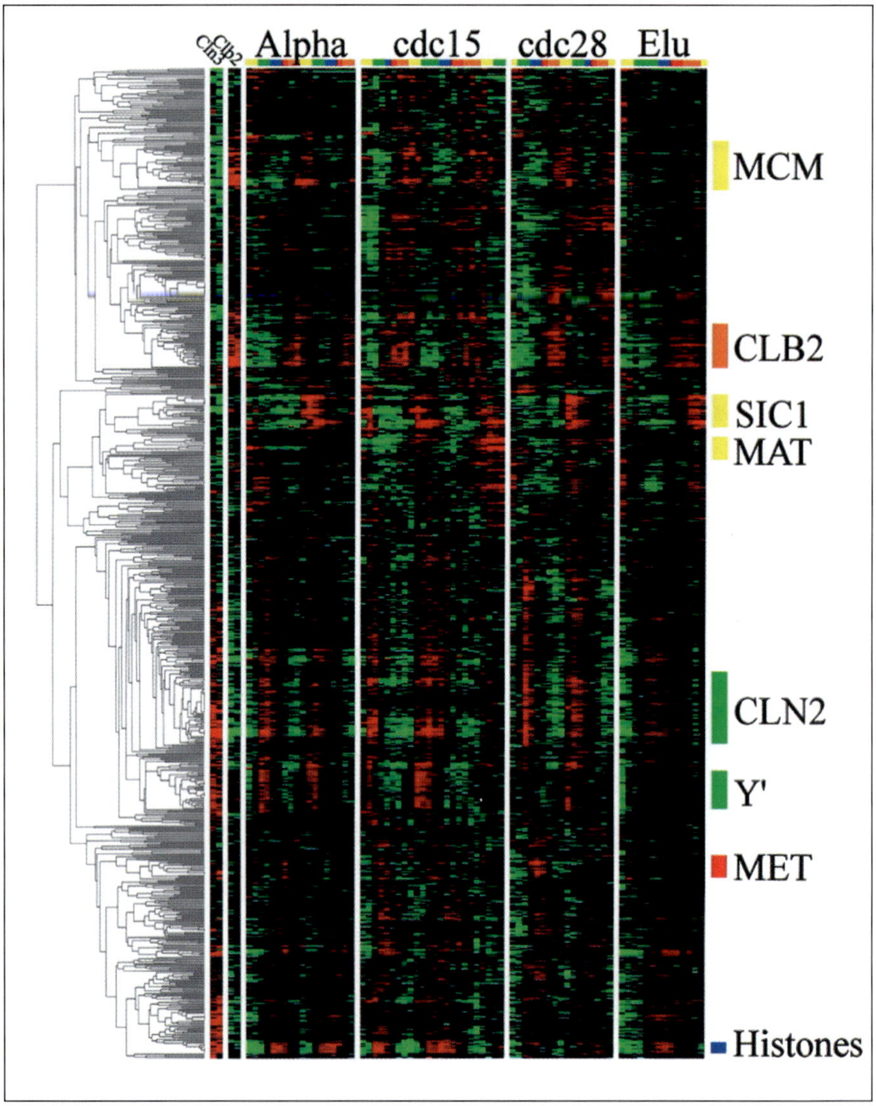

Chapter 16, Figure 5. Gene expression during the yeast cell cycle. This figure has been reproduced from Reference 53. Genes correspond to rows, and the time points of each experiment are represented by columns. The ratio of induction/repression is shown for each gene so that the magnitude is indicated by the intensity of the colors displayed. If the color is black, then the ratio of control to experimental cDNA is equal to one, whereas the brightest colors (red and green) represent a ratio of 2.8:1. Ratios greater than 2.8 are also displayed as the brightest color. In all cases, red indicates an increase in mRNA abundance, whereas green indicates a decrease in abundance compared to the control samples (stemming from asynchronous cultures of the same cells growing exponentially at the same temperature in the same medium). Gray areas (when visible) indicate absent data or data of low quality. Color bars on the right indicate the phase group to which a gene belongs (M/G1, yellow; G1, green; S, purple; G2, red; and M, brown). These same colors indicate cell cycle phase along the top of the figure. Genes that share similar expression profiles are grouped. The dendrogram on the left shows the structure of the cluster. Reprinted from Molecular Biology of the Cell, (1988 Dec; 9(12):3273-3297), with permission by the American Society for Cell Biology.

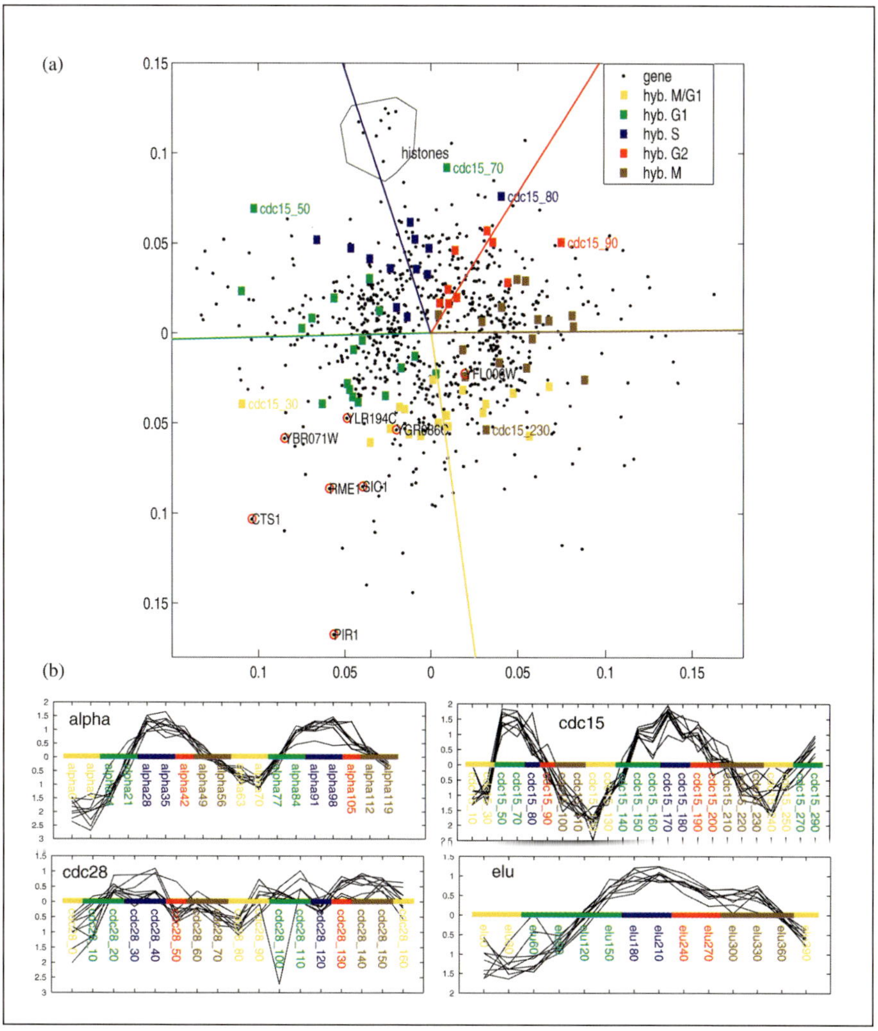

Chapter 16, Figure 6. Cell cycle synchronization data by Spellman et al. The data set composed of 800 cell cycle-associated genes has been projected by CA as is. No HMS has been employed so as not to bias the resulting plot in terms of the separation of the cell cycle phases. The outlying hybridizations have been identified to be caused by a slight phase shift of the cdc15-based synchronization visible in the upper right-hand panel of Figure 6b, which shows the profiles of the nine histone genes *HHF1, HHF2, HHT1, HTB2, HHT2, HTB1, HTA1, HTA2,* and *HHO1* that are encircled in black in Figure 6a (see text for details). The abscissa represents the first principal axis, and the ordinate represents the second. Both axes are dimensionless. Reprinted from Fellenberg et al. (25), Copyright (2001) National Academy of Sciences, USA.

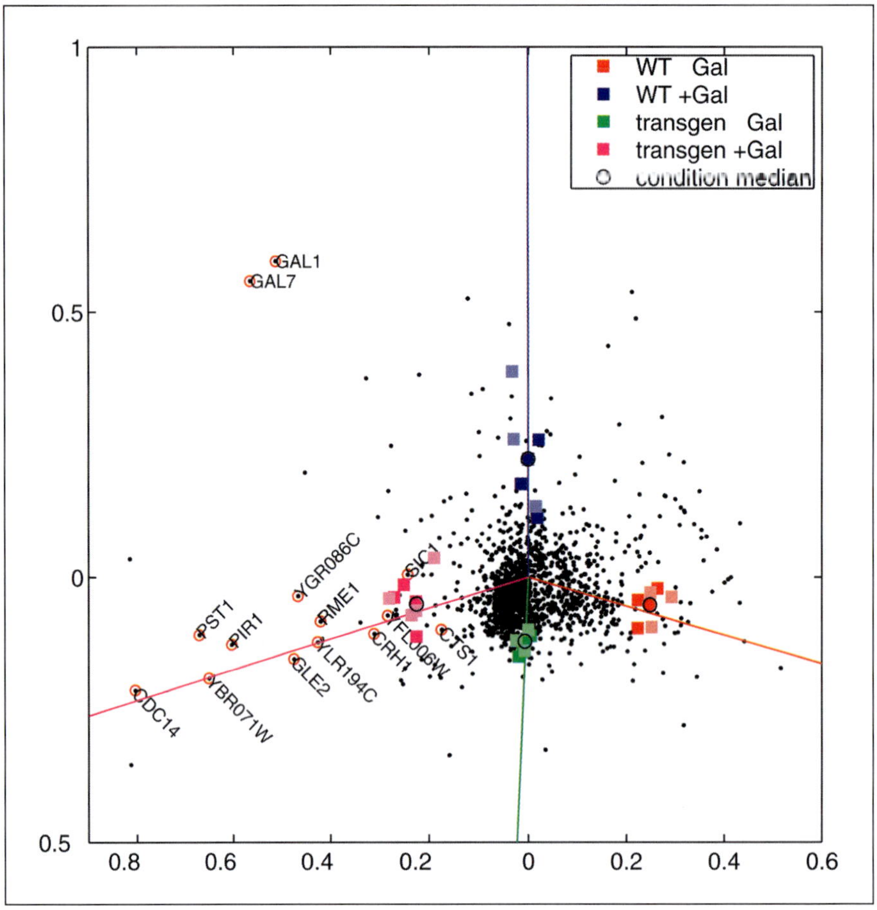

Chapter 16, Figure 7. Overexpression of Cdc14p. In arrested yeast cultures, Cdc14p expression was induced under control of the *GAL1* promoter and investigated in comparison with uninduced transgenic, uninduced wild-type, and inductor-exposed wild-type cells. From three to four hybridizations were performed for each experimental condition. Both spot sets of the array are drawn separately for each hybridization, primary and secondary spot sets are depicted in light and dark colors, respectively. The conditions are color-coded, and their hybridization medians are marked according to the legend in the upper right-hand corner. Lines are drawn in the direction of the standard coordinates of the condition medians in appropriate colors: genes such as *GAL7* and *GAL1* are associated with both the wild-type and the transgenic strain grown in the presence of galactose to an equal share. *CDC14* is associated with the induced transgenic strain only. The abscissa represents the first principal axis, and the ordinate represents the second. Both axes are dimensionless. Reprinted from Fellenberg et al. (25), Copyright (2001) National Academy of Sciences, USA. WT, wild-type; transgen, transgenic.

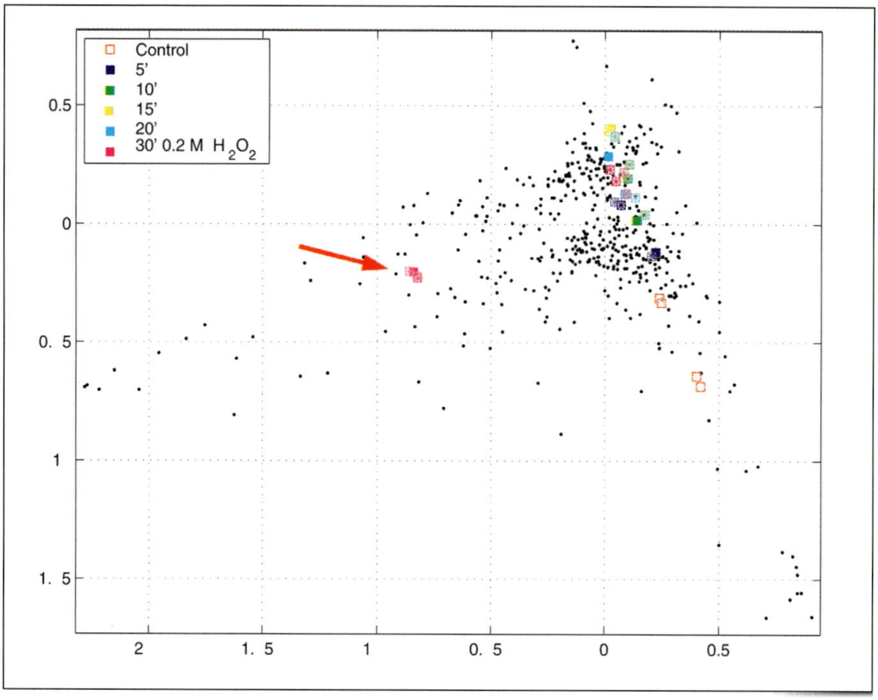

Chapter 16, Figure 8. Oxidative stress. The CA plot shows a data set recorded from wild-type yeast cells responding to 0.2 M hydrogen peroxide in the medium. Genes are depicted as black dots, and measurements (in this case monochannel hybridizations) are shown as squares that are color-coded according to the experimental condition (here time point) to which they belong (see text for details). The outlying cluster of pink-labeled measurements (red arrow) is further characterized by experiment annotation values that are over-represented or underrepresented in this cluster, as shown in Table 1. The abscissa represents the first principal axis, and the ordinate represents the second. Both axes are dimensionless. Modified from Fellenberg et al. (24), with permission by Oxford University Press.

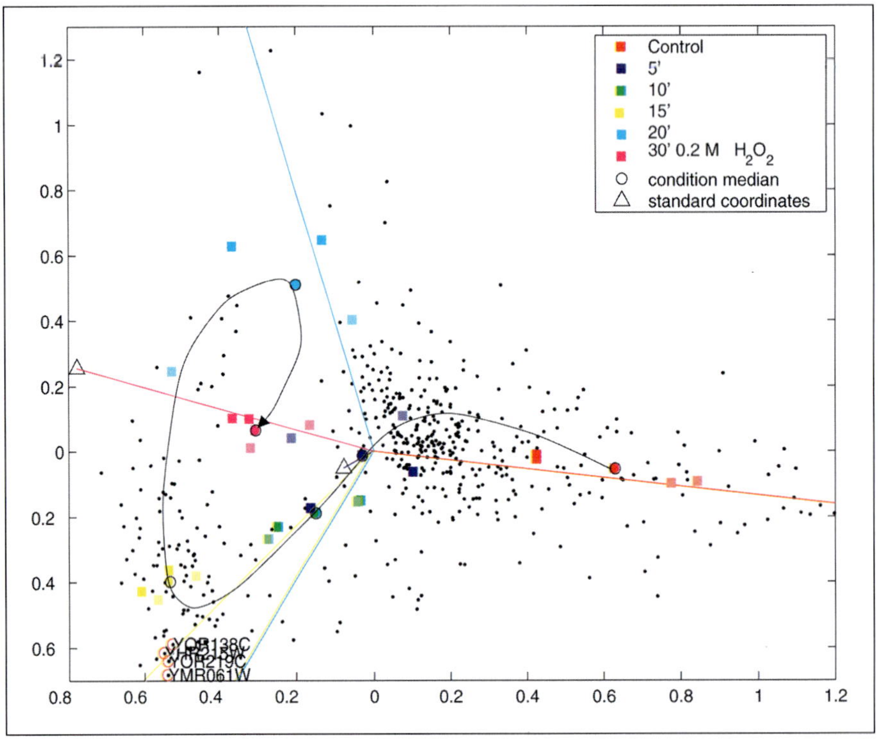

Chapter 16, Figure 9. Oxidative stress. Wild-type yeast was grown in the presence of 0.2 M hydrogen peroxide and sampled in 5-min intervals (10 min at end). The experiment was performed twice, which yielded two hybridizations per experimental condition (time point). These conditions are color-coded according to the legend, double-spotting in dark and light colors. A black arrow shows the course of the experiment. Four genes are tagged that are almost exclusively associated with the 15-min time point. Their profiles are given in Figure 10. The abscissa represents the first principal axis, and the ordinate represents the second. Both axes are dimensionless. Modified from the supplemental material to Reference 24.

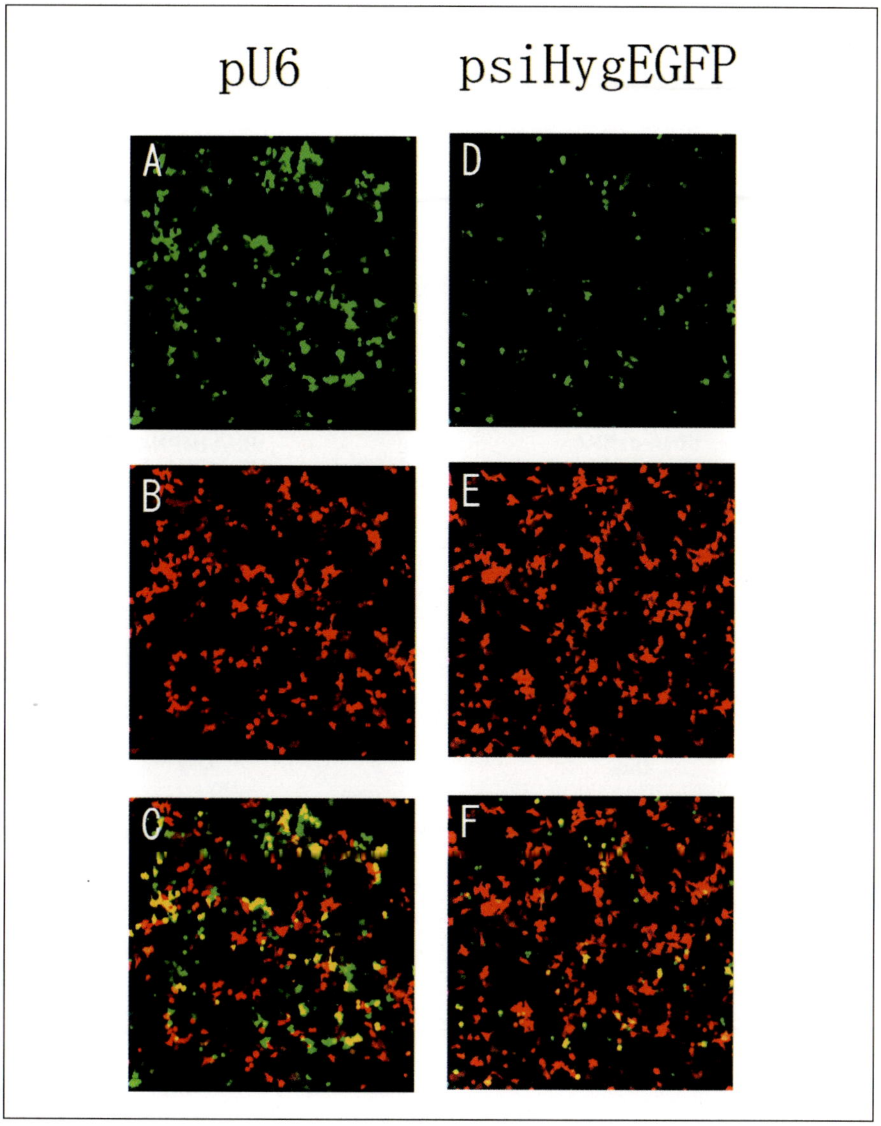

Chapter 18, Figure 4. Effects of an siRNA expression vector targeted against the gene for a hygromycin/EGFP fusion gene. The top panels (A and D) show the expression of hygromycin/EGFP, and the middle panels (B and E) show the expression of DsRed. The lower panels (C and F) show overlay images caused by hygromycin/EGFP and DsRed. Cells that co-expressed hygromycin/EGFP and DsRed appear as yellow cells.

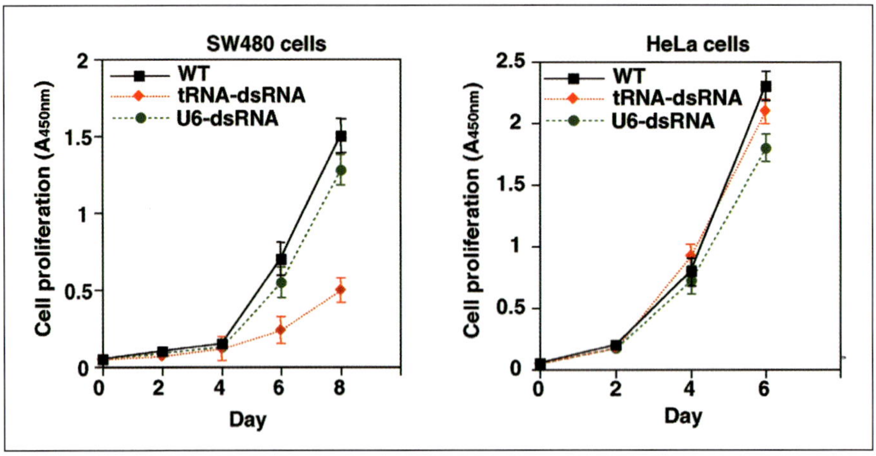

Chapter 19, Figure 7. Effects on the proliferation of cells of tRNA-dsRNA and of U6-dsRNA. The rates of proliferation were determined as described in the text. In each case, the values are the means ± SD of the results from three replicates. WT, wild-type.

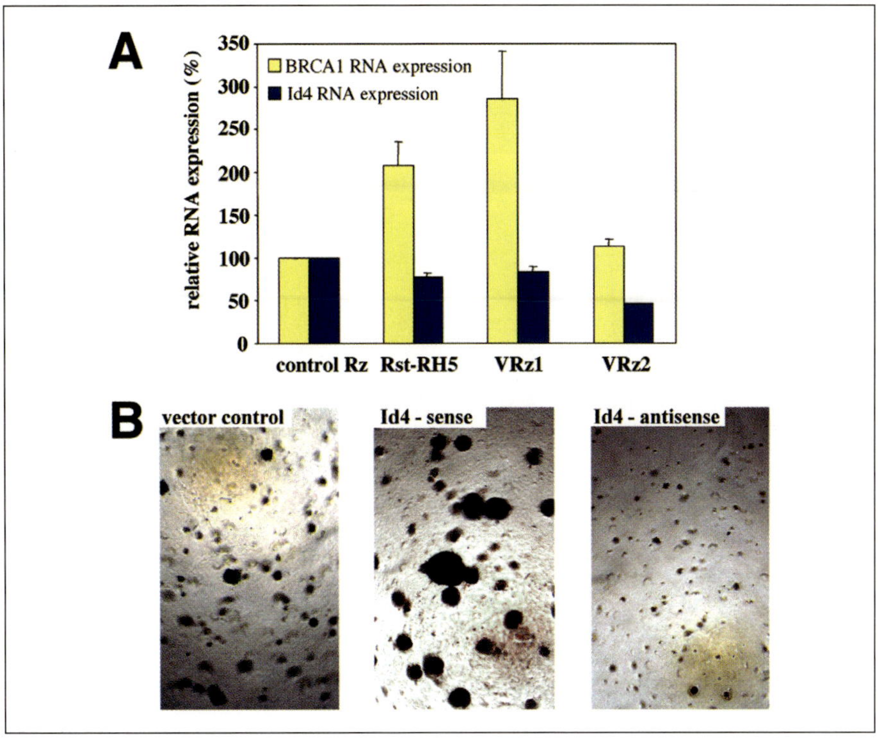

Chapter 20, Figure 7. Validation of *Id4* as a negative regulator of the BRCA1 promoter. (A) Effects of Rst-RH5 and two validation ribozymes (VRz1 and VRz2) on endogenous *BRCA1* and *Id4* expression as measured by real-time quantitative RT-PCR. Expression of *BRCA1* or *Id4* mRNA was normalized to GAPDH mRNA. Numbers are given as relative RNA levels in percent compared to the control ribozyme (mean ± SEM). (B) Anchorage-independent growth of P8EGBR3 cells transduced with control vector, *Id4*-sense or *Id4*-antisense expression vectors. Adapted with permission from Beger et al. (5).

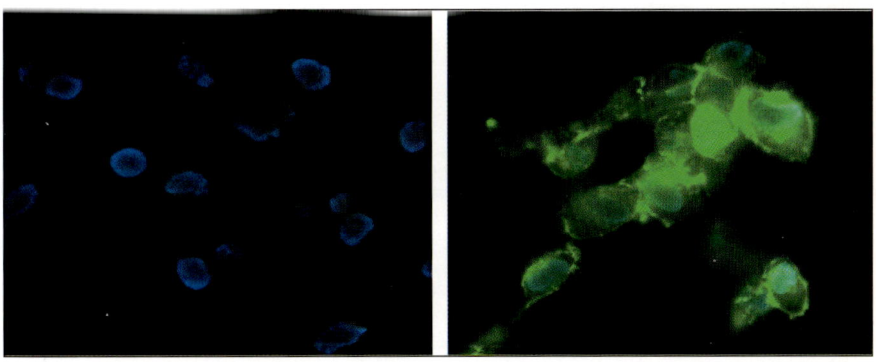

Chapter 21, Figure 4. Creation of a RAGE library expressing known and novel integral membrane proteins. A RAGE library was created using the vector shown in Figure 3. The cells were sorted using FACS and an antibody specific for the cell surface epitope encoded by the RAGE vector. Control cells used to make the RAGE library are shown on the left, and a heterogeneous population of RAGE-activated cells expressing chimeric integral membrane proteins is shown on the right (both obtained after sequential FACS sorts).

Chapter 22, Figure 1. Ten distinguishable emission colors of ZnS-capped CdSe QDs excited with a near-UV lamp. From left to right (blue to red), the emission maxima are located at 443, 473, 481, 500, 518, 543, 565, 587, 610, and 655 nm. (Adapted with permission from Han, Gao, Su, and Nie, Nat. Biotechnol. 19:631, 2001).

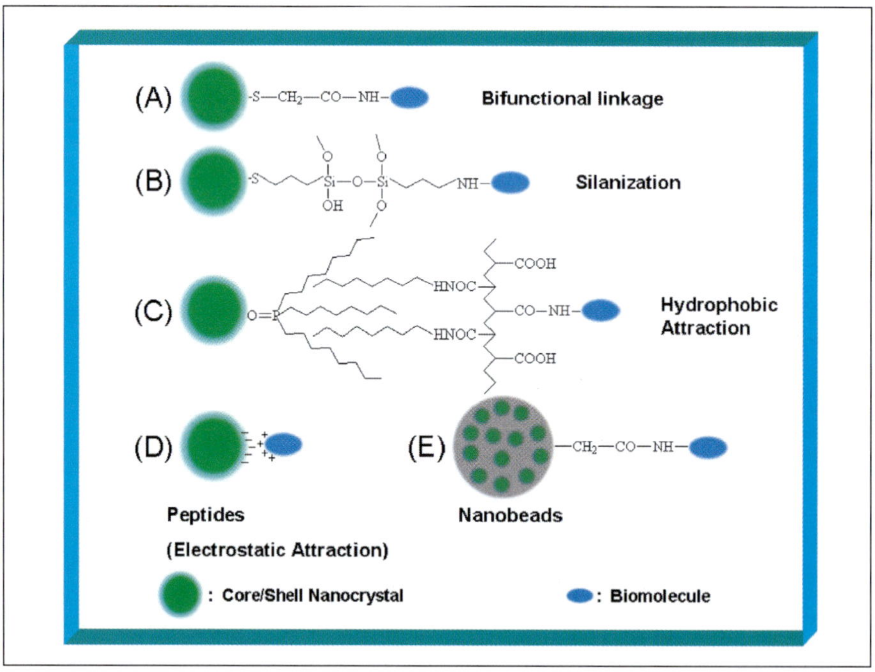

Chapter 22, Figure 3. Schematic illustration of bioconjugation methods. (Courtesy of Dr. Mingyong Han).

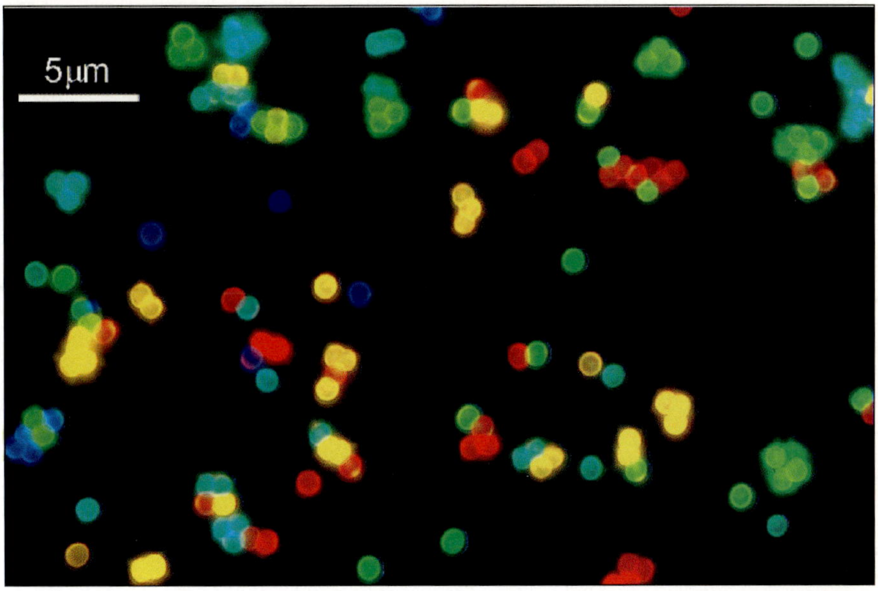

Chapter 22, Figure 4. Fluorescence micrograph of a mixture of CdSe/ZnS QD-tagged beads emitting single-color signals at 484, 508, 547, 575, and 611 nm. The beads were spread and immobilized on a poly-lysine-coated glass slide that caused a slight clustering effect. (Reproduced with permission from Han, Gao, Su, and Nie, Nat. Biotechnol. 19, 631, 2001).

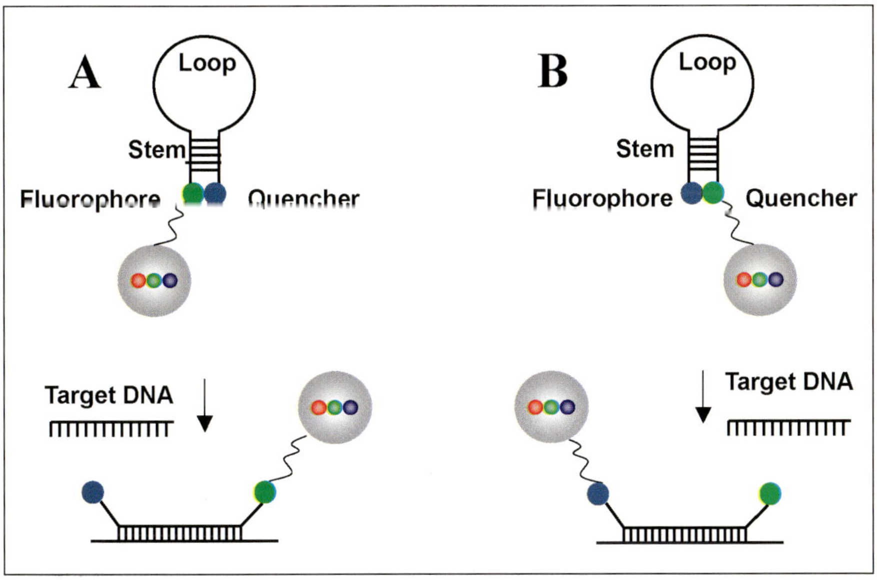

Chapter 22, Figure 5. Schematic illustration of gene screening experiments using molecular beacons and multicolor QD-coded microbeads. The bead is conjugated to either a fluorophore (A) or a quenching moiety (B) at the ends of the hairpin loop. When the beacons encounter their complementary sequences, the loop opens up and restores fluorescence, which indicates the presence of particular sequences.

coding technology is likely to be spectral overlap with the coding signals. To minimize this spectral interference, it will be necessary to separate the coding and target signals as far as possible; for example, we are currently using visible QDs for coding and near-infrared QDs for target labeling (R. Bailey, J. Strausburg, and S. Nie; unpublished data). Under complex biological conditions, we expect the performance (e.g., specificity and sensitivity) of our QD-tagged beads to be similar to that reported elsewhere (20,21,58). In a recent paper, Ferguson et al. used 3.1-µm encoded beads to study 25 sequences (including cancer and cystic fibrosis genes) and achieved a detection sensitivity of 10–100 fM target DNA (21). The underlying principles of nucleic acid hybridization and fluorescence detection are essentially the same, but multicolor QD coding should provide important advantages and applications unavailable with organic dyes.

For the rapid screening of unlabeled genetic materials, the encoded beads could be combined with molecular beacons (unpublished data). A schematic illustration is shown in Figure 5. The hairpin-shaped beacons (with both a fluorophore and a quencher) are covalently attached to the surface of the encoded beads. The beacon stem keeps the fluorescing and quenching moieties in close contact, resulting in fluorescence quenching by energy transfer and complex formation. When the probe encounters a target molecule, the molecular beacon undergoes a spontaneous conformational reorganization that forces the stem apart,

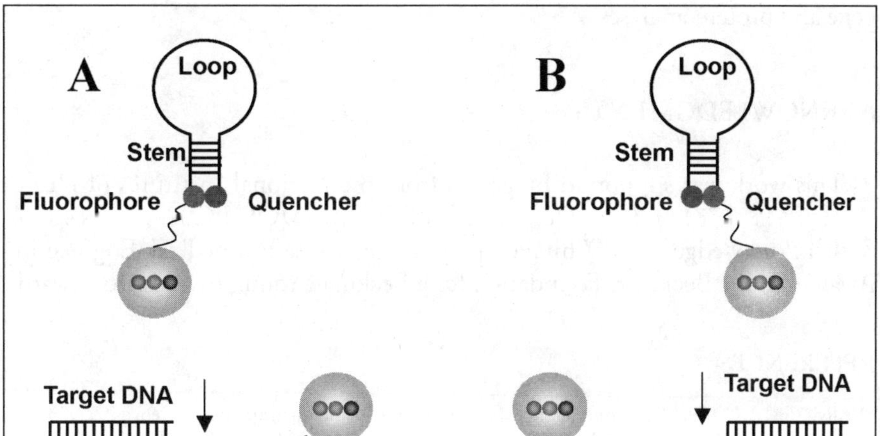

Figure 5. Schematic illustration of gene screening experiments using molecular beacons and multicolor QD-coded microbeads. The bead is conjugated to either a fluorophore (A) or a quenching moiety (B) at the ends of the hairpin loop. When the beacons encounter their complementary sequences, the loop opens up and restores fluorescence, which indicates the presence of particular sequences. (**See color plate A39.**)

leading to the restoration of fluorescence. The most important advantage of this approach is that the target molecules do not need to be labeled.

We envision integrated bead-based assays using microfluidic devices in a serial fashion or parallel binding assays using high-density microplates. In both formats, a library of encoded beads and recognition molecules (e.g., oligonucleotide probes and antibodies) will need to be developed, similar to the preparation of individual DNA clones for printing microarrays. While this involves a significant amount of "front-end" work, such encoded beads should offer tremendous flexibility for practical use in multiplexed diagnostics, drug screening, and gene expression profiling.

CONCLUSION

The development of new technologies for gene expression analysis is essential to the molecular profiling of cells, tissues, and organs. Rapid detection and quantification of a large number of genes and proteins in complex mixtures are playing an important role in both drug discovery and medical diagnostics. In the next decade, we envision that novel platforms based on multicolor QDs and QD-encoded "smart" beads will be developed for massively parallel biosensing and biodetection. These multiplexing technologies will likely combine the advantages of QDs with those of microfluidics and microarrays for simultaneous gene and protein analyses.

ACKNOWLEDGMENTS

This work was supported by grants from the National Institutes of Health (R01 GM58173) and the Department of Energy (DOE FG02-98ER14873). S.N. acknowledges the Whitaker Foundation for a Biomedical Engineering Award and the Beckman Foundation for a Beckman Young Investigator Award.

REFERENCES

1. **Alivisatos, A.P.** 1996. Semiconductor clusters, nanocrystals, and quantum dots. Science *271*:933-937.
2. **Banin, U., M. Bruchez, A.P. Alivisatos, T. Ha, S. Weiss, and D.S. Chemla.** 1999. Evidence for a thermal contribution to emission intermittency in single CdSe/CdS core/shell nanocrystals. J. Chem. Phys. *110*:1195-1201.
3. **Baral, S., A. Fojtik, H. Weller, and A. Henglein.** 1986. Photochemistry and radiation chemistry of colloidal semiconductors: intermediates of the oxidation of extremely small particles of CdS, ZnS, and Cd_3P_2 and size quantization effects (a pulse radiolysis study). J. Am. Chem. Soc. *108*:375-378.
4. **Becerra, L.R., C.B. Murray, R.G. Griffin, and M.G. Bawendi.** 1994. Investigation of the surface morphology of capped CdSe nanocrystallites by P-31 nuclear magnetic resonance. J. Chem. Phys. *100*:3297-3300.
5. **Blanton, S.A., M.A. Hines, and P. Guyot-Sionnest.** 1996. Photoluminescence wandering in single CdSe nanocrystals. Appl. Phys. Lett. *69*:3905-3907.

6. **Bruchez, M., M. Moronne, P. Gin, S. Weiss, and A.P. Alivisatos.** 1998. Semiconductor nanocrystals as fluorescent biological labels. Science *281*:2013-2016.
7. **Brus, L.E.** 1983. A simple model for the ionization potential, electron affinity, and aqueous redox potentials of small semiconductor crystallites. J. Chem. Phys. *79*:5566-5571.
8. **Chan, W.C.W.** 2001. Semiconductor quantum dots for biological detection and imaging [Ph.D. dissertation]. Department of Chemistry, Indiana University, Bloomington, Indiana.
9. **Chan, W.C.W., D.J. Maxwell, X. Gao, R.E. Bailey, M.-Y. Han, and S. Nie.** 2002. Luminescent quantum dots for multiplexed biological detection and imaging. Curr. Opin. Biotechnol. *13*:40-46.
10. **Chan, W.C.W. and S. Nie.** 1998. Quantum dot bioconjugates for ultrasensitive nonisotopic detection. Science *281*:2016-2018.
11. **Chen, J.W., M.A. Iannone, M.S. Li, J.D. Taylor, P. Rivers, A.J. Nelson, K.A. Slentz-Kesler, A. Roses, and M.P. Weiner.** 2000. A microsphere-based assay for multiplexed single nucleotide polymorphism analysis using single base chain extension. Genome Res. *10*:549-557.
12. **Dabbousi, B.O., J. Rodriguez-Viejo, F.V. Mikulec, J.R. Heine, H. Mattoussi, R. Ober, K.F. Jensen, and M.G. Bawendi.** 1997. (CdSe)ZnS core-shell quantum dots: synthesis and characterization of a size series of highly luminescent nanocrystallites. J. Phys. Chem. B *101*:9463-9475.
13. **Dahan, M., T. Laurence, F. Pinaud, D.S. Chemla, A.P. Alivisatos, M. Sauer, and S. Weiss.** 2001. Time-gated biological imaging by use of colloidal quantum dots. Opt. Lett. *26*:825-827.
14. **Dahan, M., T. Laurence, A. Schumacher, D.S. Chemla, A.P. Alivisatos, M. Sauer, and S. Weiss.** 2000. Fluorescence lifetime study of single qdots. Biophys. J. *78*:2270.
15. **DeSmet, Y., L. Deriemaeker, and R.A. Finsy.** 1997. A simple computer simulation of Ostwald ripening. Langmuir *13*:6884-6888.
16. **DeSmet, Y., L. Deriemaeker, E. Parloo, and R.A. Finsy.** 1999. On the determination of Ostwald ripening rates from dynamic light scattering measurements. Langmuir *15*:2327-2332.
17. **Efros, A.L.** 1992. Luminescence polarization of CdSe microcrystals. Phys. Rev. B *46*:7448-7458.
18. **Empedocles, S.A. and M.G. Bawendi.** 1999. Spectroscopy of single CdSe nanocrystallites. Acc. Chem. Res. *32*:389-396.
19. **Empedocles, S.A., R. Neuhauser, and M.G. Bawendi.** 1999. Three-dimensional orientation measurements of symmetric single chromophores using polarization microscopy. Nature *399*:126-130.
20. **Ferguson, J.A., T.C. Boles, C.P. Adams, and D.R. Walt.** 1996. A fiber-optic DNA biosensor microarray for the analysis of gene expression. Nat. Biotechnol. *14*:1681-1684.
21. **Ferguson, J.A., F.J. Steemers, and D.R. Walt.** 2000. High-density fiber-optic DNA random microsphere arrays. Anal. Chem. *72*:5618-5624.
22. **Fodor, S.P.A., J.L. Read, M.C. Pirrung, L. Stryer, A.T. Lu, and D. Solas.** 1991. Light-directed, spatially addressable parallel chemical synthesis. Science *251*:767-773.
23. **Gao, X., W.C.W. Chan, and S. Nie.** 2002. Quantum-dot nanocrystals for ultrasensitive biological labeling and multicolor optical encoding. J. Biomed. Optics *7*:532-537
24. **Gerion, D., F. Pinaud, S.C. Williamas, W.J. Parak, D. Zanchet, S. Weiss, and A.P. Alivisatos.** 2001. Synthesis and properties of biocompatible water-soluble silica-coated CdSe/ZnS semiconductor quantum dots. J. Phys. Chem. B *105*:8861-8871.
25. **Guzelian, A.A., J.E.B. Katari, A.V. Kadavanich, U. Banin, K. Hamad, E. Juban, A.P. Alivisatos, R.H. Wolters, et al.** 1996. Synthesis of size-selected, surface-passivated InP nanocrystals. J. Phys. Chem. *100*:7212-7219.
26. **Han, M.Y., X. Gao, J.Z. Su, and S. Nie.** 2001. Quantum-dot-tagged microbeads for multiplexed optical coding of biomolecules. Nat. Biotechnol. *19*:631-635.
27. **Hao, E., H. Sun, Z. Zhou, J. Liu, B. Yang, and J. Shen.** 1999. Synthesis and optical properties of CdSe and CdSe/CdS nanoparticles. Chem. Mater. *11*:3096-3102.
28. **Haugland, R.P.** 1995. Handbook of Fluorescent Probes and Research Chemicals, Molecular Probes, Eugene, OR.
29. **Henglein, A.** 1982. Photo-degradation and fluroescence of colloidal-cadmium sulfide in aqueous solution. Ber. Bunsenges. Phys. Chem. *86*:301-305.
30. **Hermanson, G.T.** 1996. Bioconjugate Techniques. Academic Press, New York.
31. **Hines, M.A. and P. Guyot-Sionnest.** 1996. Synthesis of strongly luminescing ZnS-capped CdSe nanocrystals. J. Phys. Chem. B *100*:468-471.
32. **Hines, M.A. and P. Guyot-Sionnest.** 1998. Bright UV-blue luminescent colloidal ZnSe nanocrystals. J. Phys. Chem. B *102*:3655-3657.

33. **Holland, P.A., R.A. Abramson, R. Watson, and D.H. Gelfand.** 1991. Detection of specific polymerase chain reaction product by utilizing the 5′ to 3′ exonuclease activity of *Thermus acquaticus* DNA polymerase. Proc. Natl. Acad. Sci. USA *88*:7276-7280.
34. **Hu, J.T., L. Li, W. Yang, L. Manna, L. Wang, and A.P. Alivisatos.** 2001. Linearly polarized emission from colloidal semiconductor quantum rods. Science *292*:2060-2063.
35. **Kapitonov, A.M., A.P. Stupak, S.V. Gaponenko, E.P. Petrov, A.L. Rogach, and A. Eychmuller.** 1999. Luminescence properties of thiol-stabilized CdTe nanocrystals. J. Phys. Chem. B *103*:10109-10113.
36. **Katari, J.E.B., V.L. Colvin, and A.P. Alivisatos.** 1994. X-ray photoelectron spectroscopy of CdSe nanocrystals with applications to studies of the nanocrystal surface. J. Phys. Chem. *98*:4109-4117.
37. **Lacoste, T.D., X. Michalet, F. Pinaud, D.S. Chemla, A.P. Alivisatos, and S. Weiss.** 2000. Ultra high-resolution multicolor colocalization of single fluorescent probes. Proc. Natl. Acad. Sci. USA *97*:9461-9466.
38. **Lie, Y.S. and C.J. Petropoulos.** 1998. Advances in quantitative PCR technology: 5′ nuclease assays. Curr. Opin. Biotechnol. *9*:43-48.
39. **Lorenz, J.K. and A.B. Ellis.** 1998. Surfactant-semiconductor interfaces: perturbation of the photoluminescence of bulk cadmium selenide by adsorption of tri-n-octylphosphine oxide as a probe of solution aggregation with relevance to nanocrystal stabilization. J. Am. Chem. Soc. *120*:10970-10975.
40. **Madelung, O.** 1996. Semiconductors—Basic Data. Springer Publishing, New York.
41. **Mahtab, R., J.P. Rogers, and C.J. Murphy.** 1995. Protein-sized quantum dot luminescence can distinguish between straight, bent, and kinked oligonucleotides. J. Am. Chem. Soc. *117*:9099-9100.
42. **Mahtab, R., J.P. Rogers, C.P. Singleton, and C.J. Murphy.** 1996. Preferential adsorption of a kinked DNA to a neutral curved surface—comparisons to and implications for nonspecific DNA-protein interactions. J. Am. Chem. Soc. *118*:7028-7032.
43. **Mattoussi, H., J.M. Mauro, E.R. Goldman, G.P. Anderson, V.C. Sundar, F.V. Mikulec, and M.G. Bawendi.** 2000. Self-assembly of CdSe-ZnS quantum dots bioconjugates using an engineered recombinant protein. J. Am. Chem. Soc. *122*:12142-12150.
44. **Mitchell, G.P., C.A. Mirkin, and R.L. Letsinger.** 1999. Programmed assembly of DNA functionalized quantum dots. J. Am. Chem. Soc. *121*:8122-8123.
45. **Micic, O.I., H.M. Cheong, H. Fu, A. Zunger, J.R. Sprague, A. Mascarenhas, and A.J. Nozik.** 19976. Size-dependent spectroscopy of InP quantum dots. J. Phys. Chem. B *101*:4904-4912.
46. **Murray, C.B., D.J. Norris, and M.G. Bawendi.** 1993. Synthesis and characterization of nearly monodisperse CdE (E = S, Se, Te) semiconductor nanocrystallites. J. Am. Chem. Soc. *115*:8706-8715.
47. **Nirmal, M., B.O. Dabbousi, M.G. Bawendi, J.J. Macklin, and L.E. Brus.** 1996. Fluorescence intermittency in single sadmium selenide nanocrystals. Nature *383*:802-804.
48. **O'Neil, M., J. Marohn, and G. McLendon.** 1990. Dynamics of electron-hole pair recombination in semiconductor clusters. J. Phys. Chem. *94*:4356-4363.
49. **Peng, X.G., M.C. Schlamp, A.V. Kadavanich, and A.P. Alivisatos.** 1997. Epitaxial growth of highly luminescent CdSe/CdS core/shell nanocrystals with photostability and electronic accessibility. J. Am. Chem. Soc. *119*:7019-7029
50. **Peng, Z.A. and X.G. Peng.** 2001. Formation of high-quality CdTe, CdSe, and CdS nanocrystals using CdO as precursor. J. Am. Chem. Soc. *123*:183-184.
51. **Peng, X.G., L. Manna, W.D. Yang, J. Wickham, E. Scher, A. Kadavanich, and A.P. Alivisatos.** 2000. Shape control of CdSe nanocrystals. Nature *404*:59-61.
52. **Peng, Z.A. and X.G. Peng.** 2001. Mechanisms of the shape evolution of CdSe nanocrystals. J. Am. Chem. Soc. *123*:1389-1395.
53. **Peng, X.G., J. Wickham, and A.P. Alivisatos.** 1998. Kinetics of II-VI and III-V colloidal semiconductor nanocrystal growth: "focusing" of size distributions. J. Am. Chem. Soc. *120*:5343-5344.
54. **Qu, L., Z.A. Peng, and X.G. Peng.** 2001. Alternative routes toward high quality CdSe nanocrystals. Nano Lett. *1*:333-337.
55. **Rogach, A.L., A. Kornowski, M. Gao, A. Eychmuller, and H. Weller.** 1999. Synthesis and characterization of a size series of extremely small thiol-stabilized CdSe nanocrystals. J. Phys. Chem. B *103*:3065-3069.
56. **Schena, M., D. Shalon, R.W. Davis, and P.O. Brown.** 1995. Quantitative monitoring of gene-expression patterns with a complementary-DNA microarray. Science *270*:467-470.

57. **Schreder, B., T. Schmidt, V. Ptatschek, U. Winker, A. Materny, E. Umback, M. Lerch, G. Muller, et al.** 2000. CdTe/CdS clusters with "core-shell" structure in colloids and films: the path of formation and thermal breakup. J. Phys. Chem. B *104*:1677-1685.

58. **Steemers, F.J., J.A. Ferguson, and D.R. Walt.** 2000. Screening unlabeled DNA targets with randomly ordered fiber-optic gene arrays. Nat. Biotechnol. *18*:91-94.

59. **Talapin, D.V., A.L. Rogach, A. Kornowski, M. Haase, and H. Weller.** 2001. Highly luminescent monodisperse CdSe and CdSe/ZnS nanocrystals synthesized in a hexadecylamine-triocytlphosphine oxide-trioctylphosphine mixture. Nano Lett. *1*:207-211.

60. **Trindade, T. and P. O'Brien.** 1996. A single source approach to the synthesis of CdSe nanocrystallites. Adv. Mater. *8*:161-163.

61. **Tyagi, S., D.P. Bratu, and F.R. Kramer.** 1998. Multicolor molecular beacons for allele discrimination. Nat. Biotechnol. *16*:49-53.

62. **Tyagi, S. and F.R. Kramer.** 1996. Molecular beacons: probes that fluoresce upon hybridization. Nat. Biotechnol. *14*:303-308.

63. **Weiss, S.** 2000. Shattering the diffraction limit of light: a revolution in fluorescence microscopy. Proc. Natl. Acad. Sci. USA *97*:8747-8749.

64. **Weller, H.** 1993. Quantized semiconductor particles—a novel state of matter for materials science. Adv. Mat. *5*:88-95.

INDEX

C

D

E

F

G

H

I

J

K

L

M

N

O

P

S

T

U

V

W

X

Y

Z